Handbook of Climate Modelling and Planning

Handbook of Climate Modelling and Planning

Edited by Bruce Mullan

SYRAWOOD
PUBLISHING HOUSE
New York

Published by Syrawood Publishing House,
750 Third Avenue, 9th Floor,
New York, NY 10017, USA
www.syrawoodpublishinghouse.com

Handbook of Climate Modelling and Planning
Edited by Bruce Mullan

International Standard Book Number: 978-1-68286-447-0 (Hardback)

Cataloging-in-publication Data

Handbook of climate modelling and planning / edited by Bruce Mullan.
p. cm.
Includes bibliographical references and index.
ISBN 978-1-68286-447-0
1. Climatology. 2. Climatic changes. 3. Climatic changes--Risk assessment. 4. Climate change mitigation.
I. Mullan, Bruce.
QC981 .H36 2017
551.6--dc23

Printed in the United States of America.

TABLE OF CONTENTS

PREFACE

Climate modelling is the design of artificial models using quantitative methods for climate prediction. Climate modelling is a sophisticated science that takes into account the transfer of energy between ecological bodies, changes in temperature levels in the atmosphere and at surface levels and fluid and hydrological motion that is observed every day. Special consideration is given to the changes caused due to greenhouse gas emissions as well as the rapid deterioration of the ozone. This book will be of great help to students, experts and professionals in the fields of climatology, meteorology and geography. For all those who are interested in climate modelling, this book can prove to be an essential guide. This book traces the progress of this field and highlights some of its key concepts and applications. The topics included in this book on climate modeling are of utmost significance and bound to provide incredible insights to readers. It consists of contributions made by international experts.

Various studies have approached the subject by analyzing it with a single perspective, but the present book provides diverse methodologies and techniques to address this field. This book contains theories and applications needed for understanding the subject from different perspectives. The aim is to keep the readers informed about the progress in the field; therefore, the contributions were carefully examined to compile novel researches by specialists from across the globe.

Indeed, the job of the editor is the most crucial and challenging in compiling all chapters into a single book. In the end, I would extend my sincere thanks to the chapter authors for their profound work. I am also thankful for the support provided by my family and colleagues during the compilation of this book.

Editor

1

Predicting the Current and Future Potential Distributions of Lymphatic Filariasis in Africa Using Maximum Entropy Ecological Niche Modelling

Hannah Slater, Edwin Michael*¤

Department of Infectious Disease Epidemiology, Imperial College London, London, United Kingdom

Abstract

Modelling the spatial distributions of human parasite species is crucial to understanding the environmental determinants of infection as well as for guiding the planning of control programmes. Here, we use ecological niche modelling to map the current potential distribution of the macroparasitic disease, lymphatic filariasis (LF), in Africa, and to estimate how future changes in climate and population could affect its spread and burden across the continent. We used 508 community-specific infection presence data collated from the published literature in conjunction with five predictive environmental/climatic and demographic variables, and a maximum entropy niche modelling method to construct the first ecological niche maps describing potential distribution and burden of LF in Africa. We also ran the best-fit model against climate projections made by the HADCM3 and CCCMA models for 2050 under A2a and B2a scenarios to simulate the likely distribution of LF under future climate and population changes. We predict a broad geographic distribution of LF in Africa extending from the west to the east across the middle region of the continent, with high probabilities of occurrence in the Western Africa compared to large areas of medium probability interspersed with smaller areas of high probability in Central and Eastern Africa and in Madagascar. We uncovered complex relationships between predictor ecological niche variables and the probability of LF occurrence. We show for the first time that predicted climate change and population growth will expand both the range and risk of LF infection (and ultimately disease) in an endemic region. We estimate that populations at risk to LF may range from 543 and 804 million currently, and that this could rise to between 1.65 to 1.86 billion in the future depending on the climate scenario used and thresholds applied to signify infection presence.

Editor: Matthew Baylis, University of Liverpool, United Kingdom

Funding: The authors acknowledge the Natural Environment Research Council, UK, for a NERC/ESRC Interdisciplinary Research Studentship to HS. The authors also acknowledge the Grantham Institute for Climate Change at Imperial College London and the National Institutes of Health, USA (under grant number RO1 AI069387-01A1), for partial financial support of this work. No additional external funding was received for this study. The funders had no role in study design, data collection and analysis, decision to publish, or preparation of the manuscript.

Competing Interests: The authors have declared that no competing interests exist.

* E-mail: emichael@nd.edu

¤ Current address: Department of Biological Sciences, University of Notre Dame, Notre Dame, Indiana, United States of America

Introduction

The role of risk mapping in describing the spatial patterns of infection and guiding the planning of parasite control is now well-established, and has been demonstrated for a range of major parasitic diseases, including malaria [1,2], trypanosomiasis [3,4], schistosomiasis [5,6], onchocerciasis [7], and lymphatic filariasis [8,9,10]. It has also led to an increased understanding of the climatic and environmental ecology of parasitic infections [8,11], including improving appreciation of species thermal tolerances and the impact of key environmental variables on ecological traits that affect transmission, such as parasite development and survival rates. More recently, focus in parasite distribution modeling has expanded to evaluating the potential for the establishment and spread of invasive vector species [12,13,14] and assessing parasite or vector species responses to global climate change [15].

Lymphatic filariasis (LF) is a vector-borne infectious disease endemic in the tropics, including sub-Saharan Africa, and is thought to present the second largest public health burden of any disease worldwide [16]. The disease is transmitted to humans by infective mosquitoes that release parasitic filarial worms into the blood stream when taking a blood meal. Many patients are asymptomatic, but infection can lead to major debilitating conditions, including lymphedema, which causes swelling of arms, legs, breasts and genitalia, and hydrocele, which causes swelling of the scrotum in males [16,17]. It has been estimated that approximately 13% of infected people suffer from the first condition while up to 21% of males living in endemic areas may experience hydrocele. As a result, and following the conclusion by an independent International Task Force for Disease Eradication that lymphatic filariasis may be one of only six infectious diseases that can be considered to be "eradicable" or "potentially eradicable" [18], the World Health Assembly in 1997 adopted Resolution WHA50.29 calling for the elimination of LF as a public health problem globally.

Although attempts have been made in the past to map the geographic distribution of LF in Africa, this has either been based on simply displaying infected sites as points or as ranges interpolated between such points on local-level maps [19,20,21,22,23,24,25], geostatistical modelling of point prevalence data at regional levels [10,26] or mapping of aggregated levels of infection at various within and between country or regional levels [9,10,27]. The exception has been the work of Lindsay and Thomas [8], who used published community LF prevalence data in conjunction with climate layers and a logistic regression model to predict the distribution and refine the first estimates of the population at risk for LF across sub-Saharan Africa [9].

These statistical modelling approaches have been important in describing and delimiting geographic ranges of species distributions; however, recent studies have highlighted several limitations of applying these models to mapping parasite distributions. First, simple statistical models, such as logistic regression, are restricted because they often fit linear functions between environmental variables and presence/absence data, when it is most likely that such associations are highly complex and non-linear [11,28]. Second, it is also difficult using these methods to address complex interactions between such variables [29,30]. Finally, using absence data in logistic regression modelling of LF distribution is complicated by the unreliability of such data owning to the use of variable blood volumes for diagnosing mf infection [31]. The key problem here is that any "absence" record may either represent a true absence of infection (implying non-suitability of location) or arise as a limitation of parasite detectability, whereas if infection is recorded as being present in a location, it is fairly certain that it occurs there.

Here we adopt a machine learning approach that allows flexible modelling of complex non-linear dependencies between infection presence and predictor variables in multidimensional space. This allows us to better understand the ecological niche and to construct a more reliable map of the potential spatial distribution of LF [30,32,33,34]. Such ecological niche models predict the geographic range of a disease or species by: (1) extracting associations between presence data and environmental covariates, (2) using these relationships to characterise the environmental requirements of the species, and (3) deploying this information to predict suitable habitats over unsurveyed areas. This approach has traditionally been used to predict the geographic range of species [34,35], but more recently it has been used to model the distribution of diseases [36,37,38].

There are currently a wide array of algorithms that can be used to model species' ecological niches using machine learning approaches [39,40]. In this study, we evaluated Maxent, a presence-only maximum entropy-based niche modelling technique [41], to describe the ecological requirement and current potential distribution of LF in Africa, and to determine for the first time how future climate change may affect the distribution and burden of this disease on the continent so that better prevention and control efforts could be directed to mitigate against the effects of such change.

Methods

LF Occurrence Data

Point data for LF occurrence or presence were collated from community surveys published in the research literature dating from 1940 to 2009, using the online and manual search procedures described in Michael *et al.* [17]. Studies were selected if the surveys described the number of people surveyed, the number positive for microfilaraemia, and were conducted at a specific community site. We found a total of 664 community-specific datapoints of which 508 comprised presence data. These were used in the present analysis (see details of selected studies in Appendix S1 in Supporting Information). Geo-coordinates for each chosen datapoint were either referenced from information given in the literature or by using Google Earth (see Figure 1). We were unable to find latitude and longitude details for 19 of these data points, while geo-coordinates for approximately 21% of the data locations used were only expressed to 2 decimal places.

Environmental Layers

We initially selected ten environmental and demographic variables, believed to influence the transmission of LF in this analysis [8]. Population density has not normally been employed as a predictor in most previous studies of pathogen distribution modelling; however, we view it as a key determinant of the potential distribution of LF for two reasons: 1) it is a component of the basic reproduction number for vector-borne diseases, such as LF, which determines the extend of spread and prevalence of such diseases, and 2) LF can only occur in inhabited places as the humans are the only host reservoir of the LF parasite in Africa [37,42,43,44].

The use of interpolated climate data or remote sensing data in combination with advanced statistical techniques to map the distribution of vector-borne diseases has accelerated greatly over the last 25 years [45,46]. Interpolated climate data layers are created by collecting large amounts of weather station data which are then processed to produce continuous climate maps using various smoothing algorithms. One of the most commonly used interpolated global climate data resource is WorldClim (www.worldclim.org) [47]. The WorldClim data are a set of climate data layers of the whole world available at resolutions of around 1 km, 5 km, 9 km or 18 km. The variables available are monthly mean, minimum and maximum temperature and monthly precipitation, and 19 derived bioclimatic variables. The WorldClim layers representing current climate conditions are smooth maps of averaged monthly climate data obtained over the period 1950–2000 from thousands of weather stations (47,554 locations for precipitation data, 24,542 for mean temperature, and 14,835 for minimum and maximum temperature – www.worldclim.org). The data have been interpolated down to a 30 arc-second high resolution grid (often referred to as "1 km^2" resolution) using a second-order thin plate smoothing spline with altitude, longitude and latitude as independent variables (Hijmans et al. 2005). Uncertainty in the data can arise from inaccurate weather station data or from the interpolation method – this second effect will be magnified in areas with sparse weather station data. For example, while precipitation data are fairly densely distributed in Africa, temperature data is much sparser. There are also very few data points in areas with low population density, particularly in the Sahara and Central Africa (Hijmans et al. 2005). These heterogeneities mean that such data and modelling uncertainties must be taken into consideration when assessing the accuracy of the predictions from the Maxent model.

The worldclim dataset is useful for infection mapping as the data are freely available on a small spatial scale. The data can be used to create new data layers, for example minimum temperature in the coldest month, or maximum temperature in the hottest month, to represent the temperature extremes in a region that could be important for vector and parasite dynamics [48,49]. One major drawback is that the climate surfaces represent average temperature or precipitation over a period of time, and hence there is no indication of the annual variability which could have a major impact on transmission dynamics.

Altitude data for this study were also obtained from www.worldclim.org – these data were collected by http://www2.jpl.nasa.gov/srtm/ and produced from data collected by a radar system circulating the earth to create a high resolution map of the globe. Similarly, NDVI data were downloaded from http://edit.

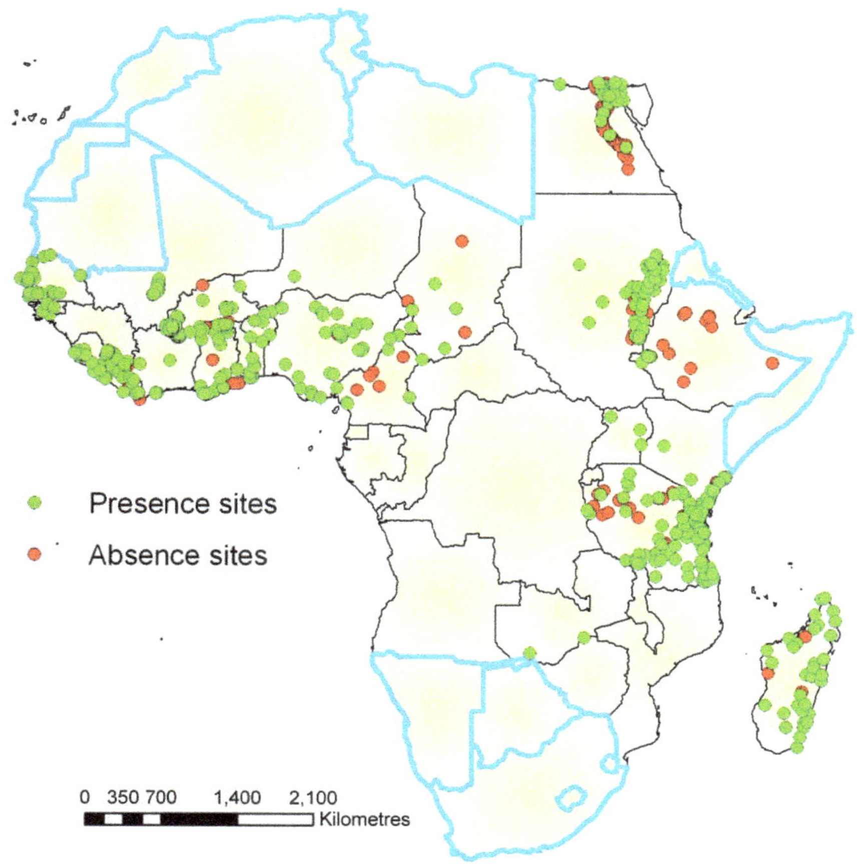

Figure 1. Locations of study sites. Green points show sites where LF infection were found to be present and red sites show sites where it was absent. Non-endemic countries are outlined in blue.

csic.es/GISdownloads.html; these maps were originally obtained from satellite images (NOAA-AVHRR) over the entire globe. Twelve monthly NDVI maps are available, each of which represents the mean monthly NDVI over an 18 year period from 1982 to 2000. We averaged these maps to produce an annual mean NDVI map. Population density data was created using data from, amongst others, the Socioeconomic Data and Applications Center (SEDAC) at Columbia University (http://sedac.ciesin.columbia.edu/gpw/). These data are created by interpolating global census data to create smooth population maps which are then scaled to match United Nations totals.

The data had slightly different spatial scales (worldclim data ~9 km^2, NDVI ~12 km^2 and population density ~5 km^2), and so were resampled using ArcGIS to give all the layers the same grid size. This resulted in a scale of around 12 km^2.

Ecological Niche Modelling

The ecological niche of a species can be defined as those ecological conditions under which it can maintain populations without immigration [50]. Ecological niches and associated potential geographic ranges can be approximated using correlative algorithms that by relating known point-occurrence data to digital GIS data layers, summarize spatial variations in these layers in multidimensional environmental space [51]. Here, we used the maximum entropy method as implemented by the Maxent software to derive the ecological niche for LF occurrence in Africa. We initially compared the performance of Maxent with another widely used modelling package GARP [52]. Maxent was selected for further use in this study as it performed better in tests of model predictive ability (Appendix S2).

Maxent is a general-purpose machine learning programme and has been widely used to predict species distributions [33,41,53,54]. The maxent algorithm essentially builds ecological niche models by quantifying the unknown probability distribution defining the occurrence of a species across a study area without inferring any unfounded information about the observed distribution. The approach aims to find the probability distribution of maximum entropy (that which is closest to uniform) subject to constraints imposed by the observed spatial distributions of the species and environmental conditions. Maxent thus outputs the maximum entropy distribution that satisfies these constraints, thereby providing the least biased description for a given dataset [41,55]. We implemented Maxent models using version 3.3.1 of the software developed by S. Phillips and colleagues (http://www.cs.princeton.edu/~schapire/maxent/). Selection of the convergence threshold and regularization values was carried out following default rules and the number of iterations was chosen such that all models converged. The default logistic model was used to ensure that predictions gave estimates between 0 and 1 of the probability of infection presence per map pixel.

Performance Measures

The performance of a model predicting the potential distribution of species presence is traditionally assessed by calculating the area under the curve (AUC) of the receiver operator characteristic (ROC) [56]. This is a plot of the sensitivity (the proportion of correctly predicted known presences, also known as absence of omission error) vs. 1-specificity (the proportion of incorrectly predicted known absences or the commission error) over the whole range of threshold values between 0 and 1. The model AUC thus calculated is compared to the null model which is an entirely random predictive model with AUC = 0.5, and models with an AUC above 0.75 are normally considered useful [57]. Using this method, the commission and omission errors are therefore weighted with equal importance for determining the performance of the model. However, for a presence only ecological niche model this method may be unsuitable for two key reasons [58,59]: 1) we are less interested in the performance of the model over all the whole of the ROC space, for example, where the omission or commission error is very high, and 2) as we do not have absence data, Maxent simulates pseudo absence data which are drawn at random from the training region. Since these do not represent true absences, mispredicting a known presence may be a more serious failing of the model than mispredicting a possible absence because while the presences are known, the absences are 'guessed'. In addition omission error has been shown to provide a better metric than commission errors for assessing model fit [60].

For these reasons, we carried out the analysis of model performance using the partial AUC procedure as described in Peterson *et al.* [59]. The criticisms raised above are answered using this method by: 1) using only presence data (not pseudo absence data) and 2) introducing a user defined variable E which refers to the maximum allowable level of omission error. The ROC curve is now a plot of sensitivity versus the proportion of the study area predicted as present. Only the region where the omission error is less than E is considered. The partial AUC is then a ratio of the AUC of the restricted ROC curve to the AUC of the restricted null model line (see Figure 2 and Peterson *et al.* [59] for full details of this method). The partial AUC was calculated using Simpsons trapezium rule via routines implemented in R. We closely examined two levels of omission error, E = 100 which is essentially a traditional ROC plot as we are assessing the model over all levels of omission error, and E = 10 where we assume that 10% of the positive predictions are actually negative, ie., we are only concerned with assessing models where omission error is less than 10%. Note that overlooking specificity could have significant effects on model accuracy as well as the predicted prevalence of infection (the overall proportion of locations where infection is predicted to be present). This outcome, however, is unlikely to be a major problem for the present study given that 76% of the surveys in our overall dataset (see Methods) reported positive LF infection, with analytical studies showing that at this moderately high level of prevalence specificity issues may have low significance for binary classification [58–63].

Model Implementation

The data were split into two groups: 75% was used to construct the model and form the functional relationships between presence and the environmental variables, and the remaining 25% was used to test the predictive ability of the model. The training region was chosen to be all the countries that are thought to be LF endemic, and the resulting model was projected over the whole of Africa. We assessed model performance by considering the partial AUC values of the testing data. We estimated the error associated with these values by performing a bootstrap algorithm, where we sample with replacement from the testing data 200 times and calculate the partial AUC for each sample.

Maxent has five feature classes (linear (L), quadratic (Q), product (P), threshold (T) and hinge (H)) that can be used to model the functional response of presence probability to changes in the environmental variables [41,54]. We experimented with using different combinations of features to produce the best performing model. Some of the explanatory layers are also likely to be more predictive than others. We thus aimed to find a set of variables that are predictively powerful and independent as possible. We employed two techniques to determine the most important variables: 1) by considering the percentage contribution that each variable made to the total test gain; and 2) by determining which

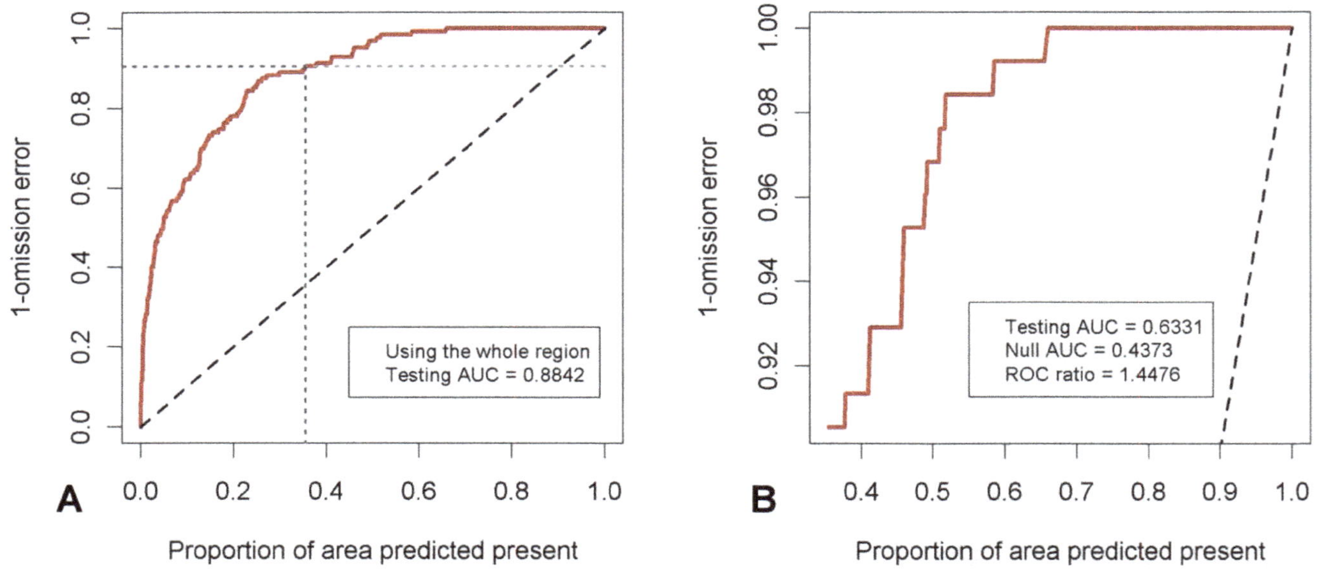

Figure 2. Comparison of traditional versus partial receiver operating characteristic (ROC) curves for the Maxent A model. A) Traditional ROC curve. The horizontal dash line indicates the region of interest for the partial AUC plot – where the omission error is less than E (in this case E = 10). B) The partial AUC plot. The dashed line indicates the null model.

variables caused the biggest lost in AUC when the data was resampled using a jackknife procedure where one variable was excluded at a time.

In addition, a quadratic discriminant analysis (QDA) was carried out in R to explore how interactions between the identified climate variables determine areas of LF presence or absence. Discriminant analysis essentially seeks to assign data into a series of discrete groups or classes based on the characteristics (**X**) of each data point, such that the probability of correct classification is maximised. QDA extends simple Linear Discriminant Analysis by allowing the intraclass covariance matrices to differ between classes, so that discrimination is based on quadratic rather than linear functions of **X**. In our case, we used QDA to classify presence and absence data correctly based on the climatic conditions of each point.

Estimating populations at risk

We estimated the number of individuals at risk by overlaying a LF binary risk map on a population map and calculating the population in the 'positive' at-risk cells. The SEDAC 2010 population layer for Africa was used for calculating the current at-risk population (http://sedac.ciesin.columbia.edu/gpw/global.jsp). Note that climate data for 2010 was unavailable, and we were therefore forced to use the data averaged between 1950 and 2000 for making these estimations. We constructed the LF binary layer by converting the continuous risk maps produced by Maxent into areas that are suitable and unsuitable by defining thresholds below which the probability of LF occurrence is considered to be zero and above which the probability is considered to be one. Traditionally these classification thresholds are determined by selecting the value that a) maximises the sum of sensitivity and specificity [61], b) where commission error = omission error [62] or c) is equal to the lowest predicted probability at a training presence site [41,63]. However, methods a) and b), as noted above, assume equal importance of omission and commission errors, and method c) is not suitable when we have an accepted level of omission error. When E = 100 we adopt the lowest training presence threshold approach, and when E = 10 we use a slightly modified version of c) suggested by A.T. Peterson (personal communication), where we take the threshold to be the value of the predicted probability from the Eth quantile of the values at training data sites (ie. when E = 10 we use the 10th percentile training presence value).

Future LF Predictions

The future potential distribution of LF was estimated by using the current Maxent model to make projections over projected climate and population density for 2050. The future climate data were downloaded from www.worldclim.org. These layers were constructed using data from general circulation models (GCMs). The IPCC report [64] considers around 25 GCMs and several emissions scenarios. The temperature projections amongst all the climate models are fairly consistent, however, there is much more uncertainty regarding precipitation. In this study, we consider just two of these GCMs - the Hadley Centre global climate model HADCM3 and the Canadian Centre for Climate Modelling and Analysis model CCCMA under two IPCC climate scenarios – A2a and B2a [65]. A2a is a more extreme scenario, assuming massive disparities between regions in high population growth and energy use, whereas B2a aims to capture a less disparate world with efforts focused towards social equity; this scenario also assumes lower population and economic growth than A2a. To account for differences in population growth between the two climate scenarios we multiplied the 2000 population data by country specific UN medium variant population growth rate predictions for the B2a scenario and by the high variant growth rate predictions for the A2a scenario (http://esa.un.org/unpp/).

Note that WorldClim provides projected future climate data (for years 2020, 2050 and 2080) at four spatial resolutions; 30 seconds (~1 km^2 spatial resolution), 2.5 minutes, 5 minutes, and 10 minutes (~344 km^2 resolution). These data have been produced with a simple downscaling technique from the coarser resolution predictions of climate models. In this procedure, projected changes in a climate variable, specifically the absolute or relative differences between outputs of a GCM simulation for the baseline years (typically 1960–1990 for future climate studies) and the simulated target years (eg. 2050), are first developed. Then, these changes are interpolated to grid cells with 30 arc-second resolution, with the assumption made that the change in climate is relatively stable over space (ie. has high spatial autocorrelation). Finally, these high resolution changes are applied/calibrated against interpolated observed climate data of the current period (WorldClim data set) to get high resolution projected climate data of the target year.

Results

Model Selection

Maxent models can be run with any combination of five feauture classes or real-valued functions, $f_1,\ldots f_n$ on environmental variables, X (viz. linear, quadratic, product, threshold and hinge. We initially ran a series of models using different combinations of these feature classes (L,Q,P,T,H) and selected three candidate models with the highest testing partial AUC values to investigate further. Model A employed the quadratic and threshold features, model B used the linear and threshold features, and model C used all the feature classes.

The relative importance and contribution of the original ten environmental, altitude and population density variables to the initially selected three niche models of LF occurrence, assessed by considering the percentage contribution that each variable made to the total test gain and by using a jackknife procedure to determine which of these variables caused the biggest lost in model AUC when each was excluded one at a time, resulted in the selection of the following five variables: population density, mean maximum temperature, mean temperature in the coldest month, mean annual precipitation and altitude. Together they accounted for more than 88% of the total test gain. Specifically, these were selected by firstly excluding the variables which performed poorly using both methods: NDVI, annual mean temperature, mean temperature in the warmest month, and secondly, by identifying the most correlated variables (mean temperature in the coldest month and mean minimum temperature (0.92), and precipitation in the wettest month and mean annual precipitation (0.95)), and selecting the best performing variable from each pair. These were mean temperature in the coldest month as it contributed more than twice as much to the test gain and performed similarly using the jackknife test, and mean annual precipitation as it added slightly more to the test gain and caused a bigger loss in AUC when excluded using the jackknife test.

The three selected models were rerun with the new set of five explanatory layers and model performance was assessed using two different levels of acceptable omission error. This showed that model A, which uses quadratic and threshold features (Table 1), has a slightly higher combined testing partial AUC and the highest entropy. Figure 2 compares the partial AUC plot (E = 10) for model A (Figure 2b) against the whole AUC plot (E = 100) (Figure 2a) with 1- omission error depicted on the y-axis

Table 1. Summary results from the three Maxent models tested in the present analyses.

Model (Features)	E = 10	E = 100	Entropy	Environmental Variables (% contribution)	Features
A (Q+T)			**8.2938**	Population density 57.3	Q,T
	1.4656	**1.7622**		Altitude 26	Q,T
	1.3331, 1.5927	1.6934, 1.8336		Mean temp in coldest month 6.7	Q,T
				Mean annual precipitation 5.2	Q,T
				Mean max temp 4.8	Q,T
B (L+T)			**8.2921**	Population density 57.3	L,T
	1.4562	**1.7646**		Altitude 25.9	L,T
	1.3315, 1.6057	1.6762, 1.8322		Mean temp in coldest month 6.8	L,T
				Mean annual precipitation 5.3	L,T
				Mean max temp 4.8	L,T
C (All)			**8.2577**	Population density 57	All
	1.4227	**1.7455**		Altitude 25.4	Q,P,T,H
	1.3278, 1.5600	1.6461, 1.8249		Mean temp in coldest month 6.8	L,P,T,H
				Mean annual precipitation 5.9	L,P,T,H
				Mean max temp 4.7	L,P,T,H

See text for explanations of terms. Model A with quadratic and threshold features was selected as it performed the best using the two E – the acceptable level of omission error – thresholds used in this study.

and the proportion of area predicted positive on the x-axis for both plots.

The relative contributions of the explanatory variables to the different Maxent models (assessed using the jackknife procedure) is shown in Table 1. The results indicate that population density contributed the most (up to 57%) to each of the tested models, followed by altitude (around 26%) as the next most significant factor. For model A, the three climatic layers contributed in total to around 17% of the overall prediction of LF occurrence. All our final models performed significantly better than the null model (all partial AUC's >1.42), re-emphasizing the high predictability that can result from ecological niche modelling using the Maxent programme [41,66].

Model Predictions

The distribution of LF occurrence in Africa predicted by the best performing Maxent model (A) is shown in Figure 3. The map shows that LF in Africa occurs over a large area extending from the west to the east primarily across the middle region of the continent. The results also depict a high degree of heterogeneity in the probability of LF occurrence on the continent. There appears to be a large zone exhibiting a high probability of LF occurrence in the Western Africa region, whereas in Central and Eastern Africa and in Madagascar, large areas of medium probability are interspersed with smaller areas of high probability, especially along the coasts. Importantly, all LF-free countries (as shown in Figure 1) are shown to have fairly low probabilities of infection. Most of the training data are located in west and east Africa and there are very few datapoints covering central Africa. Little is known about the state of LF in many of these countries, meaning we have no way to validate the model in these regions. For this reason, we need to be cautious when interpreting the results from these countries compared to more densely sampled countries.

Individual response curves (marginal responses obtained by keeping all other variables at their average sample value) of the relationships between each environmental variable and the probability of disease occurrence as estimated by model A are portrayed in Figure 4. The results clearly exhibit complex but quadratic relationships between each of the best five environmental/population drivers and probability of LF occurrence. In general, however, there is an overall negative response observed between altitude and LF occurrence and nonlinear positive responses observed for the rest of the variables. There also appears to be evidence for threshold effects in each of the estimated relationships (most clearly observed for the association between mean temperature in coldest month and probability of LF infection (Figure 4e)), wherein the probability of LF occurrence begins to increase only after about 10°C).

To visualise the LF ecological niche in Africa, the Maxent predictions were further related to environmental conditions at both presence sites and areas where the disease is known not to exist (Figure 5). The two-dimensional plots in the figure show that differences in the identified ecological conditions may strongly influence the probability of LF infection presence and absence. These results indicate that LF occurs mainly in the hot and wet regions of Africa, with non-endemic areas all having an annual rainfall level below around 100 mm. The mean maximum temperature and mean temperature in the coldest month both need to be relatively high for the disease to occur, with no presence sites occurring when the temperature in the coldest month is 3.7 degrees and the mean maximum temperature in 22.4 degrees.

Results from the quadratic discriminant analysis of the contribution of key environmental variables to LF occurrence are shown in Figure 6. These highlight not only that different regions of each variable space can determine where LF is likely to occur and not occur, but also the dependency of such classification on variable interaction. Thus, the levels of rainfall and temperature required for the disease to occur are dependent on each other, whereby in warmer regions, less rainfall is needed to sustain parasite transmission. However, a key finding is that the minimum threshold for mean temperature in the coldest month is around 11 degrees with apparently little variation in this value with increasing mean maximum temperature (Figure 6c).

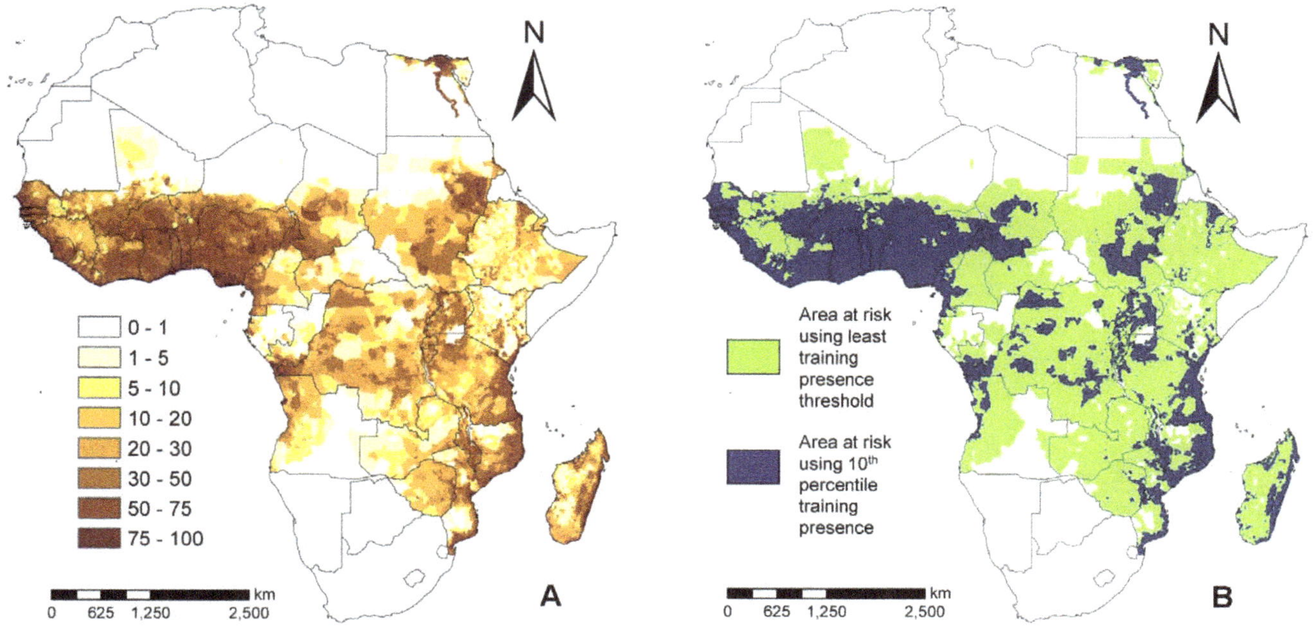

Figure 3. Probability and infection binary maps of the current occurrence of LF in Africa as predicted by the final Maxent A model containing the variables: mean annual precipitation, mean maximum temperature, and mean temperature in coldest month, altitude and population density, and quadratic and threshold features. A) Probability map where probability of occurrence is depicted in the form of percentages. B) LF binary map showing areas with and without infection presence for E = 100 (ie. using classification value from the least training presence threshold) and E = 10.

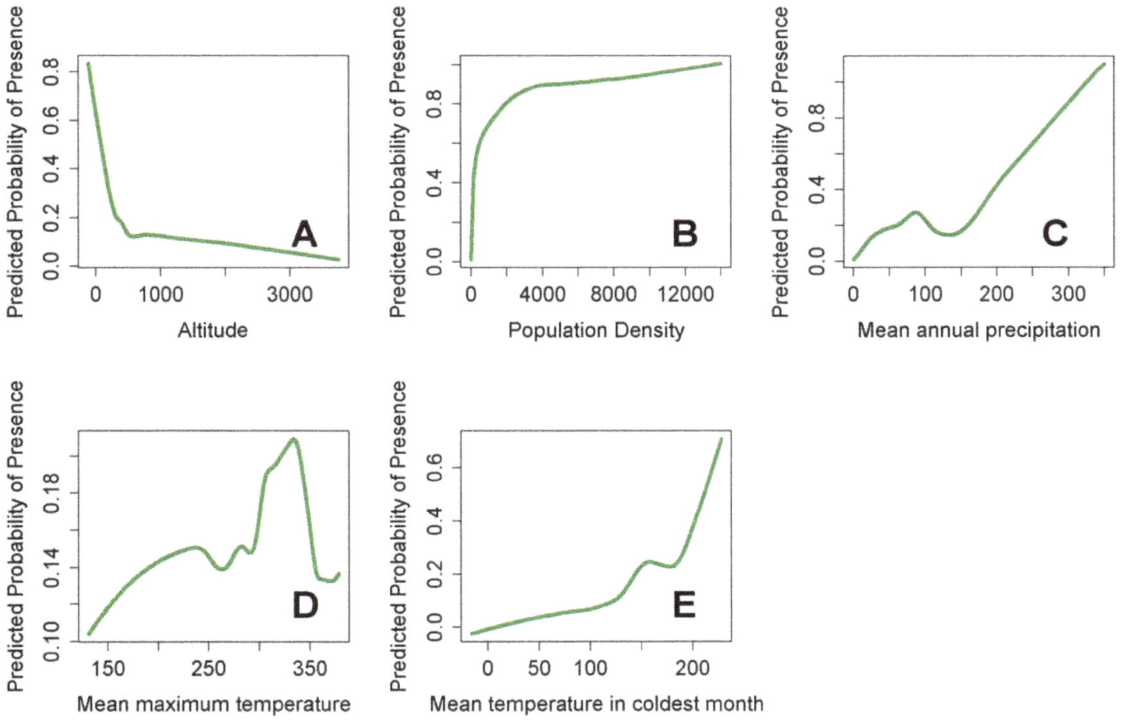

Figure 4. Graphs showing the marginal relationships between each environmental variable and the probability of LF occurrence. Temperature values are expressed in ×10°C, precipitation is in *mm* per month and altitude is in metres.

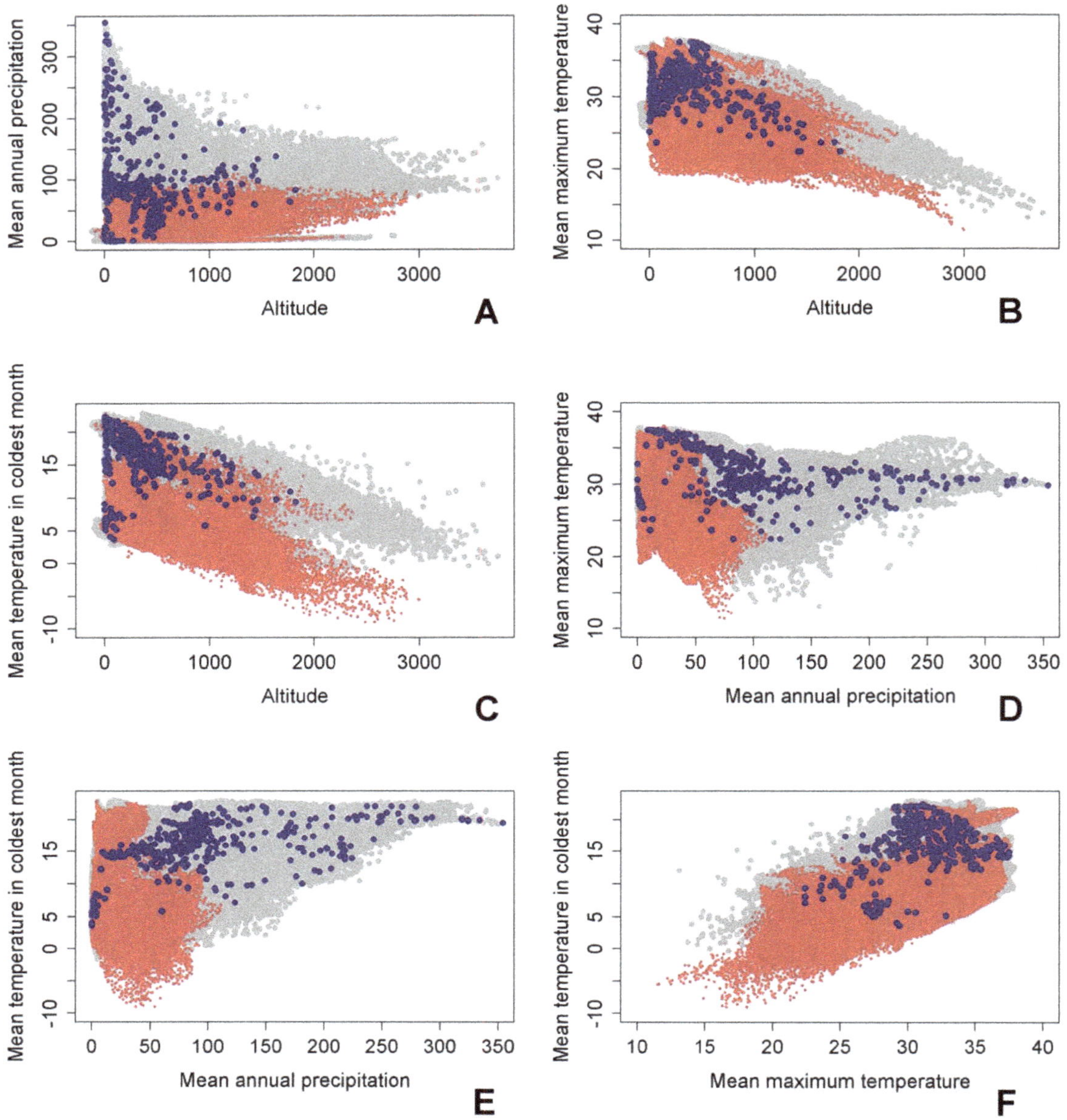

Figure 5. Visualizations of the modelled LF ecological niche in Africa. The grey points are the environmental conditions for every cell in Africa, the red points represent the conditions in non-endemic sites while the conditions underlying presence sites are shown in blue. Temperature values are expressed in ×10°C, precipitation is in *mm* per month and altitude is in metres.

Future Climate Predictions

We used model A in conjunction with the four climate change projections and their associated population growth estimates outlined earlier to investigate how the potential distribution of LF could change between now and 2050, assuming that no control measures are implemented. Our model predictions shown in Figure 7 indicate that LF occurrence could increase in large parts of Africa with the highest increases expected in areas bordering the current northern extent of the disease, particularly across regions of Mauritania, Sudan, and Somalia. LF occurrence is also predicted to increase in countries in the southern parts of the continent. The probability of disease occurrence could, however, decrease in other areas, mainly in the west near Ivory Coast and Nigeria and also the Democratic Republic of Congo (Figure 7). Overall, the mean change in probability of LF occurrence over the whole continent was found to be 0.1, suggesting that LF transmission is likely to increase in Africa as climate changes.

Estimating current and future populations at risk to LF

The populations at risk were estimated in this study by converting the Maxent prediction from model A into a binary map using two thresholds – the value of the least training presence (LTP) prediction which was 1.9% and the value of the 10th percentile of the training presence (10% TP) predictions, which was 29.8%. For each threshold, each cell in the map with a value above these values was deemed to as having LF present. The threshold map for 2010 is shown in Figure 3b and for 2050 in Figures 7c and 7d. The current (2010) population at risk to LF in

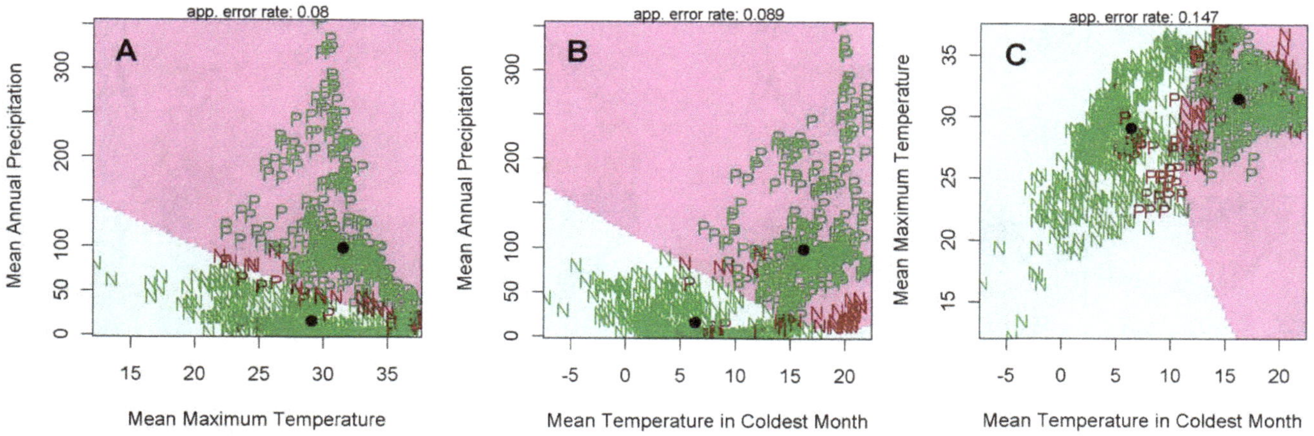

Figure 6. Results from the quadratic discriminant analysis used in this study to determine the importance of the climatic variables included in the final Maxent A model in classifying positive and negative LF occurrence sites in Africa. The purple area of the graph shows the area where infection is present, while the blue portion shows where it is absent. The green points represent correctly classified datapoints and the dark red points show incorrectly classified data points. 91% of the points were classified correctly.

Africa is calculated to be 804 million using LTP threshold (E = 100) and 542 million using the 10% TP (E = 10) threshold. The 2050 estimates range from 1.86 billion to 1.46 billion using the LTP threshold and from 1.65 billion to 1.30 billion using the 10% TP threshold (Table 2). On average, the A2a scenarios predict a larger at-risk population, indicating that 13% more people would potentially live in at-risk areas when compared with the effects of the B2a scenario.

Discussion

We have used an ecological niche modelling approach based on infection presence-only data to firstly reveal the spatial distribution of LF in Africa, and the environmental determinants that underlie this pattern, and secondly to investigate how climate change may affect the future potential distribution and burden of this important parasitic disease on that continent. The performance of the Maxent models developed here were assessed using the partial AUC measure, a modification of the usual AUC tool used for evaluating the accuracy of ecological niche models. The benefits of this method over a traditional AUC approach are that it: 1) eliminates the used of pseudo absence data in accuracy measurements, and 2) allows the user to define an acceptable level of omission error.

The advantages of using machine learning approaches, such as the maximum entropy modelling algorithm implemented in the Maxent programme, over simpler statistical tools, such as logistic regression, for species distribution modelling have been thoroughly reviewed previously [30,39,41]. Here, we highlight that two chief benefits of applying such methods to parasitic infection mapping arise from their flexibility in specifically accounting for: (1) the complex non-linear associations of infection occurrence with individual explanatory variables, and (2) the impact that interactions occurring among these variables may have on infection presence. This flexibility has provided new insights as to how climate variables may functionally influence LF presence in Africa.

Thus, for example, although the relationship between the probability of LF presence and mean annual precipitation was the least non-linear (Figure 4c), its impact on infection probability is found to be low below a threshold of around 150 mm per year. Biologically, this may be because a certain amount of water is needed to provide suitable laying sites for LF vectors. However, it has been suggested that vector survival can also be affected if there is too much rainfall as egg laying sites can get washed away [67]. If this is true, then our result might imply that such washouts will occur only at precipitation levels above 350 mm. Similarly, the LF occurrence - mean maximum temperature response curve (Figure 4d), is found to increase until it peaks between 25°C and 32.5°C, after which it begins to decrease suggesting this temperature range is the most suitable for LF transmission. This result is consistent with experimental findings that both mosquito survival [68] and the development of LF larvae within the mosquito [49] peak around 22–34°C. Although different measures of temperature were used, it is also consistent with the previous findings of Lindsay and Thomas [8], who found that the temperatures of sites in Africa with microfilaraemic individuals lie within the range between 22 to 30°C. However, our results also indicate that mean temperature in the coldest month (Fig. 4e) could induce the most non-linear effect on LF presence, showing that at temperatures <5°C, the probability of disease presence is almost zero but above this threshold to at least 22°C, a dramatic positive impact on parasite occurrence may occur. These findings suggest that fluctuations in temperature limits rather than mean temperature may represent the key temperature-related bioclimatic thresholds important for supporting LF transmission. In contrast to the effect of climate variables, the relationship between altitude and LF occurrence was found to be negative, although again the association was distinctly non-linear (Figure 4a). Such negative correlations between infection presence and altitude have been recorded previously in field studies [70,71], and most likely reflect the negative effect of falling temperature with increasing altitude (ie. the lapse rate) on mosquito survival rate and the rate at which the parasite develops within the vector [49,70].

Exploration of the Maxent modelling results has also allowed a first depiction of how subtle interactions between key climatic variables may govern the suitability of a geographic region for LF transmission to occur. The key finding here is that levels of precipitation and temperature in particular could interact strongly to define the multivariate space required for the disease to occur, with generally less rainfall needed in warmer regions to sustain parasite transmission and vice versa (Figure 5 and 6). The

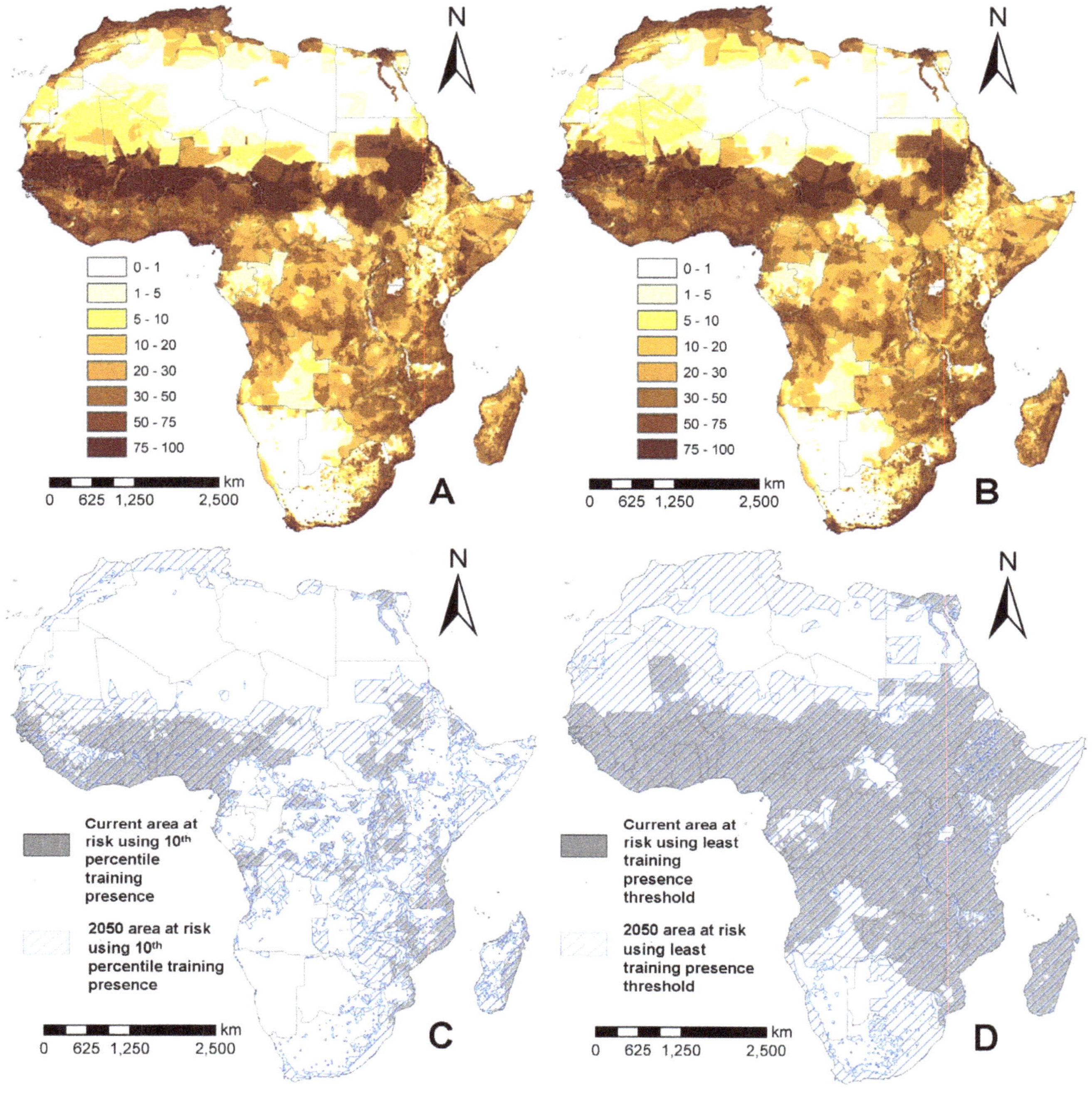

Figure 7. 2050 predictions of the distribution of LF using HADCM3 A2a and medium population increase scenario (A), and HADCM3 B2a and high population increase scenario (B). Model predictions for the CCCMA climate predictions are very similar to those shown here. (C) shows the predicted at-risk areas in 2000 (grey) and in 2050 (stripped) for the a2a scenario using a threshold value of 29.8% given by setting the acceptable omission error to E = 10. (D) shows the 2000 (grey) and 2050 (stripped) areas at risk from LF obtained with the a2a climate scenario and using the LTP threshold, with an associated threshold value of 1.9% (see text).

biological significance of this finding is that such interactive effects could result in compensatory responses among vector and parasite ecological traits (vector birth, survival and biting rates, and larval development rate in the vector) that would not only dampen the effects of variations in individual key climatic variables but also allow the transmission of LF to occur over a much wider area than would be the case if habitat suitability is defined solely by each single variable. However, the results also show that an important absolute limiting factor is that the minimum temperature threshold for mean temperature in the coldest month needs to be around 5°C for transmission to occur.

A major finding of this study is that human population density was by far the most significant variable that may influence LF occurrence in Africa. This supports not only theoretical expectations that host population density (and the attendant mosquito density) is a key driver of the transmission of vector-borne

Table 2. Population at risk from LF in endemic countries in Africa estimated applying two different threshold values based on different levels of acceptable omission error, E, to the final Maxent model A predictions.

	Population increase 2010–2050		Least training presence threshold			10th percentile training presence threshold		
Country name	medium	high	2010	2050 a2a	2050 b2a	2010	2050 a2a	2050 b2a
Angola	2.23	2.51	15.77 m	42.34 m	37.29 m	4.27 m	18.86 m	17.56 m
Benin	2.39	2.69	7.42 m	20.86 m	18.54 m	7.38 m	20.8 m	18.53 m
Burkina Faso	2.51	2.81	15.71 m	45.07 m	40.22 m	15.26 m	45.06 m	40.22 m
Burundi	1.74	1.97	8.76 m	16.76 m	14.8 m	5.35 m	9.16 m	9.08 m
Cameroon	1.84	2.1	18.13 m	38.48 m	33.79 m	13.3 m	31.29 m	27.61 m
Central African Rep.	1.69	1.95	4.04 m	8.19 m	7.1 m	0.82 m	4.13 m	3.41 m
Chad	2.41	2.7	10.51 m	28.91 m	25.78 m	7.01 m	28.28 m	24.93 m
Congo	1.83	2.08	3.96 m	7.77 m	6.94 m	3.02 m	6.14 m	5.5 m
Djibouti	1.67	1.89	0.57 m	1.2 m	1.07 m	0.42 m	1.01 m	0.89 m
Egypt	1.53	1.77	74.71 m	136.27 m	118.09 m	74.11 m	135.39 m	117.34 m
Equatorial Guinea	2.09	2.36	0.53 m	1.23 m	1.09 m	0.37 m	1.23 m	1.09 m
Ethiopia	2.05	2.31	78.08 m	173.5 m	156.7 m	6 m	29.9 m	25.7 m
Gabon	1.65	1.9	1.19 m	2.57 m	2.23 m	0.68 m	1.37 m	1.21 m
Gambia, The	2.15	2.44	1.36 m	3.24 m	2.88 m	1.36 m	3.24 m	2.88 m
Ghana	1.86	2.1	23.67 m	47.64 m	43.27 m	23.67 m	47.33 m	43.07 m
Guinea	2.32	2.62	9.83 m	25.57 m	22.68 m	4.96 m	22.61 m	18.36 m
Guinea-Bissau	2.16	2.42	1.43 m	2.91 m	2.65 m	1.41 m	2.88 m	2.63 m
Ivory Coast	2.01	2.29	18.3 m	41.48 m	36.99 m	17.39 m	40.37 m	36.2 m
Kenya	2.09	2.39	36.23 m	93.34 m	81.86 m	12.21 m	37.78 m	36 m
Liberia	2.16	2.45	4.52 m	9.35 m	8.41 m	4.33 m	9.28 m	8.34 m
Madagascar	2.12	2.42	20.44 m	48.95 m	42.91 m	13.47 m	41.57 m	38.25 m
Malawi	2.33	2.64	13.95 m	39.34 m	34.7 m	12.68 m	38.96 m	34.25 m
Mali	2.12	2.39	15.02 m	34.32 m	30.5 m	11.52 m	33.45 m	29.29 m
Mozambique	1.89	2.16	20.7 m	47.2 m	41.65 m	16.71 m	45.08 m	39.86 m
Niger	3.66	4.04	14.91 m	62.34 m	56.3 m	10.57 m	60.1 m	52.75 m
Nigeria	1.83	2.06	143.97 m	288.68 m	257.45 m	139.88 m	286.8 m	255.36 m
Rwanda	2.15	2.42	9.44 m	23.8 m	21.16 m	3.7 m	11.74 m	11.67 m
Senegal	2.03	2.3	9.75 m	20.71 m	19.03 m	9.16 m	20.71 m	19.02 m
Sierra Leone	2.13	2.42	5.9 m	13.12 m	11.67 m	5.71 m	13.03 m	11.65 m
Sudan	1.76	2	37.47 m	75.98 m	66.58 m	20.24 m	65.67 m	53.55 m
Tanzania	2.43	2.75	41.79 m	120.35 m	106.21 m	24.05 m	98.13 m	88.77 m
Togo	1.95	2.22	5.57 m	12.39 m	10.93 m	5.57 m	12.36 m	10.93 m
Uganda	2.7	3.04	32.52 m	97.25 m	86.75 m	25.32 m	83.56 m	76.17 m
Dem. Rep. of Congo	2.17	2.45	70.73 m	165.66 m	147.09 m	33.99 m	113.17 m	97.77 m
Zambia	2.18	2.48	12.67 m	32.73 m	28.84 m	2.25 m	21.99 m	18.53 m
Zimbabwe	1.75	2.04	14.9 m	26.04 m	22.44 m	4.76 m	21.38 m	19.16 m
Total			804.42 m	1,855.52 m	1,646.58 m	542.88 m	1,463.8 m	1,297.54 m

2010 populations at risk are calculated by projecting the derived Maxent model against 2000 climate data (but using 2010 population data (see text)) while the 2050 populations at risk are derived using the predicted 2050 environmental/population density data. All figures are expressed in 000's. Also given are the UN predicted increases in population, shown in terms of the factor s by which national populations are expected to increase.

infections [42], but is also biologically intuitive given that the adult parasites live in the host and mosquito vectors have a preference for human blood meals to oviposit and reproduce. This result would suggest that climate variables *per se* may play a smaller role in determining the ecological niche and hence the potential distribution of LF. However, given that the best-fitting Maxent model predicts low probabilities of infection occurrence in the known non-endemic and high altitude regions of the continent despite the presence of significant human populations in these areas (most indubitably via effects on both vector and parasites), it is clear that both population density and environmental variables will need to be included together in any study attempting to model the potential geographic distribution of this or any other parasitic infection [37,43,44].

The Maxent model for LF occurrence across Africa generated a risk map giving a probability of infection presence in each location between 0 and 1, with a probability close to 1 indicating sites with the highest risk and possibly levels of infection. Thus, the map shown in Figure 2 provides not only information regarding the potential spatial extend of LF endemicity across Africa but also crudely data on variations in the intensity of transmission that can be expected in different parts of the continent. Based on the variation in relative risk shown in the figure, the highest LF transmission therefore appears to occur in the Western Africa region, whereas infection levels in large parts of Central and Eastern Africa and in Madagascar are predicted to be considerably more heterogeneous, with moderate levels interspersed with smaller areas of high infection occurring along the coasts. Despite the fact that the present Maxent results are based only on presence data, this conclusion is surprisingly well supported by the actual national LF prevalence values estimated for the endemic countries of Africa given in Michael *et al.* [17] and Michael and Bundy [9]. This represents an important technical insight as it suggests that modelling of presence-only data may provide a good approximation to actual levels of parasite transmission intensity in an area [8], possibly due to climate-derived variations in the abundance of the relevant transmitting vector species. This is supported by the remarkable similarity of the African ecological niche maps for *Anopheles funestus* and *A. gambiae*, the two principal anopheline vectors of LF in Africa, developed by Moffett and colleagues [37,72], with the LF risk map shown in Figure 2.

Estimations of the at-risk population for LF in Africa have varied significantly between previous studies, with recently reported figures appearing to increase over time possibly due to the effects of both increasing population and surveillance. Thus, in 1992, it was estimated that some 113 million individuals lived in endemic areas [73], which increased to 212 million [74] and 396 million by 2009 [75]. These estimates, which are normally made by identifying infection positive districts and calculating the number of people in each of these districts, not only take little account of the spatial variation that exists at the local level [22,26,76] but are also highly dependent upon the existence of field surveys covering all relevant endemic areas. By contrast, machine learning-based ecological niche models using presence-only data approximate the realised distribution of a disease [54], and via the derivation of a continuous potential distribution map may offer a more accurate method to determine the true extent of infection and hence actual populations at risk. The additional prospect of being able to use different cut-off disease presence thresholds with this method means that we can also explore the implications of error in the data for quantifying risk and disease burdens. Thus, using a low presence threshold equal to the least training presence, we estimate that 804 m people in Africa may be living in at risk areas, whereas assuming a more stringent 10% omission error, we estimate that some 542 m people may be at risk. Given that the average prevalence of LF (infection and disease) in African endemic countries has been estimated to be around 11% [8,17], our estimate of the populations at risk also thus suggests that we can expect between 60 and 89 million cases of LF in 2010 on that continent compared to the 51 million and 47 million estimated by Michael and Bundy [9] and Lindsay and Thomas [8] respectively for the years 1990 and 2000.

The future potential distribution and burden of LF in Africa as a result of predicted changes in climate and population growth were produced by using the best-fit Maxent model derived for 2000 and projecting the functional relationships therein onto the two 2050 climate scenarios, i.e. we consider that niche dynamics are static and that climate change will not affect either the form of the biotic relationships governing the vector and the parasite population dynamics or any adaptation by these populations to the new environment [32,77,78]. We examined the impact of two 2050 climate scenarios from two different global climate models. The more extreme scenario, A2a, predicted on average 13% more people living in at-risk areas than the B2a scenario. Predictions of the 2050 population at risk range from 1.30 to 1.86 billion (Table 2); although a large component of this increase is a result of population growth, changes in climate are also shown to increase the area of Africa that is suitable for LF transmission. In particular, large regions below the Sahara desert and in Zambia, Zimbabwe and Angola are predicted to have increased probability of LF (Figure 7c,d), suggesting that the ecological niche of infection could increase and extend both northwards and southwards. When interpreting these results attention must be paid to the uncertainty and error associated with the future climate data – both from the GCMs and the downscaling procedures adopted, and the fact that we are only considering two GCMs and two emissions scenarios. These results obviously do not take into account increases in disease control activities on the continent of Africa, which has accelerated greatly since the Global Alliance to Eliminate Lymphatic Filariasis was created in 2000. It also does not take account of the increase in vector control on that continent, primarily targeted at malaria, which will have an impact on LF infection via reductions in vector biting rates and lifespans [79]. Indeed, our predictions of the likely future increase in LF burden argues strongly for strengthening and expanding these interventions even further as an important mitigation strategy to counter the predicted spread and intensifying of this debilitating disease in Africa as population density increases and climate changes.

Although our study has yielded several important and novel insights into the determinants and structuring of the ecological niche and the present and future spatial occurrence of LF in Africa, there are several limitations that need to be borne in mind when interpreting the present results. First, even though ecological niche modelling approaches based on occurrence data alone, such as the Maxent algorithm used in this study, are optimized for predicting the realised or actual (rather than the fundamental) distribution of a species [41,54], predictions of presence will still be dependent on the sample locations of the available data with any deficiency in sample coverage of all suitable areas able to bias the results. Second, the crude scale of the environmental layers used to construct the Maxent model means that the validity of predictions on small focal spatial scales is questionable. Third, we have no error estimates associated with our predictions – in reality we would expect a heterogeneous error map of model predictions in Africa caused by different levels of error associated with the climate data and model fit, and from the biased distribution of presence data. However, it is hoped that our predictions are fairly robust on the district or country wide scales that are typically used in policy decisions regarding disease control and eradication strategies, especially in countries with more accurate climate data and more LF survey data. We have also used 50-year averaged climate layers to approximate a phenomenon that might have changed in the past decade or so to characterise "current" climate in our analyses.

The above caveats indicate that our application is likely to be at the lower limit of the usefulness of the available data. Although it might be possible to use remote-sensed data to overcome a part of this limitation [45], forward projection of such data to future climates is clearly not possible. Combining correlative spatial modelling approaches with mechanistic models linking climate/environmental and population variables to parasite transmission processes in conjunction with regional climate models, may, on the other hand, provide a more useful solution to improving the detail of spatial predictions [69]. The practical modelling frameworks and tools required for successfully achieving this synthesis is, however, still largely indeterminate. We suggest that resolving these conceptual and methodological issues represents the next major challenge in species, including parasite, distribution modelling.

Supporting Information

Appendix S1 Details of published data used in the Maxent analysis. The number of data points from each study or review is given in brackets. The list of study references for the data used are given below the table.

Appendix S2 Comparison of GARP and Maxent model fits. Partial AUC ratios taken across 200 bootstrap replications are shown for omission errors, E, of 1 and 5.

Author Contributions

Conceived and designed the experiments: EM HS. Analyzed the data: HS EM. Wrote the paper: HS EM.

References

1. Hay SI, Omumbo JA, Craig MH, Snow RW (2000) Earth observation, geographical information systems and *Plasmodium falciparum* malaria in sub-Saharan Africa. Adv Parasitol 47: 174–215.
2. Rogers DJ, Randolph SE, Snow RW, Hay SI (2002) Satellite imagery in the study and forecast of malaria. Nature 415: 710–715.
3. Rogers DJ, Randolph SE (1993) Distribution of tsetse and ticks in Africa, past, present and future. Parasitol Today 9: 266–271.
4. Robinson TP (1998) Geographic Information Systems and the selection of priority areas for control of tsetse-transmitted trypanosomiasis in Africa. Parasitol Today 14: 457–461.
5. Brooker S (2007) Spatial epidemiology of human schistosomiasis in Africa: risk models, transmission dynamics and control. Trans R Soc Trop Med Hyg 101: 1–8.
6. Clements AC, Brooker S, Nyandindi U, Fenwick A, Blair L (2008) Bayesian spatial analysis of a national urinary schistosomiasis questionnaire to assist geographic targeting of schistosomiasis control in Tanzania, East Africa. Int J Parasitol 38(3–4).
7. Richards FO (1993) Use of geographic information systems in control programs for onchocerciasis in Guatemala. Bull Pan Am Health Organ 27: 52–55.
8. Lindsay SW, Thomas CJ (2000) Mapping and estimating the population at risk from lymphatic filariasis in Africa. Trans R Soc Trop Med Hyg 94: 37–45.
9. Michael E, Bundy DA (1997) Global mapping of lymphatic filariasis. Parasitol Today 13: 472–476.
10. Sabesan S, Palaniyandi M, Michael E, Das PK (2000) Mapping lymphatic filariasis at the district-level in India. Ann Trop Med Parasitol 94: 591–606.
11. Brooker S, Michael E (2000) The potential of geographical information systems and remote sensing in the epidemiology and control of human helminth infections. Adv Parasitol 47: 246–288.
12. Hartley S, Harris R, Lester PJ (2006) Quantifying uncertainty in the potential distribution of an invasive species: climate and the Argentine ant. Ecol Lett 9: 1068–1079.
13. Roura-Pascual N, Suarez AV (2008) The utility of species distribution models to predict the spread of invasive ants (hymenoptera: Formicidae) and to anticipate changes in their ranges in the face of global climate change. Myrmecol News 11: 67–77.
14. Ward DF (2007) Modelling the potential geographic distribution of invasive ant species in New Zealand. Biol Invasions 9: 723–735.
15. González C, Wang O, Strutz SE, González-Salazar C, Sánchez-Cordero V, et al. (2010) Climate Change and Risk of Leishmaniasis in North America: Predictions from Ecological Niche Models of Vector and Reservoir Species. PLoS Negl Trop Dis 4(1): e585.
16. WHO (2004) Lymphatic filariasis elimination in the African region: progress report. WHO, Congo Brazaville.
17. Michael E, Bundy DA, Grenfell BT (1996) Re-assessing the global prevalence and distribution of lymphatic filariasis. Parasitology 112: 409–428.
18. CDC (1993) Recommendations of the International Task Force for Disease Eradication. MMWR Morbid Mortal Wkly Rep 42: 1–38.
19. Brinkmann UK (1976) Epidemiological investigations of Bancroftian filariasis in the Coastal Zone of Liberia Tropenmed. Parasitol 28: 71–76.
20. Casaca VMR (1966) Contribuicao para o estudo da filariase *bancrofti* em Angola. An Inst Hig Med Trop (Lisb) 23: 127–132.
21. Sowilem MM, Bahgat IM, el-Kady GA, el-Sawaf BM (2006) Spectral and landscape characterization of filarious and non-filarious villages in Egypt. J Egyt Soc Parasitol 36: 373–388.
22. Thompson DF, Malone JB, Harb M, Faris R, Huh OK, et al. (1996) Bancroftian filariasis distribution and diurnal temperature differences in the southern Nile delta. Emerg Infect Dis 2: 234–235.
23. Wijers DJB (1977) Bancroftian filariasis in Kenya. I. Prevalence survey among adult males in the Coast Province. Ann Trop Med Parasitol 71: 313–331.
24. Juminer B, Diallo S, Diagne S (1971) Le foyer de filariose lymphatique du secteur de Sandiara (Senegal). 1. Evaluation de l'endemicite. Ach de l'Inst Pasteur de Tunis 48: 231–246.
25. Lamontellerie M (1972) Resultats d'enquetes sur les filarioses dans l'Ouest de la Huate-Volta (Cerle de Banfora). Annal Parasitol Hum Comp (Paris) 47: 743–838.
26. Srividya A, Michael E, Palaniyandi M, Pani SP, Das PK (2002) A geostatistical analysis of the geographic distribution of filariasis infection prevalence in Southern India. Am J Trop Med Hyg 67: 480–489.
27. Meyrowitsch DW, Nguyen DT, Hoang TH, Nguyen TD, Michael E (1998) A review of the present status of lymphatic filariasis in Vietnam. Acta Trop 70: 335–347.
28. Lawler JJ, White D, Neilson RP, Blaustein AR (2006) Predicting climate-induced range shifts: model differences and model reliability. Global Change Biol 12: 1568–1584.
29. Austin MP (2002) Spatial prediction of species distribution: an interface between ecological theory and statistical modelling. Ecol Model 157: 101–118.
30. Pearce JL, Boyce MS (2006) Modelling distribution and abundance with presence-only data. J Anim Ecol 43: 405–412.
31. Michael E, Malecela MN, Zervos M, Kazura JW (2008) Global eradication of lymphatic filariasis: the value of chronic disease control in parasite elimination programmes. PLoS One 3: e2936.
32. Guisan A, Thuiller W (2005) Predicting species distribution: offering more than simple habitat models. Ecol Lett 8: 993–1009.
33. Peterson AT, Papes M, Eaton M (2007) Transferability and model evaluation in ecological niche modeling: a comparison of GARP and Maxent. Ecography 30: 550–560.
34. Peterson AT (2006) Ecologic niche modeling and spatial patterns of disease transmission. Emerg Infect Dis 12: 1822–1826.
35. Soberon J, Peterson AT (2005) Interpretation of models of fundamental ecological niches and species' distributional areas. Biodivers Infor 2: 1–10.
36. Holt A, Salkeld D, Fritz C, Tucker J, Gong P (2009) Spatial analysis of plague in California: niche modeling predictions of the current distribution and potential response to climate change. Int J Health Georgr 8: 38.
37. Moffett A, Shackelford N, Sarkar S (2007) Malaria in Africa: vector species' niche models and relative risk maps. PLoS One 2: e824.
38. Peterson AT, Shaw J (2003) Lutzomyia vectors for cutaneous leishmaniasis in Southern Brazil: ecological niche models, predicted geographic distributions, and climate change effects. Int J Parasitol 33: 919–931.
39. Olden JD, Lawler JJ, Poff NL (2008) Machine Learning Methods Without Tears: A Primer for Ecologists. Quart Rev Biol 83: 171–193.
40. Segurado P, Araújo MB (2004) An evaluation of methods for modelling species distributions. J Biogeogr 31: 1555–1568.
41. Phillips SJ, Anderson RP, Schapire RE (2006) Maximum entropy modeling of species geographic distributions. Ecol Model 190: 231–259.
42. Anderson RM, May RM (1992) Infectious Diseases of Humans. Dynamics and Control. Oxford: Oxford University Press.
43. Hales S, de Welt N, Maindonald J, Woodward A (2002) Potential effect of population and climate changes on global distribution of dengue fever: an empirical model. Lancet 360: 830–834.
44. Mills JN, Gage KL, Khan AS (2010) Potential influence of climate change on vector-borne and zoonotic diseases: a review and proposed research plan. Env Health Persp 118(110): 1507–1514.
45. Kalluri S, Gilruth P, Rogers D, Szczur M (2007) Surveillance of Arthropod Vector-Borne Infectious Diseases Using Remote Sensing Techniques: A Review. PLoS Path 3: e116.
46. Beck LR, Lobitz BM, Wood BL (2000) Remote sensing and human health: new sensors and new opportunities. Emerging Infectious Diseases 6(3): 217–222.
47. Hijmans RJ, Cameron SE, Parra JL, Jones PG, Jarvis A (2005) Very high resolution interpolated climate surfaces for global land areas. Int J Climatol 25: 1965–1978.
48. Craig MH, Snow RW, le Sueur D (1999) A climate-based distribution model of malaria transmission in sub-Saharan Africa. Parasitology today (Personal ed.) 15(3): 105–11.
49. Lardeux F, Cheffort J (1997) Temperature thresholds and statistical modelling of larval Wuchereria bancrofti (Filariidea:Onchocercidae) developmental rates. Parasitology 114: 123–134.
50. Grinnell J (1917) The niche-relationships of the California Thrasher. Auk 34: 427–433.
51. Guisan A, Zimmermann NE (2000) Predictive habitat distribution models in ecology. Ecol Model 135: 147–186.

52. Stockwell DRB, Peters DP (1999) The GARP modelling system: problems and solutions to automated spatial prediction. Int J Geogr Infor Sci 13: 143–158.
53. Phillips SJ, Dudik M, Schapire RE (2004) A maximum entropy approach to species distribution modeling. Proceedings of the 21st International Conference on Machine Learning. Banff, Alberta, Canada: ACM.
54. Phillips SJ, Dudik M (2008) Modeling of species distributions with Maxent: new extensions and a comprehensive evaluation. Ecography 31: 161–175.
55. Jaynes ET (1957) Information theory and statistical mechanics. Physics Rev 106: 620–630.
56. Delong ER, Delong DM, Clarke-Pearson DL (1988) Comparing the areas under two or more correlated receiver operating characteristic curves: a nonparametric approach. Biometrics 44: 837–845.
57. Elith J (2006) Novel methods improve prediction of species' distributions from occurrence data. Ecography 29: 129–151.
58. Lobo JM, Jimenez-Valverde A, Real R (2007) AUC: a misleading measure of the performance of predictive distribution models. Global Ecol Biogeogr 17: 145–151.
59. Peterson AT, Papes M, Soberon J (2008) Rethinking receiver operating characteristic analysis applications in ecological niche modeling. Ecol Model 213: 63–72.
60. Anderson RP, Lew D, Peterson AT (2003) Evaluating predictive models of species' distributions: criteria for selecting optimal models. Ecol Model 162: 211–232.
61. Zweig MH, Campbell G (1993) Receiver-operating characteristic (ROC) plots: a fundamental evaluation tool in clinical medicine. Clin Chem 39: 561–577.
62. Jimenez-Valverde A, Lobo JM (2007) Threshold criteria for conversion of probability of species presence to either -or- presence-absence. Acta Oecol 31: 361–369.
63. Pearson RG, Raxworthy CJ, Nakamura M, Peterson AT (2007) Predicting species distributions from small numbers of occurrence records: a test case using cryptic geckos in Madagascar. Journal of Biogeography 34(1): 102–117.
64. Boko M, Niang I, Nyong A, Vogel C, Githeko A, et al. (2007) Africa: Climate Change 2007: Impacts, Adaptation and Vulnerability. Contribution of Working Group II to the Fourth Assessment Report of the Intergovernmental Panel on Climate Change, Cambridge University Press, Cambridge UK.
65. IPCC (2007) Climate Change 2007: The Physical Basis. Contribution of Working Group I to the Fourth Assessment Report of the Intergovernmental Panel on Climate Change. Cambridge, UK: Cambridge University Press. 996 p.
66. Elith J, Graham C (2009) Do they? How do they? WHY do they differ? On finding reasons for differing performances of species distribution models. Ecography 32: 1–12.
67. McMichael AJ (2003) Global climate change: will it affect vector-borne infectious diseases? Int Med J 33: 554–555.
68. Martens WJM, Jetten TH, Focks DA (1997) Sensitivity of malaria, schistosomiasis and dengue to global warming. Clim Change 35: 145–156.
69. Kearney M, Porter W (2009) Mechanistic niche modelling: combining physiological and spatial data to predict species' ranges. Ecol Lett 12: 334–350.
70. Ngwira B, Tambala P, Perez AM, Bowie C, Molyneux D (2007) The geographical distribution of lymphatic filariasis infection in Malawi. Filaria J 6: 12.
71. Onapa AW, Simonsen PE, Baehr I, Pedersen EM (2005) Rapid assessment of the geographical distribution of lymphatic filariasis in Uganda, by screening of schoolchildren for circulating filarial antigens. Ann Trop Med Parasitol 99: 141–153.
72. Levine RS, Peterson AT, Benedict MQ (2004) Geographic and ecologic distributions of the *Anopheles gambiae* complex predicted using a genetic algorithm. Am J Trop Med Hyg 70(2): 105–109.
73. WHO (1992) Lymphatic Filariasis: The Disease and Its Control. Geneva: WHO.
74. WHO (2005) Global Programme to Eliminate Lymphatic Filariasis: Progress report for 2004. Wkly Epidemiol Rec. pp 202–212.
75. WHO (2009) Global programme to eliminate lymphatic filariasis. Progress report on mass drug administration in 2008. Wkly Epidemiol Rec 84: 437–444.
76. Gyapong JO, Kyelem D, Kleinschmidt I, Agbo K, Ahouandogbo F, et al. (2002) The use of spatial analysis in mapping the distribution of bancroftian filariasis in four West African countries. Ann Trop Med Parasitol 96: 695–705.
77. Pearman PB, Guisan A, Broennimann O, Randin CF (2007) Niche dynamics in space and time. Trends Ecol Evol 23: 149–158.
78. Lafferty KD (2009) The ecology of climate change and infectious diseases. Ecology 90: 888–900.
79. Manga L (2002) Vector-control synergies between 'Roll Back Malaria' and the Global Programme to Eliminate Lymphatic Filariasis in the African Region. Ann Trop Med Parasitol 96: S129–S132.

2

Wind Speed during Migration Influences the Survival, Timing of Breeding, and Productivity of a Neotropical Migrant, *Setophaga petechia*

Anna Drake*, Christine A. Rock, Sam P. Quinlan, Michaela Martin, David J. Green

Center for Wildlife Ecology, Department of Biological Sciences, Simon Fraser University, Burnaby, British Columbia, Canada

Abstract

Over the course of the annual cycle, migratory bird populations can be impacted by environmental conditions in regions separated by thousands of kilometers. We examine how climatic conditions during discrete periods of the annual cycle influence the demography of a nearctic-neotropical migrant population of yellow warblers (*Setophaga petechia*), that breed in western Canada and overwinter in Mexico. We demonstrate that wind conditions during spring migration are the best predictor of apparent annual adult survival, male arrival date, female clutch initiation date and, via these timing effects, annual productivity. We find little evidence that conditions during the wintering period influence breeding phenology and apparent annual survival. Our study emphasizes the importance of climatic conditions experienced by migrants during the migratory period and indicates that geography may play a role in which period most strongly impacts migrant populations.

Editor: Nicola Saino, University of Milan, Italy

Funding: Research support came from The Fish and Wildlife Compensation Program (WSI4639) (www.bchydro.com/bcrp), core funding to the CWE provided by Environment Canada (www.sfu.ca/biology/wildberg), a Natural Sciences and Engineering Research Council of Canada (NSERC) Discovery grant to D.J.G. (RGPIN261899) and NSERC fellowships to A.D., C.R., S.Q. and M.M. (www.nserc-crsng.gc.ca). The funders had no role in study design, data collection and analysis, decision to publish, or preparation of the manuscript.

Competing Interests: The authors have declared that no competing interests exist.

* E-mail: aedrake@sfu.ca

Introduction

Every year, it is estimated that six billion songbirds leave wintering regions in Africa and Central and South America and redistribute themselves across Europe and North America [1]. These migratory species, which pollenate plants, disperse seeds, or consume insects and small mammals, quadruple bird abundance in the north and play a critical role in northern ecosystems [2]. Population declines among many migratory songbirds have led to a renewed interest in when and where factors influence their demography [3–5]. However, the geographic scale of migratory movement and our limited knowledge of the linkage between wintering and breeding populations, pose a challenge for ecologists and conservation biologists [5].

Climatic conditions are capable of impacting the population dynamics of migratory songbirds at all stages of the annual cycle. Conditions during the breeding period can alter migrant productivity by impacting total food availability and the timing of breeding events (e.g. [6,7]). Within wintering regions, climate can directly influence songbird survival [1] and, via shifts in the timing of spring migration, indirectly influence productivity [8–11]. Conditions experienced during migration can influence stopover decisions and survival [4,12,13], although fewer studies exist and this period remains the most difficult to examine [5]. Understanding the relative importance of conditions at each stage of the annual cycle in explaining changes in population numbers becomes increasingly important as climate change is expected to impact regions differently [14].

In this study, we evaluate the relative importance of conditions during the winter, spring migratory, and breeding season in influencing the survival and breeding phenology of yellow warblers in western Canada, explicitly recognizing that events at one period may impact processes occurring in periods that follow. We developed climate models that considered the impact of: (1) rainfall patterns on wintering grounds, (2) wind speed and rainfall on the migration route in spring, and (3) temperature conditions on the breeding grounds in May. Rainfall in overwintering areas, through its influence on primary productivity and insect abundance, was expected to increase survival and allow birds to migrate earlier [15,16]. Strong winds and precipitation impact the energetic cost and speed of movement during migration, and hostile conditions were expected to reduce survival and delay arrival [12]. Finally, local temperatures at the beginning of the breeding season impact the timing of insect development and therefore food availability [17] and migratory birds may adjust arrival and/or clutch initiation in response to these local/regional conditions [6].

Materials and Methods

Ethics statement

Targeted mist-netting and banding of individuals was carried out in accordance with Canadian Council on Animal Care recommendations and under permits issued by Environment Canada (CWS Banding Permit: 10759H; Scientific Permits: BC-SCI-59-04-0335; 59-05-0328; 59-06-0347; 59-07-0331; 59-08-0388; and BC-09-0296; 10-0022; 11-0037; 12-0010). Field protocols were approved by the University Animal Care Committee at Simon Fraser University (Protocol # 709B-04, 869B-04, and 1038B-04).

Study species

Yellow warblers (*Setophaga petechia*) are small, insectivorous neotropical migrants with a broad breeding distribution in North America. Genetic-isotopic work indicates eastern and western lineages have parallel migration systems and differing wintering distributions [18]. Birds that breed in the northwestern region of the continent winter primarily in lowland areas within the occidental and isthmus region of Mexico [18] where they occupy coastal scrub, riparian corridors, agricultural habitat, second growth, and tropical evergreen forest [19,20]. Migratory yellow warblers are present on the western and central flyways between March and May [21].

Study system

We have studied a population of yellow warblers breeding in Revelstoke, British Columbia (50.97°N, -118.20°W) for nine years (2004-2012). Colour-banded and monitored individuals breed within three, 30-39 ha plots of seasonally flooded grassland interspersed with isolated willow thickets (*Salix* sp.) and black cottonwood *(Populus trichocarpa)* forest along the northern section of the Upper Arrow Lakes Reservoir (elevation 435-441 m). All banded individuals are aged as either "young" (SY; first breeding season) or "older" (ASY; at least 2 years old) birds based on plumage characteristics [22].

Survival and breeding phenology

Our 2004-2012 mark-resight dataset contained 279 individuals (141 females, 138 males) and 460 annual encounters. Detailed breeding data was collected from 2005-2006 and 2008-2012. In these years, the study plots were surveyed every 1-2 days from early-May to late-June to determine male arrival dates. Breeding pairs were then monitored every 3 days until late-July in order to determine when females initiated their first clutch, and document the fate of all nesting attempts. Nestlings were banded seven days post-hatch and the number of nestlings present at this date was assumed to be the number of young fledged from nests where fledging was subsequently confirmed and evidence of predation was absent. Annual productivity was defined as the total number of young fledged across all nesting attempts made by a given individual (for further detail see [20]).

Climate models

Climate conditions on the wintering grounds. We used standardized monthly Southern Oscillation Index (SOI) values [23] to describe climatic conditions on the wintering grounds because the El Niño Southern Oscillation (ENSO) impacts rainfall patterns in Mexico [24]. Within the occidental and isthmus regions of Mexico, the majority of rainfall (>60%; [24]) occurs during the summer monsoon (May-August). Weak monsoon years are associated with El Niño phases of ENSO [24] and likely result in reduced food availability for birds wintering in this region. We therefore predicted that negative mean SOI values (dry, El Niño conditions) in the May-August period would be associated with low survivorship and delayed phenology in our population the following spring (Model 1: $SOI_{MAY\text{-}AUG}$). It is possible that late winter rainfall is disproportionately important to neotropical migrants. Such rainfall contributes significantly less moisture to over-winter habitat than the monsoon but occurs at a critical time, when regions are experiencing drought conditions. In winter El Niño conditions (negative SOI) promote greater precipitation [24]. We therefore predicted that negative SOI values in the December-March period would be associated with favorable conditions that would improve survivorship and advance breeding phenology (Model 2: $SOI_{DEC\text{-}MAR}$).

Climate conditions on spring migration. Conditions experienced during spring migration could alter both timing and survival rates. We used mean nighttime wind vectors (westerly (U-) and southerly (V-) components (m/s)) and precipitation ($Kg/m^2/s$) between March and May as measures of migration costs. Variables were derived from modeled climate data extracted from the National Center of Environmental Prediction (NCEP) Reanalysis 1 data archives at the NOAA-CIRES Climate Diagnostics Center at Boulder, Colorado, USA [25] using the RNCEP program [26]. These data have a spatial resolution of 2.5° latitude and longitude and temporal resolution of six hours. We defined the western flyway for our population as the overland region west of the easternmost portion of the continental divide (107°W), beginning at the northern extent of the yellow warbler wintering range (25°N) and ending at the latitude of our study site (50°N) (figure 1). U- and V-wind speed components were averaged from the 850 mb (~1500 m AMSL) and 925 mb (~700 m AMSL) level, and thereby encompassed conditions within much of the altitudinal range of migrant songbirds (e.g. [27,28]). Rain and wind vectors were then averaged for the March-May period, dropping noon values so that the series represented conditions encountered during nighttime migration (between 18:00h and 6:00h). Within our defined migratory region during the March-May period, averaged U-winds were westerly in all years of our study and V-winds were southerly in all years except 2008. We expected high wind speeds (westerly (Model 3: U-wind), southerly (Model 4: V-wind), and combined (Model 5: U-wind + V-wind)) and greater precipitation (Model 6: Migration Rain) to be associated with reduced survivorship and delayed breeding phenology.

Climate conditions during breeding. We used averaged daily mean temperatures (°C) at our study site during the month of May to parameterize breeding conditions. This metric has been shown to be correlated with when yellow warblers initiate breeding events, likely due to its influence on prey availability [6]. Warmer springs may also result in greater food availability and earlier fledge dates, affording breeding pairs additional energy and time to moult and fatten for fall migration. Temperatures were drawn from "Revelstoke A" weather station (ID: 1176749; 50°57′40N, -118°11′00W, elevation 444.7 m [29]). Higher mean May temperatures at our study site were expected to be associated with earlier breeding phenology in the same year, and improved adult survivorship in the following year (Model 7: May°C).

Analysis

We estimated apparent annual adult survival for the period 2004-2012 using the Cormack-Jolly-Seber model. We calculated the probability of an adult returning to the study site (φ) after adjusting for the re-sighting probability of banded individuals (p) using program MARK (5.1) [30,31]. Probability-of-return (φ) reflects both survival and emigration, thus our apparent annual survival estimates do not include surviving individuals who permanently emigrate from the study site. The global model that allowed adult survival to vary as a function of gender, age and year, and re-sighting probability to vary as a function of gender and years where detailed breeding data was collected (2005–2006 and 2008–2012 vs. 2007), fit the data well (median procedure, $\hat{c} = 1.09$).

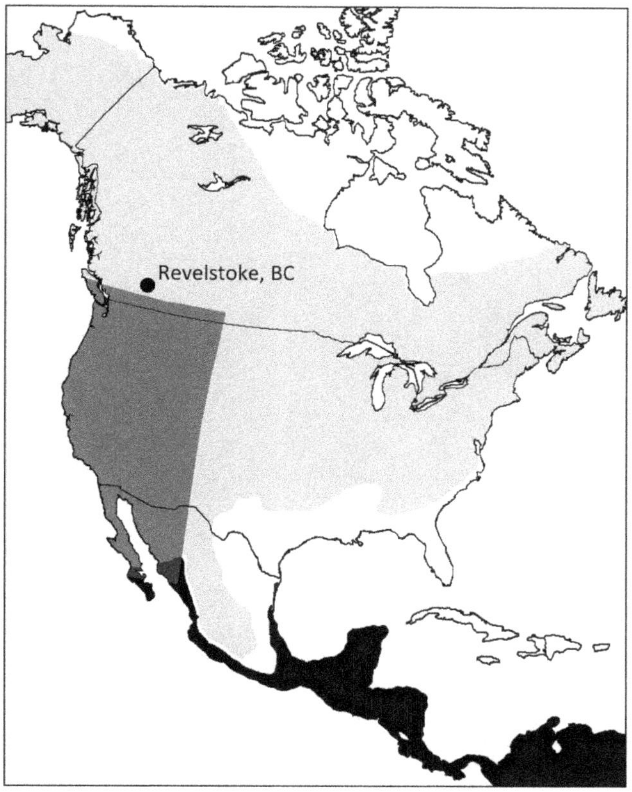

Figure 1. The location of our study, the yellow warbler breeding range (light grey) and wintering range (black), and the area used to calculate wind speed values during migration (dark grey). Base map: Wikimedia Commons.

We first determined the best model structure for the re-sighting rate, and then modeled survival rates with candidate models containing gender, age, year, and all possible interactions. We used Akaike's Information Criterion corrected for small sample sizes and over-dispersion ($QAIC_c$), to rank competing models. Re-sighting rates were best modeled with only a gender term; males were estimated to have higher re-sighting probabilities than females (male $= 0.92 \pm 0.04$, female $= 0.65 \pm 0.08$). Apparent annual survival rates were best described by a model that included year and age (AIC weight $= 0.53$) or a model that included year, age and gender (AIC weight $= 0.41$). The influence of climate variables were therefore investigated using a candidate model set that included models with: (1) only age and climate variables as main effects, (2) age, gender, and climate variables as main effects, and (3) main effects, age×climate interactions and/or gender× climate interactions (n = 47 individual models, table S1). Models were ranked using QAICc [31].

To assess regional climate effects on male arrival and female clutch initiation, we created a candidate model set that included linear models with age and climate variables as main effects, and linear models that included 'age×climate' interactions (n = 15 individual models, table S2). Models were run in R version 2.14.1 (2011, The R Foundation for Statistical Computing) and ranked using AICc. We subsequently created a data subset restricted to older individuals (2+ years) that returned to the study site in more than one year (males, n = 58; females, n = 26). We used this dataset to investigate whether climate variables that influenced breeding phenology at the population level also explained year-to-year variation in the breeding phenology of individuals.

Three of the climate variables we examined exhibited a temporal trend (table S3). We report our analysis of non-de-trended climate variables here [32] because we were interested in how absolute variation in climate influences population parameters, not in how year-to-year deviation from short-term trends account for unexplained variance in our population parameters. De-trended survival analyses had greater model uncertainty (table S4), but de-trended breeding phenology analyses yielded similar results to those obtained using non-de-trended climate variables (table S5).

We used path analysis to estimate how breeding delays associated with climate conditions influenced the annual productivity of young and older females [33]. For each age-class we first calculated the total effect (TE) as the product of the two path coefficients (standardized partial regression coefficients (β)) in the pathway between climate, clutch initiation date, and annual productivity. We then used the TE score, the standard deviation of the climate variable, and the standard deviation in the age-specific annual productivity of females in our population to predict annual productivity across the observed range of climatic conditions [33].

We report the relationship between mean SOI values (March-May; [23]) and wind speed (March-May) over a 50-year period (1963–2012) to allow comparison with other studies that have examined El Niño climate cycle effects on migrant survival and breeding phenology. We additionally use this dataset to assess patterns in wind speed over the March to May period.

Results

Survival and breeding phenology

Inter-annual variation in the apparent annual adult survival of yellow warblers breeding in Revelstoke was best described by a model including migration wind speed (table 1). Wind speed models (3, 4, 5) received 79% of the total model support (table S1). Wintering models (1, 2) and the breeding period model (7) received 13% and 3% of the total model support, respectively. The top model (3a) indicated that apparent annual survival varied with age and declined as westerly wind speeds increased (figure 2).

Annual variation in male arrival and female clutch initiation dates was also best described by models including migration wind speed (table 2). Wind speed models describing male arrival date received 66% of the total model support and wind speed models describing female clutch initiation date received 88% of total model support (table S2). The top model in both candidate model sets was (3b) which included westerly wind speed, age and a 'wind speed×age' interaction (table 2). The arrival of older males was delayed as westerly wind speed on migration increased, whereas the arrival of young males – who arrived later than older males – did not vary with wind speed (figure 3). Females of both age classes were delayed in initiating their first clutch in years with stronger westerly winds on migration but older females were less sensitive to variation in wind speed than younger females (figure 3).

Differences in the arrival date of individual males and the clutch initiation date of individual females were also associated with inter-annual variation in westerly wind speed. Individual males exhibited a median delay of 2 days ($t_{57} = 3.66$, $P < 0.001$) and individual females exhibited a median delay of 4 days ($t_{25} = 3.22$, $P = 0.002$) in years where westerly winds were stronger (figure 4).

Carry-over Effect on Female Productivity

Annual productivity declined with later clutch initiation dates (older females: $F_{(1,103)} = 8.25$, $P = 0.005$, $\beta = -0.27$; young

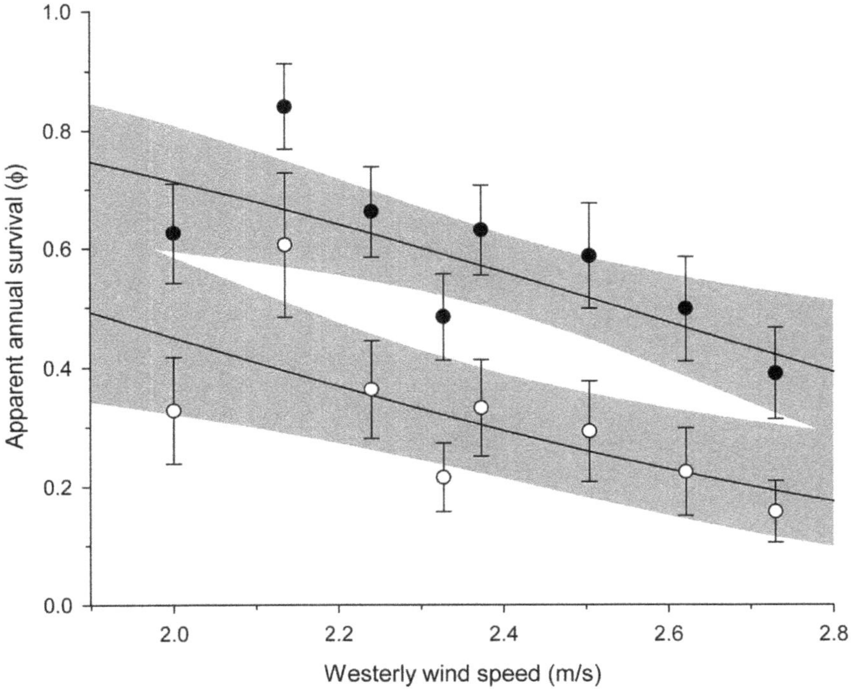

Figure 2. Relationship between westerly wind speed during migration and apparent annual survival of yellow warblers. Points are apparent annual survival (± SE) of young (1 yr, open points) and older (≥2 yrs, filled points) birds for 2004 to 2012. Solid lines and shading represent predicted apparent annual survival (ϕ) ±95% CI from the top model assuming an average southerly wind vector.

females: $F_{(1,68)} = 14.51$, $P<0.001$, $\beta = -0.42$). The total effect of westerly wind speed on annual productivity was consequently −0.03 for older females and −0.16 for young females. Based on these scores, older females would be expected to produce 0.2 fewer fledglings and younger females would be expected to produce 0.8 fewer fledglings in our highest versus lowest wind speed years.

Wind speed variation and ENSO

Within the western flyway region (figure 1), average westerly wind speeds and SOI between March and May were correlated positively over the 50-year period we examined ($r = 0.61$, $n = 50$; Spearman's $\rho = 0.55$; $P<0.0001$). Thus, stronger westerlies on the Pacific side of the Rocky Mountains are associated with La Niña conditions in the central Pacific. We did not observe a relationship between southerly wind speeds and SOI within the same period and region ($r = -0.12$, $n = 50$, Spearman's $\rho = -0.17$; $P = 0.24$).

All wind speed vectors declined from March to May (westerly wind: $F_{(2,147)} = 3.38$, $P = 0.04$; southerly wind: $F_{(2,147)} = 14.04$, $P< 0.001$). Westerly wind speeds were lower in May than in March and April while southerly winds showed a steady decline over the three-month period.

Discussion

Research conducted on neotropical migrants that over-winter in the Caribbean and breed in eastern North America and on palaearctic-African migrants indicate that populations are limited by climatic conditions during the wintering and/or breeding period [1,7,34,35]. In contrast, within our western North American yellow warbler population we find only weak evidence that climatic conditions during the wintering and breeding periods influence survival and phenology. Our findings strongly support

Table 1. Climate models obtaining substantial support (ΔQAICc ≤2) and the base temporal model ('Year + Age') describing apparent annual survival of yellow warblers breeding in Revelstoke, British Columbia (n = 279 individuals, 460 encounters).

Model #	Period	Variables	K	QAICc	ΔQAICc	ω_i
3a	Migration	U-WIND + AGE	5	613.61	0	0.197
5a	Migration	U-WIND + V-WIND + AGE	6	614.28	0.67	0.141
3b	Migration	U-WIND + AGE + SEX	6	615.08	1.47	0.094
5e	Migration	U-WIND + V-WIND + AGE + SEX + U-WIND*SEX + V-WIND*SEX	9	615.33	1.72	0.083
3c	Migration	U-WIND + AGE + U-WIND*AGE	6	615.52	1.91	0.076
-	-	YEAR + AGE	11	616.94	3.33	0.037

Model numbers match those described in the text. Age was included in all models as a covariate (*see* Methods).

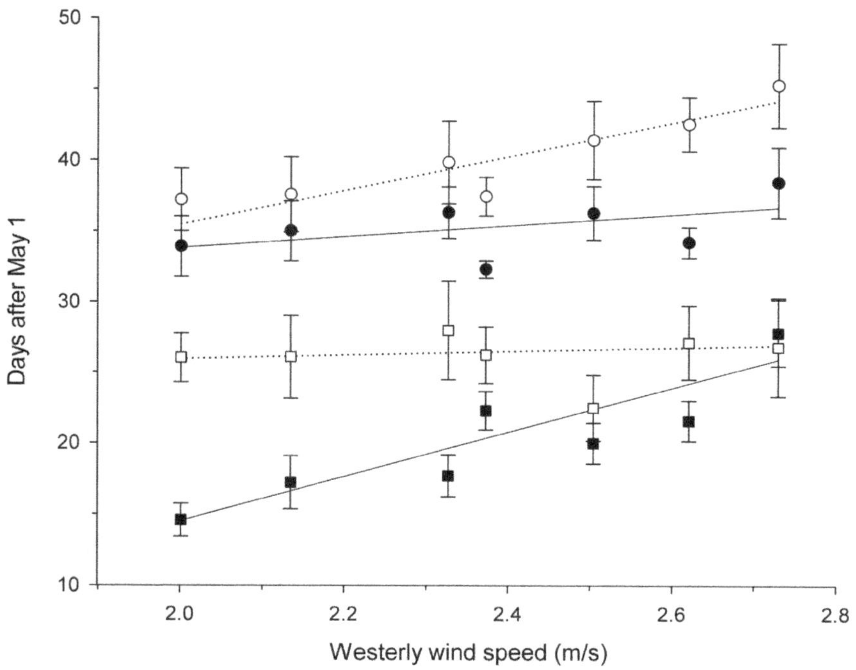

Figure 3. Male arrival date (squares) and female clutch initiation date (circles) for yellow warblers as a function of westerly wind speed during migration. Points represent mean dates ± SE for young (1 yr, open points) and older (≥2 yrs, filled points) birds in 2005–2006 and 2008–2012.

the argument that avian populations can be limited during the migratory period [4]. Not only did conditions on migration best describe variation in annual survival within our population but these conditions also best described the timing of breeding, which in turn strongly predicts annual productivity within our population. To our knowledge, this is the first study to demonstrate that conditions experienced by birds on migration can impact both survival and productivity.

Migration is known to be a period of high mortality for migratory birds [36] and climate conditions on migration can theoretically limit migrant numbers [4,37]. Higher westerly wind speeds during migration may be associated with lower apparent annual survival in our study because they reflect storm events that increase the risk of in-flight mortality [4]. High wind speeds that oppose the direction of spring movement should also increase flight costs and refueling requirements, and delay departure from stopover sites [12]. Resulting increases in migrant density, feeding effort, and competition at stopover sites may consequently increase the risk of starvation and predation *en route* [4]. However, declines in apparent annual survival may not necessarily indicate mortality. Changes in this metric associated with shifts in wind speed could also be a product of individuals being blown off course or "dropping out" of migration earlier in years when migration conditions are hostile. These individuals would then breed further east or further south and not return to our study site in subsequent years. As younger birds are less likely to have bred successfully at our study location, they may be more willing to forgo the benefits of philopatry [38,39]. If climate effects on apparent annual survival are a product of dropping out, then we might expect young individuals to show a stronger response to wind speed. However, we found relatively little support for survival models that included a 'wind speed×age' interaction term (*see also* figure 2).

Increased wind speed on migration was associated with later male arrival dates and later female clutch initiation dates. This relationship was also supported by shifts in the phenology of individuals who returned to our study site in multiple years. The effect of wind speed on breeding phenology varied with age in both sexes. Older males arrived at our study site 12 days before younger males when wind conditions on migration were favorable and at the same time as young males when conditions were more hostile (figure 3). Given that wind speeds are lower later in the migratory period, adult male arrival dates may reflect a trade-off between the advantages of early arrival and the costs of early migration. Consistently "late" arrival dates among young males may reflect cost-minimization alone. In contrast to males, older females appeared to be less sensitive to variation in wind speed than young females (exhibiting a 3 day vs. 9 day delay in clutch initiation dates across the observed range of wind speeds). Older females may be better than young females at compensating for arrival delays if experience allows them to pair and initiate breeding more rapidly.

Conditions on migration that delay reproduction can decrease productivity because later clutch initiation dates are associated with reduced fledging success [40]. We found that wind-induced delays in breeding phenology in our population could reduce the annual productivity of young and older females by as much as 0.8 and 0.2 fledglings, respectively. These declines are significant as yellow warblers typically only raise a single brood of three – five nestlings and average productivity in our population is low (2.2±1.9 fledglings per year [38]).

We found no consistent evidence that wintering climate impacted our population. In males, the mean SOI index that predicts monsoon rainfall ($SOI_{MAY\text{-}AUG}$) was associated with the timing of arrival on the breeding grounds. However, counter to our *a priori* prediction, this model indicated that males arrived

Table 2. Climate models obtaining substantial support (ΔAICc≤2) and the base temporal model ('Year + Age') describing male arrival date (n = 210) and female clutch initiation dates (n = 177) for yellow warblers in Revelstoke, British Columbia.

Model #	Period	Variables	r^2	K	AIC.c	ΔAICc	ω_i
Male Arrival:							
3b	Migration	U-WIND + AGE + U-WIND*AGE	0.21	5	1452.88	0	0.535
1b	Winter	$SOI_{MAY-AUG}$ + AGE + $SOI_{MAY-AUG}$*AGE	0.20	5	1453.85	0.98	0.328
-	-	YEAR+AGE	0.18	9	1464.32	11.44	0.002
Female clutch initiation date:							
3b	Migration	U-WIND + AGE + U-WIND*AGE	0.16	5	1194.89	0	0.289
5a	Migration	U-WIND + V-WIND + AGE	0.15	5	1195.00	0.11	0.273
3a	Migration	U-WIND + AGE	0.14	4	1195.82	0.93	0.181
5b	Migration	U-WIND + V-WIND + AGE + U-WIND*AGE + V-WIND*AGE	0.16	7	1196.47	1.58	0.131
-	-	YEAR + AGE	0.16	9	1197.90	3.00	0.064

Age was included in all models as a covariate (*see* Methods). Wintering model support for male arrival date was counter to prediction (*see* Discussion).

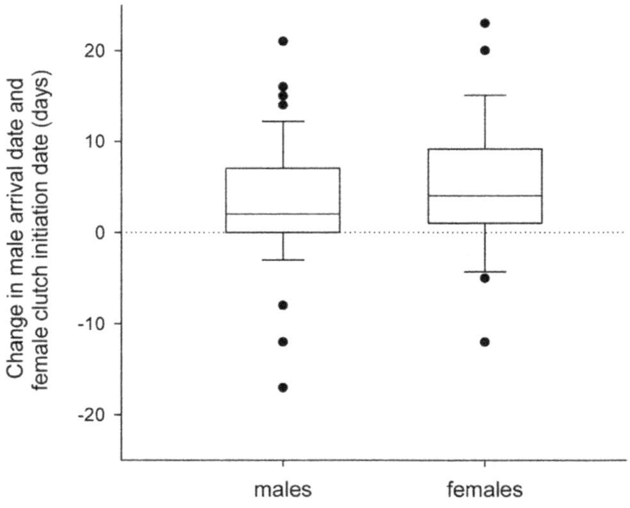

Figure 4. Within-individual changes in breeding phenology (male n = 58, female n = 26) associated with inter-annual variation in average westerly wind vectors. Box-plots show the median change, 10th and 90th percentile, and outliers. Positive values represent delayed breeding phenology in years with stronger winds on migration. Data was restricted to individuals that were ≥2 years of age in both years.

earlier in years with a more negative SOI $_{MAY-AUG}$ index, when there is less monsoon rain and conditions on wintering territories should be worse than in years with a positive SOI $_{MAY-AUG}$ index (e.g. [41]). Food availability immediately prior to migration could be higher in years with less monsoon rain if aquatic insects emerge as ephemeral water-bodies disappear [42]. This might allow males to leave earlier. However, in the climate zone typical of the occidental and isthmus region of Mexico, moist habitats make up a small percentage of landscape and water-limitation is more likely to reduce food availability for yellow warblers [16]. We therefore believe that support for this model instead reflects the positive correlation between $SOI_{MAY-AUG}$ and westerly wind speed on migration (U-wind) (table S3).

Previous work on the western flyway by Macmynowski et al. [43] and Nott et al. [44] indicate that the movement of migrants through California is delayed in La Niña years while migrant productivity in the Pacific Northwest is higher in El Niño years. These patterns were attributed to increased spring rainfall in the northern region of the Pacific slope of Mexico as well as more favorable southerly winds between March and May in El Niño years [44]. However, our analysis of ENSO and wind vectors within the "western flyway" (including California and the Pacific Northwest; figure 1) indicate ENSO is poorly correlated with southerly winds, but that La Niña conditions (+ve SOI values) are associated with strong westerly winds blowing off the Pacific during migration, while El Niño conditions (-ve SOI values) are associated with weaker westerlies. Our findings therefore corroborate those of Macmynowski et al. [43] and Nott et al. [44] but implicate westerlies as a possible additional explanation for their observations.

For our yellow warbler population, apparent annual survival models that included pre-migration rain ($SOI_{DEC-MAR}$) or migration rainfall received little support (table S1). In contrast, apparent annual survival of Swainson's thrush (*Catharus ustulatus*) in the northwestern United States is positively correlated with higher spring rainfall in Sinaloa, Sonora and southwestern California

[45]. Lack of support for migration rain in our study may be due to differences in the period and region over which our rainfall values were calculated. Rainfall in the southwestern portion of our migratory region during the pre-migratory period may be important with respect to food availability *en route* [44,45]. As higher spring rainfall in this region is associated with El Niño and therefore reduced wind speed years, further work is needed to assess the independent effect of each variable on migrants.

The majority of studies conducted in western Europe and eastern North America that have examined the impact winter conditions on migrant songbird survival or reproduction have found support for wintering effects (e.g. [34,35,46] *but see* [7]). In contrast, only three western and central North American studies have found evidence of wintering conditions influencing breeding populations [11,20,44] while four have found either no support for wintering effects, or greater support for non-wintering effects ([6,34,45], this study). Migrant populations that breed in western Europe and eastern North America encounter large ecological barriers during migration (the Mediterranean sea, Sahara desert, the Gulf of Mexico and the Caribbean sea). Songbirds using migratory routes in western North America only cross the smaller deserts of the American southwest and those using central routes may avoid barriers altogether. Such geographical differences may be important in determining when in the annual cycle migrants are most impacted by climatic conditions. In regions where birds must cross large barriers, population responses linked to overwintering conditions may instead be the product of conditions encountered at pre-barrier stopover sites as these sites often overlap with wintering areas (for example, areas encompassed in [4,8,11]). Populations that are not required to make long, uninterrupted flights may compensate for poor departure conditions by shortening flight distance and increasing stopover number [47]; weaker support for wintering effects among these populations, including our study population, may be the result of a weaker dependency on any one wintering or stopover region to fuel migration.

Pacific wind patterns may explain the strong relationship between migratory conditions and demography in our study. These winds oppose the north/northwesterly course of spring migration and therefore represent a climatological barrier to spring movement. In contrast, winds encountered in central and eastern regions of the USA facilitate the northeasterly movement of migratory birds in those regions [48]. We suggest that, whereas pre-migration and, potentially, pre-barrier fattening in wintering regions may be essential for the movement and survival migrant populations in western Europe and eastern North America, *en route* conditions and habitats are more important for migrants in western North America (*see also* [45]). Our findings indicate that migration is the most costly period of the annual cycle for western neotropical migrants, impacting both survival and productivity. If so, stopover habitats in the southwestern USA may play a significant role in maintaining western populations [49,50]. These habitats, limited and threatened by development [50], may need to be prioritized for conservation.

Supporting Information

Table S1 All climate models describing apparent annual survival of yellow warblers breeding in Revelstoke, British Columbia (n = 279 individuals, 460 encounters). Models are ranked using QAICc. Model number refers to regionally specific climate variables described in the text. Age was included in all models as a covariate (*see* Methods). The number of parameters in the model (K), Akaike's information criterion ($QAIC_c$), $QAIC_c$ difference from the top model ($\Delta QAIC_c$), and Akaike weight (ω_i) are reported.

Table S2 Ranked summary of AICc support for all candidate climate models and a null model ('Year + Age') describing (A) male arrival date (n = 210) and (B) female clutch initiation dates (n = 177) for yellow warblers in Revelstoke, British Columbia. Age (young = 1 yr; older ≥2 yrs) was included in all models as a covariate (*see* Methods). Model adjusted r^2, the number of parameters in the model (K), Akaike's information criterion adjusted for small sample size (AIC_c), AIC_c difference from the top model (ΔAIC_c), and Akaike weight ($\boldsymbol{\omega_i}$) are reported.

Table S3 Correlation (r) matrix of explanatory climate variables and time (year) (n = 8). Significant relationships (Spearman's ρ) are starred ($P<0.05$ = *, $P<0.001$ = **).

Table S4 De-trended models describing apparent annual survival of yellow warblers breeding in Revelstoke, British Columbia (n = 279 individuals, 460 encounters). Models are ranked using QAICc. Model number refers to regionally specific climate variables described in the text. De-trended variables, entered in the model with the trend labeled as "(Year)", appear in italics. Age was included in all models as a covariate (*see* Methods). The number of parameters in the model (K), Akaike's information criterion ($QAIC_c$), $QAIC_c$ difference from the top model ($\Delta QAIC_c$), and Akaike weight (ω_i) are reported.

Table S5 Ranked summary of AICc support for de-trended climate models and the null model ('Year + Age') predicting (A) male arrival (n = 210) and (B) female clutch initiation dates (n = 177) for yellow warblers in Revelstoke, British Columbia. De-trended variables, entered in the model with the trend labeled as "(Year)", appear in italics. Age (young = 1 yr; older ≥2 yrs) was included in all models as a covariate (*see* Methods). Model adjusted r^2, the number of parameters in the model (K), Akaike's information criterion adjusted for small sample size (AIC_c), AIC_c difference from the top model (ΔAIC_c), and Akaike weight ($\boldsymbol{\omega_i}$) are reported.

Acknowledgments

We thank M. Hepp, C. Grande, M. Pennell, P. Levesque, A. Potvin, D. O'Farrell, and P. Ridings for their work in the field and S. Bearhop, D. Lank, E. Krebs, R. Ydenberg, P.P. Marra and C. Krebs for insight and feedback on earlier drafts. Thank you also to W. Vansteelant and Lund University's CAnMove for inspiring the RNCEP approach.

Author Contributions

Conceived and designed the experiments: AD DJG. Performed the experiments: CAR SPQ MM. Analyzed the data: AD DJG. Contributed reagents/materials/analysis tools: AD CAR SPQ MM. Wrote the paper: AD DJG.

References

1. Newton I, Brockie K (2008) The migration ecology of birds. Amsterdam; London: Elsevier-Academic Press.
2. Lundberg J, Moberg F (2003) Mobile link organisms and ecosystem functioning: implications for ecosystem resilience and management. Ecosystems 6: 0087–0098. (doi:10.1007/s10021-002-0150-4)

3. Newton I (2004) Population limitation in migrants. Ibis 146: 197–226. (doi:10.1111/j.1474-919X.2004.00293.x)
4. Newton I (2006) Can conditions experienced during migration limit the population levels of birds? J Ornithol 147: 146–166. (doi:10.1007/s10336-006-0058-4)
5. Faaborg J, Holmes RT, Anders AD, Bildstein KL, Dugger KM, et al. (2010) Recent advances in understanding migration systems of New World land birds. Ecol Monogr 80: 3–48. (doi: 10.1890/09-0395.1)
6. Mazerolle DF, Sealy SG, Hobson KA (2011) Interannual flexibility in breeding phenology of a Neotropical migrant songbird in response to weather conditions at breeding and wintering areas. Ecoscience 18: 18–25. (doi:10.2980/18-1-3345)
7. Sillett TS, Holmes RT, Sherry TW (2000) Impacts of a global climate cycle on population dynamics of a migratory songbird. Science 288: 2040–2042. (doi:10.1126/science.288.5473.2040)
8. Saino N, Szép T, Ambrosini R, Romano M, Møller AP (2004) Ecological conditions during winter affect sexual selection and breeding in a migratory bird. Proc R Soc B Biol Sci 271: 681–686. (doi: 10.1098/rspb.2003.2656)
9. Gordo O, Brotons L, Ferrer X, Comas P (2005) Do changes in climate patterns in wintering areas affect the timing of the spring arrival of trans-Saharan migrant birds? Glob Change Biol 11: 12–21. (doi:10.1111/j.1365-2486.2004.00875.x)
10. Rockwell SM, Bocetti CI, Marra PP (2012) Carry-over effects of winter climate on spring arrival date and reproductive success in an endangered migratory bird, Kirtland's warbler (*Setophaga kirtlandii*). Auk 129: 744–752. (doi:10.1525/auk.2012.12003)
11. McKellar AE, Marra PP, Hannon SJ, Studds CE, Ratcliffe LM (2013) Winter rainfall predicts phenology in widely separated populations of a migrant songbird. Oecologia 172: 595–605. (doi:10.1007/s00442-012-2520-8)
12. Liechti F (2006) Birds: blowin' by the wind? J Ornithol 147: 202–211. (doi:10.1007/s10336-006-0061-9)
13. Marra PP, Francis CM, Mulvihill RS, Moore FR (2005) The influence of climate on the timing and rate of spring bird migration. Oecologia 142: 307–315. (doi:10.1007/s00442-004-1725-x)
14. Giorgi F, Whetton PH, Jones RG, Christensen JH, Mearns LO, et al. (2001) Emerging patterns of simulated regional climatic changes for the 21st century due to anthropogenic forcings. Geophys Res Lett 28: 3317–3320. (doi:10.1029/2001GL013150)
15. Dugger KM, Faaborg J, Arendt WJ, Hobson KA (2004) Understanding survival and abundance of overwintering warblers: does rainfall matter? Condor 106: 744–760. (doi:10.1650/7632)
16. Studds CE, Marra PP (2007) Linking fluctuations in rainfall to nonbreeding season performance in a long-distance migratory bird, *Setophaga ruticilla*. Clim Res 35: 115–122. (doi:10.3354/cr00718)
17. Visser M, Holleman L, Gienapp P (2006) Shifts in caterpillar biomass phenology due to climate change and its impact on the breeding biology of an insectivorous bird. Oecologia 147: 164–172. (doi:10.1007/s00442-005-0299-6)
18. Boulet M, Gibbs HL, Hobson KA (2006) Integrated analysis of genetic, stable isotope, and banding data reveal migratory connectivity and flyways in the Northern yellow warbler (*Dendroica petechia*; Aestiva group). Ornithol Monogr 61: 29–78. (doi: 10.2307/40166837)
19. Hutto RL (1980) Winter habitat distribution of migratory land birds in western Mexico, with special reference to small foliage-gleaning insectivores. In: Keast A, Morton ES, editors. Migrant Birds in the Neotropics: Ecology, Behaviour, Distribution and Conservation. Washington DC: Smithsonian Institution Press. pp. 181–203.
20. Drake A, Rock C, Quinlan SP, Green DJ (2013) Carry-over effects of winter habitat vary with age and sex in yellow warblers (*Setophaga petechia*). J Avian Biol 44: 321–330. (doi:10.1111/j.1600-048X.2013.05828.x)
21. eBird (2013) eBird: An online database of bird distribution and abundance (http://www.ebird.org). Ithaca NY. Accessed: March 20, 2013.
22. Pyle P (1997) Identification Guide to North American Birds, Part I: Columbidae to Ploceidae. Bolinas CA: Slate Creek Press. 732 p.
23. www.cpc.ncep.noaa.gov/data/indices/soi.
24. Caso M, González-Abraham C, Ezcurra E (2007) Divergent ecological effects of oceanographic anomalies on terrestrial ecosystems of the Mexican Pacific coast. Proc Natl Acad Sci 104: 10530–10535. (doi:10.1073/pnas.0701862104)
25. www.esrl.noaa.gov/psd/data/gridded/data.ncep.reanalysis.html.
26. Kemp MU, Emiel van Loon E, Shamoun-Baranes J, Bouten W (2012) RNCEP: global weather and climate data at your fingertips. Methods Ecol Evol 3: 65–70. (doi:10.1111/j.2041-210X.2011.00138.x)
27. Alerstam T, Chapman JW, Bäckman J, Smith AD, Karlsson H, et al. (2011) Convergent patterns of long-distance nocturnal migration in noctuid moths and passerine birds. Proc R Soc B Biol Sci 278: 3074–3080. (doi:10.1098/rspb.2011.0058)
28. Felix RK Jr., Diehl RH, Ruth JM (2008) Seasonal passerine migratory movements over the arid southwest. Stud Avian Biol 37: 126–137
29. www.climate.weatheroffice.gc.ca/climateData.
30. Lebreton J-D, Burnham KP, Clobert J, Anderson DR (1992) Modeling survival and testing biological hypotheses using marked animals: a unified approach with case studies. Ecol Monogr 62: 67–118. (doi:10.2307/2937171)
31. White GC, Burnham KP (1999) Program MARK: survival estimation from populations of marked animals. Bird Study 46: S120–S139. (doi:10.1080/00063659909477239)
32. Grosbois V, Gimenez O, Gaillard J-M, Pradel R, Barbraud C, et al. (2008) Assessing the impact of climate variation on survival in vertebrate populations. Biol Rev 83: 357–399. (doi:10.1111/j.1469-185X.2008.00047.x)
33. Norris DR, Marra PP, Kyser TK, Sherry TW, Ratcliffe LM (2004) Tropical winter habitat limits reproductive success on the temperate breeding grounds in a migratory bird. Proc R Soc Lond B Biol Sci 271: 59–64. (doi:10.1098/rspb.2003.2569)
34. Wilson S, Ladeau SL, Tøttrup AP, Marra PP (2011) Range-wide effects of breeding- and nonbreeding-season climate on the abundance of a Neotropical migrant songbird. Ecology 92: 1789–1798. (doi: 10.1890/10-1757.1)
35. Baillie SR, Peach WJ (1992) Population limitation in Palaearctic-African migrant passerines. Ibis 134: 120–132. (doi:10.1111/j.1474-919X.1992.tb04742.x)
36. Sillett TS, Holmes RT (2002) Variation in survivorship of a migratory songbird throughout its annual cycle. J Anim Ecol 71: 296–308. (doi: 10.1046/j.1365-2656.2002.00599.x)
37. Erni B, Liechti F, Bruderer B (2005) The role of wind in passerine autumn migration between Europe and Africa. Behav Ecol 16: 732–740. (doi:10.1093/beheco/ari046)
38. Rock CA, Quinlan SP, Green DJ (2013) Age-dependent costs of cowbird parasitism in yellow warblers (*Setophaga petechia*). Can J Zool 91: 505–511. (doi:10.1139/cjz-2013-0014)
39. Hoover JP (2003) Decision rules for site fidelity in a migratory bird, the Prothonotary warbler. Ecology 84: 416–430. (doi:10.1890/0012-9658(2003)084[0416:DRFSFI]2.0.CO;2)
40. Perrins CM (1970) The timing of birds' breeding seasons. Ibis 112: 242–255. (doi: 10.1111/j.1474-919X.1970.tb00096.x)
41. Renton K, Salinas-Melgoza A (2004) Climatic variability, nest predation, and reproductive output of Lilac-crowned Parrots (*Amazona finschi*) in tropical dry forest of Western Mexico. Auk 121: 1214–1225. (doi: 10.2307/4090489)
42. Leeper DA, Taylor BE (1998) Insect Emergence from a South Carolina (USA) Temporary Wetland Pond, with Emphasis on the Chironomidae (Diptera). J North Am Benthol Soc 17: 54–72. (doi: 10.2307/1468051)
43. Macmynowski DP, Root TL, Ballard G, Geupel GR (2007) Changes in spring arrival of Nearctic-Neotropical migrants attributed to multiscalar climate. Glob Change Biol 13: 2239–2251. (doi:10.1111/j.1365-2486.2007.01448.x)
44. Nott MP, Desante DF, Siegel RB, Pyle P (2002) Influences of the El Niño/Southern Oscillation and the North Atlantic Oscillation on avian productivity in forests of the Pacific Northwest of North America. Glob Ecol Biogeogr 11: 333–342. (doi:10.1046/j.1466-822X.2002.00296.x)
45. LaManna JA, George TL, Saracco JF, Nott MP, Desante DF (2012) El Niño-Southern Oscillation influences annual survival of a migratory songbird at a regional scale. Auk 129: 734–743. (doi: 10.1525/auk.2012.12017)
46. Reudink MW, Marra PP, Kyser TK, Boag PT, Langin KM, et al. (2009) Non-breeding season events influence sexual selection in a long-distance migratory bird. Proc R Soc B Biol Sci 276: 1619–1626. (doi:10.1098/rspb.2008.1452)
47. Alerstam T (1979) Wind as selective agent in bird migration. Ornis Scand 10: 76–93. (doi: 10.2307/3676347)
48. Gauthreaux SA (1980) The influence of global climatological factors on the evolution of bird migratory pathways. Proc XVII Int Ornithol Congr 17: 517–525).
49. Kelly JF, Hutto RL (2005) An east-west comparison of migration in North American wood warblers. Condor 107: 197–211. (doi:10.2307/4096504)
50. Skagen SK, Kelly JF, van Riper C, Hutto RL, Finch DM, et al. (2005) Geography of spring landbird migration through riparian habitats in southwestern North America. Condor 107: 212–227. (doi:10.1650/7807)

3

Updating Known Distribution Models for Forecasting Climate Change Impact on Endangered Species

Antonio-Román Muñoz[1,2,3]*, Ana Luz Márquez[1], Raimundo Real[1]

1 Biogeography, Diversity and Conservation Research Team, Dept. of Animal Biology, Faculty of Sciences, University of Malaga, Malaga, Spain, **2** Fundación Migres, N-340, Km. 96 Huerta Grande, Pelayo, Algeciras, Spain, **3** Depto. de Didáctica de las Matemáticas, de las Ciencias Sociales y de las Ciencias Experimentales, Faculty of Education Sciences, University of Malaga, Malaga, Spain

Abstract

To plan endangered species conservation and to design adequate management programmes, it is necessary to predict their distributional response to climate change, especially under the current situation of rapid change. However, these predictions are customarily done by relating *de novo* the distribution of the species with climatic conditions with no regard of previously available knowledge about the factors affecting the species distribution. We propose to take advantage of known species distribution models, but proceeding to update them with the variables yielded by climatic models before projecting them to the future. To exemplify our proposal, the availability of suitable habitat across Spain for the endangered Bonelli's Eagle (*Aquila fasciata*) was modelled by updating a pre-existing model based on current climate and topography to a combination of different general circulation models and Special Report on Emissions Scenarios. Our results suggested that the main threat for this endangered species would not be climate change, since all forecasting models show that its distribution will be maintained and increased in mainland Spain for all the XXI century. We remark on the importance of linking conservation biology with distribution modelling by updating existing models, frequently available for endangered species, considering all the known factors conditioning the species' distribution, instead of building new models that are based on climate change variables only.

Editor: Matt Hayward, Bangor University, United Kingdom

Funding: This work was partially financed by the Ministerio de Ciencia e Innovación of Spain and FEDER (project CGL2009-11316/BOS). The funder had no role in study design, data collection and analysis, decision to publish, or preparation of the manuscript.

Competing Interests: The authors have declared that no competing interests exist.

* E-mail: roman@uma.es

Introduction

At present there are evidences suggesting that climate is warming globally and fast, partially in response to the increased output of greenhouse gases. The Report of the Intergovernmental Panel on Climate Change [1] concluded that past, present and future emissions of greenhouse gases are expected to warm the global climate between 1.4 and 5.8°C by 2100, what is a projected rate of warming much larger than the observed changes during the 20th century [2], and likely without precedent during the last 10,000 years, according to palaeoclimate data [3]. These climatic changes are already altering some physical and biological systems and have already affected the distribution and population dynamics of a number of taxa across a broad range of geographical locations and habitats [4–9], and are expected to have even more severe consequences over the coming century [10]. Climate is one of the main determinant factors affecting the geographical range of species [4,11–13], and birds, a well-studied group of organisms, may respond to climate change changing wintering areas, migration routes and breeding grounds [14,15], undergoing changes in their phenology [16–21] and their local abundances [22], and also changing their overall distributions [23–26]. In this way, being able to anticipate the effects of climate change on the distribution of species could improve their management and conservation policy.

A frequently used method to assess the potential impact of climate change on species is to model species distributions, relating observations to a series of environmental variables [27]. However, these predictions normally do not take into account previous knowledge about the historical, geographical, ecological and human-related factors that are known to condition the species distribution, which tend to be available for endangered species [28]. On the other hand, this knowledge is difficult to incorporate into climate change models, as the variables involved in them are not the same as those produced by the climate change scenarios. To take advantage of known species distribution models, a promising approach is to update them to the variables yielded by climatic models before projecting to the future.

An explanatory model was described for the distribution of the endangered Bonelli's Eagle (*Aquila fasciata*) in Spain based on three variables: slope, mean temperature of July and mean annual precipitation [29]. Consequently, expected modifications of the temperature in July and annual precipitation due to climate change may affect the distribution of this species along this century. The most fundamental measure of the Earth's climate is surface temperature, and precipitation is also a key element of climate [30], so this explanatory model can be used to evaluate the possible effect of climate change on the distribution of this species.

According to the predictions of the different Atmosphere-Ocean General Circulation Models (AOGCMs) and Special Report on

Emissions Scenarios (SRESs) of the IPCC, in Spain there will be a decrease in precipitation and an increase in temperature through the present century. The Agencia Estatal de Meteorología (AEMET) of Spain regionalized to Spain several climate change models produced by the Intergovernmental Panel on Climate Change (IPCC), but the resulting variables of mean temperature of July and mean annual precipitation for the present were numerically different (although nominally equivalent) from those used in the existing explanatory model about Bonelli's Eagle distribution in Spain [29], which derived from actual readings of meteorological data. On the other hand, the known Bonelli's Eagle distribution model cannot be transferred to the future at face value, as the correlation among the explanatory variables is different from that existing among the AEMET variables, which affect the parameterization process and, consequently, the value of the parameters in the model. Therefore, the explanatory model needs to be updated to the AEMET variables before being fit for transference to the future scenarios.

In the present study, we modelled the future potential distribution of Bonelli's Eagle in Spain under several future climatic scenarios by updating the existing distribution model involving both climate and topography. Our aim was also to evaluate the effect of climate in relation with topography in the updated model, which could either inflate or obscure the pure effect of climate on the distribution of this cliff-nesting species, before projecting the models to the future.

Methods

Study area

The study area is mainland Spain, an area of 493,518 km^2 characterized by a heterogeneous climate, which makes it particularly appropriate for analyzing different climate change scenarios. There is a mainly eastward and southward decreasing gradient of precipitation and a mainly northward-decreasing gradient of temperature [31]. Annual precipitation varies from less than 200 mm to more than 2000 mm, whereas mean annual temperatures vary from less than 6°C to more than 18°C.

Peninsular Spain has important mountain ranges, which reach a maximum altitude of 3478 m, many of them in the coastal areas contributing to isolate the central plateau from sea influences. Mainland Spain may be divided into three climatic areas: Atlantic, Mediterranean and Interior. Mild winters are found in the Atlantic area, together with cool summers, and the precipitation is abundant and regular. The Mediterranean part is characterized by hot summers and mild winters; rainfall rarely exceeds 500 mm annually and occurs mainly during spring and autumn. In the Interior, the temperatures are high in summer and low in winter, and precipitation is irregular and scarce [32].

Target species

Bonelli's Eagle is one of the rarest raptors in Europe and is now listed as endangered [33–34]. During the 70 s and 80 s European populations of the species suffered a severe population decline of 20–50% [35–37], although in recent years the population appears to have stabilized [38], with a current estimated population of 920–1100 pairs [33]. Because of this, it is a priority-target species for conservation in Europe (Council Directive 79/409/EEC). The majority of the European population (aprox. 80%) is concentrated in the Iberian Peninsula, where this raptor has experienced a population decline of 50% over the last three eagle generations [39]. Consequently the Bonelli's Eagle is also a priority-target species for special conservation measures in Spain (Real Decreto 439/1990). Main factors involved in the decline were primarily a high mortality rate in adults and sub-adults [40,41], and the loss of suitable habitat caused by alterations in land-use [34,42]. Interspecific competition with other raptors for breeding sites and home-ranges could also have had an effect [43,44].

Bonelli's eagle is a long-lived, with deferred maturity and sedentary species, with adult birds typically tied to a specific territory throughout the year [34]. Young eagles normally settle in dispersal areas during the period preceding sexual maturity that are clearly separated from the breeding range [34,45]. The home-range size for Bonelli's eagles in the study area normally varies from 20 to 110 km^2 [39]. Its distribution ranges from India and Southern China to the Iberian Peninsula and NW Africa [46]. In the western limit of its distribution area it occupies mainly the Mediterranean area, which is considered highly responsive to climate change because of its geographical situation between the temperate central Europe and the arid northern Africa [47,48]. Suitable areas for this species are mountainous with a Mediterranean climate, characterized by hot summers and low precipitation [29], although human disturbance may also affect at a local scale [49,50]. We obtained presence and absence data for a UTM of 10×10 km (5167 squares) from the last national survey conducted in 2005, which was produced with high accuracy and completeness [38].

Updating the known species distribution model

To forecast species distributions it is necessary first to balance the impact of climate change against the effects promoted by other influential factors [13]. The ideal way to balance these different effects is to consider actual climatic data, if available, rather than fictitious climatic variables derived from AOGCM-SRES combinations, together with other drivers of species distribution. This assessment was actually done for Bonelli's Eagle in mainland Spain [29], and yielded a parsimonious model including climate and topography as main drivers of the species distribution. This model, however, is not directly transferable to the future using climate change scenarios, as the future climatic variables reflect a simulated variation of climatic conditions in relation to the modeled present climate rather than the actual present climate. The best approach in this situation is to take advantage of the known model by updating it to the simulated climate provided by the AOGCM-SRES combinations. Updating methods are re-calibration procedures that have been used to adjust previously developed models to contemporary and/or local circumstances when a new sample is available [51].

The original explanatory model [29] was updated for each combination of AOGCM and SRES, by performing the updating method 4 used in [52] -corresponding to the updating method 5 of [51]-. Consequently, we fitted a new logistic regression of the most recent distribution data published in [38] on the invariant slope *(Slop)*, and the projected mean July temperature *(Tjul)* and mean annual precipitations *(Prec)* for the period 1961–1990 by the AOGCM-SRES combination, re-estimating all the coefficients. From these logistic regressions we obtained the corresponding updated favourability functions, which represented the present updated favourability (F_p) for the species in each cell,

$$F = \frac{e^y}{\frac{n_1}{n_0} + e^y}$$

where F is the logit link of the favourability function, e is the Neperian number, y is the logistic regression model equation, and n_1 and n_0 are the numbers of presences and absences, respectively [53].

Some authors argue in favor of using only climatic variables in this type of models [54], given that climate is strongly correlated with topography. However, at least for mountain species, topography is an influential factor in the species distribution, not a mere surrogate of climate. We have included topography to better understand the relationship of habitat structure with the potential distribution of the species, which may balance the impact of climate change against the inertia induced by other not changing influential factor. This is especially important when dealing with a species intimately linked to cliffs. It is already known that the true effect of topography is obscured by climate in the case of Bonelli's Eagle [29], and other mountain species [13]. In the case of Golden Eagle (*Aquila chrysaetos*), another cliff-nesting raptor, topographic variables are involved in those models better explaining its occurrence [55].

To assess the extent to which climatic and non-climatic variables explain the species distribution we differentiated, in each updated favourability model, the contribution of the climatic variables from that of slope using a variation partitioning procedure, following the approach described in [29,56]. In this way, we distinguished the Pure Climatic Factor (PCF, measured with R^2_{pClim}), i.e., the pure effect of climate on the model variation not affected by the collinearity with slope; the Pure Non-Climatic Factor (PNCF, measured with R^2_{pNClim}), i.e., the variation in the model that was due to the pure effect of slope not affected by the collinearity with precipitation and temperature; and the Shared Climatic Factor (SCF, measured with $R^2_{ClimNClim}$), i.e., the proportion which was assignable to their shared effect [57–61]. The part of the variation in the model explained by each factor (i.e. R^2_{Clim}, R^2_{NClim}) was obtained by a linear regression of the logit function of the model with the variables of each factor. Then, the pure effect of each factor was assessed by subtracting from 1 the variation of the model explained by the other factor ($R^2_{pClim} = 1 - R^2_{NClim}$; $R^2_{pNClim} = 1 - R^2_{Clim}$; and $R^2_{ClimNClim} = 1 - R^2_{pClim} - R^2_{pNClim}$). We also estimated the proportion of the climatic factor represented by the pure climate (ρ) for each climatic model ($\rho = R^2_{pClim}/R^2_{Clim}$).

The updated favourability models were projected to the future by replacing the values of *Tjul* and *Prec* by their corresponding values in the periods 2011–2040, 2041–2070 and 2071–2100 while maintaining the coefficients and the values of *Slop* (Table 1), which will not change substantially in the near future. The digital slope (*Slop*) was obtained according to the method described in [29,62], and the climatic variables were obtained from data supplied by the Agencia Estatal de Meteorología (AEMET) of Spain and digitalized using the method explained by [63]. These data resulted from the regionalization to Spain of the climate change models produced by the Intergovernmental Panel on Climate Change (IPCC).

Table 1. Variables used to model the species distribution.

Code	Variables
Slop	Slope (°) (calculated from altitude)(1)
Prec	Annual precipitation (mm)(2)
TJul	July mean temperature(2)

(1)US Geological Survey (GTOPO30) (http://edcdaac.usgs.gov/gtopo30/gtopo30.asp);
(2)Agencia Estatal de Meteorología of Spain (AEMET), Ministerio de Medio Ambiente (http://www.aemet.es/es/elclima/cambio_climat/escenarios).

Climate change scenarios

We used four different AOGCMs: CGCM2 from the Canadian Climate Centre for Modeling and Analysis, ECHAM4 from the Max Planck Institut für Meteorologie, and HadAM3H and HadCM2SUL from the Hadley Centre (U.K.), which differ in horizontal and vertical resolutions and in the parameterizations of physical processes (convection, land surface processes, cloud cover, and radiation, among others). According to the data obtained from the AEMET the circulation models CGCM2 and ECHAM4 were run with the conditions forecasted by the SRES A2 and B2 [64], HadAM3H was run with the scenario A2, and HadCM2SUL was run with the scenario IS92a, as they are the scenarios regionalized for Spain [65] (See Table 2). Scenarios A2 and B2 represent an intermediate position of the range of projected temperature change scenarios for Spain, A2 being medium-high and B2 medium-low [66]. The A2 storyline describes a very heterogeneous world with a regionally oriented economic development preserving local identities, and assumes modest reductions in overall population growth. The B2 storyline describes a world in which the emphasis is on environmental sustainability and local solutions to economic and social issues, and assumes more substantial reductions in overall population growth.

All the climatic models were run for the periods: 1961–1990, 2011–2040, 2041–2070, 2071–2100, with the exception of the HadAM3 which only had data for 1961–1990 and 2071–2100 (Table 2), obtaining in each cell a value of expected future favourability (F_f) according to each AOGCM-SRES combination.

Applying the expression $F_{f1} = F_p + \rho(F_f - F_p)$ we calculated the minimum or the maximum climatic effect over the species distribution, i.e., F_f and F_{f1} represent the limits of the forecasted effects of climate change on the spatial distribution of the favourability for the species.

As favourability values may be interpreted as the degree of membership of the sites to the fuzzy set of localities favourable to the species [53,63], we used some fuzzy logic operations [67] to calculate, for each future projection, the IOMS features of the forecasted effect of climate change of the species favourability proposed by [63], namely the increment in favourability (I), the favourability overlap (O), the favourability maintenance (M), and the forecasted shift in favourability (S) with respect to the 1961–1990 period:

$$I = \frac{c(F_f) - c(F_p)}{c(F_p)}$$

$$O = \frac{c(F_f \cap F_p)}{c(F_f \cup F_p)}$$

Table 2. The combination of AOGCMs and scenarios used in this study.

	SRE		
AOGCM	A2	B2	IS92a
CGCM2	x	x	
ECHAM4	x	x	
HAdAM3	x		
HadCM2SUL			x

$$M = \frac{c(F_f \cap F_p)}{c(F_p)}$$

$$S = \frac{Min[c(F_p) - c(F_f \cap F_p), c(F_f) - c(F_f \cap F_p)]}{c(F_p)}$$

where,

- $c(X)$ is the cardinality of the X fuzzy set, that is, the sum of all cells' membership degrees to the fuzzy set X.
- F_f is the fuzzy set of future favourable areas for the species, and the membership degree of each cell to F_f is defined by the future favourability value for the species in the cell.
- F_p is the fuzzy set of present favourable areas for the species, and the membership degree of each cell to F_p is defined by the present favourability value for the species in the cell.
- $F_f \cap F_p$ is the intersection between future and present favourabilities, and the membership degree of each cell to $F_f \cap F_p$ is defined by the minimum of the two favourability value for the species in the cell.
- $F_f \cup F_p$ is the union between future and present favourabilities, and the membership degree of each cell to $F_f \cup F_p$ is defined by the maximum of the two favourability values for the species in the cell.

We proceeded analogously for obtaining the IOMS features comparing F_p with F_{f1}.

Positive values of increment (I) indicate a favourability expansion for the species, that is, a gain of favourable areas, whereas negative values of I mean a net loss of favourability areas for the species. High values of overlap (O) indicate that the distributions of future local favourability values are predicted to be similar to that shown at present. Maintenance (M) indicates the degree to which current local favourability values are predicted to persist in the future, so that low values of M are of more conservation concern that high M values. Favourability shift (S) measure the proportion of the present favourability that is predicted to be lost in the future but may be compensated with new favourability opportunities elsewhere.

Results

Coefficients of the logit function of the favourability models for the period 1961–1990 are shown in Table 3. Table 4 shows the results of the variation partitioning of the favourability model, specifying the percentages of variation explained by the Pure Non-Climatic Factor (PNCF), the Pure Climatic Factor (PCF), the interaction that is due to Share Climatic Factor (SCF) and the proportion of pure climatic factor in relation to the whole climatic factor (ρ). In all favourability models climate had a more important effect than topography on the distribution of the species. All SCF values were negative, which indicates that topography tends to obscure the effect of climate on the species distribution.

Figure 1 shows the future favourability for *A. fasciata* according to the climatic conditions forecasted for every time period by each AOGCM and SRES combination, including the minimum and maximum expected change in favourability for every case. Minimum and maximum values of increment (I), overlap (O), maintenance (M), and shift (S) in favourability between the 1961–1990 period and the forecasted future favourability are shown in

Table 3. Coefficients in the logit function (y) of the favourability models for the period 1961–1990.

AOGCM	*y*
CGCM2-A2	0.319 * *Slop* - 0.0023 * *Prec*+0.366 * *TJul* - 9.51
CGCM2-B2	0.319 * *Slop* - 0.0023 * *Prec*+0.366 * *TJul* - 9.51
ECHAM4-A2	0.375 * *Slop* - 0.0035 * *Prec*+0.428 * *TJul* - 12.25
ECHAM4-B2	0.375 * *Slop* - 0.0035 * *Prec*+0.428 * *TJul* - 12.25
HAdAM3-A2	0.348 * *Slop* - 0.0043 * *Prec*+0.296 * *TJul* - 8.29
HadCM2-IS92a	0.312 * *Slop* - 0.0017 * *Prec*+0.394 * *TJul* - 10.18

For each AOGCM *Prec* and *TJul* are the forecasted for them.

Table 5. All the climatic models forecasted the maintenance of Bonelli's Eagle present favourability areas, as well as its expansion (positive increment) during all the XXI century and especially in the last period (2071–2100).

Discussion

The modelling approach

Generalised Linear Models (GLMs) formulate the relationship between distribution and environmental variables explicitly, and thus are appropriate tools to generate hypotheses about how species respond to spatial and environmental variability and to provide insights into the potential response to regional climate change [68]. These methods have the advantage of modelling both presence and absence data, which is critical for threatened species [69]. Although some authors recommend the use of profile modelling techniques that supposedly only require presence data, and thus are thought not to be affected by false absences [70–72], these methods are equally affected by missing presences (i.e., false absences) while not paying due attention to the specific causes of absences. We modelled absences explicitly because the true absence of a species from an area may be due to ecological, historical, or anthropogenic reasons, all of which are relevant factors in biogeography and conservation [56,73,74]. When consistent absence data are available, the explicit consideration of absences in the regression analysis improve the quality of the models, as they provide more explicit information about less favourable locations or unfavourable conditions for the species. This is why assessing the quality of the absence data (for example,

Table 4. Results of the variation partitioning of combined favourability model.

	CGCM2		ECHAM4	HadAM3H	HadCM2SUL
	A2	B2	A2/B2	A2	IS92a
PNCF	30.6	30.6	32.1	33.1	31.4
PCF	92	92	94.7	93.5	90.3
SCF	−22.6	−22.6	−26.8	−26.6	−21.7
ρ	1.326	1.326	1.395	1.398	1.316

Values shown are the percentages of variation explained by the Pure Non-climatic Factor (PNCF), the Pure Climatic Factor (PCF) and the interaction that is the Share Climatic Factor (SCF). (ρ).: Proportion of pure climatic factor in relation to whole climatic factor.

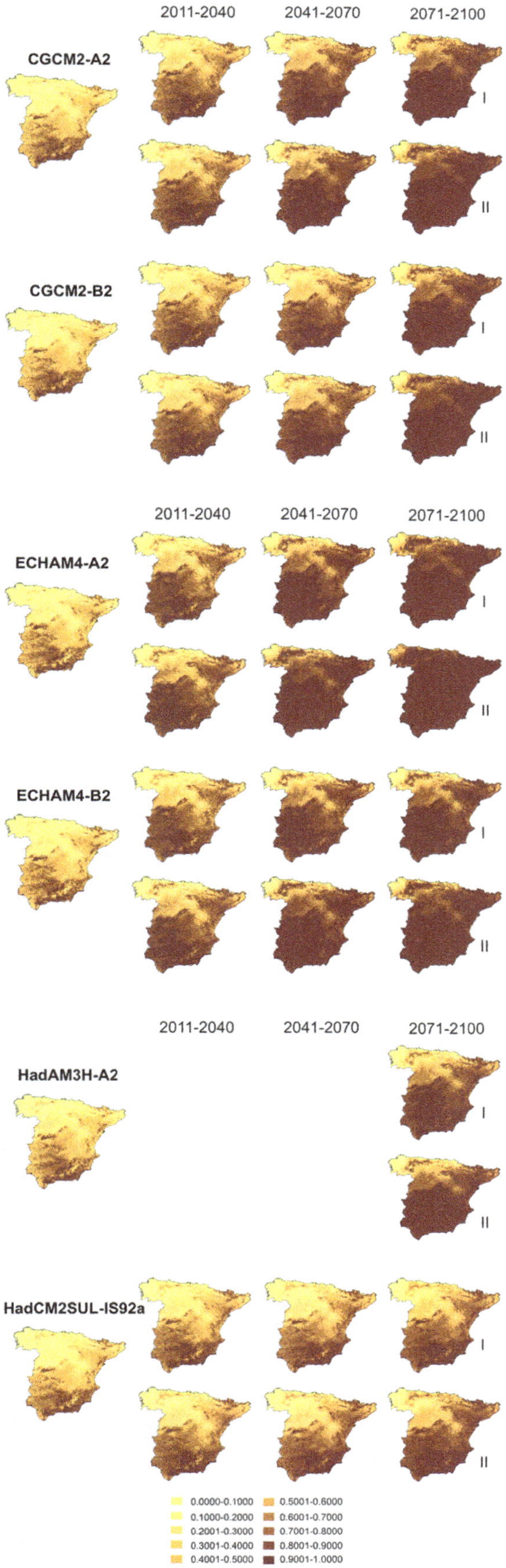

Figure 1. Favourability values forecasted at each 10 km×10 km UTM square of mainland Spain for Bonelli's Eagle, according to each climatic model and for each considered period. I and II indicate the minimum and maximum expected change in favourability, respectively.

measuring specificity) should be considered as important as the assessment of presence data in modelling procedures [13,75].

Variation partitioning

Accurate predictions about future species distributions and responses to future climate largely depend on the combination of the causal factors involved. Our models for Bonelli's Eagle included a climatic and a topographic factor, and depending on the importance of the former the future projections of the species distribution will be more or less affected by climate change. By using variation partitioning and weighting the effect of climate in relation to topography, we have evaluated the pure contribution of climate, not affected by the covariation with topography, in making a given area favourable for this mountain species. The effect of temperature and precipitation (i.e. the pure climatic factor) is obscured by slope (i.e. the non-climatic factor) in the amount expressed by the negative shared effect shown in Table 5. The pure effect of climate in the models is roughly three times that of topography (see Table 5) in all combined favourability models. This is probably the reason why the species is absent from Iberian mountains outside Mediterranean areas.

Management in a changing climate

Over the last century, mean annual temperatures have increased by 0.8°C in Europe, at the same time as annual precipitation has increased by 10–40% in northern Europe and decreased by up to 20% in parts of southern Europe [76]. Climate change is expected to have a noticeable effect on bird populations across a variety of habitats, as both ambient temperatures and levels of precipitation have a direct influence on the distribution, survival rates and productivity of individual species, and thus on population sizes [29,77–79].

Generalist species are thought to deal with rapid environmental change, while it is likely that species with more specialized ecological niches will face more severe challenges [80]. In Europe there are examples of bird species with more northerly geographic distributions that have declined, populations with more southerly distributions that have remained relatively stable or increased [22,81], and even cases in which African species have recently colonized southern Europe [82,83].

In the Mediterranean context Spain is a highly important area for bird conservation. It is the European member state with the largest surface area devoted to SPA (Special Protection Areas) for birds [84], and it is among the most responsive areas to global climate change due to its geographic situation [47,48]. In general, climate change implies a challenge for the current conservation policy, which generally assumes static species ranges, and do not consider the dynamism of the reserve borders nor the natural system dynamics caused by a changing world. In the case of Bonelli's Eagle in Andalusia (South of Spain), which is one of the most important strongholds for the species in Europe, 52.4% of the breeding territories are currently in protected areas [85], but most of the new favourable areas are predicted to occur outside the network of Andalusia's reserves, and thus the percentage of "unprotected" eagles is expected to increase. In Eastern Spain it has been demonstrated that the current network of special protected areas becomes insufficient to protect Bonelli's Eagle

Table 5. Values of the rates of increment (I), overlap (O), maintenance (M) and shifting (S) of favourability forecasted for each future projection with respect to the 1961–1990 period.

			I		O		M		S		CFf	
			I	*II*	*I*	*II*	*I*	*II*	*I*	*II*	*I*	*II*
CGCM2	**A2**	**2011–2040**	0,392	0,512	1,392	1,512	1,0	1,0	0,0	0,0	2869,0	3116,1
		2041–2070	0,716	0,899	1,716	1,899	1,0	1,0	0,0	0,0	3536,8	3914,3
		2071–2100	0,995	1,182	1,995	2,182	1,0	1,0	0,0	0,0	4113,0	4497,5
	B2	**2011–2040**	0,362	0,472	1,362	1,472	1,0	1,0	0,0	0,0	2811,0	3038,2
		2041–2070	0,560	0,719	1,560	1,719	1,0	1,0	0,0	0,0	3220,8	3548,9
		2071–2100	1,010	1,194	2,010	2,194	1,0	1,0	0,0	0,0	4149,6	4528,8
ECHAM4	**A2**	**2011–2040**	0,529	0,714	1,529	1,714	1,0	1,0	0,0	0,0	3143,3	3524,2
		2041–2070	0,936	1,177	1,936	2,177	1,0	1,0	0,0	0,0	3980,2	4476,5
		2071–2100	1,213	1,399	2,213	2,399	1,0	1,0	0,0	0,0	4551,0	4933,0
	B2	**2011–2040**	0,485	0,656	1,485	1,656	1,0	1,0	0,0	0,0	3053,1	3406,1
		2041–2070	0,817	1,053	1,817	2,053	1,0	1,0	0,0	0,0	3735,4	4221,9
		2071–2100	0,966	1,200	1,966	2,200	1,0	1,0	0,0	0,0	4042,8	4524,8
HadAM3	**A2**	**2011–2040**	0,236	0,329	1,236	1,329	1,0	1,0	0,0	0,0	2542,5	2735,1
		2041–2070	0,321	0,445	1,321	1,445	1,0	1,0	0,0	0,0	2717,1	2972,3
		2071–2100	0,517	0,704	1,517	1,704	1,0	1,0	0,0	0,0	3121,0	3505,1
HadCM2SUL	**IS92a**	**2071–2100**	0,784	0,956	1,784	1,956	1,0	1,0	0,0	0,0	3679,8	4034,4

cF_f is the cardinality of the fuzzy set of favourable areas forecasted for the respective future period. *I*: considering the apparent climatic effect, and *II*: considering the pure climatic effect, at present and in each future period for the four climatic models considered.

[86]. Species are likely to change their distributions, adjusting it to the emergence of new favourable and unfavourable areas, and therefore their representation levels in static reserves are prone to be altered [87,88]. Therefore, an effort should be made to spatially coordinate reserve management to capture these biological dynamics among multiple protected areas and across the landscape [89].

The capacity to simulate the potential changes in the distribution range of Bonelli's Eagle in Spain as precisely as possible is important to favour the conservation of the species, especially taking into account that Spain concentrates most of the European population. Changes in temperatures and precipitation patterns may have direct and indirect effects on the survival rates and productivity of the species [90], thus influencing the viability of its populations.

Our analyses indicate that the favourable areas for Bonelli's Eagle, according to all the AOGCM and scenarios used, will increase during the XXI century in Spain. The impact of climate change on this species in our study area will not be negative as it occurs for other bird species that are expected to suffer important decreases in their distribution area [26,91]. We recommend that to model species distributions in the future, multiple climatic models, i.e. the combination of AOGCMs and SRESs, should be used.

Predicting the future favorability and potential distribution

It is widely acknowledged that species distribution models provide a simplified representation of the processes governing the geographic distributions of species [27,92]. Although it is difficult to fully explore uncertainties arising from the large number of AOGCMs that are currently being generated, our approach and results are consistent in predicting an increase in climatic favourability for all the scenarios used. However, the intensity of the forecasted increment in favourability differ for the AOGCMs used, ranging from the more drastic changes predicted according to ECHAM4 to the more conservative predictions of HadCM2-SUL. Our impression after visual inspection of Figure 1 is that predictions of HadCM2SUL seem to be more reasonable, but this could be affected by an ill-founded expectation of moderate changes in nature.

The uncertainty associated to the differences in AOGCMs and SRESs has already been measured [63]. In this work we assessed a new source of uncertainty associated to the models, which derives from not knowing the exact role of climate in the biogeographical response of the species. At least, our approach allows putting limits to the minimum and maximum expected influence of climate on the species distribution and, consequently, forecasting minimum and maximum future changes in environmental favourability. Another possible source of uncertainty, especially in those species with a projected increase in distribution, as is our case, is the overestimation due to the truncated response curves [93]. Considering new environmental conditions that are outside of the calibration range could lead to erroneously predict the new conditions as favourable overlooking the fact that warmer temperatures and lower values of precipitation could be unsuitable for the species (e.g. physiological limitations or new conditions of competence) [54]. In our case, 99,1% of the predicted new favourable squares are within the range of the function y.

It is necessary to consider that an increase in the existence of favourable areas does not necessarily mean an increase in the species distribution. Human interaction will probably prevent Bonelli's Eagles from establishing in many climatically-favourable zones. Although this species may tolerate high levels of human disturbance [29,94–96], the main causes of mortality for Bonelli's eagle in Spain are human induced, mostly due to power lines casualties and also direct persecution [41]. It is remarkable that a significant proportion of the new favourable areas are predicted in flat or undulating landscapes, which lack natural perching sites for the eagles and would favor the use of electric pylons, making them more vulnerable to electrocution. Therefore, in order to enhance the conservation of the species, mitigation measures to prevent power lines-induced mortality might be accordingly contemplated in these areas, considering that management actions normally require long temporal scales. In the case of the endangered Spanish Imperial Eagle (*Aquila adalberti*) it has already been demonstrated that eagle electrocution is an affordable problem whenever there is political interest and financial support [97].

Since Bonelli's Eagle is a species of conservation concern in Europe and Spain, we take advantage of the pros provided by a regional pre-existing distribution model, and the most recent distribution data, together with the simulated climate change variables. This could be of particular interest to the species in the European context because mainland Spain includes approximately 80% of European Bonelli's Eagles [33].

This paper predicts an increase in environmental favourability for the species in the Iberian Peninsula, but many of these new favourable areas are outside mountain ranges and have little or no availability of cliffs, which currently are the usual nesting areas. In Spain Bonelli's Eagle breeds mainly in rocky substrates, since 95.5% of the nests are found in this substrate, while trees and power lines are occasionally used, 4% and 0.5% respectively. Our models can be considered as realistic only if the nesting behaviour of the species in Spain changes significantly to use trees or power lines much more than currently. On a global scale Bonelli's eagle occupies mountains, cliffs, crags, gorges, hills and plains with forest or woodland [98,99], although in some areas built their nest on lofty trees, as in southern India [99] and Portugal. In the case of neighboring Portugal the proportion of pairs nesting on trees is completely different from that found in Spain. There 64% of the population nest in trees [100], like Cork Oaks, Pines and large Eucalyptus, and in the south of the country 61 out of the 65 pairs (94%) are tree-nesters. This demonstrates the plasticity of the species to choose nesting substrate and to breed in trees in those favorable regions with no mountains, which could mean a future increase in the range of Bonelli's Eagle if it starts to breed also in trees in Spain. Interestingly, the African Hawk-eagle (*Aquila spilogaster*), the sister species of Bonelli's Eagle, is distributed in tropical Africa south of the Sahara, lives in woodlands and breeds exclusively on trees, mainly in *Acacia* riparian woodland, *Baikiaea* and mixed woodland [101].

Species may respond to global climate change by shifting their geographical distribution in absence of any evolutionary change [5] but, as it has been already pointed out, evolutionary adaptation can be rapid helping species achieve new ecological opportunities arising from climate change [102]. Another source of uncertainty when projecting species distribution models to the future is our inability to forecast how species might express phenotypic plasticity to changing environmental conditions [103]. For this reason, to consider the evolutionary potential of species and including the possibility of evolution in distribution modelling would provide detailed information of great interest in order to better determine the effect of climate change on species. It would also allow incorporating this information into better informed management programs designed to prevent biodiversity loss under rapid climate change. Long-term monitoring and observations, especially in long-lived territorial raptors characterised by deferred maturity [104], and future updating of pre-existing models, considering new distribution ranges known in the future and also new variables, are needed to provide an assessment of the predictions about climate change and about Bonelli's Eagle response by the possible changing of nesting behaviour and increase of its distribution range.

In general, global distribution models are preferable to regional models, as predicting the future distribution of a species from a part of its range could be oblivious to the variation in climate tolerance that is not present in the studied area. We updated a regional distribution model because this was the only pre-existing model available. Nevertheless, if a global niche model of the species were at hand, we would recommend it to be updated to the target region anyway, because in a global model the relationship between climate and non-climatic factors, and their separate and combined effects on the species distribution, are averaged throughout the species range, in this case from Portugal to China, including Africa and Indonesia, while some factors may be more, or less, critical than average in specific zones of the species range. The territory we analyzed in this study is characterized by a heterogeneous climate and hosts the core of Bonelli's Eagle European population, which makes it particularly appropriate for analyzing different climate change scenarios. Additionally, mainland Spain encompasses the whole variability of breeding behaviours known for the species in its entire range. Another possible advantage of focusing the study of the effect of climate change on a specific and discrete part of its distribution area is that allopatric distribution of Bonelli's Eagle (e.g. China, the Indian subcontinent and Indonesian populations), probably represent relatively different natural histories, and presumably different responses to environmental conditions. Thus, the updating of pre-existing models allows retaining the potential of these models, either regional or global, while recalibrating them to optimize their performance in specific situations.

Conclusions

To perform good species distribution models is time consuming. When working with many species simultaneously modelling may become a routine task which does not allow paying the necessary attention to the uncertainty related to each species. In this article we showed the value of using already existing models in well studied species to forecast climate change impacts, remarking the importance of linking conservation biology with distribution modelling by updating existing models, since conservation objectives are more likely to be achieved when knowledge informs actions. Models of this kind are scarce, but they are sometimes available for species of conservation concern and it is preferable to update them considering all the known factors conditioning the species' distribution to better infer climate change effects, instead of building new models that are based on climate change variables only.

Acknowledgments

Juan Carlos del Moral (SEO/BirdLife) provided the digital data of the national survey. The Agencia Estatal de Meteorología (AEMET) of Spain provided the climatic data. This paper is part of Antonio-Román Muñoz PhD thesis.

Author Contributions

Conceived and designed the experiments: ARM RR. Analyzed the data: ARM ALM. Wrote the paper: ARM RR. Edited the manuscript: ARM.

References

1. IPCC (2001) Climate Change 2001: The Scientific Basis. Technical Summary. Intergovernmental Panel on Climate Change. Cambridge: Cambridge University Press.
2. IPCC (2008) Climate Change 2007: Synthesis Report. Contribution of Working Groups I, II and III to the Fourth Assessment Report of the Intergovernmental Panel on Climate Change. Geneva: IPCC.
3. IPCC (2001) Climate Change 2001: Impacts Adaptation and Vulnerability. Summary for Policymakers. Intergovernmental Panel on Climate Change. Cambridge: Cambridge University Press.
4. Walther GR, Post E, Convey P, Menzel A, Parmesan C, et al. (2002) Ecological responses to recent climate change. Nature 416: 389–395.
5. Parmesan C, Yohe G (2003) A globally coherent fingerprint of climate change impacts across natural systems. Nature 421: 37–42.
6. Root TL, Price JT, Hall KR, Schneider SH, Rosenzweig C, et al. (2003) Fingerprints of global warming on wild animals and plants. Nature 421: 57–60.
7. Pauli H, Gottfried M, Reiter K, Klettner C, Grabherr G (2007) Signals of range expansions and contractions of vascular plants in the high Alps: observations (1994–2004) at the GLORIA*master site Schrankogel, Tyrol, Austria. Global Change Biol 13: 147–156.
8. Buisson L, Thuiller W, Lek S, Lim P, Grenouillet G (2008) Climate change hastens the turnover of stream fish assemblages. Global Change Biol 14: 2232–2248.
9. Rivalan P, Barbraud C, Inchausti P, Weimerskirch H (2010) Combined impacts of longline fisheries and climate on the persistence of the Amsterdam Albatross *Diomedia amsterdamensis*. Ibis 152: 6–18.
10. Loarie SR, Duffy PB, Hamilton H, Asner GP, Field CB, et al. (2009) The velocity of climate change. Nature 462: 1052–1055.
11. Walther GR, Burga CA, Edwards PJ (2001) Fingerprints of Climate Change. Adapted Behaviour and Shifting Species Ranges. New York: Kluwer Academic/Plenum.
12. Pearson RG, Dawson TP (2003) Predicting the impacts of climate change on the distribution of species: are bioclimate envelope models useful? Global Ecol Biogeogr 12: 361–371.
13. Márquez AL, Real R, Olivero J, Estrada A (2011) Combining climate with other influential factors for modelling climate change impact on species distribution. Climatic Change 108: 135–157.
14. Ahola M, Laaksonen T, Sippola K, Eeva T, Rainio K, et al. (2004) Variation in climate warming along the migration route uncouples arrival and breeding dates. Global Change Biol 10: 1–8.
15. Gordo O, Brotons L, Ferrer X, Comas P (2005) Do changes in climate patterns in wintering areas affect the timing of the spring arrival of trans-Saharan migrant birds? Global Change Biol 11: 12–21.
16. Ptaszyk J, Kosicki J, Sparks TH, Tryjanowski P (2003) Changes in the timing and pattern of arrival of the White Stork *(Ciconia ciconia)* in western Poland. J Ornithol 144: 323–329.
17. Sanz JJ, Potti J, Moreno J, Merino S, Frías O (2003) Climate change and fitness components of migratory bird in the Mediterranean region. Global Change Biol 9: 461–472.
18. Tryjanowski P, Kuzniak S, Sparks TH (2005) What affects the magnitude of change in first arrival dates of migrant birds? J Ornithol 146: 200–205.
19. Zalakevicius M, Bartkeviciene G, Raudonikis L, Janulaitis J (2006) Spring arrival response to climate change in birds: a case study from eastern Europe. J Ornithol 147: 326–343.
20. Gordo O, Sanz JJ (2006) Climate change and bird phenology: a long-term study in the Iberian Peninsula. Global Change Biol 12: 1993–2004.
21. Halupka L, Dyrcz A, Borowiec M (2008) Climate change affects breeding of reed warblers *Acrocephalus scirpaceus*. J Avian Biol 39: 95–100.
22. Lemoine N, Bauer HG, Peintinger M, Böhning-Gaese K (2007) Effects of climate and land-use change on species abundance in a central European bird community. Conserv Biol 21: 495–503.
23. Hughes L (2000) Biological consequences of global warming: is the signal already apparent? Trends Ecol Evol 15: 56–61.
24. McCarty JP (2001) Ecological consequences of recent climate change. Conserv Biol 15: 320–331.
25. Virkkala R, Heikkinen RK, Leikola N, Luoto M (2008) Projected large-scale range reductions of northern-boreal land bird species due to climate change. Biol Conserv 141: 1343–1353.
26. Barbet-Massin M, Walther BA, Thuiller W, Rahbek C, Jiguet F (2009) Potential impacts of climate change on the winter distribution of Afro-Palaearctic migrant passerines. Biol Letters 5: 248–251.
27. Guisan A, Thuiller W (2005) Predicting species distribution: offering more than simple habitat models. Ecol Lett 8: 993–1009.
28. Niamir A, Skidmore AK, Toxopeus AG, Muñoz AR, Real R (2011) Finessing atlas data for species distribution models. Divers Distrib 17: 1173–1185.
29. Muñoz AR, Real R, Barbosa AM, Vargas JM (2005) Modelling the distribution of Bonelli's Eagle in Spain: implications for conservation planning. Divers Distrib 11: 477–486.
30. Serreze MC (2010) Understanding Recent Climate Change. Conserv Biol 24: 10–17.
31. Font I (2000) Climatología de España y Portugal. Salamanca: Ediciones Universidad de Salamanca.
32. Capel JJ (1981) Los climas de España. Barcelona: Oikos-Tau SA Ediciones.
33. BirdLife International (2004) Birds in Europe: population estimates, trends and conservation status. BirdLife Conservation Series n° 12. Cambridge: Bird-Life International. 374 p.
34. Real J (2004) Aguila azor-perdicera, Hieraaetus fasciatus. In: Madroño A., González C, Atienza JC, editors. Libro Rojo de las Aves de España. Madrid: Dirección General de Conservación de la naturaleza-Sociedad Española de Ornitología. pp. 154–157.
35. Rocamora G (1994) Bonelli's Eagle *Hieraaetus fasciatus*. In: Tucker GM, Heath MF, editors. Birds in Europe, their conservation status. BirdLife Conservation Series n° 3. Cambridge: BirdLife International. pp. 184–185.
36. BirdLife International/EBCC (2000) European bird populations: estimates and trends. BirdLife Conservation Series n° 10. Cambridge: BirdLife International. 160 p.
37. Real J (2003) Águila-azor perdicera. In: Marti R, del Moral JC, editors. Atlas de las aves reproductoras de España. Madrid: Dirección General de Conservación de la naturaleza-Sociedad Española de Ornitología. pp. 192–193.
38. Del Moral JC (2006) El águila perdicera en España. Población en 2005 y método de censo. Madrid: SEO/BirdLife.
39. Bosch R, Real J, Tintó A, Zozaya EL, Castell C (2010) Home-ranges and patterns of spatial use in territorial Bonelli's Eagles *Aquila fasciata*. Ibis 152: 105–117.
40. Real J, Mañosa S (1997) Demography and conservation of Western European Bonelli's Eagle (*Hieraaetus fasciatus*) populations. Biol Cons 79: 59–66.
41. Real J, Grande JM, Mañosa S, Sánchez-Zapata JA (2001) Causes of death in different areas for Bonelli's Eagle *Hieraaetus fasciatus* in Spain. Bird Study 48: 221–228.
42. Balbontín J (2005) Identifying suitable habitat for dispersal in Bonelli's eagle: an important issue in halting its decline in Europe. Biol Conserv 126: 74–83.
43. Fernández C, Insausti JA (1990) Golden Eagles take up territories abandoned by Bonelli's Eagles. J Raptor Res 24: 124–125.
44. Carrete M, Sánchez-Zapata JA, Tella JL, Gil-Sánchez JM, Moleón M (2006) Components of breeding performance in two competing species: habitat heterogeneity, individual quality and density-dependence. Oikos 112: 680–690.
45. Moleón M, Bautista J, Madero A (2011) Communal roosting in young Bonelli's Eagles (*Aquila fasciata*). J Raptor Res 45: 353–356.
46. Del Hoyo J, Elliot A, Sargatal J (1994) Handbook of the birds of the world, Vol. II. New world vultures to guineafowl. Barcelona: Lynx Edicions.
47. Sánchez E, Gallardo C, Gaertner MA, Arribas A, Castro M (2004) Future climate extreme events in the Mediterranean simulated by a regional climate model: a first approach. Global Planet Change 44: 163–180.
48. Giorgi F, Lionello P (2008) Climate change projections for the Mediterranean region. Global Planet Change 63: 90–104.
49. López-López P, García-Ripollés C, Aguilar JM, García-López F, Verdejo J. (2006) Modelling breeding habitat preferences of Bonelli's Eagle (*Hieraaetus fasciatus*) in relation to topography, disturbance, climate and land use at different spatial scales. J Ornithol 147: 97–107.
50. Muñoz AR, Real R (2013) Factors determining the distribution of Bonelli's Eagle *Aquila fasciata* in southern Spain: scale may matter. Acta Ornithol. 48: *in press*.
51. Steyerberg EW, Borsboom GJJM, van Houwelingen HC, Eijkemans MJC, Habbema JDF (2004) Validation and updating of predictive logistic regression models: a study on sample size and shrinkage. Statist Med 23: 2567–2586.
52. Gastón A, García-Viñas JI (2010) Updating coarse-scale species distribution models using small fine-scale samples. Ecol Model 221: 2576–2581.
53. Real R, Barbosa AM, Vargas JM (2006) Obtaining environmental favourability functions from logistic regression. Environ Ecol Stat 13: 237–245.
54. Engler R, Randin C, Thuiller W, Dullinger S, Zimmermann NE, et al. (2011) 21st century climate change threatens mountain flora unequally across Europe. Global Change Biol 17: 2330–2341.
55. López-López P, García-Ripollés C, Soutullo A, Cadahía L, Urios V (2007) Identifying potentially suitable nesting habitat for golden eagles applied to 'important bird areas' design. Anim Conserv 10: 208–218.
56. Muñoz AR, Real R (2006) Assessing the potential range expansion of the exotic monk parakeet in Spain. Divers Distrib 12: 656–665.
57. Legendre P (1993) Spatial autocorrelation: trouble or new paradigm? Ecology 74: 1659–1673.
58. Legendre P, Legendre L (1998) Numerical ecology. Second English edition. Amsterdam: Elsevier Science.
59. Bjorholm S, Svenning JC, Skov F, Balslev H (2005) Environmental and spatial controls of palm (Arecaceae) species richness across the Americas. Global Ecol Biogeogr 14: 423–429.
60. Farfán MA, Vargas JM, Guerrero JC, Barbosa AM, Duarte J, et al. (2008) Distribution modelling of wild rabbit hunting yields in its original area (S Iberian Peninsula). Ital J Zool 75: 161–172.
61. Randin CF, Jaccard H, Vittoz P, Yoccoz NG, Guisan A (2009) Land use improves spatial predictions of mountain plant abundance but not presence-absence. J Veg Sci 20: 996–1008.

62. Barbosa AM, Real R, Olivero J, Vargas JM (2003) Otter (*Lutra lutra*) distribution modeling at two resolution scales suited to conservation planning in the Iberian Peninsula. Biol Conserv 114: 377–387.
63. Real R, Márquez AL, Olivero J, Estrada A (2010) Species distribution models in climate change scenarios are still not useful for informing policy planning: an uncertainty assessment using fuzzy logic. Ecography 33: 304–314.
64. Nakicenovic N, Alcamo J, Davis G, de Vries B, Fenhann J, et al. (2000) IPCC Special Report on Emissions Scenarios. Cambridge: Cambridge University Press.
65. Leggett J, Pepper WJ, Swart RJ (1992) Emissions Scenarios for the IPCC: An Update. In: Houghton JT, Callander BA, Varney SK, editors. Climate Change 1992: The Supplementary Report to the IPCC Scientific Assessment. Cambridge: Cambridge University Press. pp. 69–95.
66. Brunet M, Casado MJ, de Castro M, Galán P, Lopez JA, et al. (2007) Generación de escenarios de cambio climático para España. Madrid: Ministerio de Medio Ambiente.
67. Kuncheva LI (2001) Using measures of similarity and inclusion for multiple classifier fusion by decision templates. Fuzzy Set Syst 122:401–407.
68. Calef MP, McGuire AD, Epstein HE, Rupp TS, Shugart HH (2005) Analysis of vegetation distribution in Interior Alaska and sensitivity to climate change using a logistic regression approach. J Biogeogr 32: 863–878.
69. Estes LD, Mwangi AG, Reillo PR, Shugart HH (2011) Predictive distribution modeling with enhanced remote sensing and multiple validation techniques to support mountain bongo antelope recovery. Anim Conserv 14: 521–532.
70. Peterson AT, Vieglais DA (2001) Predicting species invasions using ecological niche modeling. BioScience 51: 363–371.
71. Hirzel AH, Helfer V, Metral F (2001) Assessing habitat suitability models with a virtual species. Ecol Model 145: 111–121.
72. Phillips SJ, Anderson RP, Schapire RE (2006) Maximum entropy modeling of species geographic distributions. Ecol Model 190: 231–259.
73. Castro A, Muñoz AR, Real R (2008) Modelling the spatial distribution of the Tengmalm's owl *Aegolius funereus* in its southwestern Palaeartic limit (NE Spain). Ardeola: 55, 71–85.
74. Aragón PA, Lobo JM, Olalla-Tárraga MA, Rodríguez MA (2010) The contribution of contemporary climate to ectothermic and endothermic vertebrate distributions in a glacial refuge. Global Ecol Biogeogr 19: 40–49.
75. Jiménez-Valverde A, Lobo JM, Hortal J (2008) Not as good as they seem: the importance of concepts in species distribution modelling. Divers Distrib 14: 885–890.
76. Parry ML (2000) Assessment of Potential Effects and Adaptations for Climate Change in Europe: The Europe ACACIA project. Norwich: Jackson Environment Institute, University of East Anglia.
77. Thomson DL, Baillie SR, Peach WJ (1997) The demography and age-specific annual survival of song thrushes during periods of population stability and decline. J Anim Ecol 66: 414–424.
78. Freeman SN, Crick HQP (2003) The decline of the Spotted Flycatcher *Muscicapa striata* in the UK: an integrated population model. Ibis 145: 400–412.
79. Robinson RA, Green RE, Baillie SR, Peach WJ, Thomson DL (2004) Demographic mechanisms of the population decline of the song thrush *Turdus philomelos* in Britain. J Anim Ecol 73: 670–682.
80. Huntley B, Collingham YC, Green RE, Hilton GM, Rahbek C, et al. (2006) Potential impacts of climatic change upon geographical distributions of birds. Ibis 148: 8–28.
81. Julliard R, Jiguet F, Couvet D (2004) Common birds facing global changes: what makes a species at risk? Global Change Biol 10: 148–154.
82. Elorriaga J, Muñoz AR (2010) First breeding record of North African Long-legged Buzzard *Buteo rufinus cirtensis* in continental Europe. British Birds 103: 396–404.
83. Elorriaga J, Muñoz AR (2013) Hybridisation between the Common Buzzard *Buteo buteo buteo* and the North African race of Long-legged Buzzard *Buteo rufinus cirtensis* in the Strait of Gibraltar: prelude or preclude to colonisation? Ostrich 84: 41–45.
84. Morillo C, Gómez-Campo C (2000) Conservation in Spain, 1980–2000. Biol Conserv 95: 165–174.
85. Moleón M (2006) El águila perdicera en Andalucía. In: del Moral JC, editor. El águila perdicera en España. Población en 2005 y método de censo. Madrid: SEO/BirdLife. pp. 24–30.
86. López-López P, García-Ripollés C, Soutullo A, Cadahía L, Urios V (2007) Are important bird areas and special protected areas enough for conservation?: the case of Bonelli's eagle in a Mediterranean area. Biod Cons 16: 3755–3780.
87. Heller NE, Zavaleta ES (2009) Biodiversity management in the face of climate change: a review of 22 years of recommendations. Biol Conserv 142: 14–32.
88. Alagador D, Martins MJ, Cerdeira JO, Cabeza M, Araújo MB (2011) A probability-based approach to match species with reserves when data are at different resolutions. Biol Conserv 144: 811–820.
89. Hannah L, Salm R (2005) Protected areas management in a changing climate. In: Lovejoy TE, Hannah L, editors. Climate Change and Biodiversity. Yale University Press. pp. 363–371.
90. Gil-Sánchez JM, Moleón M, Otero M, Bautista J (2004) A nine-year study of successful breeding in a Bonelli's eagle population in southeast Spain: a basis for conservation. Biol Conserv 118: 685–694.
91. Huntley B, Green RE, Collingham YC, Willis SG (2007) A Climatic atlas of European Breeding Birds. Barcelona: Durham University, The RSPB and Lynx Edicions.
92. Diniz-Filho JAF, Bini LM, Rangel TF, Loyola RD, Hof C, et al. (2009) Partitioning and mapping uncertainties in ensembles of forecasts of species turnover under climate change. Ecography 32: 897–906.
93. Thuiller W, Brotons L, Araujo MB, Lavorel S (2004) Effects of restricting environmental range of data to project current and future species distributions. Ecography 27: 165–172.
94. Gil-Sánchez JM, Molino F, Valenzuela G (1996) Selección de hábitat de nidificación por el águila perdicera (*Hieraaetus fasciatus*) en Granada (SE de España). Ardeola 43: 189–197.
95. López-López P, García-Ripollés C, García-López F, Aguilar J, Verdejo J (2004) Distribution pattern among Golden Eagle *Aquila chrysaetos* and Bonelli's Eagle *Hieraaetus fasciatus* in the Castellon province. Ardeola 51: 275–283.
96. Carrascal LM, Seoane J (2009) Factors affecting large-scale distribution of the Bonelli's eagle *Aquila fasciata* in Spain. Ecol Res 24: 565–573.
97. López-López P, Ferrer M, Madero A, Casado E, McGrady M (2011) Solving man-induced large-scale conservation problems: the Spanish Imperial Eagle and power lines. Plos One 6(3): e17196. doi:10.1371/journal.pone.0017196.
98. Cramp S, Simmons KEL (1980) Handbook of the birds of Europe, the Middle East and North Africa, Vol. II. Oxford: Oxford University Press.
99. Ali S, Ripley SD (2001) Handbook of the birds of India and Pakistan, Vol. I. New Delhi: Oxford University Press.
100. European Raptors, Biology and Conservation website. Available: http://www.europeanraptors.org/raptors/bonellis_eagle.html. Accessed 2013 Jun 5.
101. Hustlers K, Howells WW (1988) The effect of primary production on breeding success and habitat selection in the African Hawk-eagle. Condor 90: 583–587.
102. Hoffmann AA, Sgro CM (2011) Climate change and evolutionary adaptation. Nature 470: 479–485.
103. Theurillat JP, Guisan A (2001) Potential impact of climate change on vegetation in the European Alps: a review. Climatic Change 50: 77–109.
104. Balbontín J, Penteriani V, Ferrer M (2003) Variations in the age of mates as an early warning signal of changes in population trends? The case of Bonelli's eagle in Andalusia. Biol Conserv 109: 417–423.

4

Natural Hazards in a Changing World: A Case for Ecosystem-Based Management

Jeanne L. Nel[1]*, David C. Le Maitre[1], Deon C. Nel[2], Belinda Reyers[1], Sally Archibald[3,4], Brian W. van Wilgen[1,5], Greg G. Forsyth[1], Andre K. Theron[1], Patrick J. O'Farrell[1], Jean-Marc Mwenge Kahinda[4], Francois A. Engelbrecht[4], Evison Kapangaziwiri[4], Lara van Niekerk[1], Laurie Barwell[1]

1 Natural Resources and the Environment, Council for Scientific and Industrial Research (CSIR), Stellenbosch, Western Cape, South Africa, **2** World Wide Fund for Nature (WWF), Cape Town, Western Cape, South Africa, **3** School of Animal, Plant and Environmental Sciences, University of the Witwatersrand, Johannesburg, Gauteng, South Africa, **4** Natural Resources and the Environment, Council for Scientific and Industrial Research (CSIR), Pretoria, Gauteng, South Africa, **5** Centre for Invasion Biology, Department of Botany and Zoology, Stellenbosch University, Stellenbosch, Western Cape, South Africa

Abstract

Communities worldwide are increasingly affected by natural hazards such as floods, droughts, wildfires and storm-waves. However, the causes of these increases remain underexplored, often attributed to climate changes or changes in the patterns of human exposure. This paper aims to quantify the effect of climate change, as well as land cover change, on a suite of natural hazards. Changes to four natural hazards (floods, droughts, wildfires and storm-waves) were investigated through scenario-based models using land cover and climate change drivers as inputs. Findings showed that human-induced land cover changes are likely to increase natural hazards, in some cases quite substantially. Of the drivers explored, the uncontrolled spread of invasive alien trees was estimated to halve the monthly flows experienced during extremely dry periods, and also to double fire intensities. Changes to plantation forestry management shifted the 1:100 year flood event to a 1:80 year return period in the most extreme scenario. Severe 1:100 year storm-waves were estimated to occur on an annual basis with only modest human-induced coastal hardening, predominantly from removal of coastal foredunes and infrastructure development. This study suggests that through appropriate land use management (e.g. clearing invasive alien trees, re-vegetating clear-felled forests, and restoring coastal foredunes), it would be possible to reduce the impacts of natural hazards to a large degree. It also highlights the value of intact and well-managed landscapes and their role in reducing the probabilities and impacts of extreme climate events.

Editor: Vanesa Magar, Centro de Investigacion Cientifica y Educacion Superior de Ensenada, Mexico

Funding: Financial support was provided by the CSIR (www.csir.co.za), Santam Limited (santam.co.za) and the Global Environmental Facility under the auspices of UNEP's Project for Ecosystem Services (www.unep.org). The funders had no role in study design, data collection and analysis, decision to publish, or preparation of the manuscript.

Competing Interests: The authors bring to your attention to their co-funded collaboration in this study with Santam insurance – a commercial enterprise. Santam's role in this study was to help select a suitable study site and then to mainstream the knowledge the authors generated into their organisation and the broader insurance sector. The authors would like to emphasize that, once the study site was chosen, Santam had no role in study design, data collection and analysis, decision to publish, or preparation of the manuscript. Santam's interest in this study was not for direct financial gain, but rather to use as 'proof-of-concept' to inform further in-house business research on how to improve risk assessment techniques in a changing world. Both organizations in the collaboration (Santam and CSIR) committed similar amounts of funding, and the relationship with Santam insurance and the authors can be more correctly described as a 'knowledge partnership' than a 'client-customer relationship'.

* E-mail: jnel@csir.co.za

Introduction

Since the turn of the century, several major natural disasters have attracted international attention, including hurricanes Katrina and Sandy, floods in Thailand and the Indian Ocean tsunami. These disasters, along with countless more frequent disasters of smaller magnitude, have been responsible for the loss of at least a million lives over the last decade, with recovery often taking years and financial losses estimated to be in the trillions of US dollars [1]. Such disasters occur when extreme physical events – or 'natural hazards' – impact adversely on vulnerable and exposed communities and infrastructure, which are the human elements of disaster [2]. As a result, disaster risk is affected by changes in the incidence of natural hazard, as well as alterations in the patterns of societal exposure and vulnerability. This paper deals specifically with changes in the incidence of natural hazards, which are expected to increase into the future. Changes in climate are expected to result in higher sea levels and increased hurricane activity, bringing more frequent and severe storm-waves that flood and erode coastal areas [2,3]. In many regions of the world, climate-induced shifts in the water cycle will result in more frequent and intense periods of flooding and drought [4,5]. Wildfires are also expected to become more widespread and frequent, being closely linked to hot, dry weather conditions and drought [6].

The anticipated increased incidence of natural hazards is not only attributed to climate change. There is a growing concern that rapid and widespread land cover change is leading to the loss of the buffering capacity that healthy ecosystems provide against these natural hazards [7]. For example, healthy mangrove ecosystems, coral reefs and coastal foredunes are able to dissipate wave energy, reducing the impacts of coastal flooding and erosion during storms [8,9]. Areas with intact mangroves were much less affected by the 1999 and 2004 Indian Ocean tsunamis, than areas where mangroves had been removed [10]. Likewise, healthy inland wetland ecosystems and riparian zones help to absorb peak flows and sediment during extreme rainfall events, reducing flooding and sedimentation hazards to downstream areas [11,12]. Invasion of natural vegetation by alien trees that use more water than the indigenous vegetation that they replace, exacerbates the effects of water scarcity in drought-prone regions [13,14]. Invasive alien trees can also increase fuel loads and thus wildfire hazard [15], or can turn an ecosystem that was not prone to burning into a flammable landscape [16].

This growing body of evidence that land cover change influences the frequency and severity of natural hazards presents new opportunities for managing and reducing risks faced by society, and forms the foundation of ecosystem-based approaches to adaptation. These approaches seek to manage, conserve or restore ecosystems and their associated ecosystem services to help people cope with the impacts of climate change [17]. Ecosystem-based adaptation approaches can be used to complement or replace technological or engineering solutions to adaptation, and often present more cost-effective, self-sustaining and flexible alternatives in the long term [18,19,20]. Apart from enhancing the buffering capacity of ecosystems, these approaches often come with multiple co-benefits to humans, such as improved fisheries production [21], timber harvesting [22], biodiversity conservation [23], and recreational value [24].

A key challenge to incorporating ecosystem-based approaches into disaster risk reduction is quantifying the extent to which land cover changes influence the occurrence and consequences of natural hazards. Clear examples are needed to quantify the benefits of ecosystem-based approaches to disaster risk reduction. Such assessments will require, *inter alia*, understanding the multiple land cover and climate drivers of natural hazards, quantifying how ecosystems will respond to changes in these drivers, and examining trade-offs between ecosystem-based management relative to alternative solutions [19]. This paper describes a 'proof of concept' study which aimed to address the first two of these challenges, exploring how land cover and climate change might affect four natural hazards – floods, droughts, wildfires and storm-waves – in the south coastal region of South Africa. These natural hazards frequently affect southern Africa's emerging economies and vulnerable communities. We limit the scope of this paper to natural hazards, but in the discussion reflect on translating natural hazards to disaster risk, which is the product of the likelihood of a natural hazard event occurring and its consequence on society. The intention of this proof of concept was to inform local authorities and businesses about local land cover drivers in their region and their potential for reducing the risk posed by natural hazards, thus contributing to comprehensive strategies for disaster risk reduction and climate change adaptation. But beyond informing local and national stakeholders in South Africa, the methods and lessons developed in this study have broader implications for ecosystem-based approaches and their evidence-base globally [19].

Methods

Study Area

The study area, hereafter 'Eden' (Figure 1), is 3 820 km^2 in size and comprises the local authorities along the southern Cape coast of South Africa within the Eden District Municipality (viz. Mossel Bay, George, Knysna and Bitou municipalities). Eden regularly experiences extremely heavy rainfall events which, combined with the small, steep catchments in the coastal areas, often results in high runoff and flash flooding [25]. The most severe incidences of flooding generally coincide with large storm-waves associated with cut-off low atmospheric pressure systems over southern Africa [26]. Floods are also interspersed with prolonged periods of extremely low rainfall, and Eden is frequently declared a disaster area due to persistent drought conditions [27]. The area also naturally experiences moderately frequent fires (every 10–13 years) due to the co-occurrence of its indigenous, flammable vegetation ('fynbos'), periods of hot, dry weather, and readily available sources of ignition [28]. Intense wildfires pose significant risks when associated with high population densities, and the intensity of wildfires is further exacerbated by invasive alien trees, which increase the fuel loads [29,30]. Natural hazards in Eden coincide with diverse socio-economic contexts, which results in inequalities to prepare for, cope with and adapt to disasters. Direct damage costs between 2003 and 2008 were estimated to be more than 3.5 times higher than the average annual household income in the most vulnerable and exposed communities, providing an indication of the vulnerability of some resident communities [25].

Approximately 68% of Eden's surface area is covered by indigenous vegetation, with agriculture (lucerne, vegetables and hops) and timber plantations respectively comprising 17% and 12% of Eden [31]. Urban areas are concentrated along the coast, and estimated to occupy just over 2% of the area. Eden has a 26% population growth rate [32], which is well above the national growth statistic of 15%, and which places considerable pressure on both natural resources and built infrastructure in the region [27]. Almost all the remaining indigenous fynbos (98%) is invaded by alien plants to some extent, which translates to approximately 12% of Eden being invaded at 100% density [31].

Climatically, Eden is located in a transition zone between a winter and summer rainfall regime [33]. Rainfall occurs throughout the year, peaking in autumn (March) and spring (October) with the lowest monthly rainfall occurring in June. The spring and autumn rainfall peaks coincide with increased frequency of cut-off low pressure systems over southern Africa [26], and about one of every five of these brings flooding and damage to the coastal areas [34]. Being a transition zone makes the area particularly susceptible to climate change, because the climate is influenced by changes in the circulation systems of both climatic regimes. An increase in annual maximum temperatures of about 1.2°C has been observed since 1960 [35]. While trends in projected future annual rainfall are weak relative to temperature [36], the combined effects of temperature, rainfall and evapotranspiration are amplified in the hydrological cycle and can have profound impacts on water resources in southern Africa [37].

Climate Data and Natural Hazard Models

Four models were developed to examine how land cover and climate change influence the natural hazards of floods, droughts, wildfires and storm-waves in Eden. Particular sites and catchments within Eden were selected to facilitate the modelling of each hazard (Figure 1). We identified drivers of each hazard based on literature reviews and expert consultations, highlighting those which were identified as important, relevant, plausible, and for which adequate data were available. The climate change scenario was extracted from the A2 scenario of the IPCC Special Report on Emission Scenarios (SRES) [38], while land cover change scenarios were based on published information and expert knowledge of the region. We then modelled the effect of the scenarios of change on each natural hazard (Table 1; Table S1).

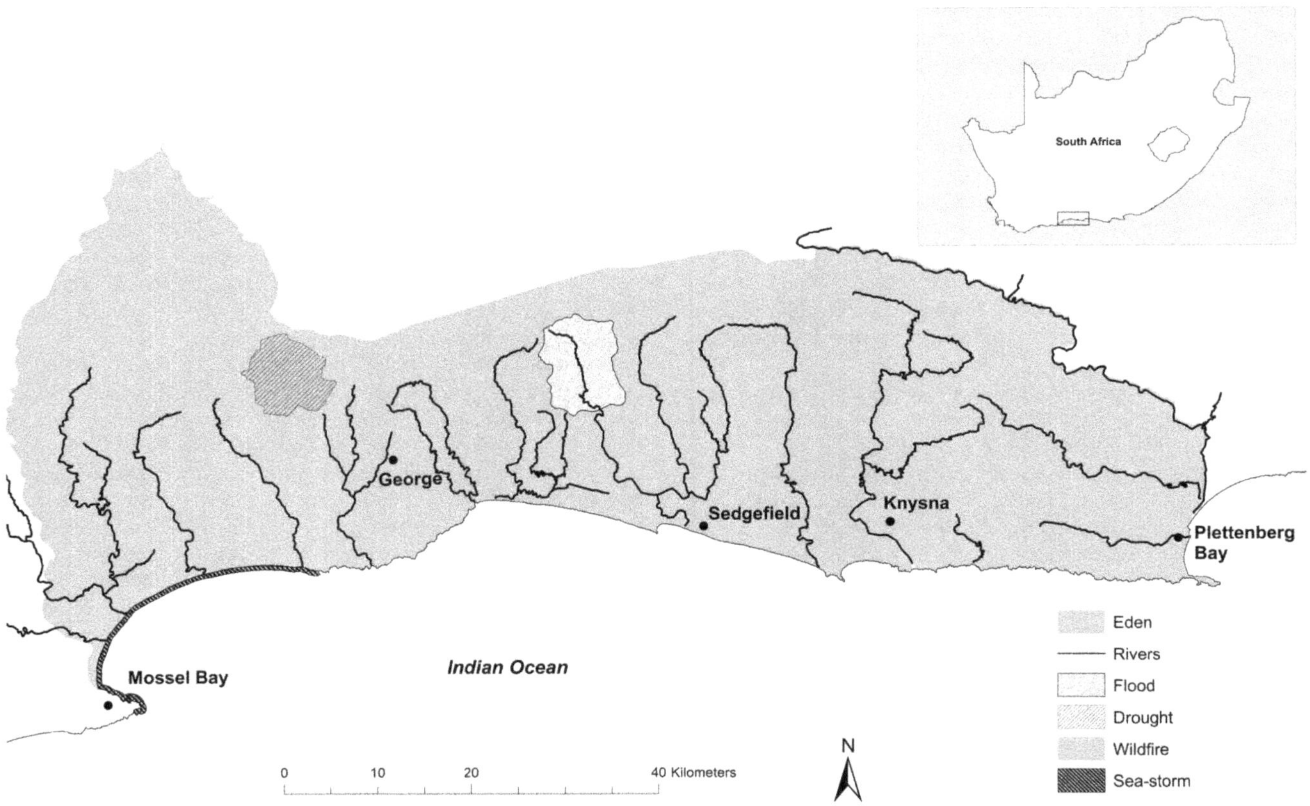

Figure 1. The location of the hazard model study areas within Eden. The inset showing the location of Eden in South Africa.

To explore the changes in flood, drought and storm-wave hazards, we produced simulations that respectively estimated return periods of extreme peak flow events, low flows and wave run-up events. The change in wildfire hazard was examined by calculating the change in fire intensities with future changes in fuel loads and climate.

Climate projections. Projections of future climate, used by the flood, drought and wildfire models, were obtained using a regional climate model, the conformal-cubic atmospheric model (CCAM) [39]. A multiple downscaling procedure was followed to obtain high resolution simulations of future climate change over southern Africa. In the first phase of the downscaling procedure, the sea-ice and bias-corrected sea-surface temperatures of the CSIRO Mark3.5 global climate model [3] was used as lower-boundary forcing in CCAM simulations performed globally at a quasi-uniform resolution of approximately 200 km [3,40]. The simulations used the SRES A2 scenario [38], which was the only detailed downscaled data for the southern African region. The A2 scenario represents a low mitigation, or high emissions, scenario implying rapid continued growth in greenhouse gas concentrations in the atmosphere. It is thus appropriate in examining the impacts of extreme climate events, although we do note throughout the limitations of only having one climate scenario available. CCAM was subsequently applied in a stretched-grid mode, to obtain simulations of approximately 60 km resolution in the horizontal, over southern Africa. This CCAM modelling procedure has been shown to provide satisfactory simulations of annual rainfall and temperature distributions, as well as the intra-annual cycle in rainfall and circulation over the southern African region [40]. It also realistically simulates observed daily climate statistics over southern Africa, such as the frequency of occurrence of extreme precipitation events, and the cut-off low atmospheric pressure systems and tropical cyclones [40,41]. Climate data required for each natural hazard model were extracted for the period 1961–2050, which included daily data for: rainfall, maximum temperature, wind speed, and relative humidity (Dataset S1).

Modelling flood hazard. We developed a hydrological model for an upper catchment within Eden (Figure 1) that drains into a coastal lake near the low-lying town of Sedgefield, which has in the past been highly susceptible to flooding during heavy rains. The model explored how inflows to the lake are affected by catchment land cover and climate change. This upper catchment was chosen because it had reliable gauging weir data for calibrating simulated flows, and included the typical land cover found in Eden.

The agrohydrological modelling system (ACRU) was used to simulate daily flows based on climate data, physical catchment characteristics (particularly soils) and land cover [42]. ACRU contains a soils and vegetation database that was developed specifically for South African conditions, and has been widely used, both locally and elsewhere [43,44]. We used this database, together with 30 m resolution land cover data [45], to identify five 'hydrological response units': pine plantation, clear-felled pine plantation, wattle plantation, indigenous forest and fynbos. We constructed a model for baseline conditions using daily temperature for 1961–1990 from our climate model, and daily rainfall [46] for the same period (Table 1). The latter was considered to better reflect the rainfall gradients within the catchment than the relatively coarse resolution of our climate model. All hydrological response units fed directly into the outflow point at the gauging

Table 1. Scenarios of land cover and climate change used to quantify changes to flood, drought, wildfire and storm-wave hazards.

Hazard	Scenario
Flood	1. BASELINE: current land cover and climate
Flood	2. Clear-felling and non-replanting of the plantation; current climate
Flood	3. Burning down of the plantation by a moderate wildfire with replacement by degraded fynbos; current climate
Flood	4. Burning down of the plantation by a severe wildfire with replacement by degraded fynbos; current climate
Flood	5. Current land cover; future climate
Drought	6. BASELINE: Indigenous natural vegetation with no invasive alien plants or human activities; current climate
Drought	7. Alien trees invade to maximum potential; current climate
Drought	8. Indigenous natural vegetation with no invasive alien plants or human activities; future climate
Wildfire	9. BASELINE: Current levels of invasion by alien trees; current climate
Wildfire	10. Alien trees invade to maximum potential; current climate
Wildfire	11. Current levels of invasion by alien trees; future climate
Wildfire	12. Alien trees invade to maximum potential; future climate
Wildfire	13. Alien trees and shrubs are cleared and maintained at levels below 5% cover; future climate
Storm-wave	14. BASELINE: Current beach slope; current climate
Storm-wave	15. 3° increase in beach slope; current climate
Storm-wave	16. Current beach slope; future climate

More detailed descriptions of each scenario and the associated data used are available in Table S1.

weir, so that simulated flows could be calibrated against observed flows (Table 1). The initial settings in the ACRU model were based on default values in the model for land cover classes and soils [47]. The simulated flows were much greater than the observed flows and too responsive to rainfall events so: (i) soil and effective rooting depths were increased to accommodate regional variations in soils and greater rooting depths for fynbos and pine trees [43]; and (ii) quickflow and baseflow response coefficients were altered according to Royappen et al. [48] for this catchment. These adjustments brought the mean simulated flows to within 10% of the observed daily flows, which were used to describe the baseline for runoff events.

We used this calibrated model to explore the implications of different forestry management practices and climate change on flood events (Table 1). Future daily temperature data for 2021–2050 were extracted from our climate model. To account for the inadequate rainfall resolution, future daily rainfall was calculated using a proportional adjustment to the current daily rainfall data [46], based on regional trends in current and future daily rainfall from the climate model used here. Regional trends were calculated as proportional differences for modelled daily rainfall percentiles and then applied to the corresponding percentiles of the observed data to generate the future rainfall for 1991–2020 and 2021–2050. The exceedance probabilities for extreme flows for both current (1961–1990) and future (2021–2050) climate conditions were calculated from the simulated flow record using the Log Pearson III distribution, which is widely used for calculating extreme values and their return intervals [49]. An extreme rainfall event in the climate model dataset was defined as 25 mm of rain falling within 24 hours over a grid cell of 0.5° latitude×0.5° longitude.

Modelling drought hazard. Many definitions of drought exist – e.g. 'meteorological', 'atmospheric', 'hydrological', 'agricultural' and 'water management' droughts [50] – each differing in their emphasis on describing the characteristics and causes of drought, vulnerability to drought, or impacts of drought. Here, we study the effects of land cover and climate on hydrological drought, which is defined as a persistently low discharge or volume of water in streams or reservoirs, lasting months or years. Although hydrological drought is a natural phenomenon underpinned by dry climate, it is greatly exacerbated by human land use activities, which often affect the magnitude and frequency of the drought [50]. We used flow duration curves to explore changes in hydrological patterns. Flow duration curves plot the percentage of time a flow exceeds a certain threshold, with flows between 70–99% exceedance depicting low flows [51], and flows with >90% exceedance describing extreme low flows and taken here to represent drought.

Flow simulations used to develop flow duration curves were derived for a headwater sub-catchment within Eden (Figure 1) using the Pitman model [52,53,54]. This sub-catchment supplies water for high-value hops farming, and inadequate flow, particularly during the dry season, is frequently problematic [27]. The Pitman model is a conceptual monthly time-step rainfall-runoff model that has been frequently used for water resource assessments in southern Africa for many years and has become a standard method used by many practitioners. It includes explicit routines to simulate interception, infiltration, excess surface runoff, soil moisture (or unsaturated zone) runoff, groundwater recharge and drainage to stream flow, as well evaporative losses from the unsaturated zone and the groundwater storage in the vicinity of the river channel. We used a physically-based parameter estimation procedure described previously [55,56], which uses physical property data at the sub-basin scale (typically 50–1000 km^2). The agricultural land types GIS layer for South Africa [57] provides much of the physical property data required, including soil depths for different parts of the sub-basin (hilltop, valley sides and valley bottoms), soil texture (translated into soil hydraulic properties), topographic slope, and sub-surface geological conditions. Five land types were identified in the study area (Db30, Db32, Fc42, Lb139 and Lb141), based on spatial land cover data [45], and these were used to estimate runoff generating, soil moisture accounting and groundwater parameters for the Pitman model. The model was then run to generate time-series of mean monthly runoff under baseline conditions (Table 1).

Resulting output ensembles were within the 90% confidence intervals of the regionalised mean annual runoff ratios, the gradient of the monthly flow duration curve, and within range of the three recharge estimates of the groundwater resource assessment study [58].

Using the calibrated Pitman model, we then explored changes in flow resulting from maximum potential invasion by alien trees and from climate change (Table 1). An ecological module was used to estimate invasive alien tree water use [59], which links back to the Pitman model to predict changes in streamflow [60]. Alien trees were only allowed to invade untransformed vegetation, which were calculated as those areas not classified as urban, agriculture, forestry plantation or waterbodies in the land cover GIS layer for South Africa [45]. We used monthly temperature and rainfall data from our climate model. The inadequate resolution of the rainfall data was addressed by scaling the projected future rainfall according to the monthly distribution statistics of historical rainfall data as outlined previously [61,62]. Changes in flow for each scenario were compared using flow duration curves, focusing on flows produced at >90% exceedance range.

Modelling wildfire hazard. The damage caused by a wildfire is directly related to Byram's fireline intensity [63], which measures the rate of energy released along the fire front. The relative effort required to control a fire, and the damage it does, are both strongly correlated with fireline intensity [64]. The higher the fireline intensity, the more difficult a fire is to control, which has important implications for human safety and fire-sensitive assets [65]. Wildfires are a regular occurrence in the fire-prone indigenous fynbos in Eden [28]. However, invasion by alien trees have the effect of increasing the above-ground biomass or fuel, and therefore fire intensity [29]. If uncontrolled, invasive alien trees will continue to spread leading to further increases in fire intensity. In addition, climate change is predicted to result in hotter, drier and windier weather, which will further increase the intensity of wildfires.

We used Byram's [63] equation (Equation 1) to estimate changes in fireline intensity across 106 spatial assessment units in Eden under different scenarios. These assessment units are similar-sized irregular areas (mean size = 49 km^2) that are nested within administrative and physiographic boundaries [66].

$$I = Hwr \tag{1}$$

Where I = fireline intensity in kW m^{-1};

H = heat yield (assumed to be constant at 20000 J g^{-1});

w = fuel loads in g m^{-2}; and

r = rate of fire spread in m s^{-1}

For each assessment unit, we assumed that the area covered by untransformed fynbos vegetation (as calculated for the drought model) would be available to burn in wildfires. We considered five scenarios, each with a unique combination of climate and invasive alien trees (Table 1).

We used recent spatial data [31] on the area invaded by pine trees and hakea shrubs (*Pinus* and *Hakea* species) to estimate fuel loads in each assessment unit under different scenarios. Pines have annual spread rates of 3.75–20.6% in the fynbos biome [67,68,69], and hakea can spread at 8% annually [70]. We assumed that pines and hakea would spread at a conservative rate of 4% per year into the available area within each assessment unit (to a maximum of 100% of untransformed vegetation only), and used this to estimate the area that would become invaded by 2050.

Fuel loads in fynbos vegetation are approximately 1 800 g m^{-2} [29]. Invasion by hakea and pines can increase these fuel loads to 3 900, and to 20 000 g m^{-2}, respectively [29,30]. The proportional mix of invasive species in Eden as a whole was 72.5% pine and 27.5% hakea, so we assumed a mean fuel load for assessment units of 15 572 g m^{-2}. Based on these biomass estimates, we used the relative proportions occupied by uninvaded and invaded fynbos respectively to estimate a mean fuel load for each assessment unit under the different scenarios.

We used data on fire weather conditions and associated rates of fire spread in fynbos [71] to establish the relationships between McArthur's Forest Fire Danger Index (FDI) [72] and rates of fire spread (Equation 2). McArthur's FDI is based on observed relationships between the behaviour of fires and the environmental conditions (air temperature, relative humidity and wind speed) under which they burn. The FDI provides an index of the degree of difficulty of suppressing a fire.

$$r = -3.2\left(1.1^{-0.2(FDI-8)}\right) + 3.5 \tag{2}$$

Where r = rate of fire spread in m s^{-1} and FDI = McArthur's Forest Fire Danger Index

Equation 2 was used to convert FDIs to estimates of current (1961–1990) and future (2011–2050) rates of fire spread using daily data from our climate model for temperature, wind speed, rainfall, and relative humidity. We then used the resultant mean rates of spread for current and future conditions (1.1 m s^{-1} and 1.2 m s^{-1} respectively) to estimate the fire intensity under different scenarios (Table 1).

Modelling storm-wave hazard. Storm-waves, in the southern African context, refer to extreme offshore wave events which result in wave impacts that are experienced at the shoreline. The severity of storm-wave impacts is highly dependent on wave run-up height (the maximum point that storm waves can reach on land) and coastal erosion potential [73]. Two models were developed to examine the spatial and temporal variation of wave run-up and coastal erosion potential within the study area and how these are likely to change in the future.

First, we used a numerical wave model, SWAN [74], to translate offshore wave data to inshore wave conditions. Offshore data on wave height, period and direction, from the National Centre for Environmental Prediction (NCEP), were used as input variables [75,76]. Simulations were run to determine inshore wave heights for various return periods. Simulations included offshore wave conditions that result from 1:10 year south-south-westerly swells, as these conditions result in severe inshore conditions in the study area. Second, we used the inshore wave height and period from the SWAN wave model simulations as input into the Nielsen and Hanslow model [77] to calculate wave run-up elevations at 0.5 km points along the coastline, together with corresponding return periods. This model requires information on inshore wave height and period, sea water level and beach slope, and has previously shown an acceptable prediction accuracy ($R^2 = 0.79$) for local South African conditions [73]. We used wave heights from the SWAN model simulations as inputs for the inshore wave height and period, and sea water level was based on spring high tidal level predictions. This is a relatively extreme but realistic scenario as spring high tides occur every 14 days along the coastline, making the chances of storm waves coinciding with spring high tides relatively high. Beach and inshore slopes were calculated using the distance to the 20 m depth contour obtained from the South African Navy's bathymetric charts and available 5 m contour intervals. Using these outputs, baseline conditions of wave run-up elevations and corresponding return periods were plotted for each 0.5 km location along the coast (Table 1; Figure 1).

These models were then used to examine the change in wave run-up elevations resulting from the future potential influence of anthropogenic effects on coastal erosion (resulting in a steeper slope), and the change in future wave climate and sea-level rise (Table 1). Anthropogenic effects on coastal erosion were simulated by assuming a 3° increase in beach slope which was used as input into the Nielsen and Hanslow run-up model [77]. This is a conservative increase in beach slope that local coastal engineers considered to be a realistic effect of coastal erosion, which occurs as a consequence of human activities, e.g. hardening of the coastline through urban and industrial development, removal of coastal foredunes. Future climate conditions were modelled by assuming a 0.5 m rise in sea-level based on reviews of recent publications [78,79], and by applying a 6% increase to offshore extreme waves based on regional projections from metocean climate modelling [80]. We calculated expected changes in wave run-up elevation and return period up to the year 2100 and compared these to baseline conditions.

Results

The flood model for all land cover and future A2 climate scenarios showed an increase in extreme daily flows for equivalent return periods compared to the current baseline condition (Figure 2). Estimated extreme daily flows for a two-year return period showed a 16% increase from the baseline condition for the scenario of clear-felling without replanting. Similarly, the scenarios describing the burning down of the timber plantation under moderate and severe fires show an increase of approximately 24% and 32% respectively. The A2 scenario of future climate change also shows an increased trend in extreme daily flows, estimated to increase by over 24% of the baseline for the 50-year return period. Under the A2 scenario of climate change, flood return intervals are substantially reduced, doubling the frequency of 1:100 year floods to 1:50 years. Land cover changes also resulted in increases in flood frequency, shifting the 1:100 year flood event to about a 1:80 year return period in the most extreme scenario in which timber plantations were burnt down by a severe fire and replaced with degraded fynbos (Figure 2).

The full range of river discharges in the drought study is displayed in Figure 3, from the low flows to the high flows (broadly defined as flows respectively above or below 50% exceedance). Figure 3 shows that both invasion by alien trees and future climate change (under the A2 scenario) will exacerbate extreme low flows (flows >90% exceedance). Flow for alien trees at maximum potential invasion were consistently lower than baseline flows, halving the expected flow at 90% exceedance (0.02 million m^3 under baseline compared to 0.01 million m^3 under full invasion; Figure 3). While the A2 climate change scenario shows an overall trend of higher flows, its flow duration curve drops sharply below the baseline at 90% exceedance, suggesting that under this scenario of climate change much lower flows will be experienced during drought periods.

Mean fireline intensities of around 80 000 kW m^{-1} were estimated under current baseline conditions (Figure 4). This could increase by about 88% (to 150 000 kW m^{-1}) under a scenario in which invasive alien trees continue to spread unhindered, without any climate change, and could potentially more than double (to over 180 000 kW m^{-1}) if the combined effects of alien tree spread and the A2 scenario of climate change are taken into account. Under a hypothetical scenario in which climate change took place, but alien tree invasions remained at current levels, mean fireline intensities would increase by about 9% (to 87 000 kW m^{-1}) under the A2 scenario of climate change compared to current conditions. However, under a scenario in which alien trees are brought under control by reducing and maintaining them at below 5% cover, the estimated mean intensity of fires would be reduced by almost half (to 43 200 kW m^{-1}) when compared to the current situation.

Using a sandy beach location as an example of a typical area prone to storm-waves in Eden, simulations of wave run-up elevation for spring high tide and south-south-westerly swell conditions are currently predicted to range between approximately 5.7 m for a 1:1 year return period to 6.5 m for a relatively extreme event with a 1:50 year return (Figure 5). Under future wave climate and sea-level rise predictions, the 1:50 year wave run-up elevation will be reduced to a 1:3 year return interval at this location. Increasing beach slope by 3° to simulate conservative anthropogenic effects on coastal erosion, produced a substantial increase in wave run-up elevation for respective return periods. In this scenario, the wave heights of a 1:100 year return period are likely to occur on an annual basis.

Discussion

Natural Hazards are Increasing

Our results show that climate change in Eden will increase the frequency of all natural hazard events examined, substantially so in the case of floods, droughts and storm-waves (Figures 2–5), and to a lesser extent for wildfires (Figure 4). When looking beyond climate change impacts, drivers of land cover change appear to have a similar effect on increasing the incidence of natural hazards (Figures 2–5). Allowing the spread of invasive alien trees into untransformed vegetation was estimated to halve the monthly river flows experienced during drought and double fire-line intensities (Figures 3 and 4 respectively). In the case of wildfire (Figure 4), the fireline intensities for invaded fynbos are all orders of magnitude greater than the limits for effective fire control or suppression [81]. This poses significant threats to life and infrastructure [65], and can have significant hydrological impacts [82]. Similarly, severe 1:100 year storm-waves are estimated to occur on an annual basis with only modest human-induced coastal hardening and the removal of coastal foredunes (Figure 5). The impacts of land cover change on floods also attest to the impacts of land management practices on natural hazards by reducing the return time between large flood events by nearly 20% (Figure 2).

In interpreting these findings, we focus on broad trends rather than quantitative measurements because of the inherent uncertainty, gaps in data and models, and limited scenarios explored in predicting land cover and climate change impacts. With only one climate change scenario (focusing on a high emission, high risk scenario), we cannot compare land cover and climate change impacts with great certainty. However, we can point out that even under this extreme climate change scenario, land cover changes were shown to have as great and sometimes greater impacts on natural hazards, highlighting the importance of land cover management in reducing natural hazards. When considering the land cover scenarios, our focus was usually on single drivers of land cover change (e.g. spread of alien trees or forestry management practices); however, even within this small set of land cover drivers, the impacts were substantial on natural hazard incidence. As more climate scenarios become available for southern Africa, as well as data on other land cover and land use drivers relevant to natural hazards (e.g. agricultural practices, ground water abstraction), the additional impacts of these on natural hazards can be incorporated into our models and integrated into other ecosystem-based management options. In moving beyond this study's focus on general trends and evidence of importance of land cover and climate change drivers on natural hazards, future work on new

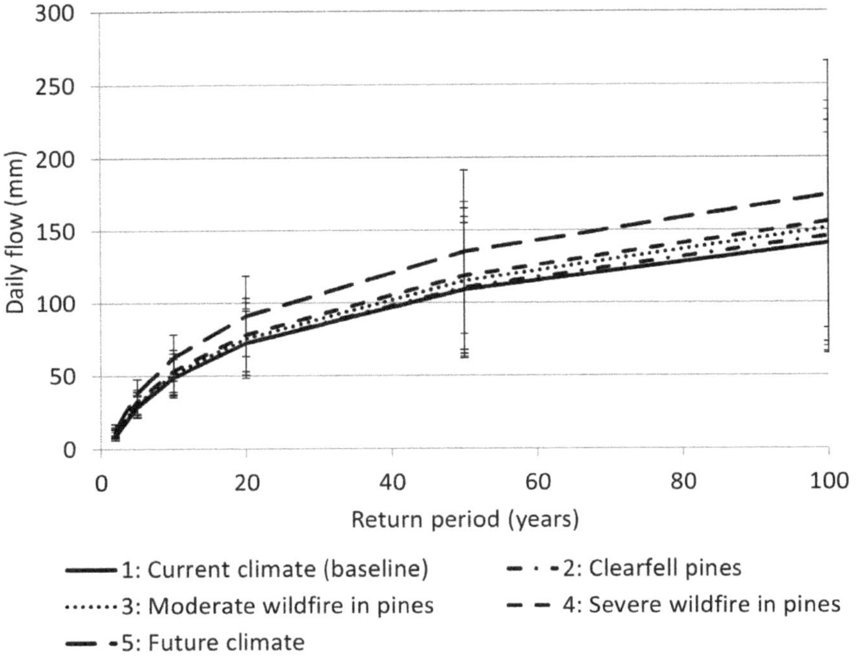

Figure 2. Flood return intervals for different scenarios of land cover and climate change. The numbers prefixing the annotated description of each scenario provides a reference to Table 1, which describes each scenario in more detail. The changes in the values for each return interval illustrate the potential changes in the likelihood of extreme flow events under the different scenarios. For example, the return period of a flood with a daily flow of 150 mm (similar to the May 1981 flood in this area) would decrease from a baseline of more than 100 years to 70 years under future climate (scenario 5).

drivers, data and scenarios would help in better reflecting the ranges of impacts, and the uncertainty associated with the predictions. This in turn will improve the ability to examine trade-offs of alternative solutions to disaster risk reduction and climate change adaptation.

Data and modelling techniques are also continuously evolving, and several improvements can be made to increase the confidence of the hazard models. For example, within specific natural hazard models (e.g. wildfire) the limited availability of fine scale data resulted in the use of a relatively coarse grid scale, possibly underestimating the impacts found. An analysis of extremes (rather than means) at a finer resolution may show higher climate-related change (see Figure S1). Moreover, models that explore interactions between drivers and between natural hazards would be more informative than the separate models produced here. For example, droughts and wildfire are closely linked hazards with similar climate drivers (warmer and drier conditions). Droughts can exacerbate wildfires because they usually increase dry, highly flammable standing biomass [6]. Severe wildfires also exacerbate floods, especially when they precede the onset of rains. This is particularly the case when wildfires occur in densely invaded areas, which burn more intensely than indigenous vegetation, and which result in increased runoff caused by resin-induced water-repellency and associated reductions in infiltration [84].

Natural Hazards can be Reduced by Appropriate Land Management

This study makes a case for the incorporation of ecosystem-based management approaches, in tandem with other approaches (e.g. mitigation or engineering responses), into disaster risk reduction and climate change adaptation. Our findings show that through appropriate and pro-active land use management, it would be possible to reduce the impacts of natural hazards to a large degree. Clear-felling of timber plantations should ideally be associated with rehabilitation and re-vegetation to avoid increasing the flood hazard. Because the timber plantations (mostly *Pinus* sp) are similar to dense stands of invasive alien trees, the flood model also supports the clearing of invasive alien trees to reduce the hazard posed by flood events soon after a wildfire. Clearing invasive alien trees and restoring the natural fynbos vegetation is also an effective tool for reducing wildfire and drought hazards. The wildfire model shows that the impacts of climate change can be substantially reduced by clearing invaded areas and maintaining a healthy cover of indigenous vegetation, lowering the fireline intensity to half that of current levels (Figure 4).

The costs to clear existing invasive alien trees in Eden are much smaller than the estimated losses caused by damaging wildfires to, for example, timber plantations. Estimates of the cost of clearing of invasive trees range from US\$ 100 ha^{-1} [85] to US\$ 800 ha^{-1} [86], giving estimates of between US\$ 2.3–19.2 million to clear in the Eden area. A single wildfire in 2007 caused losses of US\$ 200 million to the local timber industry [87]. Similarly, estimated costs of clearing were estimated to be lower than the economic impact invasive alien trees have on the hops industry in Eden (c.a. US\$ 250 000 per year for 15 years, compared to US\$ 350 000 per year for perpetuity to cover the additional groundwater pumping costs) [88]. Similar invasive alien tree clearing initiatives have been proposed to lower the long term economic impacts of natural hazards to the fruit industry in Eden [27].

Human activities that harden the coastline exacerbate beach erosion, thereby increasing beach slope and wave run-up [8]. Coastal foredunes are the South African equivalent of salt marsh and mangrove wetlands that offer protection from hurricane storm-surge or tsunamis elsewhere in the world [8,21]. Maintain-

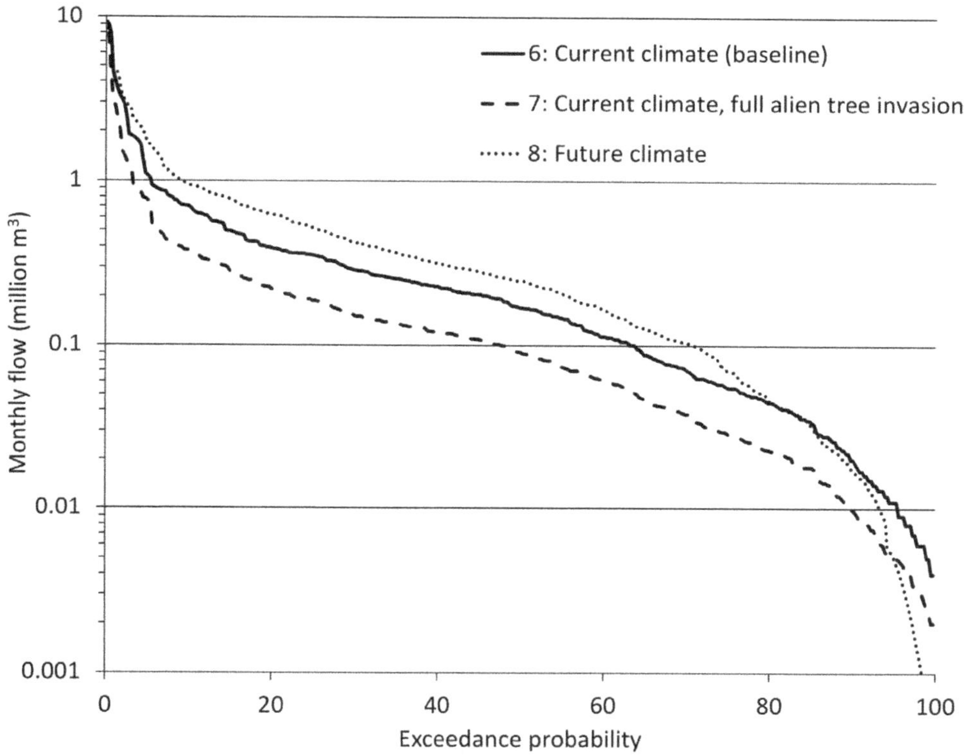

Figure 3. Flow duration curve for different scenarios of land cover and climate change. This shows the cumulative proportion of the months where a flow exceeded a given discharge for the different scenarios. The numbers prefixing the annotated description of each scenario provides a reference to Table 1, which describes each scenario in more detail. Extreme low flows were defined as those with >90% exceedance, which were used in this study to represent severe drought conditions. A log-normal probability curve was used to allow the low and high flow ends of the plot to be more clearly displayed.

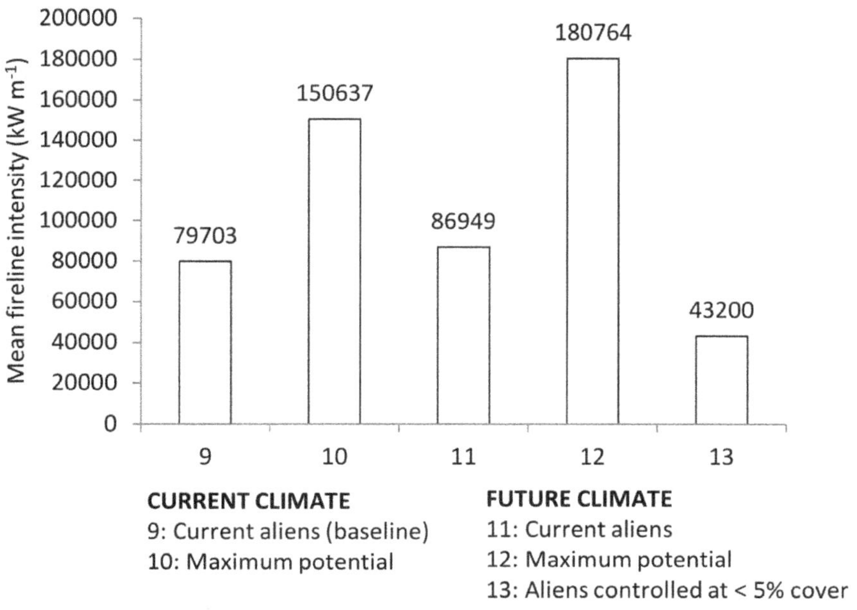

Figure 4. Estimated mean fireline intensity experienced in 106 assessment units for different scenarios of climate change and alien tree management. The numbers prefixing the annotated description of each scenario provides a reference to Table 1, which describes each scenario in more detail.

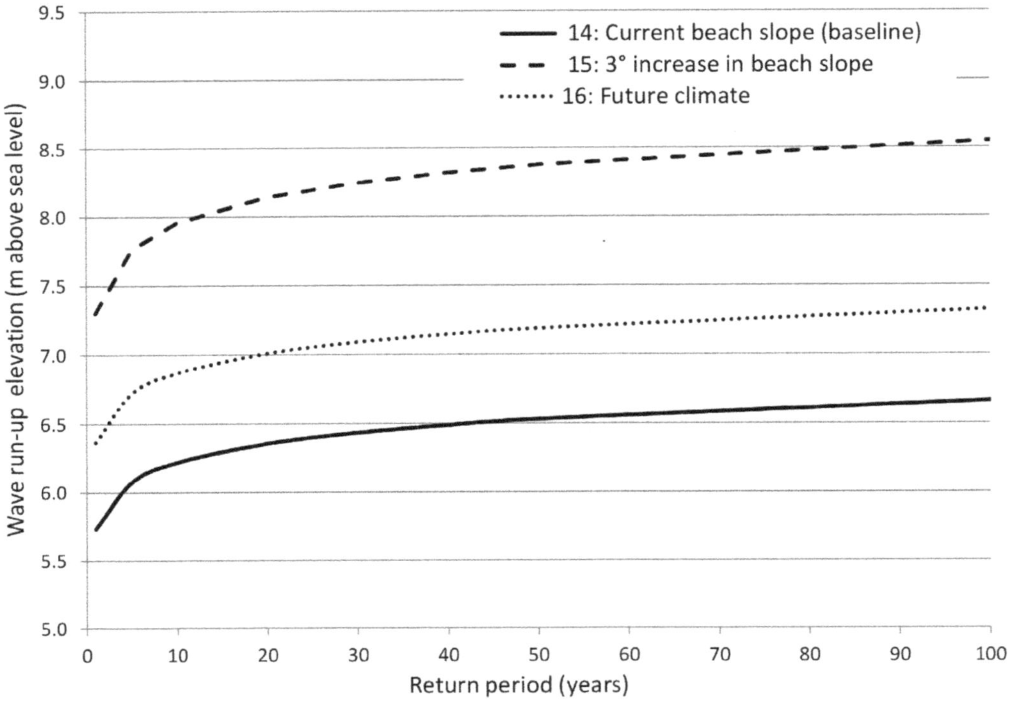

Figure 5. Wave run-up elevations for various storm-wave return intervals for different scenarios of beach slope and climate change. Simulations used here are for a typical sandy beach location in Eden (Tergniet, near Mossel Bay), which is prone to storm-wave damage. Return periods were based on the simulated wave run-up elevations for a south-south westerly swell, and spring high tide levels. The numbers prefixing the annotated description of each scenario provides a reference to Table 1, which describes each scenario in more detail.

ing natural vegetation, sand volume, and natural sediment movement, and restricting developments on foredunes can reduce impacts of wave run-up. Given that human-induced coastal erosion substantially influences the impacts of extreme storm-wave events (Figure 5), rehabilitation of foredunes should also be seriously considered as a means of reducing the impacts of climate change. Indeed, this is the strategy that has seen the US government investing billions of dollars in restoring the coastal salt marshes to protect the Gulf of Mexico from extreme events such as Hurricane Katrina in 2005 [21,89].

Way Forward in Eden and in Ecosystem-based Adaptation

The broad trends from this 'proof of concept' study have provided sufficient evidence to mobilise action within Eden, and several public-private initiatives have been launched to clear invasive alien trees, and restore catchments and foredunes. Implementation mechanisms have also been established to facilitate local action, including the appointment of a catchment manager by the hops farming industry, and the establishment of the 'Business Adopt a Municipality' forum in Eden, which explores how best the insurance sector can support local authorities and communities to manage their natural hazards and environmental risk.

While refining the hazard models is clearly important in identifying on the ground actions, perhaps more important is the need to understand how these natural hazards and associated trends express themselves as risk to communities in Eden, so that they can prepare for, cope with, and recover from disasters [90,91,92]. Disaster risk is widely accepted as the product of a natural hazard and its consequences on society [93]. The latter depends on the exposure and vulnerability of communities to natural hazards [3]. Eden has a naturally high relative exposure, and this coincides with particularly vulnerable communities living in the area [25]. Farmers are highly affected by repeated setbacks from droughts, floods and wildfires, and this has severe ripple effects on rural farm labourers and the entire local economy. This, in turn, increases the vulnerability of communities in Eden, making them more susceptible to subsequent disasters.

Disaster risk reduction is an activity that seeks systemic ways to reduce the severity or occurrence of natural hazards, as well as the consequences that such events have on people [93]. Eden has a relatively well-capacitated Disaster Risk Reduction unit compared to other districts in South Africa. However, much like in many parts of the world, efforts are still very much focussed on recovery from disaster (e.g. through providing disaster relief funding), or short-term disaster preparedness (e.g. through early warning systems, or ensuring adequate supply of fire engines). Longer term efforts to reduce risk are still lacking. These could include efforts to manage drivers of risk through appropriate land use management and restoration, as well as providing opportunities for social learning that promotes individual, collective and institutional capacity to manage risk [91]. Although the disaster risk framework in South Africa (Disaster Management Act No. 57 of 2002), and more widely, acknowledges both short and longer term efforts, there is still a lack of explicit budget allocated to longer term efforts. This expresses itself at the local level where authorities, such as the Eden District Municipality, that are interested in piloting new approaches, have no funding to do so. Interventions at all levels of governance (local, provincial and national) will be required to remove financial implementation barriers.

Conclusions

There is a need to build an evidence-base that addresses the potential of ecosystem-based adaptation approaches [19]. Our models and findings contribute to such a call. Our study shows that land cover change is as important as climate change in influencing the effects of natural hazards. These findings offer the Eden community an empowering message. Through pro-active management of key drivers of land cover change, they will be able to reduce the impacts of floods, droughts, wildfires and storm-waves. In considering the trade-offs of such ecosystem-based approaches with alternative forms of disaster risk reduction, the multiple co-benefits of ecosystem management and restoration need to be considered. For example, clearing invasive alien trees will reduce the impacts of drought, wildfire and flood hazards, create opportunities for employment in rural poor communities, and decrease the vulnerability of agricultural production and thus the overall local economy of the region. Promoting individual, collective and institutional capacity to deal with risk through social learning networks is also increasingly being recognised as an important long term strategy in disaster risk reduction [91,92]. Indeed, a recent study found that socio-institutional interventions tended to offer the most efficient climate change adaptation options in the city of Durban, South Africa [94]. The establishment of the public-private initiatives between land-owners, businesses and government agencies in Eden is a good step in this direction, as it offers the opportunity for cross-sectoral collaboration and social learning around disaster risk reduction.

Supporting Information

Figure S1 Fire Danger Index (FDI) calculated for the current (1960–1990) and future (2010–2050) time periods using the CCAM climate data for temperature, wind speed, rainfall, and relative humidity. Although there is a strong directional change in temperature, this trend is mediated by relatively small changes to wind speed and relative humidity. However, the small change to wind speed and relative humidity may be the result of an averaging effect, because both parameters are highly variable within the relatively coarse grid cells of the CCAM climate model. An analysis of extremes (rather than means) within the grid cell may show higher climate-related change.

Table S1 Scenarios of land cover and climate change used to quantify changes to flood, drought, wildfire and storm-wave hazards, together with the associated data used in each scenario.

Dataset S1 Climate model data used in the natural hazard models for the period 1961–2050, which included daily data for rainfall in mm over a 0.5° latitude×0.5° longitude grid cell (Rain), maximum temperature in °C (MaxT), wind speed in km/h (Wind), and percentage relative humidity (RH).

Acknowledgments

We would like to thank Janis Smith for producing the study area schematic (Figure 1).

Author Contributions

Conceived and designed the experiments: DCN JLN BR SA DCLM BWvW AT JMK LvN FE. Performed the experiments: FE DCLM SA BWvW GF AT LB JMK EK. Analyzed the data: JLN BR PO DCN DCLM BWvW AT JMK EK. Wrote the paper: JLN BR DCLM BWvW PO FE.

References

1. UNISDR (2013) From Shared Risk to Shared Value – the Business Case for Disaster Risk Reduction. Global Assessment Report on Disaster Risk Reduction. Geneva, Switzerland: United Nations Office for Disaster Risk Reduction (UNISDR). 288 p.
2. IPCC (2007) Impacts, adaptation and vulnerability. In: Parry M, Canziani O, Palutikof J, van der Linden P, Hanson C, editors. Cambridge: Cambridge University Press. 982 p.
3. IPCC (2012) Managing the risks of extreme events and disasters to advance climate change adaptation. In: Field CB, Barros V, Stocker TF, Qin D, Dokken DJ, et al., editors. Special report of working groups I and II of the intergovernmental panel on climate change. Cambridge: Cambridge University Press. 582 p.
4. Bates BC, Kundzewicz ZW, Wu S, Palutikof JP (2008) Climate Change and Water. Technical Paper of the Intergovernmental Panel on Climate Change, IPCC Secretariat, Geneva.
5. Milly PCD, Betancourt J, Falkenmark M, Hirsch RM, Kundzewicz ZW, et al. (2008) Stationarity is dead: Whither water management? Science 319: 573–574.
6. Bradstock RA, Cohn JS, Gill AM, Bedward M, Lucas C (2009) Prediction of the probability of large fires in the Sydney region of south-eastern Australia using fire weather. Int J Wildland Fire 18: 932–943.
7. MA 2005. Millennium Ecosystem Assessment: Ecosystems and human well-being: Biodiversity synthesis. Available: http://www.millenniumassessment.org/documents/document.354.aspx.pdf. Accessed August 2013.
8. Shephard CC, Crain CM, Beck MW (2011) The protective role of coastal marshes: a systematic review and meta-analysis. PLoS ONE 6: e27374.
9. Day JW Jr, Boesch DF, Clairain EJ, Kemp GP, Laska SB, et al. (2007) Restoration of the Mississippi Delta: Lessons from Hurricanes Katrina and Rita. Science 315: 1679–1684.
10. Das S, Vincent JR (2009) Mangroves protected villages and reduced death toll during Indian super cyclone. PNAS 106: 7357–7360.
11. Bullock A, Acreman M (2003) The role of wetlands in the hydrological cycle. Hydrol Earth Syst Sc 7: 358–389.
12. Zedler JB, Kercher S (2005) Wetlands resources: status, trends, ecosystem services and restorability. Annu Rev Environ Resour 30: 39–74.
13. Görgens AHM, van Wilgen BW (2004) Invasive alien plants and water resources in South Africa: current understanding, predictive ability and research challenges. S Afr J Sci 100: 27–33.
14. Ehrenfeld JG (2010) Ecosystem consequences of biological invasions. Annu Rev Ecol Syst 41: 59–80.
15. Brooks ML, d'Antonio CM, Richardson DM, Grace JB, Keeley JE, et al. (2004) Effects of invasive alien plants on fire regimes. BioScience 54: 677–688.
16. d'Antonio C, Vitousek P (1992) Biological invasions by exotic grasses, the grass/fire cycle, and global change. Annu Rev Ecol Syst 23: 63–87.
17. Munang R, Thiaw I, Alverson K, Mumba M, Liu J, et al. (2013) Climate change and ecosystem-based adaptation: a new pragmatic approach to buffering climate change impacts. Curr Opin Environ Sustain 5: 67–71.
18. Aronson J, Milton SJ, Blignaut JN (2007) Restoring natural capital. Science, business and practice. Washington: Island Press.
19. Jones HP, Hole DG, Zavaleta ES (2012) Harnessing nature to help people adapt to climate change. Nature Clim Change 2: 504–509.
20. van Slobbe HJ, de Vriend S, Aarninkhof K, Lulofs M, de Vries M, et al. (2013) Building with nature: in search of resilient storm surge protection strategies. Nat Hazards 65: 947–966.
21. Barbier EB, Hacker SD, Kennedy C, Koch EW, Stier AC, et al. (2011) The value of estuarine and coastal ecosystem services. Ecol Monogr 81: 169–193.
22. Moberg F, Rönnbäck P (2003) Ecosystem services of the tropical seascape: interactions, substitutions and restoration. Ocean Coast Manage 46: 27–46.
23. Yu X, Jiang L, Li L, Wang J, Wang L, et al. (2009). Freshwater management and climate change adaptation: experiences from the central Yangtze in China. Clim Dev 1: 241–248.
24. Dubgaard A (2004) Cost-benefit analysis of wetland restoration. J Water Land Dev 8: 87–102.
25. RADAR (2010) RADAR Western Cape 2010: Risk and Development Annual Review. Disaster Mitigation for Sustainable Livelihoods Programme, University of Cape Town, South Africa: PeriPeri Publications. 104 p.
26. Singleton AT, Reason CJC (2007) Variability in the characteristics of cut-off low pressure systems over subtropical southern Africa. Int J Climatol 27: 295–310.
27. Holloway A, Fortune G, Zweig P, Barrett L, Benjamin A, et al. (2012) Eden and Central Karoo drought disaster 2009–2011: The scramble for water. Report

Number LG 10/2011/2012, Disaster Mitigation for Sustainable Livelihoods Programme, Stellenbosch University, South Africa. 132p.
28. van Wilgen BW, Forsyth GG, de Klerk H, Das S, Khuluse S, et al. (2010) Fire management in Mediterranean-climate shrublands: a case study from the Cape fynbos, South Africa. Journal of Appl Ecol 47: 631–636.
29. van Wilgen BW, Richardson DM (1985) The effect of alien shrub invasions on vegetation structure and fire behaviour in South African fynbos shrublands: a simulation study. J Appl Ecol 22: 955–966.
30. Versfeld DB, van Wilgen BW (1986) Impacts of woody aliens on ecosystem properties. In: Macdonald IAW, Kruger FJ, Ferrar AA, editors. The ecology and control of biological invasions in South Africa. Cape Town: Oxford University Press. 239–246 pp.
31. Vromans DC, Maree KS, Holness S, Job N, Brown AE (2010) The Garden Route Biodiversity Sector Plan for the George, Knysna and Bitou Municipalities: Supporting land-use planning and decision-making in Critical Biodiversity Areas and Ecological Support Areas for sustainable development. Garden Route Initiative. South African National Parks, Knysna.
32. Statistics South Africa (2012) Census 2011 Municipal Report: Western Cape Statistics South Africa Report no. 03–01–49. Pretoria. 93 p.
33. Weldon D, Reason CJC (2013) Variability of rainfall characteristics over the South Coast region of South Africa. Theor Appl Climatol DOI 10.1007/s00704-013-0882-4.
34. Taljaard JJ (1985) Cut-off lows in the South African region. South African Weather Bureau Technical Paper No 14. Pretoria. 153 p.
35. Kruger AC, Shongwe S (2004) Temperature trends in South Africa: 1960–2003. International J Climatol 24: 1929–1945.
36. Kruger AC (2006) Observed trends in daily precipitation indices in South Africa: 1910–2004. Int J Climatol 26: 2275–2285.
37. Schulze RE (2000) Modelling hydrological responses to land use and climate change: a Southern African perspective. Ambio 29: 12–22.
38. Nakicenovic N, Alcamo J, Davis G, de Vries B, Fenhann J, et al. (2000) IPCC Special Report on Emissions Scenarios. Cambridge: Cambridge University Press. 599 p.
39. Mcgregor JL,Dix MR (2008) An updated description of the Conformal-Cubic Atmospheric Model. In: Hamilton K, Ohfuchi W, editors. High Resolution Simulation of the Atmosphere and Ocean. Springer Verlag. 51–76.
40. Engelbrecht FA, Landman WA, Engelbrecht CJ, Landman S, Bopape MM, et al. (2011) Multi-scale climate modelling over Southern Africa using a variable-resolution global model. Water SA 37: 647–658.
41. Malherbe J, Engelbrecht FA, Landman WA (2013) Projected changes in tropical cyclone climatology and landfall in the Southwest Indian Ocean region under enhanced anthropogenic forcing. Clim Dynam 40: 2867–2886.
42. Jewitt GPW, Schulze RE (1999) Verification of the ACRU model for forest hydrology applications. Water SA, 25: 483–489.
43. Smithers JC, Schulze RE (1995). ACRU Agrohydrological Modelling System: User Manual Version 3.00. Department of Agricultural Engineering, University of Natal, Pietermaritzburg, South Africa.
44. Warburton ML, Schulze RE, Jewitt GP (2010) Confirmation of ACRU model results for applications in land use and climate change studies 2. Change 2: 3.
45. van den Berg EC, Plarre C, van den Berg HM, Thompson MW (2008) The South African National Land Cover 2000. Agricultural Research Council (ARC) and Council for Scientific and Industrial Research (CSIR), Report No. GW/A/2008/86, Pretoria.
46. Lynch SD (2004) Development of a raster database of annual, monthly and daily rainfall for southern Africa. Report No. 1156/1/04, Water Research Commission, Pretoria.
47. Schulze RE, Maharaj M, Warburton ML, Horan MJC, Kunz RP, et al. (2008) South African Atlas Of Climatology and Agrohydrology. Report No. 1489/1/08, Water Research Commission, Pretoria.
48. Royappen M, Dye PJ, Schulze RE, Gush MB (2002) An analysis of catchment attributes and hydrological response characteristics in a range of small catchments. Report No. 1193/1/02, Water Research Commission, Pretoria.
49. US Interagency Advisory Committee on Water Data (1982) Guidelines for Determining Flood Flow Frequency. Bulletin 17B of the Hydrology Committee, US Geological Survey, Reston, VA, USA.
50. Wilhite DA, Glantz MH (1985) Understanding the drought phenomenon: The role of definitions. Water Int 10: 111–120.
51. Smakhtin VU (2001) Low flow hydrology: a review. J Hydrol 240: 147–186.
52. Pitman WV (1973) A mathematical model for generating monthly flows from meteorological data in South Africa. Report No. 2/73, Hydrological Research Unit, University of the Witwatersrand, South Africa.
53. Midgley DC, Pitman WV, Middleton BJ (1994) Surface Water Resources of South Africa 1990: User's Manual. Report no. 298/1/94, Water Resource Commission, Pretoria.
54. Middleton BJ, Bailey AK (2009) Water Resources of South Africa 2005 Study. Water Research Commission Report Number TT 380/08, Water Research Commission, Pretoria.
55. Kapangaziwiri E, Hughes DA (2008) Towards revised physically-based parameter estimation methods for the Pitman monthly rainfall-runoff model. Water SA 32: 183–191.
56. Hughes DA, Kapangaziwiri E, Sawunyama T (2010) Hydrological model uncertainty assessment in southern Africa. J Hydrol 387: 221–232.
57. AGIS (2007) Agricultural Geo-Referenced Information System. Available: http://www.agis.agric.za. Accessed August 2013.
58. DWAF 2005. Groundwater Resource Assessment II: Recharge Literature Study Report 3A. Report of Department of Water Affairs and Forestry, Pretoria.
59. Le Maitre D, van Wilgen BW, Chapman RA, McKelly DH (1996) Invasive plants in the Western Cape, South Africa: modelling the consequences of a lack of management. J Appl Ecol 33: 161–172.
60. Mallory S. (2011) Extent of Invasive Alien Plants and the Impact of Removal on the Water Resources of Olifants River catchment. Department of Water Affairs Report No. P WMA 04/B50/00/8310/3. Department of Water Affairs, Pretoria.
61. Hughes DA, Mantel SK, Slaughter A (2011) Developing climate change adaptation measures and decision support system for selected South African water boards. Water Research Commission Report K5/2018/2, Water Research Commission, Pretoria.
62. Hughes DA (2004) Problems of estimating hydrological characteristics for small catchments based on information from the South African national surface water resource database. Water SA 30: 393–398.
63. Byram GM (1959) Combustion of forest fuels. In: Davis KP, editor. Forest Fire: Control and Use, New York: McGraw- Hill. Pp. 155–182.
64. Mercer DE, Prestemon JP (2005) Comparing production function models for wildfire risk analysis in the wildland-urban interface. Forest Policy Econ 7: 782–795.
65. Gill AM, Stephens SL, Cary GJ (2013) The worldwide "wildfire" problem. Ecol Appl 23: 438–454.
66. Naude A, Badenhorst AZH, Van Huyssteen WE, Maritz J (2007) Technical Overview Of The Mesoframe Methodology and South African Geospatial Analysis Platform. CSIR Report CSIR/BE/PSS/IR/2007/0104/B, Council for Scientific and Industrial Research, Pretoria.
67. Richardson DM, Brown PJ (1986) Invasion of mesic mountain fynbos by Pinus radiata. South Afr J Bot 52: 529–536.
68. Higgins SI, Richardson DM, Cowling RM (2000) Using a dynamic landscape model for planning the management of alien plant invasions. Ecol Appl 10: 1833–1848.
69. Moeller J (2010) Spatial Analysis of Pine Tree Invasion in the Tsitsikamma Region, Eastern Cape, South Africa: A Pilot Study. Honours dissertation, Department of Geography, Rhodes University, Grahamstown, South Africa.
70. van Wilgen BW, de Wit MP, Anderson HJ, Le Maitre, DC, Kotze IM, et al. (2004). Costs and benefits of biological control of invasive alien plants: case studies from South Africa. S Afr J Sci 100: 113–122.
71. van Wilgen BW, Le Maitre DC, Kruger FJ (1985) Fire behaviour in South African fynbos (macchia) vegetation and predictions from Rothermel's fire model. J Appl Ecol 22: 207–216.
72. Noble IR, Gill AM, Bary GAV (1980) McArthur's fire danger meters expressed as equations. Austral Ecol 5: 201–203.
73. Theron AK, Rossouw M, Barwell L, Maherry A, Diedericks G, et al. (2010) Quantification of Risks to Coastal Areas and Development: Wave Run-up and Erosion. Proceedings CSIR 3rd Biennial Conference 2010, CSIR, Pretoria, South Africa.
74. Booij N, Ris RC, Holthuijsen LH (1999) A third-generation wave model for coastal regions, Part I: Model description and validation. J Geophys Res 104: 7649–7666.
75. Tolman HL, Balasubramaniyan B, Burroughs LD, Chalikov DV, Chao YY, et al. (2002) Development and Implementation of Wind-Generated Ocean Surface Wave Models at NCEP. Weather Forecast 17: 311–333.
76. NCEP (2013) Offshore data on wave height, period and direction. Available: http://www.nco.ncep.noaa.gov/pmb/products/wave/#WW3ENS. Accessed August 2013.
77. Nielsen P, Hanslow DJ (1991) Wave run-up distributions on natural beaches. J Coastal Res 7: 1139–1152.
78. Rossouw M, Theron AK (2012) Investigation of potential climate change impacts on ports and maritime operations around the southern African coast. In: Asariotis R, Benamara H, editors. Maritime Transport and the Climate Change Challenge. Routledge: United Nations Conference on Trade and Development (UNCTAD), Earthscan. 360 p.
79. Parris A, Bromirski P, Burkett V, Cayan D, Culver M, et al. (2012). Global Sea Level Rise Scenarios for the US National Climate Assessment. NOAA Tech Memo OAR CPO-1. 37 p.
80. Mori N, Yasuda T, Mase H, Tom T, Oku Y (2010) Projection of extreme wave climate change under global warming. Hydrological Research Letters 4: 15–19.
81. McCaw L (2013) Western Australia's weather and climate: a synthesis of Indian Ocean climate initiative stage 3 research. Austral For 76: 110–112.
82. Shakesby RA, Doerr SH (2006) Wildfire as a hydrological and geomorphological agent. Earth-Sci Rev 74: 269–307.
83. Hewitson BC, Crane RG (2006) Consensus between GCM climate change projections with empirical downscaling: precipitation downscaling over South Africa. Int J Climatol 26: 1315–1337.
84. Scott DF (1993) The hydrological effects of fire in South African mountain catchments. J Hydrol 150: 409–432.
85. Marais C, van Wilgen BW, Stevens D (2004) The clearing of invasive alien plants in South Africa: a preliminary assessment of costs and progress. S Afr J Sci 100: 97–103.
86. van Wilgen BW, Forsyth GG, Le Maitre DC, Wannenburgh A, Kotzé JDF, et al. (2012) An assessment of the effectiveness of a large, national-scale invasive alien plant control strategy in South Africa. Biol Conserv 148: 28–38.

87. Godsmark R (2007) The impact of the 2007 plantation fires on the SA Forestry and Forest Products Industry. Available: http://www.forestry.co.za/industry-information-protection/. Accessed August 2013.
88. Nel DC, Colvin C, Nel JL, de Lange W, Mwenge Kahinda J, et al. (2010) SAB Hop Farms Water Risk Assessment: Project Report. CSIR Report CSIR/NRE/ECOS/2011/066/B, CSIR, Pretoria.
89. Barbier EB, Georgiou IY, Enchelmeyer B, Reed DJ (2013) The value of wetlands in protecting southeast Louisiana from hurricane storm surges. PLoS ONE 8: e58715.
90. Thomalla F, Downing T, Spanger-Siegfried E, Han G, Rockström J (2006) Reducing hazard vulnerability: towards a common approach between disaster risk reduction and climate adaptation. Disasters 30: 39–48.
91. Cutter S, Barnes L, Berry M, Burton C, Evans E, et al. (2008) A place-based model for understanding community resilience to natural disasters. Glob Environ Change 18: 598–606.
92. Brown K, Westaway E (2011) Agency, capacity, and resilience to environmental change: lessons from human development, well-being, and disasters. Annu Rev Environ Resour 36: 14.1–14.22.
93. UNISDR (2009) The 2009 UNISDR Terminology on Disaster Risk Reduction. United National International Strategy for Disaster Reduction (UNISDR), Geneva, Switzerland.
94. Cartwright A, Blignaut J, De Wit M, Goldberg K, Mander M, et al. (2013) Economics of climate change adaptation at the local scale under conditions of uncertainty and resource constraints: the case of Durban, South Africa. Environ Urban 25: 1–18.

5

Climatic Correlates of Tree Mortality in Water- and Energy-Limited Forests

Adrian J. Das[1]*, Nathan L. Stephenson[1], Alan Flint[2], Tapash Das[3], Phillip J. van Mantgem[4]

1 Western Ecological Research Center, United States Geological Survey, Three Rivers, California, United States of America, **2** California Water Science Center, United States Geological Survey, Sacramento, California, United States of America, **3** Climate Atmospheric Science and Physical Oceanography, Scripps Institution of Oceanography, La Jolla, California, United States of America, **4** Western Ecological Research Center, United States Geological Survey, Arcata, California, United States of America

Abstract

Recent increases in tree mortality rates across the western USA are correlated with increasing temperatures, but mechanisms remain unresolved. Specifically, increasing mortality could predominantly be a consequence of temperature-induced increases in either (1) drought stress, or (2) the effectiveness of tree-killing insects and pathogens. Using long-term data from California's Sierra Nevada mountain range, we found that in water-limited (low-elevation) forests mortality was unambiguously best modeled by climatic water deficit, consistent with the first mechanism. In energy-limited (high-elevation) forests deficit models were only equivocally better than temperature models, suggesting that the second mechanism is increasingly important in these forests. We could not distinguish between models predicting mortality using absolute versus relative changes in water deficit, and these two model types led to different forecasts of mortality vulnerability under future climate scenarios. Our results provide evidence for differing climatic controls of tree mortality in water- and energy-limited forests, while highlighting the need for an improved understanding of tree mortality processes.

Editor: Gil Bohrer, The Ohio State University, United States of America

Funding: Funding provided by the U.S. Geological Survey. The funders had no role in study design, data collection and analysis, decision to publish, or preparation of the manuscript.

Competing Interests: The authors have declared that no competing interests exist.

* E-mail: adas@usgs.gov

Introduction

Recent regional increases in tree mortality rates and episodes of forest die-back have been linked to rising temperatures [1,2], indicating that a potentially substantial source of biotic feedbacks to global climatic changes may already be underway [3]. Assessing the effect of such changes will require a realistic understanding of the relationships between climate and tree mortality as well as the mechanisms that underlie those relationships.

For example, increases in tree mortality with temperature might be attributed to two broad and non-mutually exclusive mechanisms: (i) increasing drought stress on trees resulting from temperature-induced increases in climatic water deficit (hereafter "deficit" : an index of evaporative demand that is not met by available water, hence drought stress [4,5]) [2,6–9], or (ii) temperature-induced increases in the reproduction, survivorship, and effectiveness of insects and pathogens that kill trees [10–12]. The first hypothesis posits that changes in mortality rate are most strongly tied to tree condition, with increasing drought stress making a given tree more susceptible to an array of mortality risks including physiological decline and attack by enemies. The second hypothesis posits that changes in mortality rate are most closely tied to the population dynamics and effectiveness of tree enemies, regardless of the condition of their hosts.

We might reasonably hypothesize that the first of these mechanisms dominates in water-limited forests: forests in which growth and other biological processes respond most strongly to changes in water availability, such as in arid regions or, in many mountain ranges, at lower elevations. The second mechanism might be expected to become more important in energy-limited forests: forests in which growth and other biological processes respond most strongly to temperature changes, such as in wet regions or at higher elevations. Although some studies have examined differences in the climatic controls of tree growth rates between water- and energy-limited forests [13], we are unaware of comparable studies of mortality rates.

Even for sites at which drought stress is clearly the best predictor of changes in mortality, the best model for relating changes in mortality to changes in deficit may not be obvious. Trees are adapted to the typical deficit within their geographical range limits, with trees in drier environments being more strongly adapted to lower water availability [14,15]. In addition, some evidence suggests that trees have some ability to acclimate to changes in water stress, in the short term by altering stomatal conductance and in the longer term by altering carbon allocation priorities or morphological traits [14,16–18].

But how is that range of adaptation defined relative to the average historical deficit at a given site? For example, are species adapted to similar absolute ranges of deficit, regardless of environment? Or are trees that are adapted to drier environments more strongly resistant to changes in deficit than trees adapted to more mesic environments? In short, does one expect mortality to be more strongly correlated with absolute or relative changes in deficit?

Arguments can be made for both possibilities. Trees in more water stressed environments have developed adaptations for survival in those environments (e.g., [8,14]), perhaps suggesting that they have increased resistance to absolute changes in water stress. Such adaptations might result in mortality having a relative relationship to changes in deficit (i.e., the more water stressed a site is, the stronger the adaptions to absolute increases in water stress). However, in some cases, severe drought results in larger increases in mortality rates at drier sites when compared to more mesic sites [19,20], perhaps indicating that absolute changes in deficit might be more predictive. While a recent global analysis of the safety margins trees maintain against drought-induced hydraulic failure provides an important first step toward distinguishing among the possibilities [15], alone it does not allow us to definitively choose between them. That work, while showing patterns of embolism resistance that might argue that tree responses to drought are best related to absolute changes in deficit, does not directly relate those safety margins to mortality and also does not address other factors related to hydraulic safety margins such as stomatal regulation.

The precise nature of these relationships (water limitation versus energy limitation, absolute versus relative changes in climatic water deficit) could result in markedly different assessments of the potential vulnerability of a given forest to climate change. For example, many high elevation forests might be substantially more vulnerable if mortality is most strongly related to relative changes in deficit (since many high elevation forests have a relatively small baseline deficit), and forests that experience relatively little change in deficit, due to abundant water and deep soils, might still be substantially vulnerable to climate change if changes in mortality rate are primarily related to changes in climatic favorability to tree enemies.

In this work, we seek to shed light on the possible mechanisms by which increasing temperature can lead to increasing tree mortality rates and to explore some implications of our findings for forecasting changes in tree mortality rates. Specifically, we use empirical data to test.

1) The hypothesis that changes in tree mortality rates in water-limited forests should best correlate (positively) with climatic water deficit, whereas mortality rates in energy-limited forests should best correlate (positively) with temperature due to temperature's direct effects on insect and pathogen populations (e.g., [21,22,23]).
2) Whether absolute or relative changes in deficit best correlate with changes in mortality rate.

With regard to #1, because deficit partly depends on temperature, deficit and temperature are correlated. However, deficit also depends on water availability, which can vary independently of temperature; for example, a warm, wet summer will have a low deficit while a warm, dry summer will have a high deficit. We therefore might reasonably expect to be able to distinguish between deficit- and temperature-related changes in tree mortality rates.

Using results of these analyses, we then explore the implications of our findings by forecasting responses of tree mortality rates to different scenarios of future climatic changes. In particular, we use our models and forecasts to provide additional insights into climatic controls of tree mortality and to highlight critical knowledge gaps. Importantly, we are not attempting to actually predict future mortality rates but rather using our forecasts to illustrate the effect that uncertainties could have on such predictions.

To accomplish these tasks we used data from long-term forest plots arrayed along a steep elevational gradient in California's Sierra Nevada mountain range. Earlier analyses of these data revealed that tree mortality rates had increased significantly over the last two decades and that, for all elevations combined, these increases were correlated with temperature-driven increases in climatic water deficit. These previous analyses also ruled out competition, fire suppression, air pollution, self-thinning, and aging as confounding factors [9,24]. This unique longitudinal data set is ideal for our current purposes because (i) samples are large and of fine temporal resolution, with the fates of >20,000 individual trees tracked annually for up to 24 years, and (ii) the forests were sampled along a 1900 m elevational gradient, ranging from water-limited forests near lower treeline to energy-limited forests at upper treeline.

Materials and Methods

Study Sites and Tree Mortality Rates

Twenty-one permanent study plots ranging in size from 0.9 to 2.5 ha were established between 1982 and 1996 in old-growth stands within the coniferous forests of Sequoia and Yosemite national parks, Sierra Nevada, California (Table S1). These forests have mixed age and size structure, with trees ranging from recent recruitment to mature canopy trees (Fig. S1). The plots have not experienced stand-replacing disturbances in at least two centuries, and probably much longer (as estimated by counting rings on increment cores or nearby stumps, or by historical records and the sizes of the largest trees), and therefore the forests contain cohorts of all ages and all sizes (Fig. S1). A few other plots in our network were excluded due to recent disturbances (fire or avalanche). The plots are arranged along a steep elevational gradient (~1900 m) from near lower to upper treeline and encompass several different forest types, including ponderosa pine-mixed conifer, white fir-mixed conifer, Jeffrey pine, red fir, and subalpine forests [25]. The sites have never been logged. Frequent, low severity fires characterized many of the forest types prior to Euro-American settlement, but the areas containing the study plots have not burned since the late 1800s [26]. The climate is montane mediterranean, with hot, dry summers and cool, wet winters in which ~25–95% of annual precipitation (which averages 1100 to 1400 mm) falls as snow, depending on elevation [27]. Mean annual temperature declines sharply with elevation (~5.2°C for every 1 km increase in elevation), ranging from roughly 11°C at the lowest plots to 1°C at the highest. Soils are relatively young (mostly inceptisols), derived from granitic parent material.

Within each plot all trees ≥1.37 m in height were tagged, mapped, measured for diameter, and identified to species. We censused all plots annually for tree mortality, and at intervals of ~5 years we remeasured diameter at breast height (dbh, 1.37 m above ground level) of living trees and recorded new recruitment. Data selection and mortality rate calculations followed van Mantgem and Stephenson [9] with the exception that mortality rates were calculated through 2006, yielding a final data set of 21,024 trees for analysis. Mortality rates were calculated only for the 86% of tree mortalities that were not the result of mechanical factors (uprooting, breaking, or being crushed), since mechanical mortalities are likely to be only indirectly associated with the climatic factors considered here [9]. This gave a total of 3788 mortalities for the period of record. Summary data are provided in Table S2.

Ethics Statement

Research was performed in Sequoia National Park and Yosemite National Park. All necessary permits were obtained from the National Park Service for this study, which complied with all relevant regulations.

Table 1. Top Ranked Models and Parameters.

Predictor Variable	Model Form	ΔAIC	Evidence Ratio	β_0	β_0 S.E.	β_1	β_1 S.E.	α	α S.E.
All Elevations (21 plots)									
D_a *(Absolute Change)*									
Current year plus two prior	Exponential	0.0	1.0	−0.0536	0.0321	0.0036	0.0005	0.1709	0.0246
Current year plus two prior	Linear	0.6	1.3	0.9726	0.0305	0.0034	0.0005	0.1711	0.0246
D_r *(Relative Change)*									
Current year plus two prior	Exponential	2.9	4.3	−1.2159	0.1774	1.1643	0.1675	0.1752	0.0249
Current year plus two prior	Linear	3.8	6.7	−0.1257	0.1500	1.0989	0.1537	0.1756	0.0250
T_a *(Absolute Change)*									
Current year plus two prior	Exponential	15.7	2565.7	−0.0823	0.0350	0.5515	0.0939	0.1870	0.0259
Current year plus two prior	Linear	14.3	1274.1	0.9388	0.0308	0.5325	0.0829	0.1856	0.0258
Low Elevation (Water-limited) (13 plots)									
D_a *(Absolute Change)*									
Current year plus two prior	Exponential	0.0	1.0	−0.0507	0.0356	0.0036	0.0006	0.1697	0.0263
Current year plus two prior	Linear	0.2	1.1	0.9748	0.0339	0.0034	0.0005	0.1697	0.0264
D_r *(Relative Change)*									
Current year plus two prior	Exponential	2.3	3.2	−1.3272	0.2170	1.2783	0.2066	0.1728	0.02659
Current year plus two prior	Linear	3.0	4.5	−0.2212	0.1802	1.1965	0.1848	0.1735	0.02670
T_a *(Absolute Change)*									
Current year plus three prior	Exponential	9.3	104.6	−0.0752	0.03843	0.6851	0.1261	0.1881	0.0279
Current year plus three prior	Linear	7.0	33.1	0.9465	0.03431	0.6695	0.1033	0.1856	0.0276
High Elevation (Energy-limited) (8 plots)									
D_a *(Absolute Change)*									
Current year plus two prior	Exponential	0.5	1.3	−0.0661	0.0751	0.0037	0.0011	0.1783	0.0686
Current year plus two prior	Linear	0.8	1.5	0.9638	0.0703	0.0035	0.0011	0.1799	0.0689
D_r *(Relative Change)*									
Current year plus two prior	Exponential	0.0	1.0	−1.0287	0.3133	0.9609	0.2863	0.1777	0.0684
Current year plus two prior	Linear	0.4	1.2	0.0648	0.2644	0.8986	0.2711	0.1793	0.0687
T_a *(Absolute Change)*									
Current year plus four prior	Exponential	3.6	6.0	−0.1481	0.0900	0.7864	0.2846	0.1829	0.06742
Current year plus four prior	Linear	3.8	6.7	0.8764	0.0734	0.7534	0.2740	0.1822	0.06766

Note: ΔAIC is the difference in AIC value between the top ranked model and the given model. Smaller values indicate better models. Evidence Ratio can be interpreted as how much stronger the evidence is for the top ranked model over the given model. Larger values indicate stronger evidence that the top ranked model is better than the given model. β_0 and β_0 S.E. are estimated intercept parameter and standard error for the model (see Eqns. 5-7). β_1 and β_1 S.E. are estimated parameter and standard error for the deficit variable for the model (see Eqns. 5-7). α is a parameter from the negative binomial distribution that quantifies over dispersion relative to a Poisson distribution, where the over dispersion factor is defined as $(1+1/\alpha)$. Therefore, smaller values of α represent larger over dispersion.

Defining Water- vs. Energy-Limited Forests

Tague *et al.* [28] used a coupled ecohydrologic model to forecast the effects of temperature changes on net primary productivity (NPP) in the central Sierra Nevada, finding that increasing temperature resulted in declining NPP below about 2200 m elevation but increasing NPP above about 2600 m elevation, suggesting that the transition from water limitation to energy limitation occurs between about 2200 m and 2600 m elevation. Similarly, the response of remotely-sensed forest greening in the Sierra Nevada to interannual variation in snowpack water content suggests that the transition from water to energy limitation occurs between about 2100 m to 2600 m elevation [29]; these results were independently corroborated by *in situ* flux tower measurements at 2015 and 2700 m elevation [29]. Thus, although the transition from predominantly water-limited (low elevation) to predominantly energy-limited (high elevation) forests is likely to be gradual, different lines of evidence are consistent in suggesting that the midpoint of the transition falls near 2400 m elevation. We chose to define the transition at 2450 m, as this elevation allowed us to cleanly segregate the plots we used in model-building by forest types: ponderosa pine-mixed conifer, white fir-mixed conifer, and Jeffrey pine forests (≤2450 m, predominantly water-limited), and red fir and subalpine forests (>2450 m, predominantly energy-limited).

Climatic Data

We obtained historical values of monthly-averaged precipitation and air temperature in a gridded map format at a 4-km spatial scale from the empirically-based Parameter-Elevation Regressions on Independent Slopes Model (PRISM) [30]. Spatial downscaling was performed on the coarse resolution grids (4 km) to produce fine resolution grids (270 m) using a model developed by Nalder and Wein [31] modified with a "nugget effect" specified as the length of the coarse resolution grid [32].

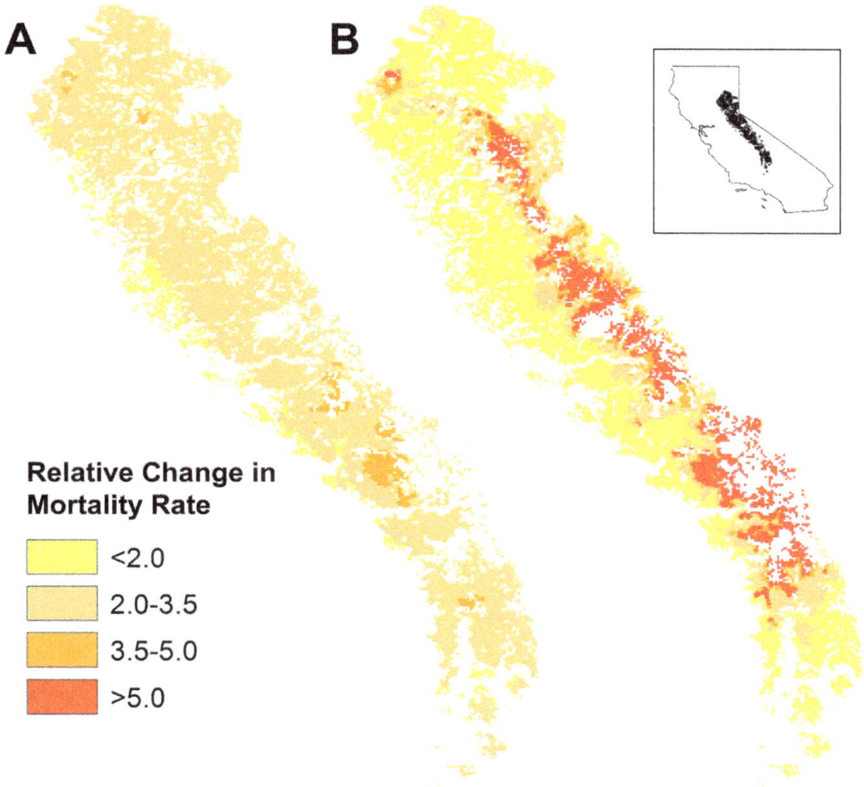

Figure 1. Projected Changes in Mortality Rate for Sierra Nevada Conifer Forests (Hypothetical). Mapped projections of average relative changes in mortality rate for the years 2090 to 2099 for coniferous forests of California's Sierra Nevada, using exponential and the GFDL A2 emissions model. Elevations generally increase from left to right. (A) Changes in mortality when absolute changes in deficit (D_a) are used as a predictor. (B) Changes in mortality when relative changes in deficit (D_r) are used as a predictor. Surfaces were interpolated from 33,594 grid points using Ordinary Kriging. Other emissions scenarios gave qualitatively similar results (Figs S3 to S5).

This technique combines a spatial Gradient and Inverse Distance Squared (GIDS) weighting to monthly point data using multiple regressions calculated for every grid cell for every month. Using the 4-km resolution digital elevation model in PRISM, parameter weighting is based on the location and elevation of the new fine resolution grid (270 m for this study) relative to existing coarse resolution grid cells [32]. The monthly maps of precipitation and minimum and maximum air temperature are used as input to a monthly water balance model (the Basin Characterization Model) to calculate potential and actual evapotranspiration and climatic water deficit for the Sierra Nevada at a grid spacing of 270 m.

Climatic Water Deficit from the Basin Characterization Model

Climatic water deficit is defined as *PET-AET*, where *PET* is potential evapotranspiration and *AET* is actual evapotranspiration. Deficit explicitly considers the seasonal interactions of temperature and precipitation, as well as the timing of snowmelt and the soil water holding capacity. Deficit is therefore a more biologically relevant measure of water stress than other commonly used indices [4,5].

To calculate deficit we used the Basin Characterization Model (BCM): a physically-based model that calculates water balance fractions based on data inputs for topography, soil composition and depth, underlying bedrock geology, and spatially-distributed values (measured or estimated) of air temperature and precipitation [33,34]. The BCM calculates monthly recharge and runoff using a deterministic water-balance approach based on the distribution of precipitation and the estimation of potential evapotranspiration [33,34] using the Priestley-Taylor equation [35]. The BCM relies on rigorous hourly energy balance calculation (used for the Priestley-Taylor equation) using topographic shading and applies available spatial maps of elevation, bedrock permeability estimated from geology, soil water storage from STATSGO and SSURGO soil databases [36], vegetation density, and PRISM maps of precipitation and minimum and maximum air temperature.

The BCM was calibrated regionally to measured potential evapotranspiration data and MODIS snow cover data [37]. Locally, the model was also calibrated to measured unimpaired streamflow data [34]. The determination of whether excess water becomes recharge or runoff is governed in part by the underlying bedrock permeability. The higher the bedrock permeability, the higher the recharge and the lower the runoff generated for a given grid cell. In small gaged basins that generate unimpaired flows, the bedrock permeability can be adjusted to calculate a total basin discharge that matches the measured basin discharge.

Tree Mortality Rate Model Development

We developed statistical models relating changes in tree mortality rate to changes in temperature and climatic water deficit. Changes were calculated relative to reference period

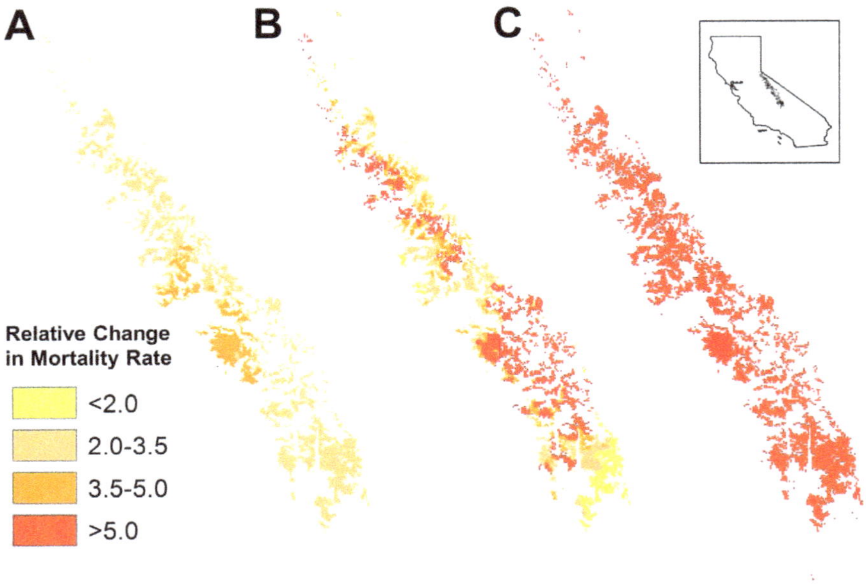

Figure 2. Projected Changes in Mortality Rate for predominantly Energy-Limited Sierra Nevada Conifer Forests (Hypothetical). Mapped projections of average relative changes in mortality rate for the years 2090 to 2099 for energy-limited coniferous forests (≥2450 m) of California's Sierra Nevada, using exponential models and the GFDL A2 emissions model. Elevations generally increase from left to right. (A) Changes in mortality when absolute changes in deficit (D_a) are used as a predictor. (B) Changes in mortality when relative changes in deficit (D_r) are used as a predictor. (C) Changes in mortality when absolute changes in temperature (T_a) are used as a predictor. Other emissions scenarios gave qualitatively similar results (Figs. S6 to S8).

averages. For mortality rate, the reference period for a given plot was the period of record for that plot (plot establishment through 2006). For climate, the reference period was the water years (October 1 to September 30) 1982 through 2006 – the full range of years of record for all plots combined. (Note that mortality rates could be calculated only for the period of record of a given plot, whereas climatic values could be calculated for longer periods).

To account for systematic variation in tree mortality rates with elevation in the Sierra Nevada [38] and to facilitate interpretation of results, we modeled relative rather than absolute changes in mortality rates, with change defined against the reference period rate. Specifically, relative change in mortality rate for each plot was calculated as

$$M_r = M_t / M_{ref} \tag{1}$$

where M_r is the relative change in mortality rate, M_t is mortality rate for the year t, and M_{ref} is the reference mortality rate (i.e., the average mortality rate for the period of record of the given plot). As a check, we also modeled absolute changes in mortality rates and found no qualitative changes to our results.

The effects of temperature on biological processes are generally expressed in terms of absolute rather than relative changes in temperature (e.g., [11,39]) so we used absolute change in temperature as an independent variable:

$$T_a = T_t - T_{ref} \tag{2}$$

where T_a is the absolute change in average annual temperature, T_t is the annual average temperature in year t, and T_{ref} is the reference average annual temperature. As a check we also modeled using relative changes in temperature [in K] and found no qualitative changes to our results.

We could identify no *a priori* biological basis for choosing between relative versus absolute changes in deficit for use as independent variables (see the Introduction), so we used both to test whether one or the other were better predictors of change in mortality rate. These were calculated as

$$D_r = D_t / D_{ref} \tag{3}$$

$$D_r = D_t - D_{ref} \tag{4}$$

where D_r and D_a are, respectively, the relative and absolute changes in climatic water deficit, D_t is the deficit for year t, and D_{ref} is the average deficit in the given plot for the reference period.

To allow for lagged and cumulative effects, we also calculated running averages of D_r, D_a, and T_a that incorporated the current year and the prior one to four years (i.e., average of current year and prior one to four years). Running averages were also calculated that excluded the current year.

The relationships between relative change in mortality and climatic variables were modeled as either linear or exponential functions:

$$M_r = \beta_0 + \beta_1 C_x \tag{5}$$

$$M_r = e^{\beta_0 + \beta_1 C_x} \tag{6}$$

where C_x is the given climatic variable (D_a, D_r, or T_a, including running averages of those variables), and β_0 and β_1 are estimated parameters. Additionally, logistic functions were tried but discarded because uncertainty in the upper asymptote resulted in unstable parameter estimations (i.e., there was not enough

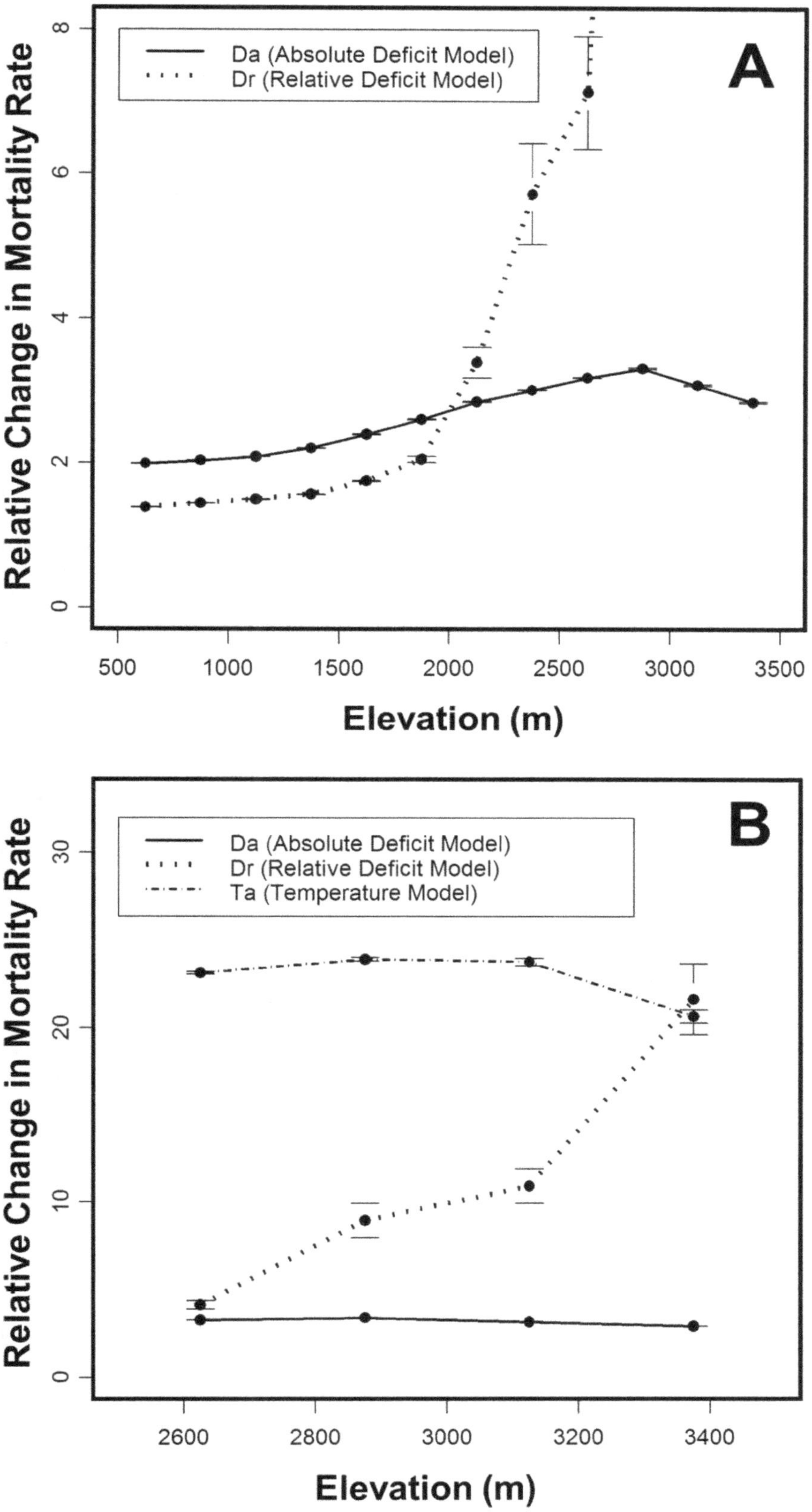

Figure 3. Projected Changes in Mortality Rate by Elevation (Hypothetical). Projected relative changes in tree mortality rate by elevation in the 2090s for the GFDL A2 climatic scenario using (A) D_a and D_r exponential models for all forests and (B) the D_a, D_r, and T_a exponential model for energy-limited forests. Rates are averaged in 250 m classes using the 33,594 Sierra Nevada coniferous forest grid points. Error bars show standard

error. Note that the scales on A and B are different. A small number of points with small baseline deficit values (<1% of all points for A and <2% for B) were excluded because the D_r models predicted exceedingly large changes in mortality rate. Projections using the PCM model and other emission scenarios as well as using linear mortality models gave qualitatively similar results.

information in the data to reliably estimate a maximum increase in mortality rate). Models were parameterized separately for all forests combined, water-limited forests (≤2450 m in elevation), and energy-limited forests (>2450 m elevation).

To account for site effects, we initially used mixed-effect models with site as a random effect. In the vast majority of cases, however, the estimated site effect was negligible and gave parameter values for the fixed effects that were statistically indistinguishable from those estimated using a fixed effects model. This was likely due to our standardization of climate and mortality values to a reference period. The mixed effects models also gave higher Akaike's Information Criterion (AIC, see below) values, indicating less support. Even in the very few cases where the mixed effects term was non-trivial, parameter estimates for the fixed effects were statistically indistinguishable from their fixed effects model counterparts and AIC model rankings were the same. In addition, maximum likelihood estimations for the mixed effects parameters tended to be highly sensitive to starting values. Therefore, given that mixed effects terms did not appear to offer any improvement to or have any meaningful effect on the analysis, we proceeded with standard fixed effects models.

Models were fit using a maximum likelihood approach. Relative changes in mortality rates were treated as counts with the previous year's total count of live trees used as an offset. For example, for linear models the count was given by:

$$M_{t,j} = n_{t-1,j} \cdot M_{ref,j} \cdot (\beta_0 + \beta_1 C_{x,t,j}) \quad (7)$$

Where $M_{t,j}$ is the count of mortalities at year t in plot j; $n_{t-1,j}$ is the count of live trees at year t-1 in plot j; $M_{ref,j}$ is the reference mortality rate in plot j; and $C_{x,t,j}$ is the given climatic variable at time t in plot j. Parameters for these equations were then fit by maximizing a negative binomial log likelihood function using SAS 9.1 (SAS Institute. 2004. SAS OnlineDoc® 9.1.3. Cary, NC: SAS Institute Inc).

Models were compared with an information theoretic approach using AIC [40] to determine which type of model (linear or exponential) and which climatic parameter (D_a, D_r, and T_a) best predicted mortality rate for each elevational class. As a guide, based upon Burnham and Anderson's [40] rules of thumb, ΔAIC values less than 2 or an evidence ratio less than 2.7 would be considered very little evidence that one model is better than another while a ΔAIC greater than 4 or an evidence ratio greater than 7.4 would be considered relatively strong evidence for one model being better than another. The highest ranked models for each elevational class were then chosen and used for forecasting change in future tree mortality rates under different climate scenarios, described below.

Mortality Forecasting

Note that the forecasts we perform here are intended to illustrate potential pitfalls in mortality forecasting given uncertainty in model choice. They are not intended nor should be interpreted as an attempt to reliably predict future mortality rates. For this illustration, changes in tree mortality rates in the coniferous forests of the Sierra Nevada were forecast through year 2100. We chose to use Sierra Nevada conifer forests and climate forecasts derived from General Circulation Models (GCM) in order to provide a biologically-relevant and biologically-plausible context for our illustration, but we might just have easily chosen a hypothetical dataset with arbitrarily chosen values.

The extent of coniferous forests was determined from the California Gap Analysis Project's primary Wildlife Habitat Relationship habitat types (WHR1) [41]. On a one km resolution grid, this resulted in the selection of 33,594 grid points within the Sierra Nevada ecoregion (as defined by Hickman [42]) encompassing an area of approximately 33,600 km^2.

Grid points were classified by elevation as containing either predominantly water- or predominantly energy-limited forest (≤2450 m or >2450 m elevation, respectively). Of course, the elevational boundary between water- and energy-limited forests almost certainly varies with latitude, topography, rain shadow effects, and other factors. However, we opted for this simple classification by elevation because our primary goal was to gain broad, qualitative insights from model comparisons, not to predict future conditions on a particular piece of ground. As an additional check, we performed forecasts in which we classified grid points as containing either water- or energy-limited forest based on forest type rather than elevation, and these forecasts showed no qualitative differences from those based on elevation that we report here.

Downscaling Future Climate Scenarios

GCM climatic forecasts are generally available for the continental US at 12 km spatial resolution. A set of these projections have been downscaled for California and its environs using the constructed analogs method of Hidalgo et al. [43], providing a basis for our further downscaling [32]. Our goal was to represent climate projections for California on the basis of global climate models that have proven capable of simulating recent historical climate, particularly the distribution of monthly temperatures and the strong seasonal cycle of precipitation that exists in the region [44–46]. In addition, models were selected to represent a range of model sensitivity to greenhouse gas forcing. On the basis of these criteria, two GCMs were selected, the Parallel Climate Model (PCM) developed by National Center for Atmospheric Research (NCAR), Department of Energy (DOE) (see [47,48]) and the National Oceanic and Atmospheric Administration (NOAA) Geophysical Fluid Dynamics Laboratory CM2.1 model (GFDL) [49,50]. The choice of greenhouse gas emissions scenarios included A2 (medium-high–essentially "business as usual") and B1 (low-essentially a "mitigated emissions" scenario), was guided by considerations presented by Nakicenovic *et al.* [51]. Thus we developed a range of climatic forecasts based on four specific scenarios; two models each driven by two emissions scenarios: "GFDL A2", "GFDL B1", "PCM A2", "PCM B1".

These four scenarios were downscaled from the 12-km grid scale to the historical PRISM data scale of 4 km for the purpose of bias correction. To make the correction possible the GCM was run for a historical forcing function to establish a baseline for modeling to match current climate. The baseline period for this study was defined as the PCM and GFDL model runs for 1950–2000, representing current (pre-2000) atmospheric greenhouse gas conditions. This baseline period was then adjusted using the PRISM data from 1950–2000, for each month and for each grid cell. Our approach to bias correction is a simple scaling of the mean and standard deviation of the projections to match those of the PRISM data following Bouwer *et al.* [52] and described in detail in Flint and Flint [32]. Once the bias correction was complete, the 4 km projections were further downscaled to 270 m spatial resolution using the GIDS spatial interpolation approach for model application.

Changes in tree mortality rate were forecast for each year for each grid point under each combination of GCM and emissions scenario by inputting forecasted climate into our best mortality models. Climate and mortality forecasts were then summarized by decade for each grid point.

Results

Model Selection and Comparison

For all elevations combined and for low elevation forest plots (≤2450 m), the strongest predictors of $M_r$were running averages ofD_a,D_r, and T_a that included the current year change in deficit or temperature plus either the prior two years or three years (the latter only for low elevation temperature models) (Table 1). Deficit models had substantially stronger support than temperature models. In contrast, the evidence was at best equivocal in distinguishing between absolute or relative deficit variables as the better predictors ofM_r. Thus, while both types of deficit variables appear to be better predictors of changes in mortality than temperature variables at all elevations combined and at low elevations, there was no strong support for one type of deficit variable over the other. In all cases, ΔAIC indicated roughly equal support for both linear and exponential models.

For high elevation forest plots (>2450 m), the strongest predictors of M_r were running averages that included the current year plus the prior two or four years (Table 1). Deficit models were still top ranked, but ΔAIC values and evidence ratios indicate that they had only marginally more support than the best T_a models. In all cases, ΔAIC indicated roughly equal support for both linear and exponential models.

Although temperature and deficit variables are correlated, the strength of that correlation (between D_a and T_a including the current plus two prior years) was indistinguishable between low and high elevations ($r = 0.68$ for low elevations, $r = 0.64$ for high elevations,). The increased relative predictive power of temperature at high elevation therefore cannot be explained by increased correlation between temperature and deficit.

In comparing the models developed for each set of data (all elevations, low elevations, and high elevations), the similarity in the parameter estimates for the deficit models, particularly the D_a models, is notable, suggesting that the relationship between changes in mortality rate and changes in deficit does not vary substantially across a very marked elevational gradient.

Forecasted Changes in Mortality Rates

Temperature increased for all combinations of GCMs and emissions scenarios, with average increases between the 2000s to the 2090s ranging from 1.2 to 4.2°C (Fig. S2). The GFDL model tended to forecast a general decrease in precipitation in the Sierra Nevada, while the PCM model showed no straightforward pattern (Fig. S2). Climatic water deficit increased in all cases, with the GFDL A2 combination showing the sharpest increase (Fig. S2).

All mortality models forecast increasing tree mortality rates throughout the coniferous forests of the Sierra Nevada, with spatially-averaged projected increases by the 2090s varying dramatically depending on the model used and position on the landscape (see below). In general, the GFDL model resulted in larger increases in mortality rates than the PCM model, and the A2 scenario resulted in larger increases than the B1 scenario.

The spatial pattern of mortality rate increases differed markedly among forecasts from the D_a, D_r, and T_a models (Fig. 1, Fig. 2, Fig. 3). In particular, the models make dramatically different forecasts about how mortality rate will change along elevational gradients. D_a models forecast gradually increasing mortality rates with elevation (Fig. 1A, Fig. 2A, Fig. 3), reflecting modestly increasing changes in absolute deficit with elevation. In contrast, D_r models forecast smaller increases in mortality rates at lower elevations but very sharply increasing rates for forests above about 2200 m (Fig. 1B, Fig. 2B, Fig. 3):a consequence of the fact that baseline deficits at higher elevations tend to be small and that relatively modest changes in absolute deficit can result in very large changes in relative deficit. Finally, in the elevational zone above 2450 m, where temperature was a relatively good predictor of changes in mortality rate, T_a models forecast large and relatively uniform increases in mortality rate at all elevations within the zone, except at the highest elevations where the magnitude of increase drops somewhat (Fig. 2C, Fig. 3B). This pattern tracks the forecasted changes in temperature at different elevations.

Discussion

Deficit versus Temperature

To our knowledge, we provide the first explicit correlative test of the effects of climate on tree mortality in water- versus energy-limited forests. As expected, changes in mortality rate in water-limited forests (low elevations) were unambiguously best modeled by deficit. In energy limited forests (high elevation) temperature also became a relevant predictor, with deficit models providing only marginally better fits.

Two factors might account for the fact that temperature variables, while stronger predictors at higher elevations, were not unambiguously the best predictors at those elevations: (i) energy and water limitation are probably more properly described as a continuum rather than as a dichotomy, and (ii) the proposed mechanisms driving increasing tree mortality rates (increased drought stress and favorability to enemies) are not mutually exclusive. Therefore, it is not surprising that deficit might still be an important driver of mortality rates at higher elevations, even if it becomes less clearly the best predictor.

On the first point, we have assumed a simple dichotomy between water- and energy-limited forests. In reality, water and energy limitation likely represent the extremes of graded variations in both space and time, with many forests being both water and energy limited (e.g., [53,54,55]). Additionally, with climatic change, as temperatures increase and winter snowpacks decrease through time, we expect that some forests will switch from being primarily energy-limited to primarily water-limited.

We also cannot eliminate the possibility that the observed difference in climatic correlates of tree mortality between our water- and energy-limited forests is a simple consequence of differences in the forests' species composition. However, we believe this is unlikely. Importantly, both sets of forests are heavily dominated by the same two tree genera, *Abies* and *Pinus*, and are attacked by a similar (and sometimes identical) suite of pathogens and herbivores, suggesting that environmental effects rather than species effects provide the most parsimonious explanation of our results (e.g., [7,8,11,12,56]).

Regardless, our finding that temperature becomes a relatively stronger predictor of mortality rate in energy-limited forests agrees with recent studies examining climatic controls on tree growth, which have found that climatic controls of both individual tree growth [13] and overall forest productivity [57] vary depending on the water or energy limitation of a given forest. Littell *et al.* [13] for example found evidence that deficit drove tree growth patterns for most of the Douglas-fir forests in their study (i.e., growth was usually water-limited) but that at higher elevation sites growth appeared to be temperature- (energy-) limited.

Our findings have implications for our understanding of how forest structure, dynamics, and carbon storage will change with a changing climate. Forests play a substantial role in the global carbon cycle [58], and changes in tree mortality can have dramatic effects on forest carbon dynamics (e.g., [1]). To date, however, most forest simulation models have relied on general mortality algorithms that do not incorporate explicit mechanisms and would therefore not be expected to adequately reflect changes in mortality processes across a gradient of water and energy limitation [59–61].

One encouraging result was the strong similarity in parameter estimations for deficit models in both low and high elevation forests. While it is likely that additional mechanisms will need to be considered across such gradients, relationships between drought and mortality may not vary substantially across fairly broad scales.

Absolute versus Relative Deficit

Our inability to clearly distinguish among models using relative versus absolute changes in deficit as predictors of mortality also has critical implications for forecasting future changes in mortality rates. The two types of models lead to substantially different predictions of changes in mortality rates across the landscape (Figs. 1,2). If future studies yield similar results across latitudinal as well as elevational gradients, we could be confronted with drastically different conclusions about which forests are most at risk from a warming climate at subcontinental scales.

We currently have no strong *a priori* reasons for selecting one deficit model over another. Although critical groundwork has been laid for mechanistic models of drought-induced tree mortality [62], current models cannot yet tell us whether mortality is likely to be best predicted by absolute or relative changes in deficit. Empirical evidence is also equivocal. While some studies in areas of chronic water stress have found increased mortality during droughts (potentially providing opportunities to study the effects of increasing deficit [19,63–65]), we do not yet have the data needed to assess whether trees in those regions were responding in a manner more consistent with absolute or relative changes in deficit. Additionally, it is not clear whether episodes of severe, short-term drought are directly comparable to the more gradual long-term increases in water stress that we are examining here. Distinguishing between the two models will likely require more definitive physiological models, larger empirical datasets that incorporate a wider range of deficits along gradients of water stress, and manipulative experiments that directly test the effect of changes in deficit.

Additional Uncertainties

Given the uncertainties noted above, we emphasize that our forecasts of future tree mortality rates in the Sierra Nevada should not be taken as quantitative predictions but rather as an illustration of the dangers of attempting to predict tree mortality rates without a solid understanding of underlying mechanisms. As noted above, we chose to use the Sierra Nevada landscape and downscaled GCM projections in order to provide a biologically-plausible and biologically-relevant context for our illustration, but we might just as easily have used a completely hypothetical climate projection and a hypothetical forest.

Additional uncertainties should be acknowledged with regard to our forecasts. Of necessity, our models were developed from a range of deficit and temperature changes that is far smaller than the range projected for the future. It is therefore unlikely that we have captured the true shape of the relationship between deficit or temperature and mortality, as is indicated by our inability to distinguish between exponential and linear models. In addition, as noted above, mechanisms may change at a given location as temperatures increase. Perhaps most importantly, our models were parameterized on background (non-catastrophic) tree mortality rates and do not capture potential insect or pathogen outbreaks, such as those seen in recent years in many parts of western North America [12]. Our assumption of a simple monotonic relationship between climate and tree mortality is undoubtedly overly simplistic (see [10,11]), and does not account for sudden threshold transitions that can result in extensive forest die-back. However, our forecasts serve their intended role of qualitatively illustrating how model choices can profoundly affect our ability to understand and predict future changes.

Conclusions

Understanding the spatial relationships between climate and tree mortality is becoming increasingly important as climate continues to change. For example, van Mantgem *et al.* [24] found a widespread increase in mortality rates in old growth forests across the western USA and showed that regional warming is likely an important contributor to that increase. We have shown here that the mechanisms relating climate to tree mortality appear to vary along an elevational gradient, suggesting that a similar pattern may hold across a broader geographic scale, with the underlying mechanisms that drive changes in mortality perhaps varying between drier (water-limited) regions and wetter (energy-limited) regions. We have also demonstrated that identifying the precise nature of the relationship between climate variables and tree mortality can be critically important for making accurate predictions of forest change. At present, our ability to correctly identify such relationships is tenuous. A more thorough grasp of mortality mechanisms will be critical as we struggle to forecast and manage forest change in the decades to come.

Supporting Information

Figure S1 Tree Size Distribution for Each Plot. Shows the size distribution of trees in each of the A) water-limited and B) energy-limited plots. The y-axis is given in a log base 10 scale.

Figure S2 Forecasted change in Sierra Nevada climate through time from different model scenarios. Climate data is averaged by decade for all 33,594 gridpoints. Gap between standard error bars for each point was too small to distinguish so they have not been included. A) temperature forecasts; B) precipitation forecasts; C) climatic water deficit forecasts.

Figure S3 Projected Changes in Mortality Rate for Sierra Nevada Conifer Forests (GFDL B1, Hypothetical). Mapped projections of average relative changes in mortality rate for the years 2090 to 2099 for coniferous forests of California's Sierra Nevada, using exponential and the GFDL B1 emissions model. Elevations generally increase from left to right. (A) Changes in mortality when absolute changes in deficit (D_a) are used as a predictor. (B) Changes in mortality when relative changes in deficit (D_r) are used as a predictor. Surfaces were interpolated from 33,594 grid points using Ordinary Kriging.

Figure S4 Projected Changes in Mortality Rate for Sierra Nevada Conifer Forests (PCM A2, Hypothetical). Mapped projections of average relative changes in mortality rate for the years 2090 to 2099 for coniferous forests of California's Sierra Nevada, using exponential and the PCM A2 emissions model. Elevations generally increase from left to right. (A) Changes in mortality when absolute changes in deficit (D_a) are used as a predictor. (B) Changes in mortality when relative changes in deficit (D_r) are used as a predictor. Surfaces were interpolated from 33,594 grid points using Ordinary Kriging.

Figure S5 Projected Changes in Mortality Rate for Sierra Nevada Conifer Forests (PCM B1, Hypothetical). Mapped projections of average relative changes in mortality rate for the years 2090 to 2099 for coniferous forests of California's Sierra Nevada, using exponential and the PCM B1 emissions model. Elevations generally increase from left to right. (A) Changes in mortality when absolute changes in deficit (D_a) are used as a predictor. (B) Changes in mortality when relative changes in deficit (D_r) are used as a predictor. Surfaces were interpolated from 33,594 grid points using Ordinary Kriging.

Figure S6 Projected Changes in Mortality Rate for predominantly Energy-Limited Sierra Nevada Conifer Forests (GFDL B1, Hypothetical). Mapped projections of average relative changes in mortality rate for the years 2090 to 2099 for energy-limited coniferous forests (≥2450 m) of California's Sierra Nevada, using exponential models and the GFDL B1 emissions model. Elevations generally increase from left to right. (A) Changes in mortality when absolute changes in deficit (D_a) are used as a predictor. (B) Changes in mortality when relative changes in deficit (D_r) are used as a predictor. (C) Changes in mortality when absolute changes in temperature (T_a) are used as a predictor.

Figure S7 Projected Changes in Mortality Rate for predominantly Energy-Limited Sierra Nevada Conifer Forests (PCM A2, Hypothetical). Mapped projections of average relative changes in mortality rate for the years 2090 to 2099 for energy-limited coniferous forests (≥2450 m) of California's Sierra Nevada, using exponential models and the PCM A2 emissions model. Elevations generally increase from left to right. (A) Changes in mortality when absolute changes in deficit (D_a) are used as a predictor. (B) Changes in mortality when relative changes in deficit (D_r) are used as a predictor. (C) Changes in mortality when absolute changes in temperature (T_a) are used as a predictor.

Figure S8 Projected Changes in Mortality Rate for predominantly Energy-Limited Sierra Nevada Conifer Forests (PCM B1, Hypothetical). Mapped projections of average relative changes in mortality rate for the years 2090 to 2099 for energy-limited coniferous forests (≥2450 m) of California's Sierra Nevada, using exponential models and the PCM B1 emissions model. Elevations generally increase from left to right. (A) Changes in mortality when absolute changes in deficit (D_a) are used as a predictor. (B) Changes in mortality when relative changes in deficit (D_r) are used as a predictor. (C) Changes in mortality when absolute changes in temperature (T_a) are used as a predictor.

Table S1 Plot Details. Characteristics of the 21 forest plots used for model development.

Table S2 Plot Mortality Data. Counts of mortalities and live trees for each plot for each census year.

Acknowledgments

We thank the many people involved in establishing and maintaining the permanent forest plots, and Sequoia and Yosemite National Parks for their invaluable cooperation and assistance. Julie Yee provided essential statistical advice. This work is a contribution of the Western Mountain Initiative, a USGS global change research project, and USGS Pacific Southwest Area Integrated Science. Any use of trade names is for descriptive purposes only and does not imply endorsement by the U.S. Government.

Author Contributions

Conceived and designed the experiments: AJD NLS PJvM AF. Performed the experiments: AJD NLS PJvM. Analyzed the data: AJD TD AF PJvM. Contributed reagents/materials/analysis tools: AF PJvM TD. Wrote the paper: AJD NLS AF.

References

1. Kurz WA, Dymond CC, Stinson G, Rampley GJ, Neilson ET, et al. (2008) Mountain pine beetle and forest carbon feedback to climate change. Nature 452: 987–990.
2. Williams AP, Allen CD, Macalady AK, Griffin D, Woodhouse CA, et al. (2012) Temperature as a potent driver of regional forest drought stress and tree mortality. Nature Climate Change Published online 30 September.
3. Adams HD, Macalady AK, Breshears DD, Allen CD, Stephenson NL, et al. (2010) Climate-Induced Tree Mortality: Earth System Consequences. EOS 91: 153–154.
4. Stephenson NL (1990) Climatic control of vegetation distribution- the role of water balance. American Naturalist 135: 649–670.
5. Stephenson NL (1998) Actual evapotranspiration and deficit: biologically meaningful correlates of vegetation distribution across spatial scales. Journal of Biogeography 25: 855–870.
6. Breshears DD, Cobb NS, Rich PM, Price KP, Allen CD, et al. (2005) Regional vegetation die-off in response to global-change-type drought. Proceedings of the National Academy of Sciences of the United States of America 102: 15144–15148.
7. Breshears DD, Myers OB, Meyer CW, Barnes FJ, Zou CB, et al. (2009) Tree die-off in response to global change-type drought: mortality insights from a decade of plant water potential measurements. Frontiers in Ecology and the Environment 7: 185–189.
8. McDowell N, White S, Pockman WT (2008) Transpiration and stomatal conductance across a steep climate gradient in the southern Rocky Mountains. Ecohydrology 1: 193–204.
9. van Mantgem PJ, Stephenson NL (2007) Apparent climatically induced increase of tree mortality rates in a temperate forest. Ecology Letters 10: 909–916.
10. Bentz BJ, Regniere J, Fettig CJ, Hansen EM, Hayes JL, et al. (2010) Climate Change and Bark Beetles of the Western United States and Canada: Direct and Indirect Effects. Bioscience 60: 602–613.
11. Hicke JA, Logan JA, Powell J, Ojima DS (2006) Changing temperatures influence suitability for modeled mountain pine beetle (Dendroctonus ponderosae) outbreaks in the western United States. Journal of Geophysical Research-Biogeosciences 111.
12. Raffa KF, Aukema BH, Bentz BJ, Carroll AL, Hicke JA, et al. (2008) Cross-scale drivers of natural disturbances prone to anthropogenic amplification: The dynamics of bark beetle eruptions. Bioscience 58: 501–517.
13. Littell JS, Peterson DL, Tjoelker M (2008) Douglas-fir growth in mountain ecosystems: Water limits tree growth from stand to region. Ecological Monographs 78: 349–368.
14. Breda N, Huc R, Granier A, Dreyer E (2006) Temperate forest trees and stands under severe drought: a review of ecophysiological responses, adaptation processes and long-term consequences. Annals of Forest Science 63: 625–644.
15. Choat B, Jansen S, Brodribb TJ, Cochard H, Delzon S, et al. (2012) Global convergence in the vulnerability of forests to drought. Nature 491: 752–756.
16. Joslin JD, Wolfe MH, Hanson PJ (2000) Effects of altered water regimes on forest root systems. New Phytologist 147: 117–129.
17. Landhausser SM, Wein RW, Lange P (1996) Gas exchange and growth of three arctic tree-line tree species under different soil temperature and drought preconditioning regimes. Canadian Journal of Botany-Revue Canadienne De Botanique 74: 686–693.
18. Metcalfe DB, Meir P, Aragao L, da Costa ACL, Braga AP, et al. (2008) The effects of water availability on root growth and morphology in an Amazon rainforest. Plant and Soil 311: 189–199.

19. Gitlin AR, Sthultz CM, Bowker MA, Stumpf S, Paxton KL, et al. (2006) Mortality gradients within and among dominant plant populations as barometers of ecosystem change during extreme drought. Conservation Biology 20: 1477–1486.
20. Mueller RC, Scudder CM, Porter ME, Trotter RT, Gehring CA, et al. (2005) Differential tree mortality in response to severe drought: evidence for long-term vegetation shifts. Journal of Ecology 93: 1085–1093.
21. Frazier MR, Huey RB, Berrigan D (2006) Thermodynamics constrains the evolution of insect population growth rates: "Warmer is better". American Naturalist 168: 512–520.
22. Bale JS, Masters GJ, Hodkinson ID, Awmack C, Bezemer TM, et al. (2002) Herbivory in global climate change research: direct effects of rising temperature on insect herbivores. Global Change Biology 8: 1–16.
23. Deutsch CA, Tewksbury JJ, Huey RB, Sheldon KS, Ghalambor CK, et al. (2008) Impacts of climate warming on terrestrial ectotherms across latitude. Proceedings of the National Academy of Sciences of the United States of America 105: 6668–6672.
24. van Mantgem PJ, Stephenson NL, Byrne JC, Daniels LD, Franklin JF, et al. (2009) Widespread Increase of Tree Mortality Rates in the Western United States. Science 323: 521–524.
25. Fites-Kaufman JA, Rundel P, Stephenson NL, Weixelman DA (2007) Montane and subalpine vegetation of the Sierra Nevada and Cascade ranges. In: Barbour MG, Keeler-Wolf T, Schoenherr AA, editors. Terrestrial Vegetation of California. Berkeley, CA: University of California Press. 456–501.
26. Caprio AC, Swetnam TW (1995) Historic fire regimes along an elevational gradient on the west slope of the Sierra Nevada, California; 1993 1995; Missoula, MT. USDA Forest Service General Techinical Report INT-GTR-320.
27. Stephenson NL (1988) Climatic Control of Vegetation Distribution: the Role of the Water Balance with Examples from North America and Sequoia National Park, California. Ithaca, N.Y.: Cornell University.
28. Tague C, Heyn K, Christensen L (2009) Topographic controls on spatial patterns of conifer transpiration and net primary productivity under climate warming in mountain ecosystems. Ecohydrology 2: 541–554.
29. Trujillo E, Molotch NP, Goulden ML, Kelly AE, Bales RC (2012) Elevation-dependent influence of snow accumulation on forest greening. Nature Geoscience 5: 705–709.
30. Daly C, Gibson W, Doggett M, Smith J, Taylor G (2004) Up-to-date monthly climate maps for the conterminous United States. 13–16.
31. Nalder IA, Wein RW (1998) Spatial interpolation of climatic Normals: test of a new method in the Canadian boreal forest. Agricultural and Forest Meteorology 92: 211–225.
32. Flint LE, Flint AL (2012) Downscaling future climate scenarios to fine scales for hydrologic and ecological modeling and analysis. Ecological Processes 1: 1–15.
33. Flint AL, Flint LE (2007) Application of the basin characterization model to estimate in-place recharge and runoff potential in the Basin and Range carbonate-rock aquifer system, White Pine County, Nevada, and adjacent areas in Nevada and Utah. U.S. Geological Survey.
34. Flint LE, Flint AL (2012) Simulation of Climate Change in San Francisco Bay Basins, California: Case Studies in the Russian River Valley and Santa Cruz Mountains. U.S. Geological Survey. 55 p.
35. Priestley C, Taylor R (1972) On the assessment of surface heat flux and evaporation using large-scale parameters. Monthly Weather Review 100: 81–92.
36. National Resources Conservation Service (NRCS) (2006) U.S. General Soil Map (STATSGO2). Soil Data Mart. Available: http://soildatamart.nrcs.usda.gov/. Accessed 2004 April.
37. Flint LE, Flint AL (2007) Regional Analysis of Ground-Water Recharge. In: Stonestrom DA, Constantz J, Ferre TPA, Leake SA, editors. Ground-Water Recharge in the Arid and Semiarid Southwestern United States. Reston, VA: U.S. Geological Survey. 29–60.
38. Stephenson NL, van Mantgem PJ (2005) Forest turnover rates follow global and regional patterns of productivity. Ecology Letters 8: 524–531.
39. Brown JH, Gillooly JF, Allen AP, Savage VM, West GB (2004) Toward a metabolic theory of ecology. Ecology 85: 1771–1789.
40. Burnham KP, Anderson DR (1998) Model selection and inference : a practical information-theoretic approach. New York: Springer. xix, 353 p. p.
41. Davis FW, Stoms DM, Hollander AD, Thomas KA, Stine PA, et al. (1998) The California Gap Analysis Project– Final Report. Santa Barbara, CA: University of California.
42. Hickman JC (1993) The Jepson Manual: Higher Plants of California; Hickman JC, editor. Berkeley, CA: University of California Press.
43. Hidalgo HG, Dettinger MD, Cayan DR (2008) Downscaling with constructed analogues: Daily precipitation and temperature fields over the United States. California Energy Commission PIER Final Project Report CEC-500-2007-123.
44. Cayan D, Tyree M, Dettinger M, Hidalgo H, Das T, et al. (2009) Climate change scenarios and sea level rise estimates for the California 2008 Climate Change Scenarios Assessment. California Climate Change Center CEC-500-2009-014-D.
45. Cayan DR, Maurer EP, Dettinger MD, Tyree M, Hayhoe K (2008) Climate change scenarios for the California region. Climatic Change 87: 21–42.
46. Knowles N, Cayan DR (2002) Potential effects of global warming on the Sacramento/San Joaquin watershed and the San Francisco estuary. Geophysical Research Letters 29: 1891.
47. Meehl GA, Washington WM, Wigley T, Arblaster JM, Dai A (2003) Solar and greenhouse gas forcing and climate response in the twentieth century. Journal of Climate 16: 426–444.
48. Washington W, Weatherly J, Meehl G, Semtner A Jr, Bettge T, et al. (2000) Parallel climate model (PCM) control and transient simulations. Climate Dynamics 16: 755–774.
49. Delworth TL, Broccoli AJ, Rosati A, Stouffer RJ, Balaji V, et al. (2006) GFDL's CM2 global coupled climate models. Part I: Formulation and simulation characteristics. Journal of Climate 19: 643–674.
50. Stouffer R, Broccoli A, Delworth T, Dixon K, Gudgel R, et al. (2006) GFDL's CM2 global coupled climate models. Part IV: Idealized climate response. Journal of Climate 19: 723–740.
51. Nakicenovic N, Alcamo J, Davis G, de Vries B, Joergen F, et al. (2000) Special Report on Emissions Scenarios : a special report of Working Group III of the Intergovernmental Panel on Climate Change. In: Pacific Northwest National Laboratory EMSL, editor. New York, NY: Cambridge University Press.
52. Bouwer LM, Aerts JC, van de Coterlet GM, van de Giesen N, Gieske A, et al. (2004) Evaluating downscaling methods for preparing global circulation model (GCM) data for hydrological impact modeling. In: Aerts JC, Droogers P, editors. Climate Change in Contrasting River Basins. London, U.K.: CAB International Publishing. 25–47.
53. Graumlich LJ (1991) Sub-alpine tree growth, climate and increasing CO_2– an assessment of recent growth trends. Ecology 72: 1–11.
54. Graumlich LJ (1993) A 1000-year record of temperature and precipitation in the Sierra Nevada. Quaternary Research 39: 249–255.
55. Lloyd AH, Graumlich LJ (1997) Holocene dynamics of treeline forests in the Sierra Nevada. Ecology 78: 1199–1210.
56. Adams HD, Guardiola-Claramonte M, Barron-Gafford GA, Villegas JC, Breshears DD, et al. (2009) Temperature sensitivity of drought-induced tree mortality portends increased regional die-off under global-change-type drought. Proceedings of the National Academy of Sciences of the United States of America 106: 7063–7066.
57. Boisvenue C, Running SW (2006) Impacts of climate change on natural forest productivity - evidence since the middle of the 20th century. Global Change Biology 12: 862–882.
58. Bonan GB (2008) Forests and climate change: Forcings, feedbacks, and the climate benefits of forests. Science 320: 1444–1449.
59. Hawkes C (2000) Woody plant mortality algorithms: description, problems and progress. Ecological Modelling 126: 225–248.
60. Bigler C, Bugmann H (2004) Assessing the performance of theoretical and empirical tree mortality models using tree-ring series of Norway spruce. Ecological Modelling 174: 225–239.
61. Keane RE, Austin M, Field C, Huth A, Lexer MJ, et al. (2001) Tree mortality in gap models: Application to climate change. Climatic Change 51: 509–540.
62. McDowell NG, Pockman WT, Allen CD, Breshears DD, Cobb N, et al. (2008) Mechanisms of plant survival and mortality during drought: why do some plants survive while others succumb to drought? New Phytologist 178: 719–739.
63. Allen CD, Breshears DD (1998) Drought-induced shift of a forest-woodland ecotone: Rapid landscape response to climate variation. Proceedings of the National Academy of Sciences of the United States of America 95: 14839–14842.
64. McDowell NG, Allen CD, Marshall L (2010) Growth, carbon-isotope discrimination, and drought-associated mortality across a Pinus ponderosa elevational transect. Global Change Biology 16: 399–415.
65. Adams HD, Kolb TE (2004) Drought responses of conifers in ecotone forests of northern Arizona: tree ring growth and leaf sigma C-13. Oecologia 140: 217–225.

6

Why Do Species Co-Occur? A Test of Alternative Hypotheses Describing Abiotic Differences in Sympatry versus Allopatry Using Spadefoot Toads

Amanda J. Chunco[1]*, Todd Jobe[2], Karin S. Pfennig[3]

1 Department of Geography, University of North Carolina at Chapel Hill, Chapel Hill, North Carolina, United States of America, 2 Signal Innovations Group, Inc., Durham, North Carolina, United States of America, 3 Department of Biology, University of North Carolina at Chapel Hill, Chapel Hill, North Carolina, United States of America

Abstract

Areas of co-occurrence between two species (sympatry) are often thought to arise in regions where abiotic conditions are conducive to both species and are therefore intermediate between regions where either species occurs alone (allopatry). Depending on historical factors or interactions between species, however, sympatry might not differ from allopatry, or, alternatively, sympatry might actually be more extreme in abiotic conditions relative to allopatry. Here, we evaluate these three hypothesized patterns for how sympatry compares to allopatry in abiotic conditions. We use two species of congeneric spadefoot toads, *Spea multiplicata* and *S. bombifrons*, as our study system. To test these hypotheses, we created ecological niche models (specifically using Maxent) for both species to create a map of the joint probability of occurrence of both species. Using the results of these models, we identified three types of locations: two where either species was predicted to occur alone (i.e., allopatry for *S. multiplicata* and allopatry for *S. bombifrons*) and one where both species were predicted to co-occur (i.e., sympatry). We then compared the abiotic environment between these three location types and found that sympatry was significantly hotter and drier than the allopatric regions. Thus, sympatry was not intermediate between the alternative allopatric sites. Instead, sympatry occurred at one extreme of the conditions occupied by both species. We hypothesize that biotic interactions in these extreme environments facilitate co-occurrence. Specifically, hybridization between *S. bombifrons* females and *S. multiplicata* males may facilitate co-occurrence by decreasing development time of tadpoles. Additionally, the presence of alternative food resources in more extreme conditions may preclude competitive exclusion of one species by the other. This work has implications for predicting how interacting species will respond to climate change, because species interactions may facilitate survival in extreme habitats.

Editor: João Pinto, Instituto de Higiene e Medicina Tropical, Portugal

Funding: This work was supported primarily by the Seeding Postdoctoral Innovators in Research and Education grant K12GM000678 from the Minority Opportunities in Research division of the National Institute of General Medical Sciences/National Institutes of Health (NIH) and the University of North Carolina (UNC) Graduate Student Opportunity Fund to AJC, with additional support provided by a grant from the National Science Foundation and a NIH Director's New Innovator Award to KP. The UNC at Cahpel Hill's Libraries provided financial support for open access publication. The funders had no role in study design, data collection and analysis, decision to publish, or preparation of the manuscript.

Competing Interests: The authors have declared that no competing interests exist.

* E-mail: chunco@email.unc.edu

Introduction

What determines whether or not closely related species co-occur? Although the forces that govern species distributions have long been a focus of ecological study [1–5], ascertaining what factors set the boundaries between closely related species is of special interest for understanding the evolutionary and ecological implications of species interactions [6–8]. Factors driving individual species ranges, let alone those driving the overlap of related species' ranges, are complex, and include abiotic factors (e.g., temperature and precipitation), and biotic factors (e.g., resource availability, competition, and predation) [3–5,9–11]. One way to evaluate why closely related species occur sympatrically in some regions but not others is to compare the abiotic conditions in sympatry versus allopatry. Doing so can provide insight into the degree to which abiotic factors, as opposed to biotic or historical factors, set the boundaries of co-occurrence between species. In particular, comparing sympatry and allopatry could support one of three hypothesized patterns of environmental variation underlying species co-occurrence. Because different types of interactions between the two species and their environment dictate each pattern, ascertaining how sympatry and allopatry differ lends insight into the types of factors driving co-occurrence of closely related species.

First, sympatry may be intermediate in abiotic environment compared to allopatry (hypothesis 1, Figure 1a). If abiotic factors are the primary drivers of species' ranges, then species should coexist wherever conditions fall within the fundamental niche of both species [4]. For example, if one species requires colder temperatures whereas the other requires warmer temperatures, then coexistence would occur at intermediate temperatures and only one species or the other will occur at more extreme temperatures (e.g. [12,13]). More generally, this pattern is expected if range margins initially arise as a response to an underlying abiotic environmental gradient [14–16].

A second pattern that can emerge in comparing the abiotic conditions of sympatry and allopatry is that sympatry may occur in

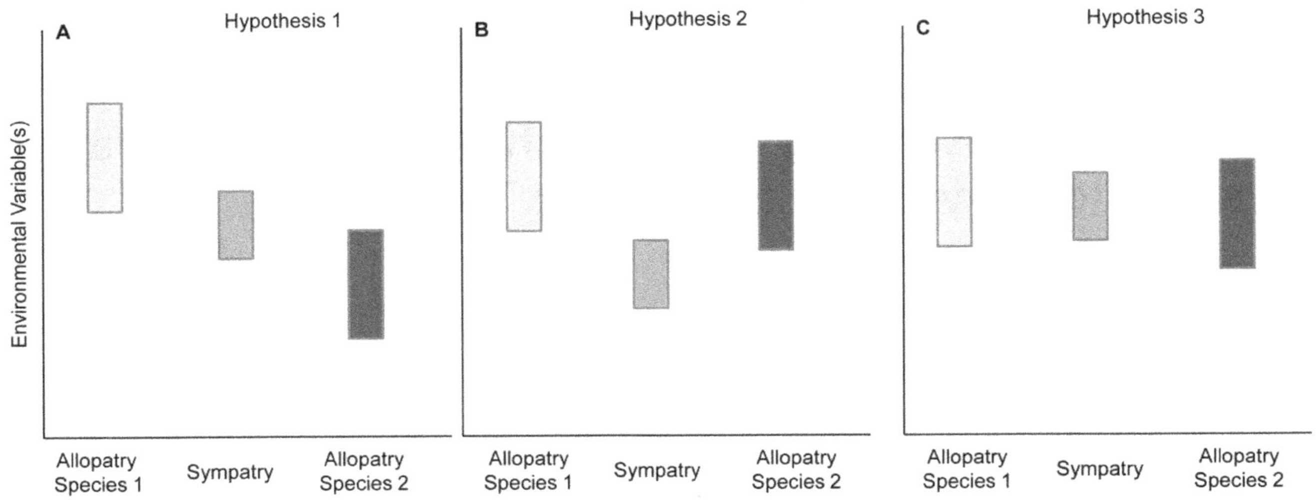

Figure 1. Alternative hypotheses for abiotic conditions underlying species distributions. A representation of three alternative patterns of environmental variation underlying sympatric and allopatric populations of two species across an environmental gradient. (a) Under hypothesis 1, species co-occur at intermediate environmental conditions where niches of the two species overlap. (b) Under hypothesis 2, biotic factors mediate co-occurrence such that species co-occur most commonly under extreme conditions. (c) Under hypothesis 3, sympatry and allopatry are governed primarily by dispersal ability, resulting in no environmental differences between sympatric and allopatric populations.

habitats that are more extreme than one of the allopatric regions (hypothesis 2, Figure 1b). In other words, sympatry may lie at one end of a continuum of environmental variable(s) as opposed to being intermediate between the two allopatric conditions. Such a pattern would emerge if *biotic* interactions mediated either, or both, species' responses to the underlying abiotic conditions in sympatry. Indeed, one species might facilitate the presence of a second species in extreme environments. This can arise with facultative mutualisms [17–19]. Similarly, hybridization may allow the colonization of extreme habitats via 'genetic facilitation' [20]. Moreover, an additional food resource may be present only in the extreme environment, or predators or parasites may be present that depress populations of one or both species, thereby precluding competitive exclusion of one species by the other [21–22]. Regardless of why sympatry occurs at one extreme of the abiotic environmental continuum, such a pattern would be primarily driven by biotic, rather than abiotic, factors.

Finally, the third pattern that could emerge in comparing abiotic conditions in sympatry and allopatry is that they do not differ (i.e. hypothesis 3, Figure 1c). Such a pattern would strongly suggest that dispersal limitation (either owing to aspects of the focal species' behaviour or physiology or due to physical barriers such as rivers or mountains) is the key factor limiting individual species' distributions within potential range boundaries [23]. Thus, sympatry and allopatry may arise owing to biogeographic history rather than an underlying environmental difference.

Distinguishing among the above patterns is important, because each hypothesis suggests how a different set of factors governs species ranges and regions of overlap between closely related species. Yet, evaluating how sympatry and allopatry differ regionally is often intractable, especially for wide ranging species. Generally, only a subset of environments within the species ranges and areas of overlap can be sampled or a small subset of variables measured. Consequently, comparing sympatry and allopatry based on field measures is often limited in the degree to which comparisons reflect range-wide patterns.

Here, we address these issues by combining niche modeling and environmental analysis to evaluate the above three alternative hypotheses for how regions of sympatry and allopatry might differ in abiotic conditions. We do so using two congeneric species of spadefoot toads, *Spea multiplicata* (Cope, 1863) and *S. bombifrons* (Cope, 1863) as our study system. In amphibians such as spadefoot toads, the abiotic environment is expected to be particularly influential on their ranges [24]. At the same time, complex biotic interactions between these species could influence their range dynamics. Indeed, as we explain in greater detail below, the tadpoles of *S. multiplicata* and *S. bombifrons* compete for resources in at least part of their range [25,26]. Consequently, the abiotic environment can also indirectly affect the distribution of these species by governing the distribution of food resources for which they compete. Moreover, as we also describe below, the two species interbreed, and hybrid fitness is determined in part by the abiotic environment [27,28]. Yet, whether such interactions scale up to affect regional patterns of sympatry and allopatry, as opposed to only affecting local distributions of the two species, is unclear. This system is therefore an excellent model for evaluating whether and how regions of sympatry and allopatry differ and can thereby provide insight into the relative importance of abiotic versus biotic forces in setting the regions of co-occurrence for closely related species.

Methods

Study system

Spea multiplicata and *S. bombifrons* inhabit arid regions of western North America [29]. Both species spend most of the year underground, and emerge to breed in ephemeral ponds that form after summer rains [30]. Because their offspring develop in ephemeral ponds, the ranges of both species should be highly sensitive to abiotic environmental conditions, such as temperature and rainfall, which affect pond duration.

Although range maps suggest a broad area of sympatry for *S. multiplicata* and *S. bombifrons* [29,31] (Figure 2), whether these species actually co-occur in the same habitat through much of their range is unclear. Within southern Arizona and New Mexico,

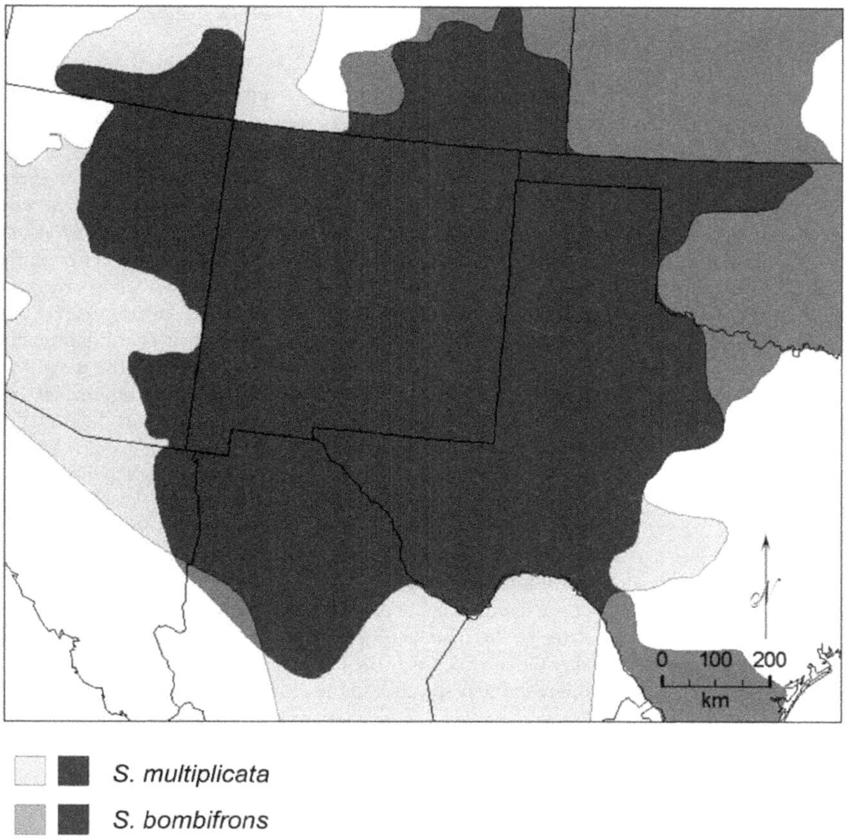

Figure 2. Range maps for *Spea multiplicata* and *S. bombifrons*. Range map showing the total range of both *S. multiplicata* and *S. bombifrons*.

for example, these species show patterns of either co-occurrence or habitat segregation along an altitudinal gradient [21,32].

Whether populations are actually sympatric or allopatric likely depends, at least in part, on interactions between species. Experimental work suggests that competition for food at the tadpole stage may, in part, drive patterns of species presence and absence [21]. Whereas adults of both species primarily feed on beetles and other small invertebrates [33], at the tadpole stage, *S. multiplicata* and *S. bombifrons* both feed on anostracan shrimp and detritus. Where both species occur together, however, *S. bombifrons* tadpoles outcompete *S. multiplicata* tadpoles for one food resource, shrimp, whereas *S. multiplicata* outcompetes *S. bombifrons* for an alternate resource, detritus [34]. When only one food resource is abundant, only the species capable of specializing on that resource is found [21]. However, when both resources are abundant, both species are also usually present [21]. Thus, these species co-occur in habitats with alternative food resources available, but competitive exclusion appears to occur when one resource is insufficient [21]. Because abiotic factors can generally influence the availability of food resources, climate could directly affect the presence and absence of the spadefoot species via physiological constraints and also indirectly affect their presence/absence because of the effects of climate on resource availability.

Reproductive interactions may also contribute to patterns of species presence and absence. These species naturally hybridize, and hybridization has been observed in areas of sympatry [32,35,36]. Hybrid tadpoles feed on detritus and anostracan fairy shrimp, as do pure species tadpoles. However, hybrids appear competitively equivalent or even superior to pure species for both resources [37]. Nevertheless, the fitness consequences of hybridization are environmentally dependent and differ for the two species [27,28]. In particular, in situations where ponds dry rapidly, hybridization is advantageous for *S. bombifrons* females, but not for *S. multiplicata* females, because hybrid offspring develop faster than pure species *S. bombifrons* offspring but slower than pure *S. multiplicata* offspring [27]. Because ponds frequently dry before tadpoles complete metamorphosis, fast development time is critical in fast drying ponds [38]. By contrast, when ponds are long lasting, hybridization is not favourable for either species because hybrid adults suffer reduced fertility [27]. Consequently, in environments where hybridization is deleterious for both species (i.e., long lasting ponds), hybridization could depress fitness and reduce population viability [39]. Alternatively, in environments where hybridization has a positive effect on *S. bombifrons*' development rate, the presence of *S. multiplicata* might actually facilitate the presence of *S. bombifrons*. Observations in the field have shown that hybridization is directional, with *S. bombifrons* females pairing with *S. multiplicata* males significantly more frequently than the reverse pairing [27]. Thus, in the spadefoots, the relative costs and benefits of hybridization in a given area—which are, in turn, determined by climate variables such a rainfall—may determine whether the two species co-occur as opposed to undergoing reproductive exclusion [10,39].

Niche modeling

To evaluate the alternative hypothesized patterns of abiotic conditions in sympatry versus allopatry (Figure 1), we first identified areas of likely co-occurrence between *S. multiplicata* and *S. bombifrons*. To do so, we independently modelled the presence of each species across their ranges using the niche modeling program MAXENT set to the default values (ver. 3.3.2, [40]). MAXENT was chosen because it demonstrates robust model performance compared to other modeling algorithms when presence-only data is available [41]. Moreover, because MAXENT requires only presence data rather than presence-absence data [40], it was ideally suited for use with museum records to identify locations of species occurrence (see below). In doing so, our aims were to: 1) identify regions where *S. multiplicata* and *S. bombifrons* are predicted to co-occur, and 2) compare environmental conditions between regions where these species are and are not predicted to coexist.

MAXENT uses species occurrence records and environmental data to build a predictive model of species distributions [40]. For species occurrence data, we compiled all *Spea bombifrons* and *S. multiplicata* records available from 29 museums throughout the United States, either via HerpNET (http:\\herpnet.org) or directly from the museums (n = 14,695; a list of the institutions that provided data is found in Table S1). Records with missing, incomplete, or inconsistent locality information or year of collection were excluded. Only one record from each unique location was used in the model, and duplicate locality records (i.e., all other records from the exact same geographic coordinates) were discarded. Each remaining record was then georeferenced, and the relative uncertainty was determined following the guidelines recommended by Chapman and Wieczorek [42]. Only museum records specific enough to identify the collection location to within one kilometer and those with a collection date between 1950 and 2000 were used in the model. This time frame corresponds with the years the climate data used to generate the WorldClim environmental layers were collected [43]. After selecting records that met the above criteria, 250 localities for *S. bombifrons* and 288 localities for *S. multiplicata* remained.

For environmental data, we initially considered 19 bioclimatic variables from WorldClim (www.worldclim.org ver. 1.4, [43]), and two hydrological variables from the U.S. Geological Survey's Hydro-1k dataset (http://eros.usgs.gov/#/Find_Data/Products_and_Data_Available/gtopo30/hydro). After removing highly correlated variables (see Methods S1), we were left with eight bioclimatic variables that were used in the final models.

Because variable selection can affect model results (e.g. [44]), we ran four separate models – using a different set of environmental variables in each model – for each species. Three models used different sets of abiotic environmental variables, whereas the fourth model used the model results for one species as the only variable in predicting the distribution of the other species (see Methods S2; Table S2). This Biotic Model was included because previous research has shown that including additional species in distribution models can improve model performance if those species interact in a biologically meaningful way [45]. After identifying the best performing model, we also performed a sensitivity analysis to evaluate the effects of the regularization multiplier (see Methods S2).

MAXENT provides a logistic output where each grid cell value is the probability of occurrence relative to a randomly selected cell based on environmental suitability. The logistic values range from 0, signifying a low probability of occurrence (i.e. low habitat suitability), to 1, signifying a high probability of occurrence (i.e. high habitat suitability) [46]. We subsequently used these values to assign records from museum specimens to being from allopatric *S. bombifrons* regions, allopatric *S. multiplicata* regions or sympatric regions (see below).

Comparing the environment in sympatry versus allopatry

We next compared the abiotic environment between regions of predicted sympatry and predicted allopatry as identified by the niche models. This allowed us to evaluate each of the three hypotheses outlined in the introduction (Figure 1), thereby providing important information about how the abiotic environment might mediate species coexistence. We were particularly interested in environmental differences between allopatric and sympatric sites in areas where species' ranges overlap (and thus areas with the potential for sympatry) rather than differences between species' range boundaries. We therefore used the localities of the museum records to construct a minimum convex polygon (i.e., a polygon described by points that fall at the outermost edge of the distribution) to define each species known range. Only museum record localities that fell within this minimum convex polygon were used in the environmental analysis. This provides a more appropriate and conservative comparison of environmental conditions between regions of sympatry and allopatry.

For each museum record that met the above criteria, we used the value of the MAXENT logistic output from the best performing abiotic model (which was the Climate-Only model; see Results S1) to designate each specific site as either: 1) predicted sympatry, 2) predicted *S. multiplicata* in allopatry, 3) predicted *S. bombifrons* in allopatry, or 4) neither species present. We used the calculations below, where P(*m*) is the logistic value for *S. multiplicata*, and P(*b*) is the logistic value for *S. bombifrons*:

1. Probability of sympatry: $P(m)xP(b)$
2. Probability of *S. multiplicata* in allopatry: $P(m)x(1-P(b))$
3. Probability of *S. bombifrons* in allopatry: $(1-P(m))xP(b)$
4. Probability of neither species present: $(1-P(m))x(1-P(b))$

Each site was assigned to one of the four location types above based on which outcome was most probable (e.g. if a site had a probability of sympatry of 0.7 while the other three categories had a probability of 0.1, the site would be assigned to sympatry). All sites were assigned to the category with the highest probability, even when the difference between two categories was small. Although this method may not correctly assign every site, it is less arbitrary than using a binomial threshold approach (e.g. assigning sites that have a logistic value of above 0.5 for both models to sympatry, [47]), and it provides a conservative measure of sympatry.

Because we were interested in the differences between regions predicted to be sympatric and those predicted to be allopatric, we restricted our analysis to sites where the niche model predicted either sympatry or allopatry. Thus, we removed locations where neither species was predicted to occur or where the locality was incorrectly assigned based on known occurrences from museum records (i.e. a locality where one species was collected that was from an area where the other species was predicted to be allopatric). *S. multiplicata* records that were identified by the model as *S. bombifrons* in allopatry were primarily found at the eastern edge of the range for *S. multiplicata*, while *S. bombifrons* records identified as *S. multiplicata* were primarily found at the western edge of the range for *S. bombifrons*. This suggests that these records might be found in sink habitats that are not favourable for long-term population persistence. Omission errors (i.e. incorrectly predicting a species' absence in areas where it is truly present) are expected in sink habitats [23]. Therefore, removing sites using the above criteria should not bias our results. Our final samples consisted of 120 *S. bombifrons* records (47 from predicted allopatry and 73 from predicted sympatry) and 170 *S. multiplicata* records (76 from predicted allopatry and 94 from predicted sympatry; the data from the 167 sympatric sites were combined for our comparisons of sympatry and allopatry below; Figure 3).

For each record designated as predicted sympatry or predicted allopatry, we extracted the value of each environmental variable used in the niche model from that location. We then compared all three location types: sympatry, allopatric *S. multiplicata*, and allopatric *S. bombifrons*. To evaluate the composite environmental differences between allopatry and sympatry, we performed a principal components analysis. We retained all principal components with an eigenvalue greater than 1 and used ANOVA to discriminate whether the location types were significantly different in terms of each retained principal component.

Finally, to further visualize the range of conditions experienced by each group, we used box-and-whisker plots to show each environmental variable individually in addition to the PCA. All statistical calculations were preformed in R v 2.10.1 [48].

Results

In generating the alternative niche models, we found that, although each model used a different subset of environmental variables, all four models showed similar areas of predicted occurrence (Figure S1, Figure S2, Figure S3). The three abiotic models showed very similar regions of moderate suitability, and these models reveal a substantially smaller region of sympatry than would be assumed based on range maps alone (Figure S1, Figure S2, Figure S3). In contrast, the Biotic Model showed a larger area of moderate habitat suitability than any of the abiotic models (Figure S1, Figure S2, Figure S3). All maps shown are the average of the ten replicate runs for each model.

The abiotic model with the best performance in terms of AUC included only the eight climate variables (i.e, the Climate-Only Model, Table S3), and the results of this model were used to evaluate sympatry and allopatry (Figure 3). The sensitivity analysis showed the effect of the regularization multiplier (Figure S4, Table S4). Although the regularization multiplier influences the extent of highly suitable habitat predicted by the model, the major areas of sympatry are the same among all models. Because Philips and Dudik [46] suggest that the default values of Maxent are appropriate for a variety of conditions, we used the default regularization multiplier value of 1.0 for our analysis of sympatry and allopatry.

As evidence that these models were good descriptors of sympatry and allopatry, nearest neighbour distances to the closest record of the other species were significantly closer in predicted sympatry (mean distance = 7.02 km) than predicted allopatry

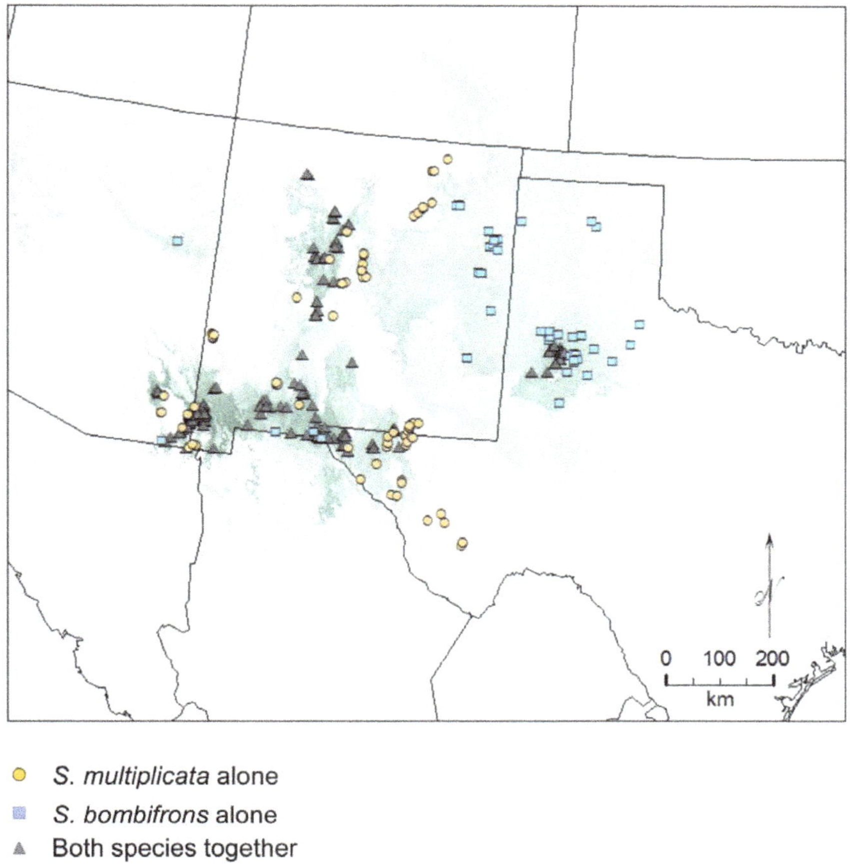

Figure 3. Range maps of predicted sympatry. Range maps of predicted sympatry between *Spea multiplicata* and *S. bombifrons*. The value for each 1 km sq pixel was calculated by multiplying the logistic value of both species, and values range from 0 (white) to 1 (dark green). Sites used in the environmental analysis are indicated by points. Specifically, blue squares represented collection locations for *S. bombifrons* that occurred in areas predicted to be allopatric for that species; orange circles represent collection locations for *S. multiplicata* records that were predicted to be allopatric for that species, whereas gray triangles represent collection locations for either species in areas predicted to be sympatric.

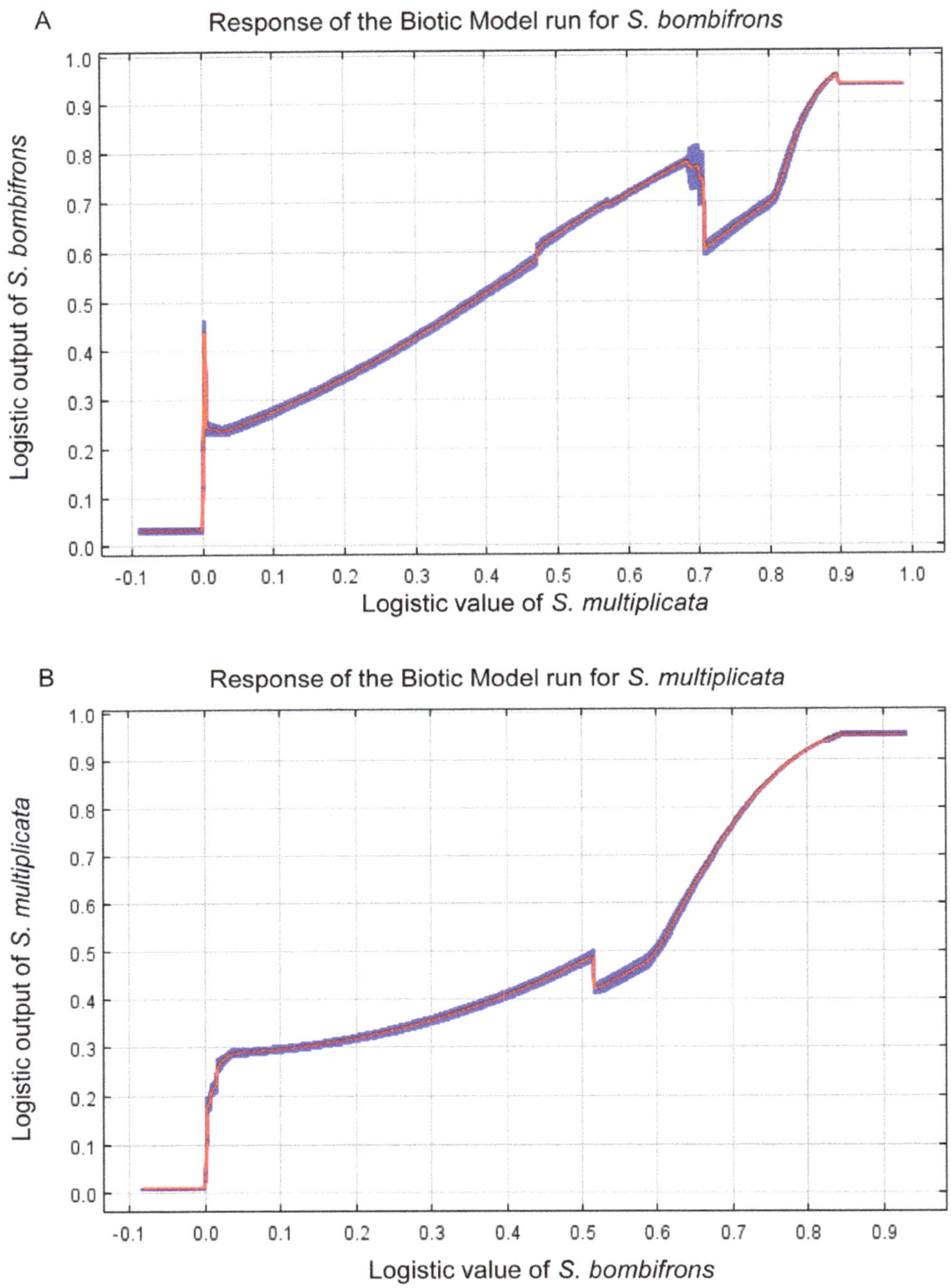

Figure 4. Maxent response curves for the Biotic Model. The response curves of the Biotic Model for (a) *S. bombifrons* and (b) *S. multiplicata*. These curves show how the logistic output changes along an 'environmental gradient'. Here, the environmental gradient is the predicted output of the other species used to create the Biotic Model. The red line shows the average of the 10 replicate runs, while the blue bands shows +/− one standard deviation. At low logistic values for one species, the other species has a low logistic value as well. Both species thus show a similar response to the environment (i.e. environments good for one species tend to be good for the other).

(mean = 20.33 km) for *S. bombifrons* ($t_{118} = 3.785$, $p<0.001$,) and, likewise, for *S. multiplicata* (mean in sympatry = 12.63 km, mean in allopatry = 51.52 km, $t_{168} = 7.659$, $p<0.001$). Moreover, most sites showed a substantial difference between the highest logistic value (i.e. the value used to assign a site to a particular geographic category) and the second highest value. The average logistic value (+/−SD) used to assign a site was 0.46 (+/− 0.13), whereas the average second highest logistic value (i.e. the next most likely geographic category) was 0.28 (+/− 0.06). Thus, assignments to predicted regions of sympatry or allopatry were not based on marginal differences in likelihood: most sites were unequivocally assigned to a particular geographic category.

The Biotic Model evaluated whether the predicted presence of one species could predict the presence of the other species. Examining the response curves for the Biotic Model shows how the logistic output changes along the environmental gradient

Table 1. Loadings on the first two principal components for the eight environmental variables used in the MAXENT model.

	PCA 1	PCA2
Mean Diurnal Range in Temperature	*	0.143
Maximum Temperature of Warmest Month	0.160	−0.176
Annual Range in Temperature	*	0.279
Mean Temperature of Driest Quarter	0.914	−0.169
Mean Temperature of Coldest Quarter	0.243	−0.158
Precipitation of Driest Quarter	*	*
Precipitation of Warmest Quarter	−0.214	−0.896
Precipitation of Coldest Quarter	0.105	−0.177

Loadings near 0 (i.e., −0.1 to 0.1) are denoted with an "*".

(where the "environmental gradient" is the predicted output of the other species). We found that the Biotic Model performed reasonably well, and that the relationship was roughly positively linear (Figure 4). At low logistic values for one species, the other has a low logistic value as well, so both species show a roughly similar response to the environment (Figure 4). These results therefore indicate that the habitat requirements between the two species are similar. Nevertheless, this Biotic Model is the poorest performing model of the four that we considered. Moreover, the predicted areas of species presence were greater than in any of the other models, suggesting some over-prediction. This indicates that, although the requirements of *S. multiplicata* and *S. bombifrons* are similar, there are important differences between them in how they respond to the environment. Thus, models that include climate variables are still a better approach than using only the presence of one species to predict the distribution of the other.

To evaluate the combined differences in abiotic variables between sympatry and allopatry, we first used a principal component analysis (PCA). We retained the first two principal components (PCs), which, together, explained 87.5% of the variance. The loadings of each environmental variable on the first two principal components are shown in Table 1. When we contrasted these principal components between sympatry, allopatry for *S. bombifrons*, and allopatry for *S. multiplicata* using an ANOVA, we found a significant effect of region on both PC scores (PC 1: $F(2, 287) = 35.522$, $p<0.001$; PC 2: $F(2, 287) = 21.678$, $p<0.001$). A Tukey HSD test further revealed that all groups were significantly different from each other for both PC 1 and PC 2 at $p<0.02$ for all group comparisons. Moreover, the mean sympatric score for both principal components was greater than the mean scores for either allopatric region, indicating that sympatry occurs in extreme, rather than intermediate, habitats relative to allopatric regions (Figure 5). Because temperature in the driest quarter and precipitation in the warmest quarter loaded most strongly on PC 1 and PC 2 respectively (Table 1), our results indicate that sympatry is warmer and drier than either allopatric region (Figure 5). This pattern of sympatry being extreme in abiotic conditions, rather than intermediate relative to allopatric regions, is most consistent with hypothesis 2 in Figure 1.

That sympatry occurs in regions at an extreme in abiotic conditions relative to allopatric regions is further emphasized when evaluating the individual variables used in the PCA. Allopatric *S. bombifrons* sites are wetter, cooler, and less variable in diurnal temperature range than allopatric *S. multiplicata* sites (Figure 6). Contrary to hypothesis 1, sympatry was not intermediate for any

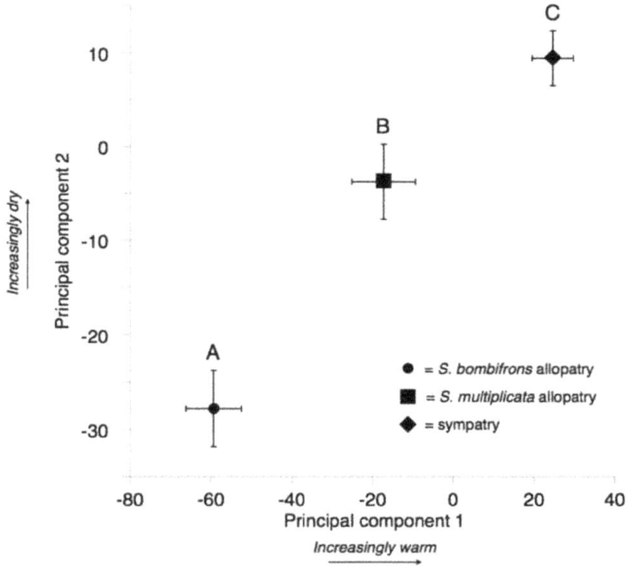

Figure 5. Principal components of the abiotic environment. Means (+/− s.e.) for the first two principal components describing variation in the eight environmental variables used to build ecological niche models. Different letters indicate significantly different means; each group (*S. multiplicata* in allopatry, *S. bombifrons* in allopatry, and sympatry), is significantly different from the other two.

of the individual environmental variables. Instead, most variables (including: mean temperature of the coldest quarter, precipitation of the driest quarter, precipitation of the warmest quarter, maximum temperature of the warmest month, and mean temperature of the driest quarter) show median values for sympatry that are at an extreme relative to either species in allopatry (Figure 6). Generally, sympatry tended to be hotter and drier than allopatric sites for either species. Of all the environmental variables, only precipitation of the coldest quarter and annual range in temperature had median values that were similar between both species in sympatry and allopatry (Figure 6). Thus, contrary to commonly held views of sympatry and allopatry, sympatric regions between *S. bombifrons* and *S. multiplicata* were at an extreme of the abiotic environment relative to regions of allopatry.

Summary

We used spadefoot toads (*Spea* spp.) as a case study for examining the factors potentially contributing to patterns of distribution and co-occurrence. We did so by constructing an independent predictive map of the entire range for *S. multiplicata* and *S. bombifrons*. We then contrasted the abiotic environment in areas where the two species were likely to co-occur (predicted sympatry) versus areas where each species was likely to occur alone (predicted allopatry).

The resulting niche models showed that sympatry is geographically interspersed within regions of allopatry for both species (Figure 3). Thus, we do not find a gradient of predominantly *S. bombifrons* habitat, then sympatry, then *S. multiplicata* habitat along a north-south axis, as might be expected in a contact zone between two species that differ latitudinally in distribution. Moreover, when we compared abiotic conditions between predicted sympatric sites and predicted allopatric sites, we found striking differences. Contrary to the common expectation that sympatric sites will be

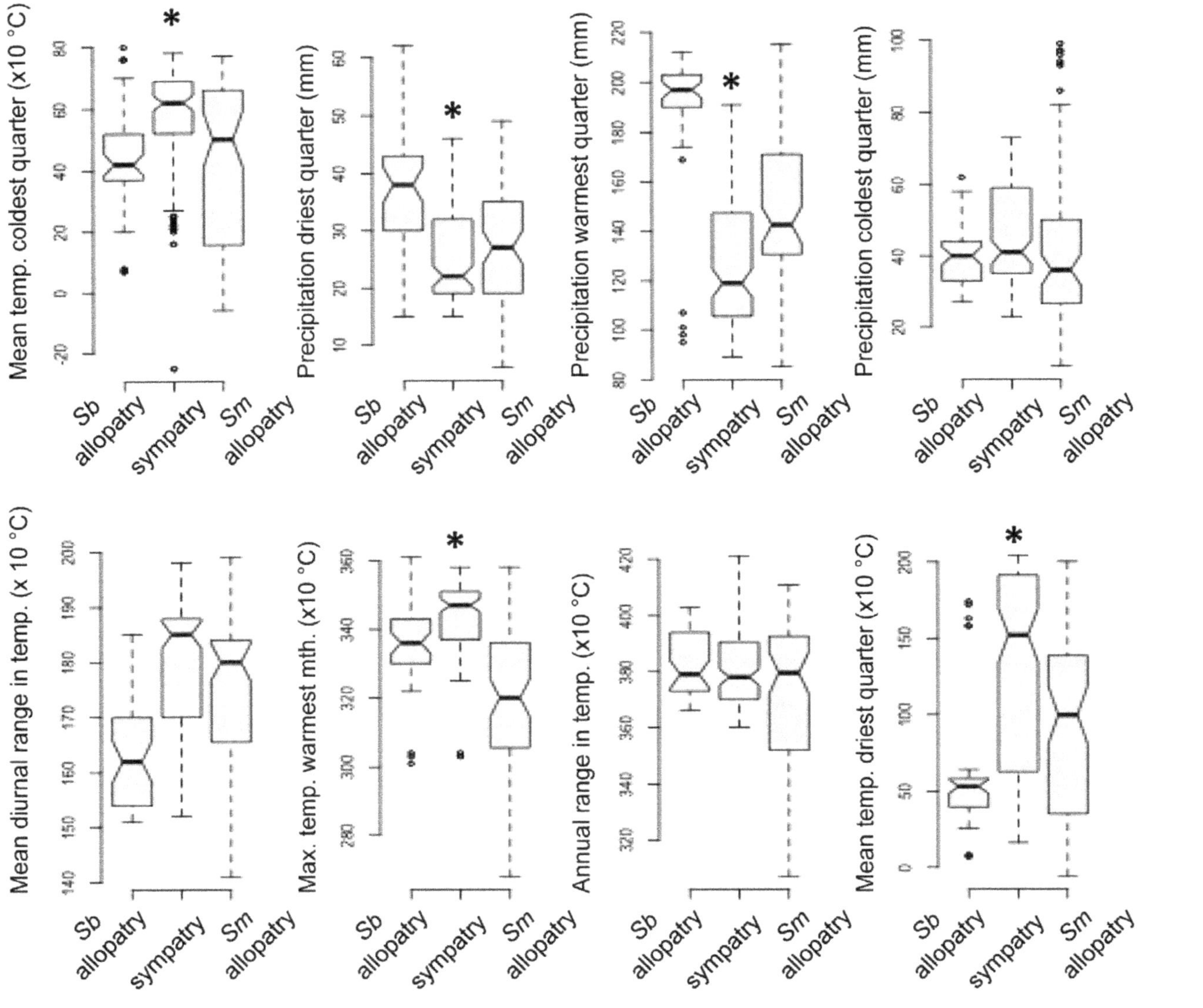

Figure 6. Environmental variation between sympatry and allopatry. Box-and-whisker plots showing environmental space occupied by predicted allopatric populations of *S. bombifrons* (abbreviated "*Sb* allopatry"), predicted allopatric populations of *S. multiplicata* (abbreviated "*Sm* allopatry"), and predicted sympatric populations of both species for each environmental variable used in the Maxent models. Non-overlapping notches are roughly equivalent to 95% confidence intervals, and therefore provide strong evidence that the medians differ [48,58]. An "*" above the sympatry box indicates variables for which sympatric sites are significantly hotter or drier than both allopatric sites.

intermediate between allopatric sites in environmental variables [14,16]; hypothesis 1 from Figure 1), we instead found sympatry to generally occupy more extreme ends of the distributions for the abiotic variables used in our study (Figures 5, 6). Specifically, sympatric sites tend to be hotter and drier than allopatric sites for either species. Allopatric *S. bombifrons* sites are substantially cooler and wetter than sympatric sites, whereas allopatric *S. multiplicata* sites appear more similar to sympatry than to allopatric *S. bombifrons* sites (Figures 5, 6).

Discussion

Although a few studies have previously used MAXENT to study patterns of species co-occurrence (e.g. [49,50]), these studies have primarily relied on a threshold approach (i.e. using a predetermined cut-off in logistic value to identify presence and absence). The specific areas of predicted presence with the threshold approach depend, however, on the specific threshold chosen [40]. Instead, using the joint probability to highlight areas of co-occurrence, as we did here, provides several advantages over a threshold approach. First, joint probability is less arbitrary than choosing a binomial threshold to predict sympatry. Second, this approach provides a continuous measure of the probability of sympatry across the region of interest rather than a binomial measure of presence or absence (e.g. Figure 2). The continuous measure provides more information about the relative likelihood of co-occurrence, and is therefore potentially more useful for targeting field surveys. It is important to note that while MAXENT highlights suitable environments for occurrence, suitable environment does not necessarily guarantee occurrence. Thus, some over-prediction is likely. However, given that these models perform well, and that multiple modeling approaches highlight similar areas of habitat suitability, these results will provide useful data for identifying new populations in the field.

The presence of sympatry in extreme rather than intermediate habitats (relative to allopatry) suggests that simple responses to the abiotic environment are not the primary factors mediating co-occurrence between *S. bombifrons* and *S. multiplicata*. Instead, our results suggest that biotic interactions within these extreme environments may be important in driving co-occurrence between *S. bombifrons* and *S. multiplicata*. In this case, the most likely biotic factor that could explain our results is hybridization. The two species are known to hybridize naturally [27,32,35], and *S. bombifrons* benefits by hybridizing in rapidly drying pools [28]. Specifically, hybrids between *S. bombifrons* and *S. multiplicata* develop faster than pure species *S. bombifrons* [27]. Beneficial hybridization may foster co-occurrence in drier, warmer habitats if hybridization allows *S. bombifrons* to maintain populations in habitats where pure species would otherwise be unable to persist. That the environment of sympatry tends to be more similar to *S. multiplicata* in allopatry than to *S. bombifrons* in allopatry is further consistent with the idea that hybridization with *S. multiplicata* facilitates the presence of *S. bombifrons* in warm, dry habitats that are dissimilar to sites found for allopatric *S. bombifrons*. Indeed, genetic data suggests that *S. bombifrons* expanded its range out of the Great Plains into the Southwestern USA [51]; our results suggest that hybridization may have fuelled this expansion into novel habitats.

An alternative (albeit not mutually exclusive) explanation for our results is that sympatry occurs where resources are available that foster coexistence. Indeed, the pattern of sympatry occurring in more extreme habitats could arise if such habitats foster the presence of alternative resources that minimize competition between ecologically similar species. Because *S. multiplicata* and *S. bombifrons* specialize on different resources (detritus and shrimp respectively) as tadpoles where they co-occur [25,34], sympatric regions might occur in a more extreme environment than one of the allopatric regions if both resources occur in those environments. For example, anostracan fairy shrimp on which the tadpoles feed are potentially more abundant in hotter, drier habitats [52]. If such habitats contain sufficient detritus, then the presence of both resources might preclude competitive exclusion of one species by the other. Consistent with this notion, field observations have shown that local co-occurrence of both *Spea* species occurs only where both shrimp and detritus are sufficiently abundant to permit co-existence rather than competitive exclusion with southern Arizona and New Mexico [21].

Whether hybridization or the presence of sufficient resources (or both factors in combination) fosters sympatry between *S. multiplicata* and *S. bombifrons* will require further investigation. Because MAXENT uses species locality data in predicting ranges, any factors that limit species ranges are indirectly incorporated into the model. Therefore, although we used climate to predict the ranges of both species, any factors (such as resources) that are themselves tightly correlated with climate will be indirectly included in the model. Using our models, we therefore cannot determine whether climate *directly* mediates co-occurrence (with hybridization facilitating *S. bombifrons*' persistence in hotter, drier habitats) or whether climate *indirectly* mediates coexistence via its effects on resources that mediate competition. Nevertheless, the results of this study can be used to guide empirical work testing specific predictions that arise from environmental comparisons of sympatry and allopatry. More generally, this approach of blending niche modeling with analysis of environmental data can guide greater insight into the factors that affect species co-occurrence.

These results also have implications for considering how the environment mediates species interactions when predicting responses to climate change. Many current models of future species' distributions assume that range limits are driven primarily by climate. Thus, although the potential for novel communities to form under future climate regimes has been well-documented [53,54], few studies have explicitly considered how species interactions will be influenced by the environment (but see [55]). Here, we show that sympatric sites occur in regions that are hotter and drier than allopatric sites of either *S. bombifrons* or *S. multiplicata*. Because the U.S. southwest, where these *Spea spp.* are found, is predicted to become hotter and drier as climate change progresses [56], sympatry may become more extensive if the environment becomes more suitable for species co-occurrence. Indeed, if hybridization fosters *S. bombifrons*' ability to invade such habitat, such changes could alter both patterns of co-occurrence and genetic exchange between the two species. Thus, climate change may alter both evolutionary processes and ecological processes in these species.

Furthermore, these kinds of non-intuitive relationships between the environment and species interactions may explain in part why many studies of recent range changes have shown higher than expected variation in the response to climate change, with as many as ~40% of species showing either no changes in range limits or a change in the opposite direction than predicted [57]. Therefore, future studies should more explicitly consider how the environment mediates species interactions when predicting responses to climate change.

Supporting Information

Methods S1 A description of the process used to select environmental layers for niche model construction.

Methods S2 A description of: 1) each of the four niche models run, 2) the methodology for identifying areas of sympatry, and 3) the sensitivity analysis.

Results S1 A description of the result of each of the four niche models run and the results of the sensitivity analysis.

Figure S1 Maxent models for *S. bombifrons*. Map of *S. bombifrons* under all 4 models. Each pixel visually represents the logistic value for *S. bombifrons* for each of the 4 models, where values range from 0 (shown in white) to 1 (shown in dark blue). The four models are: A) the Full Abiotic Model, B) the Climate-Only Model, C) the Summer Environment & Seasonality Model, and D) the Biotic Model.

Figure S2 Maxent models for *S. multiplicata*. Map of *S. multiplicata* under all 4 models. Each pixel visually represents the logistic value for *S. multiplicata* for each of the 4 models, where values range from 0 (shown in white) to 1 (shown in dark orange). The four models are: A) the Full Abiotic Model, B) the Climate-Only Model, C) the Summer Environment & Seasonality Model, and D) the Biotic Model.

Figure S3 Maxent models of sympatry. Map of predicted sympatry under all 4 models. The value for each pixel was calculated by multiplying the logistic value of both species for each model, and values range from 0 (shown in white) to 1 (shown in dark green). The four models are: A) the Full Abiotic Model, B) the Climate-Only Model, C) the Summer Environment & Seasonality Model, and D) the Biotic Model.

Figure S4 Sensitivity analysis. Maxent model results at four different regularization multipliers: A) 0.1, B) 0.5, C) 2.0, and D) 5.0. Each of these models was run based on the same environmental variables included in the Climate-Only model. Other than the regularization multiplier, all other values were held at default levels. Each map shows the average of 10 replicate runs.

Table S1 The following institutions provided locality data used in the ecological niche models.

Table S2 Environmental variables used in each MAXENT model.

Table S3 Mean and standard deviation for each of the four niche models run.

Table S4 Mean and standard deviation for each of the four niche models run for the sensitivity analysis.

Acknowledgments

We first thank the many museums and curators who were essential to this project. A list of museums that contributed data is found in Table S1. Amanda Henley provided technical help on using ARCGIS. Jennifer Costanza, Matt Simon, Anne Trainor, and Ian Breckheimer provided feedback on MAXENT results. Maria Servedio, Aaron Moody, Joel Kingsolver, and David Pfennig greatly contributed to the development of this project. Finally, David Pfennig, David Kikuchi, Juan Santos, Sumit Dhole, Antonio Serrato, Verónica Rodriguez Moncalvo, Lisa Bono, Emily Schmidt, Michelle Kovarik, Brooke Wheeler, Jill Hurst, and Martin Ferris provided comments on the manuscript.

Author Contributions

Conceived and designed the experiments: AJC KP TJ. Performed the experiments: AJC TJ. Analyzed the data: AJC TJ KP. Contributed reagents/materials/analysis tools: AJC. Wrote the paper: AJC KP.

References

1. Merriam CH (1894) Laws of temperature control the geographic distribution of terrestrial animals and plants. National Geographic 6: 229–238.
2. Griggs RF (1914) Observations on the behavior of some species at the edges of their ranges. Bulletin of the Torrey Botanical Club 41: 25–49.
3. Grinnell J (1917) Field tests of theories concerning distributional control. Am Nat 51: 115–128.
4. Hutchinson GE (1957) Population studies - animal ecology and demography - concluding remarks. Cold Spring Harb Symp Quant Biol 22: 415–427.
5. Gaston KJ (2003) The structure and dynamics of geographic ranges. Oxford: Oxford University Press.
6. Darwin C (1859) On the origin of species by means of natural selection. London: John Murry.
7. Connell JH (1961) The influence of interspecific competition and other factors on the distribution of the barnacle *Chthamalus stellatus*. Ecology 42: 710–723.
8. Brown JH (1984) On the relationship between abundance and distribution of species. Am Nat 124: 255–279.
9. Krebs CJ (1978) Ecology: The experimental analysis of distribution and abundance. New York, NY: Harper & Row. 678 p.
10. Gröning J, Hochkirch A (2008) Reproductive interference between animal species. Q Rev Biol 83: 257–282.
11. Pfennig K, Pfennig D (2009) Character displacement: Ecological and reproductive responses to a common evolutionary problem. Q Rev Biol 84: 253–276.
12. Swenson NG (2006) GIS-based niche models reveal unifying climatic mechanisms that maintain the location of avian hybrid zones in a North American suture zone. J Evol Biol 19: 717–725. 10.1111/j.1420-9101.2005.01066.x.
13. Walls SC (2009) The role of climate in the dynamics of a hybrid zone in Appalachian salamanders. Global Change Biol 15: 1903–1910. 10.1111/j.1365-2486.2009.01867.x.
14. Hewitt GM (1996) Some genetic consequences of ice ages, and their role in divergence and speciation. Biol J Linn Soc 58: 247–276. 10.1111/j.1095-8312.1996.tb01434.x.
15. Brown JH, Stevens GC, Kaufman DM (1996) The geographic range: Size, shape, boundaries, and internal structure. Annu Rev Ecol Syst 27: 597–623. 10.1146/annurev.ecolsys.27.1.597.
16. Guisan A, Theurillat J, Kienast F (1998) Predicting the potential distribution of plant species in an alpine environment. Journal of Vegetation Science 9: 65–74. 10.2307/3237224.
17. Bertness MD, Callaway R (1994) Positive interactions in communities. Trends in Ecology & Evolution 9: 191–193. 10.1016/0169-5347(94)90088-4.
18. Callaway RM, Brooker RW, Choler P, Kikvidze Z, Lortie CJ, et al. (2002) Positive interactions among alpine plants increase with stress. Nature 417: 844–848.
19. Bruno JF, Stachowicz JJ, Bertness MD (2003) Inclusion of facilitation into ecological theory. Trends in Ecology & Evolution 18: 119–125. DOI: 10.1016/S0169-5347(02)00045-9.
20. Figueroa ME, Castillo JM, Redondo S, Luque T, Castellanos EM, et al. (2003) Facilitated invasion by hybridization of *Sarcocornia* species in a salt-marsh succession. J Ecol 91(4): 616–626. 10.1046/j.1365-2745.2003.00794.x.
21. Pfennig DW, Rice AM, Martin RA (2006) Ecological opportunity and phenotypic plasticity interact to promote character displacement and species coexistence. Ecology 87: 769–779.
22. Hubbell SP (2006) Neutral theory and the evolution of ecological equivalence. Ecology 87: 1387–1398. 10.1890/0012-9658(2006)87[1387:NTATEO]2.0.CO;2.
23. Guisan A, Thuiller W (2005) Predicting species distribution: Offering more than simple habitat models. Ecol Lett 8: 993–1009. 10.1111/j.1461-0248.2005.00792.x.
24. Rodríguez MÁ, Belmontes JA, Hawkins BA (2005) Energy, water and large-scale patterns of reptile and amphibian species richness in Europe. Acta Oecol 28: 65–70. DOI: 10.1016/j.actao.2005.02.006.
25. Pfennig DW, Murphy PJ (2002) How fluctuating competition and phenotypic plasticity mediate species divergence. Evolution 56: 1217–1228.
26. Pfennig DW, Rice AM, Martin RA (2007) Field and experimental evidence for competition's role in phenotypic divergence. Evolution 61: 257–271. 10.1111/j.1558-5646.2007.00034.x.
27. Pfennig KS, Simovich MA (2002) Differential selection to avoid hybridization in two toad species. Evolution 56: 1840–1848.
28. Pfennig KS (2007) Facultative mate choice drives adaptive hybridization. Science 318: 965–967. 10.1126/science.1146035.
29. Stebbins RC (1985) A field guide to western reptiles and amphibians: Field marks of all species in western North America, including Baja California. Boston, MA: Houghton Mifflin Company. 336 p.
30. Bragg AN (1965) Gnomes of the night: The spadefoot toads. Philadelphia, Pennsylvania: University of Pennsylvania Press.
31. Elliott L, Gerhardt C, Davidson C (2009) The frogs and toads of North America: A comprehensive guide to their identification, behavior, and calls. Boston: Houghton Mifflin. 343 p.
32. Simovich MA (1985) Analysis of a hybrid zone between the spadefoot toads *Scaphiopus multiplicatus* and *Scaphiopus bombifrons*. Ph.D. Dissertation. University of California, Riverside.
33. Anderson AM, Haukos DA, Anderson JT (1999) Diet composition of three Anurans from the playa wetlands of northwest Texas. Copeia 1999: 515–520.
34. Pfennig DW, Murphy PJ (2000) Character displacement in polyphenic tadpoles. Evolution 54: 1738–1749.
35. Sattler PW (1985) Introgressive hybridization between the spadefoot toads *Scaphiopus bombifrons* and *S. multiplicatus* (Salientia: Pelobatidae). Copeia 1985: 324–332.
36. Pfennig KS (2003) A test of alternative hypotheses for the evolution of reproductive isolation between spadefoot toads: Support for the reinforcement hypothesis. Evolution 57: 2842–2851.
37. Pfennig KS, Chunco AJ, Lackey ACR (2007) Ecological selection and hybrid fitness: Hybrids succeed on parental resources. Evol Ecol Res 9: 341–354.
38. Pfennig DW (1992) Polyphenism in spadefoot toad tadpoles as a locally adjusted evolutionarily stable strategy. Evolution 46: 1408–1420.
39. Pfennig KS, Pfennig DW (2005) Character displacement as the "best of a bad situation": Fitness trade-offs resulting from selection to minimize resource and mate competition. Evolution 59: 2200–2208.
40. Phillips SJ, Anderson RP, Schapire RE (2006) Maximum entropy modeling of species geographic distributions. Ecol Model 190: 231–259. 10.1016/j.ecolmodel.2005.03.026.
41. Elith J, Graham CH, Anderson RP, Dudík M, Ferrier S, et al. (2006) Novel methods improve prediction of species' distributions from occurrence data. Ecography 29: 129–151. 10.1111/j.2006.0906-7590.04596.x.

42. Chapman AD, Wieczorek J (2006) Guide to best practices for georeferencing. Copenhagen: Global Biodiversity Information Facility.
43. Hijmans RJ, Cameron SE, Parra JL, Jones PG, Jarvis A (2005) Very high resolution interpolated climate surfaces for global land areas. Int J Climatol 25: 1965–1978. 10.1002/joc.1276.
44. Roedder D, Loetters S (2009) Niche shift versus niche conservatism? Climatic characteristics of the native and invasive ranges of the mediterranean house gecko (*Hemidactylus turcicus*). Global Ecol Biogeogr 18: 674–687. 10.1111/j.1466-8238.2009.00477.x.
45. Heikkinen RK, Luoto M, Virkkala R, Pearson RG, Körber J (2007) Biotic interactions improve prediction of boreal bird distributions at macro-scales. Global Ecology & Biogeography 16: 754–763. 10.1111/j.1466-8238.2007.00345.x.
46. Phillips SJ, Dudik M (2008) Modeling of species distributions with Maxent: New extensions and a comprehensive evaluation. Ecography 31: 161–175. 10.1111/j.0906-7590.2008.5203.x.
47. Fielding AH, Bell JF (1997) A review of methods for the assessment of prediction errors in conservation presence/absence models. Environ Conserv 24: 38–49.
48. R Development Core Team. R Foundation for Statistical Computing (2005) R: A language and environment for statistical computing. 2.11.1. ISBN 3-900051-07-0.
49. Brito JC, Acosta AL, Álvares F, Cuzin F (2009) Biogeography and conservation of taxa from remote regions: An application of ecological-niche based models and GIS to North-African canids. Biol Conserv 142: 3020–3029. DOI: 10.1016/j.biocon.2009.08.001.
50. Martínez-Freiría F, Sillero N, Lizana M, Brito JC (2008) GIS-based niche models identify environmental correlates sustaining a contact zone between three species of European vipers. Divers Distrib 14: 452–461. 10.1111/j.1472-4642.2007.00446.x.
51. Rice AM, Pfennig DW (2008) Analysis of range expansion in two species undergoing character displacement: Why might invaders generally 'win' during character displacement? J Evol Biol 21: 696–704. 10.1111/j.1420-9101.2008.01518.x.
52. Pfennig D (1990) The adaptive significance of an environmentally-cued developmental switch in an Anuran tadpole. Oecologia 85: 101–107. 10.1007/BF00317349.
53. Williams JW, Jackson ST (2007) Novel climates, no-analog communities, and ecological surprises. Frontiers in Ecology and the Environment 5: 475–482.
54. Williams JW, Jackson ST, Kutzbach JE (2007) Projected distributions of novel and disappearing climates by 2100 AD. Proc Natl Acad Sci U S A 104: 5738–5742.
55. Jankowski JE, Robinson SK, Levey DJ (2010) Squeezed at the top: Interspecific aggression may constrain elevational ranges in tropical birds. Ecology 91: 1877–1884.
56. Seager R, Ting M, Held I, Kushnir Y, Lu J, et al. (2007) Model projections of an imminent transition to a more arid climate in southwestern North America. Science 316: 1181–1184. 10.1126/science.1139601.
57. La Sorte FA, Thompson FR (2007) Poleward shifts in winter ranges of North American birds. Ecology 88: 1803–1812.
58. Chambers JM, Cleveland WS, Kleiner B, Tukey PA (1983) Graphical methods for data analysis. Wadsworth and Brooks/Cole.

7

Mapping, Bayesian Geostatistical Analysis and Spatial Prediction of Lymphatic Filariasis Prevalence in Africa

Hannah Slater[1], Edwin Michael[2]*

1 Department of Infectious Disease Epidemiology, Imperial College London, St. Mary's Campus, Norfolk Place, London, United Kingdom, 2 Department of Biological Sciences, University of Notre Dame, Notre Dame, Indiana, United States of America

Abstract

There is increasing interest to control or eradicate the major neglected tropical diseases. Accurate modelling of the geographic distributions of parasitic infections will be crucial to this endeavour. We used 664 community level infection prevalence data collated from the published literature in conjunction with eight environmental variables, altitude and population density, and a multivariate Bayesian generalized linear spatial model that allows explicit accounting for spatial autocorrelation and incorporation of uncertainty in input data and model parameters, to construct the first spatially-explicit map describing LF prevalence distribution in Africa. We also ran the best-fit model against predictions made by the HADCM3 and CCCMA climate models for 2050 to predict the likely distributions of LF under future climate and population changes. We show that LF prevalence is strongly influenced by spatial autocorrelation between locations but is only weakly associated with environmental covariates. Infection prevalence, however, is found to be related to variations in population density. All associations with key environmental/demographic variables appear to be complex and non-linear. LF prevalence is predicted to be highly heterogenous across Africa, with high prevalences (>20%) estimated to occur primarily along coastal West and East Africa, and lowest prevalences predicted for the central part of the continent. Error maps, however, indicate a need for further surveys to overcome problems with data scarcity in the latter and other regions. Analysis of future changes in prevalence indicates that population growth rather than climate change *per se* will represent the dominant factor in the predicted increase/decrease and spread of LF on the continent. We indicate that these results could play an important role in aiding the development of strategies that are best able to achieve the goals of parasite elimination locally and globally in a manner that may also account for the effects of future climate change on parasitic infection.

Editor: Michael George Roberts, Massey University, New Zealand

Funding: The authors acknowledge the Natural Environment Research Council (NERC), United Kingdom, for a NERC/Economic and Social Research Council Interdisciplinary Research Studentship to HS. The authors also acknowledge the Grantham Institute for Climate Change at Imperial College London, and the National Institutes of Health, United States of America (under grant number RO1 AI069387-01A1), for partial support of this work. The funders had no role in study design, data collection and analysis, decision to publish, or preparation of the manuscript.

Competing Interests: The authors have declared that no competing interests exist.

* E-mail: emichael@nd.edu

Introduction

Recently, there has been increasing scientific interest in acquiring a better understanding of the spatial distributions of parasitic infections [1–13]. First, such work, by detailing the distribution and severity of diseases, is important for guiding the planning of control programmes [14–17]. Maps of prevalence or intensity of infection, for example, can enable a more precise stratification of disease risk faced by communities, which can in turn allow more reliable spatial planning of intervention efforts as well as identification of the worst affected areas for prioritizing these efforts [4,8,15–16,18]. Second, mapping studies can, by combining spatial data on parasite prevalence with geographic information on biotic and abiotic ecological variables, be used to explore the underlying causes of infection risk, thus improving our understanding of the transmission ecology of parasitic infections [1,19–22]. Finally, understanding the relationships between mapped disease prevalences and environmental/climactic factors is useful for examining how climate change may affect the long-term transmission and distribution of diseases [1].

Lymphatic filariasis (LF) is a major vector-borne parasitic disease endemic to the tropics, including sub-Saharan Africa. It is thought to represent the second largest health burden of any vector-borne disease worldwide [23], and this, together with improvements in drug-based treatments and diagnostic tools [24], has led to LF being considered as one of only six infectious diseases that could be "eradicable" or "potentially eradicable" at the global level [25]. This conclusion followed by adoption of Resolution WHA50.29 by the World Health Assembly in 1997 calling for the elimination of LF as a public health problem, has resulted in the rapid implementation of large-scale national mass drug administration programmes in all endemic regions of the world [26–27]. This progress has led to impressive reductions in infection in treated communities, but has also made gaining a better understanding of the geographic distribution of LF prevalence a key requirement for more effectively guiding successful elimination activities across endemic regions.

Previous attempts to map LF in Africa have estimated infection distribution or prevalence across specific regions or countries by mapping of infection sites either as points or as ranges interpolated between such points [22,28–32], quantifying spatial patterns using geostatistical approaches [3,33] or smoothing aggregated prevalence data across areas to produce national level maps [34–37]. By contrast, Lindsay and Thomas [38] used logistic regression with environmental covariates and LF presense/absence data to predict the probability of LF presence across Africa, which was used to improve estimations of the number of people living in at-risk areas.

The importance of explicitly accounting for spatial effects when attempting to map infection data has been emphasized recently [6,39–40], and a variety of methods for dealing with spatially correlated data has been suggested, within both the frequentist [41] and Bayesian frameworks [42–44]. Generalised linear models with spatially correlated random effects are commonly used in this context [39], and such generalized linear spatial models (GLSMs) have been used to map a wide range of phenomena, including germinating seeds [45], root rot [46] and bird populations [47]. They have also been increasingly used to model and predict disease prevalences, including schistosomiasis [4,7,48], pseudorabies virus [49], trypanosomiasis [9], and malaria [18,50]. Despite this trend, no study thus far has modelled the prevalence of LF across an endemic region by addressing both environmental covariates and spatial effects together. This is in spite of the fact that 1) the spatial distribution of LF has been shown to be highly non-homogenous between and within countries [33–36,51]; and 2) both intrinsic (aggregation and dispersal) and extrinsic environmental factors affecting the demographic rates of the vector and the developmental rates of the parasite within the vector [38,52–53] are likely to vary spatially [3].

Here, we present a first attempt to develop a smooth map of LF infection prevalence across Africa using a Bayesian GLSM fitted to published community-level infection data and spatially varying demographic and environmental covariates expected to underlie the transmission of this parasitic infection in endemic regions. Recently, this analysis method has received increasing focus as a more rigorous means to obtain estimates of parameters for large classes of complicated models, including, as in the present case, for complex spatial modelling problems that require explicit accounting of spatial autocorrelation while incorporating uncertainty in input infection data and model parameters [4–7,9]. The resulting prevalence map is used to investigate the geographical limits and levels of infection prevalence, the size of the population likely to be infected, and the environmental ecology of LF infection on the continent. It is predicted that future changes in climate may have an impact on the burdens of infectious diseases [54], particularly vector-borne diseases [55–56], and thus a second major aim of this study was also to undertake a first investigation of the potential impact global warming could have on future LF spatial distribution, prevalence and burden on the continent of Africa.

Methods

LF Prevalence Data

LF prevalence data for model building were collated from published community surveys conducted across Africa from 1940 to 2009, using the online and manual search procedures described previously [51]. Studies were selected if the surveys described the number of people surveyed, the number positive for microfilaraemia (mf), and were conducted at a specific community site. We found a total of 664 community-specific datapoints providing this information and these were used in the present analysis (see details of selected studies in **Table S1** in Information S1). Since field surveys employed different blood sampling volumes for detection of mf, all prevalence values were standardized to reflect sampling of 1 ml blood volumes using a transformation factor of 1.95 and 1.15 respectively for values originally estimated using 20 μlor 100 μlblood volumes [57]. Geo-coordinates for mapping of each chosen datapoint were either referenced from coordinates given in the literature or by using Google Earth. The corresponding author may be contacted for access to these data.

Environmental Layers

A number of environmental and climatic variables, essentially related to temperature and percipitation, affect the development and survival of the *Wuchereria bancrofti* parasite and its transmitting mosquito vectors [38,52–53]. Here, we began by choosing nine environmental data layers for analyses (**Table 1**), based on both their availability and biological plausibility in reflecting these external drivers of LF transmission. In addition, a spatial data layer describing the population density of each site was also included in the analysis to take explicit account of the role that host density plays in the transmission of vector-borne diseases, including LF [12,58–59]. The climate and altitude layers were downloaded from http://www.worldclim.org on a spatial resolution of 10 km×10 km. The ground vegetation cover index (NDVI) and the population density layer for Africa were downloaded from http://edit.csic.es/GISdownloads.html and http://sedac.ciesin.columbia.edu/gpw/global.jsp respectively. Worldclim layers are smooth maps of climatic variables created using interpolated weather station data [60]. The NDVI layer was created using satellite sensor data (from NASA's NOAA AVHRR sensor) and the population layer was created using data from, amongst others, the Socioeconomic Data and Applications Center (SEDAC) at Columbia University. The two sources of data were on slightly different scales, and so were resampled using ArcGIS to give all the layers the same grid size. This resulted in a scale of around 12 km×12 km for all layers used in this study. The WorldClim layers used for developing the model were monthly and annual means of temperature and percipitation estimated for Africa for the period 1950–2000. The mean NDVI data are based on monthly values obtained over a 18-year period from 1982 to 2000. All other covariate data layers reflected data assembled for the year 2000. Covariate values were extracted from the data grids for each geographic location where a LF survey result was available.

Multivariate Bayesian Generalised Linear Spatial Model

We used a Bayesian GLSM to spatially model the community-based LF infection data in response to the multiple environmental/demographic covariates derived at each site. If we assume that the number of individuals found positive for LF at location x_i is Y_i out of the total number of people tested, M_i, then Y_i is a binomial random variable, $Y_i \sim Bin(M_i, p_i)$, where p_i is the proportion of individuals infected at each location. A multivariate GLSM can be used to fit this model to observed data with the predictor variables entered as fixed effects and the spatial structure in the residuals modelled by inclusion of a location-specific random effect, $S(x_i)$ [61]. The model is denoted $h(\mu_i)= \sum_{j=1...k} f_{ij}\beta_j + S(x_i)$ where μ_i is the expected value of infection prevalence at location x_i conditional on a random spatial process$S(x)$, f_{ij} represents the value of the jth environmental variable at the ith location and β_j is the coefficent of the jth variable. The link function is denoted by $h(\cdot)$; for binomial data we use the logit function $h(\mu)=log(\mu/(1-\mu))$. S is modelled as a stationary Gaussian process with $S \sim MVN(0,\sigma^2 R+\tau^2 I)$ where I is the identity matrix, and σ^2 and τ^2 represent the spatial variance and the nugget effect (which accounts for the additional non-spatial variation in the data) respectively. R is an $n \times n$ correlation matrix with $R_{ij}=\rho(\|x_i-x_j\|,\phi)$ with an exponential spatial correlation function dependent on the parameter ϕ.

Table 1. Details of variables used in the Bayesian geostatistical analysis.

Environmental Variable	Source	Details
Altitude	http://www.worldclim.org	-
NDVI	http://edit.csic.es/GISdownloads.html	Mean of average monthly mean temperature across all 12 months
Annual mean temperature	http://www.worldclim.org	Mean of average monthly mean temperature across all 12 months
Mean maximum temperature	http://www.worldclim.org	Mean of average monthly maximum temperature across all 12 months
Mean temperature in warmest month	http://www.worldclim.org	Maximum of the 12 average monthly maximum temperature layers
Mean minimum temperature	http://www.worldclim.org	Mean of average monthly minimum temperature across all 12 months
Mean temperature in coldest month	http://www.worldclim.org	Minimum of the average monthly minimum temperature layers
Mean annual precipitation	http://www.worldclim.org	Mean of average monthly precipitation across all 12 months
Precipitation in wettest month	http://www.worldclim.org	Maximum of the 12 average monthly precipitation layers
Population density	http://sedac.ciesin.columbia.edu/gpw/global.jsp	-

Model Estimation

Given our Bayesian modelling framework and initial uncertainty regarding parameter values, we assigned vague normal prior distributions with mean 0 and variance 5 for all the variable coefficients and the intercept. These values were selected by running a standard, *ie.* non-spatial Generalized Linear Model (GLM), examining the magnitudes of the regression coefficients and intercept, and selecting prior distributions for each parameter with a wide variability centred near the mean value. The parameter ϕ, which controls the rate of exponential decay of the spatial correlation process, is assigned a reciprocal prior distribution between 4 and 6. We assigned σ^2 a uniform prior and we assume τ^2 to be proportional to σ^2, with $\tau^2 = 0.005\sigma^2$, informed by an initial variogram analysis of the data. We modelled the spatial correlation between sites with a Matern function with $\kappa = 0.5$, which equates to the use of an exponential function for describing spatial correlation in the data.

Implementation and Convergence Diagnostics

The model was implemented using *geoRglm* [62], a software package based on the R statistical system. Model fitting was done by running 150,000 Markov Chain Monte Carlo (MCMC) iterations after a burn-in of 300,000 runs, and thereafter storing every 250th element. Stability of model parameters was assessed by examining both the within and between chain convergence of the stored MCMC runs. Within chain convergence was analysed using two methods: 1) visual checks using trace plots, and 2) examining the MCMC error as a percentage of the standard deviation, with a MCMC error of less than 5% of the standard deviation taken to indicate satisfactory convergence [63]. The between chain convergence was evaluated by running five identical models and comparing both the obtained Deviance Information Criterion (DIC) values and using the the Gelman-Rubin (GR) diagnostic [64], which assesses convergence by running parallel chains from different starting values and determining if all chains converge to the same posterior distribution. The mean values of each regression parameter and the 95% and 80% credible intervals were calculated.

Variable Selection and Model Development

We use a similar procedure as outlined in Austin and Tu [65] and Craig et al. [66] to select a set of predictive variables that are uncorrelated for developing the multivariate Bayesian GLSM. The rational behind this approach is that often due to convergence and mixing problems with the MCMC sampling approach when including all of the spatial covariates (as was the case for the present analysis), the most practical route is to reduce the list of potential correlated explanatory variables using non-spatial selection methods, before moving to a spatial context [9,66]. The approach comprised the following steps.

Step 1– We performed a non-spatial univariate logistic regression analysis relating the observed infection prevalences to each of the covariates, and recorded the Akaike Information Criterion (AIC) of each fitted model [67]. AIC is defined by: $-2L(\beta)+2k$. Here $\beta = \{\beta_0, \beta_j\}$ are the regression coefficients, L is the maximum value of the likelihood function for the model and k is the number of parameters included in the model, which indicates that AIC penalizes for the addition of parameters.

Step 2– We reduced multicollinearity and confounding effects arising from correlated variables by identifying all pairs of variables with a Spearman's rank >0.7 and eliminating the variable in each pair from an environmental theme (eg. set of temperature or precipitation variables) giving the highest AIC value as derived in step 1. We also examined scatter plots of each pair of variables for the existence of any non-linear correlations.

Step 3– Next, we determined the functional form of the variables by fitting two different functional forms, a linear versus quadratic form, to the data, and calculated the AIC of each model. The functional form with the lowest AIC was selected.

Step 4 - The GLSM described above was then run with the variables selected from step 2 and the functional forms identified in step 3 using the *binom.krige.bayes* function in the geoRglm package. We performed a manual stepwise variable selection procedure whereby we removed one variable at a time, reran the model and assessed model performance using DIC [68]. If model performance improved, the variable was retained and if not, it was eliminated. This was repeated for all variables and continued until all remaining variables contributed to model performance.

Model Validation

We assessed the predictive ability of the final model by removing 100 data points and then fitting the Bayesian spatial model to the remaining data. The fitted model was used to make predictions over the removed data locations, and model accuracy was assessed by comparing the model prediction and observed prevalence for each location. The predicted LF infection prevalence was classified as correctly predicted when the observed prevalence for a location was within the 95% Bayesian credible interval (BCI) or within the 75%, 50%, 25% and 5% BCIs resulting from the predictive posterior distribution of that location [7,69].

Spatial Predictions

For mapping, we predicted infection prevalences using the final selected model at selected grid locations covering the whole of Africa. The predicted values were posterior means realised as part of the MCMC simulations from the posterior predictive distribution [61]. Approximate standard errors were obtained by dividing the 95% credible intervals by 4. Due to computational limitations we resampled the grid to create a coarser grid with cells of around 60 km×60 km using ArcGIS. This reduced the number of prediction sites from 500,000 to just under 20,000.

Estimating the Population with LF

The number of people infected with LF in Africa was estimated by overlaying the estimated prevalence map with the population map for the year 2000. We chose to use the 2000 population data for this calculation not only to conform with the climate data, but also to estimate and present the baseline infection burden of LF in Africa prior to the initiation of large scale control programs in endemic countries, which began in ernest only after the year 2000 on that continent. The number of infected people was calculated by multiplying estimated prevalence by population on a cell-by-cell basis. The mean prevalence and the total number of people with LF in each country were then estimated using the *zonal* statistics function available in the spatial analyst package in ArcGIS (ArcMap 9.3).

Future LF Predictions

In order to estimate the impact of future climate change and population growth on the spatial distribution of LF, we applied the parameter estimations from the current model fitted to baseline data to the climate and population data from 2050, based on the simplifying initial assumption that the biological relationships governing disease transmission would remain largely unchanged

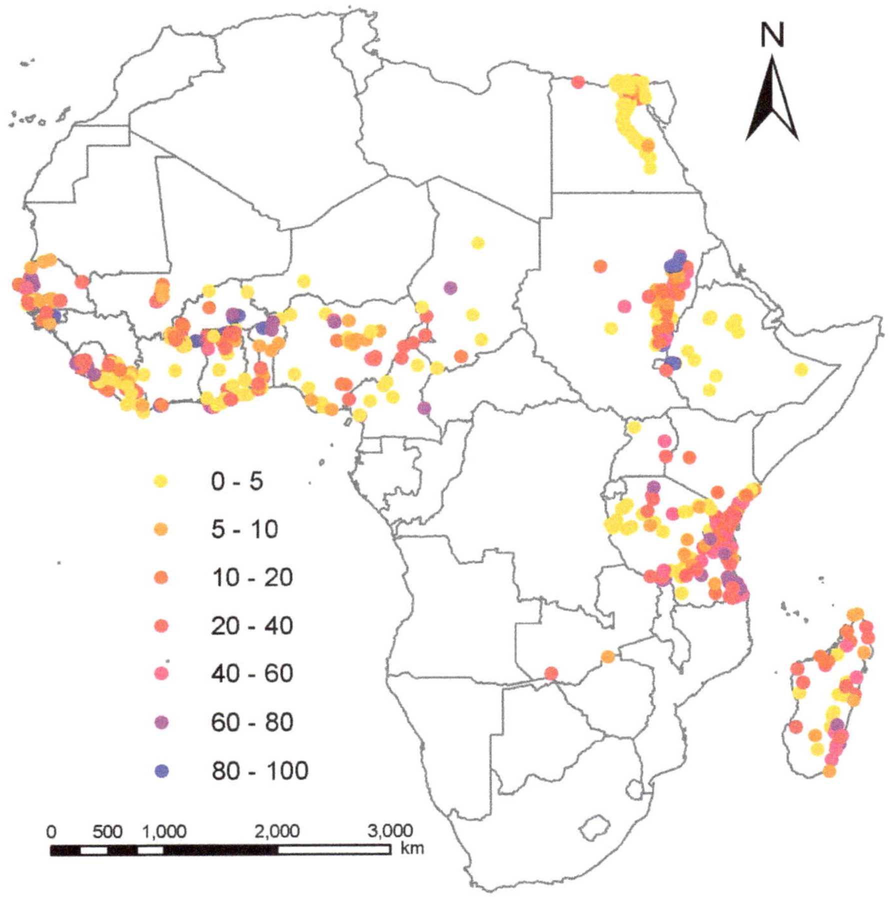

Figure 1. Locations and prevalence of LF infection for each study survey used in the present analysis. Points are coloured in relation to the percentage of survey population with mf in their blood.

over the two estimation periods. The results were used to both examine how prevalence might change from pre-intervention levels in currently defined LF endemic areas as well as whether new areas might become suitable for LF transmission in the future. We used climate predictions from two climate models – the Hadley Centre global climate model HADCM3 and the Canadian Centre for Climate Modelling and Analysis model CCCMA - under two IPCC climate scenarios – A2a and B2a [70]. A2a is a scenario assuming large disparities between regions and high population growth and energy use. B2a aims to capture a less disparate world with efforts focused towards social equity; this scenario assumes lower population and economic growth than A2a. To account for differences in population growth between the two climate scenarios we multiplied the 2000 population data by the country specific UN medium variant population growth rate predictions for the B2a scenario and by the high variant growth rate predictions for the A2a scenario (http://esa.un.org/unpd/wpp/unpp/panel_population.htm).

Results

Sample Site Locations and Observed LF Prevalence

The locations of the survey sites used in this study with their observed LF mf prevalence are shown in **Figure 1**. The depicted map shows that most LF community surveys carried out in Africa have occured along the western and eastern regions of the continent, across Sudan and Ethiopia, along the river nile in Eygpt, and on the coast of Madagascar. By contrast, very little information on infection status and prevalence are available from communities in the central regions (**Figure 1**). The observed community infection prevalences are also highly spatially heterogeneous, with sites exhibiting high prevalences occuring generally towards the central western and eastern regions of the continent, along a central gradient in the north eastern region and along the coast of Madagascar. By contrast, the surveyed community infection prevalences appear to be lower towards the southern parts of the western region, the central portion of the East-West endemic band and along the river Nile in the northern reaches of the continent (**Figure 1**).

Variable Selection and Univariate Analysis

Following steps 1–2 in the methods, we identified four pairs of variables with spearman's rank >0.7. From each correlated pair, we selected the variable with the lowest AIC from step 1 for inclusion as a predictor in the model. Non-linear correlations between variables were assessed visually, but none were identified (see **Figure S1** in Information S1). This left five out of the original 10 variables (**Table 1**) for inclusion in model development, *viz.* altitude, NDVI, population density, mean precipitation in the wettest month and mean annual temperature.

Figure 2 depicts the univariate relationship between LF prevalence and each of the selected variables or covariates. The lines show the mean fits of quadratic logistic regression models for each variable. Comparison of these fits with simple linear logistic relationships indicated that in every case the quadratic form provide a better fit to the data (not shown), suggesting that the relationship of each variable with LF prevalence was highly and significantly non-linear. Given this result, we assigned quadratic functional forms to all the six variables selected as predictors in this study.

Multivarate Bayesian GLSM Model

The results of fitting the Bayesian GLSM model to the LF prevalence data are shown in **Table 2**. The DIC values given in the table show that the full Bayesian GLSM (*ie.* with covariates and spatial component) gave the best, parsimonious, fit to the data in comparison with models containing only a spatial component and only covariates. The model with no spatial term included is the worst model, with a large DIC value of 71,246 (**Table 2**), clearly indicating the need for accounting for spatial correlation in the data. The non-spatial multivariate analysis shows that all covariates are associated with the risk of mf, but implementation of the full multivariate GLSM indicated a loss of importance for all the environmental and climatic variables. The spatial term appeared important even when 95% credible intervals are used for the full GLSM but in the case of covariates, only population density is shown to be required in the model when both the 95% and the lower 75% credible intervals were used for judging

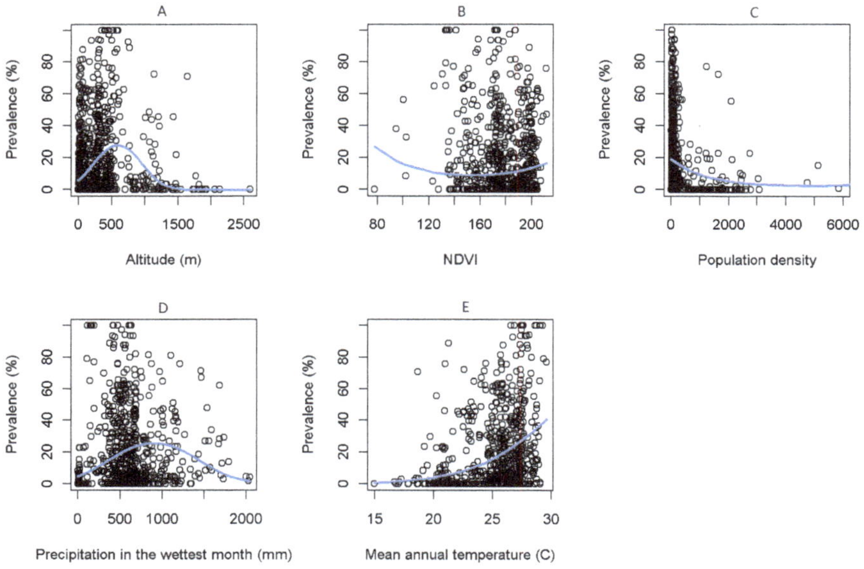

Figure 2. Scatter plots of each ecological/environmental variable against LF prevalence. The blue lines denoted the fitted quadratic functional forms for each variable.

variable importance (**Table 2**). However, running a reduced spatial model containing only this variable produced an increase in the DIC value (to 130), and also did not improve model performance in correcting predicting observed LF prevalence at validation locations (see below) compared to the full model. This suggests that considering the full set of covariates as selected in the full GLSM is required for predicting LF mf prevalence distribution across Africa.

Both trace plots and the Geldman-Rubin diagnostic indicated good convergence for all the β and σ^2 parameters of the model (see **Figure S2** and **Table S2** in Information S1). However, the results indicate that the decay parameter for spatial correlation (φ) may not have converged adequately, suggesting that the derived value for this parameter requires to be treated with caution. The posterior distributions for all parameters were normally distributed.

The functional forms of all the variables estimated using the full multivariate GLSM are portrayed in **Figure 3**. Interestingly, the forms for each variable uncovered by the multivariate model, except for precipitation in the wettest month, showed important differences with those detected for these variables using the univariate logistic regressions shown in **Figure 2**. Thus, compared to the corresponding univariate forms, LF prevalence is quantified to show a negative association with altitude, a generally positively increasing association with NDVI as well as with population density, and a monotonically increasing relationship with mean annual temperature (**Figure 3**). Given that the multivariate system facilitates parameterization of variables by taking account of the concurrent effects of other variables on the dependent response, and that the quantified functional responses are ecologically plausibile (see Discussion), we used the functions quantified by the full multivariate GLSM in all subsequent analyses in this paper.

Model Validation

Table 3 shows the mean number of test locations (from a sample of 100 randonly selected locations) with observed LF prevalences that fell into each of the selected BCIs of the posterior predictive distribution. The results show that the Bayesian spatial model with covariates performed better in predicting the number of test locations than the model without covariates across all the BCIs.

Spatial Predictions

The risk map of predicted LF prevalence created using the full Bayesian GLSM is depicted in **Figure 4**. According to the predictions of mean prevalence (**Figure 4a**), areas of high LF prevalence (>20%) are estimated to be in coastal and north West Africa, around the Sudanese region, and along the East African coast. Medium-prevalence areas (10–20%) are predicted to occur in the west, central west, and along the south eastern borders of the LF endemicity zone. Interestingly, regions of some countries

Table 2. Model parameters and the 95% and 75% credible intervals for the full model, and parameter values and 95% CIs for the model with no covariates and the model with no spatial term. Significant covariates in each model are highlighted in bold.

	Full model			No covariates		Population density model		No spatial term	
Coefficient	**Mean**	**95% CI**	**75% CI**	**Mean**	**95% CI**	**Mean**	**95% CI**	**Mean**	**95% CI**
β_0 – intercept	−16.06	(−37.035, 8.601)	(−29.160, −2.504)	−3.421	**(−6.988, −0.011)**	−2.775	(−6.585, 1.457)	1.233	**(0.943, 1.524)**
β_1 – altitude	−0.102	(−0.573, 0.318)	(−0.331, 0.146)					0.120	**(0.107, 0.131)**
β_2 – altitude^2	−0.002	(−0.022, 0.017)	(−0.014, 0.010)					−0.017	**(−0.018, −0.016)**
β_3 – NDVI	5.188	(−4.245, 14.301)	(−0.200, 10.894)					0.304	**(0.023, 0.580)**
β_4 – NDVI^2	−1.673	(−4.600, 1.419)	(−3.548, 0.092)					−0.124	**(−0.209, −0.038)**
β_5 – population density	−0.240	(−0.63, 0.114)	**(−0.463, −0.040)**			−0.262	(−0.604, 0.112)	−0.277	**(−0.298, −0.256)**
β_6 – population density^2	0.037	**(0.003, 0.075)**	**(0.017, 0.059)**			0.038	**(0.006, 0.073)**	0.023	**(0.021, 0.025)**
β_7 – prec in wet month	0.189	(−0.385, 0.733)	(−0.139, 0.540)					0.266	**(0.253, 0.279)**
β_8 – prec in wet month^2	−0.009	(−0.033, 0.015)	(−0.026, 0.006)					−0.012	**(−0.013, −0.011)**
β_9 – mean annual temp	0.698	(−1.293, 2.444)	(−0.447, 1.759)					−0.375	**(−0.408, −0.342)**
β_{10} – mean annual temp^2	−0.013	(−0.050, 0.029)	(−0.036, 0.012)					0.009	**(0.008, 0.010)**
σ^2 – spatial variance parameter	61.686	**(55.229, 68.277)**	**(57.655, 65.839)**	54.676	**(48.171, 62.542)**	62.763	**(56.217, 69.247)**		
ϕ – spatial decay parameter	5.507	**(5.488, 5.522)**	**(5.498, 5.516)**	4.875	**(4.567, 4.995)**	5.485	**(5.424, 5.545)**		
DIC	118.8			1100.7		130.3		71246	

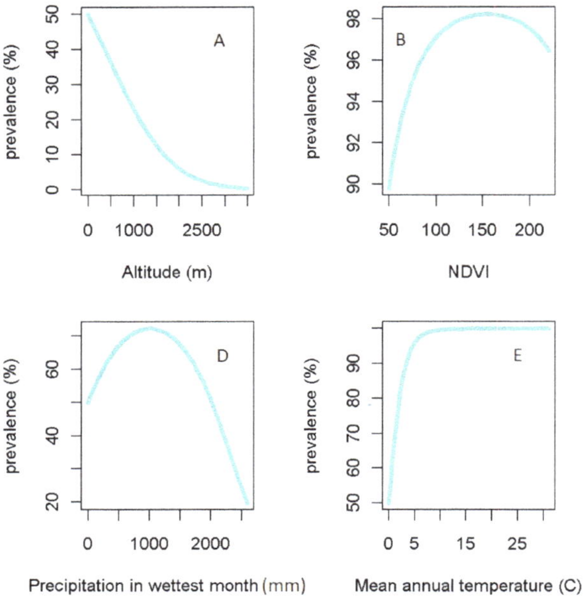

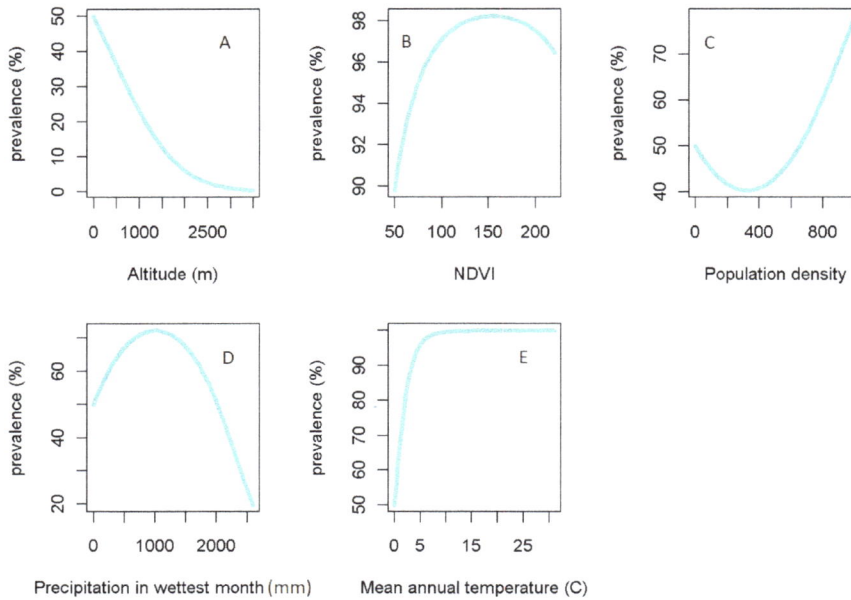

Figure 3. Functional forms for each ecological/environmental plotted using the coeficients estimated using the final multivariate bayesian spatial model. Lines show the marginal relationships estimated between each environmental variable and the prevalence of LF.

for which we had no or sparse data were found to be LF endemic, with some areas in central and southern Africa estimated to have prevalences of up to as high as 10%. However, examination of the estimation error map in Figure 4b shows that the areas with the lowest standard errors and hence uncertainty are as expected those where we have the most sample data while unsampled areas are associated with very high standard errors. This suggests that the mean prevalence predictions for the unsampled areas need to be viewed with caution. It also highlights how mapping of error estimates from running Bayesian GLSMs can be used to conduct secondary sampling of sites to fill in gaps in the currently available spatial data on community infection prevalences in an endemic region. The present results thus suggest that more community surveys of LF infection are urgently required in countries across the northern and southern most edges of the presently endemic LF zone in Africa, and in the central and south west parts of the continent (**Figure 4b**) if we are to develop more reliable maps of LF prevalence for this continent.

Number of People Infected with LF

The estimated number of people with LF and the mean prevalence in each country for 2000 prior to the initiation of large-scale national-level mass drug administration are shown in **Table 4**. The results indicate that the baseline mean LF prevalence for the whole of Africa was around 7.85%, and that some 61.55 million people were infected with *W. Bancrofti* at the start of the current control campaign. Infection prevalence and the number of people infected, however, varied a great deal between endemic countries, with prevalence ranging from as low as 0.3% in Burundi to as high as 38% in Sierra Leone (**Table 4**). We predict that prior to 2000, Nigeria had the most people infected (11.01m), with significant infected populations (up to some 27.11 million or 44.05% of the total infected African population) also quantified for Burkina Faso, Egypt, Senegal, Sudan, Tanzania, Uganda and Zaire (**Table 4**).

Table 3. Precentage of study locations with LF prevalence falling in the 5%, 25%, 50%, 75% and 95% Bayesian credible intervals for the geostatistical models without and with covariates.

	Credible Interval				
	5%	25%	50%	75%	95%
LF – with covariates	5.0±1.1	26.8±3.9	49.2±6.6	66.4±4.2	76.6±3.2
LF – without covariates	4.0±0.9	22.6±3.1	41.6±4.5	57.0±1.6	69.2±2.0

Future Changes in LF Spatial Distribution

We produced predictive LF prevalence smooth maps for Africa using the the climate and population density data for 2050 under each of the four climate change scenarios outlined in the Methods. The results were almost identical for both models and scenarios, with the mean prevalence across Africa estimated to be between 8.35 to 9.85% for all scenarios, and thus we illustrate results from the CCCMA model only in Figure 5 and Table 4. **Figure 5a** depicts the predicted prevalence in 2050, while **Figure 5b** shows the absolute change in estimated prevalences between 2000 and 2050. In constructing the latter result, we simply considered currently LF free countries to have 0% infection prevalence and used this fixed baseline value to calculate the absolute precentage change between the present and the predicted 2050 prevalence in these areas. The absolute change results sumarized in **Figure 5b** reveal firstly that the potential transmission areas of LF in Africa will significantly expand by 2050; in particular, both the northern and southern extremes of the continent will become appreciably endemic, with some currently low endemic areas in these regions predicted to reach infection prevalences >5% (**Table 4**). The second feature of the results shown in **Figure 5b** is that LF prevalence is also likely to change within the currently endemic countries/regions, significantly increasing in some areas while

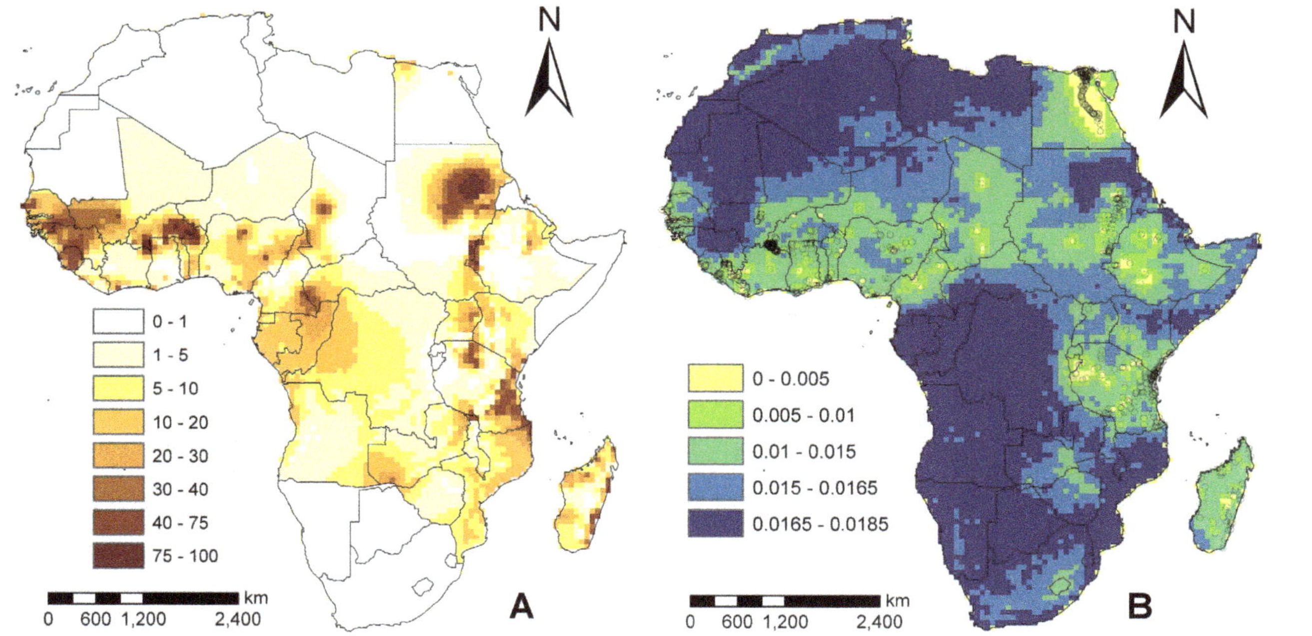

Figure 4. Predictions of the final Bayesian spatial model. (A) Estimates of LF prevalence in Africa. Known filariasis free territories have been masked out (Western Sahara, Mauritania, Morocco, Libya, Algeria, Tunisia, Eritrea, Somalia, Botswana, Namibia, South Africa, Lesotho and Swaziland – information from http://www.taskforce.org/). Values plotted are the mean of the posterior distribution from the multivariate Bayesian GLSM. (B) Map of the estimated standard errors with points showing the location of the LF prevalence data.

decreasing drastically in others (**Figure 5b** and **Table 4** - where countries with significant decreasing LF prevalence are highlighted in blue), although in the majority of these countries the change is predicted to be positive (**Table 4**). The estimated population with LF infection in 2050 based on the A2a scenario is predicted to be around 211m and 208m using the two climate models, while based on the B2a scenario, the estimated infected populations are calculated to be 184m and 179m respectively. However, as the underlying at-risk or exposed population in Africa will increase significantly by 2050 (due to both rapid population growth and spread of infection into currently non-endemic populations), the overall prevalence of LF in Africa will increase only moderately from the 2000 baseline prevalence of 7.85% to between 9.75–9.85% in 2050.

Discussion

It is important to note from the outset that our study was primarily concerned with developing a robust analytical method for predicting and mapping LF prevalence distribution and potential future change in Africa in terms of data availability, biological plausibility and model parsimony. However, despite this main objective, the methodology employed in this analysis, viz. using a novel systematically staged ecological covariate selection procedure coupled with selection of a multivariate spatial regression model containing a subset of low-correlated environmental/climatic variables that gave a good fit to the data as well as prediction of observed infection prevalence in test locations [66], has also allowed us to provide improved information regarding the likely environmental determinants of LF prevalence distribution in Africa. Thus, even though the results indicated that spatial structure in the data *per se* played the major part in the Bayesian spatial model's ability to predict LF prevalence correctly (**Table 2**), six variables belonging to three major environmental themes (viz. climate - mean annual temperature, mean minimum temperature, and mean precipitation in the wettest month; landscape - altitude and NDVI; and population density) are shown to be associated with LF prevalence distribution in Africa. Our use of a staged variable selection process to pre-select a list of predictors could be viewed as setting up the analysis for success [71], and hence could have led to the spurious selection of these covariates. However, using a reduced spatial model containing only the "important" population density variable, did not lead to a marked improvement in model fit nor improve the predictive performance of the full model. This indicates that our pre-selection process did not significantly affect the validity of the inclusion of the fuller set of six variables in our final model. Indeed, the inclusion of these covariates to the Bayesian model incorporating spatial structure only produced a near 90% reduction in DIC values (**Table 2**), indicating that both spatial structure and all six selected covariates were required for describing the observed LF prevalence spatial distribution in Africa.

Several other features of our methodology and the variables selected for use in the final multivariate Bayesian spatial model require to be noted. The first is the need to address the problem of correlation among predictors suitably when developing multivariate regression models [66]. The staged process of variable elimination employed here to produce a candidate list of low or little-correlated predictions from among a larger set of biologically plausible environmental variables as a first stage in our spatial analysis proved to be practical solution to this problem. However, although this method may address the problem of multicollinearity, it is important to note that excluding variables on the basis of low univariate correlation with the response may obscure the fact that their predictive power may be quite different in the presence of other interacting variables [66]. It is also possible that while these variables may show weak associations

Table 4. Estimated number of people with LF infection in Africa based on the full GLSM.

Country Name	2000	2050 A2a scenario	2050 B2a scenario
Algeria	0m (0%)	1.09m (2.1%)	0.91m (2%)
Angola	0.80m (6.1%)	3.63m (8.4%)	3.2m (8.3%)
Benin	0.90m (14.4%)	3.97m (18.3%)	3.45m (18%)
Botswana	0m (0%)	0.17m (5.8%)	0.14m (5.6%)
Burkina Faso	3.05m (26.5%)	11.71m (26%)	10.61m (26.4%)
Burundi	0.02m (0.3%)	0.34m (2%)	0.23m (1.6%)
Cameroon	1.52m (10.2%)	3.66m (9.4%)	3.21m (9.4%)
Central African Republic	0.40m (10.9%)	1.06m (11.8%)	0.92m (11.7%)
Chad	0.68m (8.6%)	2.27m (7.8%)	2.14m (8.2%)
Congo	0.38m (12.3%)	1.1m (13.1%)	0.96m (12.9%)
Djibouti	0.03m (4.7%)	0.06m (4.5%)	0.05m (4.5%)
Egypt	2.61m (3.9%)	11.76m (8.5%)	9.87m (8.2%)
Equatorial Guinea	0.04m (8.9%)	0.07m (5.5%)	0.05m (4.9%)
Eritrea	0m (0%)	0.71m (5.9%)	0.63m (5.9%)
Ethiopia	1.45m (2.3%)	5.22m (2.8%)	4.52m (2.7%)
Gabon	0.18m (14.9%)	0.2m (7%)	0.17m (6.8%)
Gambia, The	0.16m (14.2%)	0.22m (6.2%)	0.18m (5.8%)
Ghana	1.14m (6%)	4.28m (8.5%)	3.84m (8.7%)
Guinea	1.86m (23%)	4.67m (18%)	4.03m (17.5%)
Guinea-Bissau	0.28m (25.7%)	0.64m (18.8%)	0.49m (16.2%)
Ivory Coast	1.14m (7.2%)	3.07m (7.1%)	2.85m (7.5%)
Kenya	2.25m (7.4%)	8.53m (9.4%)	7.1m (8.9%)
Lesotho	0m (0%)	0.01m (0.2%)	0m (0.2%)
Liberia	0.29m (9.9%)	0.32m (3.2%)	0.26m (2.9%)
Libya	0m (0%)	0.81m (7.8%)	0.62m (6.7%)
Madagascar	1.73m (11%)	4.75m (9.6%)	4.36m (10.1%)
Malawi	0.59m (5.3%)	2.69m (6.9%)	2.18m (6.3%)
Mali	1.99m (17.6%)	6.62m (19.2%)	5.89m (19.2%)
Mauritania	0m (0%)	0.55m (7.9%)	0.52m (8.5%)
Morocco	0m (0%)	1.16m (2.4%)	0.95m (2.3%)
Mozambique	1.84m (10.2%)	6.02m (12.5%)	5.37m (12.8%)
Namibia	0m (0%)	0.2m (5.3%)	0.17m (5.1%)
Niger	0.29m (2.7%)	2.41m (3.8%)	2.2m (3.9%)
Nigeria	11.01m (9.7%)	31.79m (11%)	27.86m (10.8%)
Rwanda	0.04m (0.5%)	0.85m (3.6%)	0.61m (2.9%)
Senegal	2.85m (32.1%)	5.35m (20.5%)	4.73m (20.5%)
Sierra Leone	1.62m (37.6%)	3.94m (27.7%)	3.48m (27.7%)
Somalia	0m (0%)	1.11m (3.8%)	0.98m (3.7%)
South Africa	0m (0%)	1.83m (2.9%)	1.45m (2.7%)
Sudan	7.04m (22.7%)	18.09m (23.7%)	16.11m (24%)
Swaziland	0m (0%)	0.09m (5.4%)	0.08m (5.2%)
Tanzania, United Republic	5.40m (15.6%)	21.05m (17.2%)	18.45m (17%)
Togo	0.72m (16.2%)	0.85m (6.8%)	0.77m (7%)
Tunisia	0m (0%)	0.6m (4.6%)	0.44m (3.9%)
Uganda	2.97m (12.8%)	15.27m (15.6%)	13.07m (15%)
Western Sahara	0m (0%)	0.04m (3.8%)	0.04m (3.7%)
Zaire	3.19m (6.3%)	12.07m (7.3%)	10.64m (7.3%)
Zambia	0.76m (7.3%)	2.81m (8.6%)	2.4m (8.3%)
Zimbabwe	0.33m (2.6%)	1.09m (4.2%)	0.89m (4%)
Total population infected	61.55m (7.86%)	210.82m (9.85%)	184.03m (9.75%)

individually with a response, they nonetheless may interact together to causally produce the outcome of interest, especially in the case of complex outcomes [72–73]. In this study, however, inclusion of all initially evaluated ten variables into the GLSM model did not improve the fit of our final model (data not shown), confirming the relative importance of the selected predictors. Nonetheless, we indicate that further work is required to guide variable selection within a spatial framework particularly in multilevel, multicomponent systems such as parasitic-climate/environment applications, if we are to understand and manage such complex associations more accurately.

The spatial geostatistical modelling framework employed here distinguishes the correlation among observations that can attributed to the spatial proximity of data locations - perhaps generated by the transmission process governed by the flight range of LF mosquito vectors [3] - and that which can be explained by larger-scale first-order spatial correlation explained by common exposures to environmental factors. While the model is thus important for avoiding overestimation of the explanatory power of covariates (compared to a non-spatial model), another significant output of the approach is that it also allows for a better estimation and treatment of residual or unmeasured spatial variation remaining in the model [18]. For example, in the present study, such a formulation can account for the measurement error introduced by the use of prevalence data collated from different studies, each using slightly different sampling and testing methods. Such residual spatial error may also suggest that micro-scale factors, such as those related to poverty, capacity of health facilities, ongoing interventions, and other environmental factors, associated with each location may additionally influence the spatial distribution of observed LF infection. Estimation of the magnitude of such error as afforded by geostatistical analytical frameworks, such as the present model, is thus important as it not only provides a basis for determining the significance of micro-spatial effects but also for supporting further investigations aimed at identifying these additional factors.

Apart from permitting simultaneous environmental risk assessment and modelling of spatial dependence and prediction, the Bayesian modelling framework used here is also important because it allows the quantification of uncertainty in map predictions. Our results arising from such error quantification illustrated via the error map in **Figure 4b** highlight that areas that are more densely sampled for LF infection in Africa (primarily along the west and east Africa coasts and among the Ethiopian highlands) have lower errors associated with their predictions compared to areas in central and south western regions of the continent. The creation and use of such error maps are important in disease control as they indicate areas where additional surveys are required. In the present instance, we indicate that surveys for LF infection prevalence are needed in the above and other areas with large standard errors shown on the map (**Figure 4b**) if we are to develop more accurate maps of LF distribution and prevalence for this continent.

Our best-fit model, overall, indicates that several environmental factors may be associated concurrently with LF infection distribution in Africa. These results have also provided new information regarding the functional forms of their association with LF prevalence, with all the variables selected in our best-fit model – altitude, NDVI, population density, precipitation in the wettest month, mean annual temperature and mean minimum temperature – exhibiting nonlinear quadratic forms for their effects on infection (**Figure 3**). On a technical note, this work has underscored the value of using multivariate models compared to univariate analysis for reliably estimating the function forms of ecological predictors driving disease that are biologically and epidemiological interpretable. Thus, the coefficients for altitude

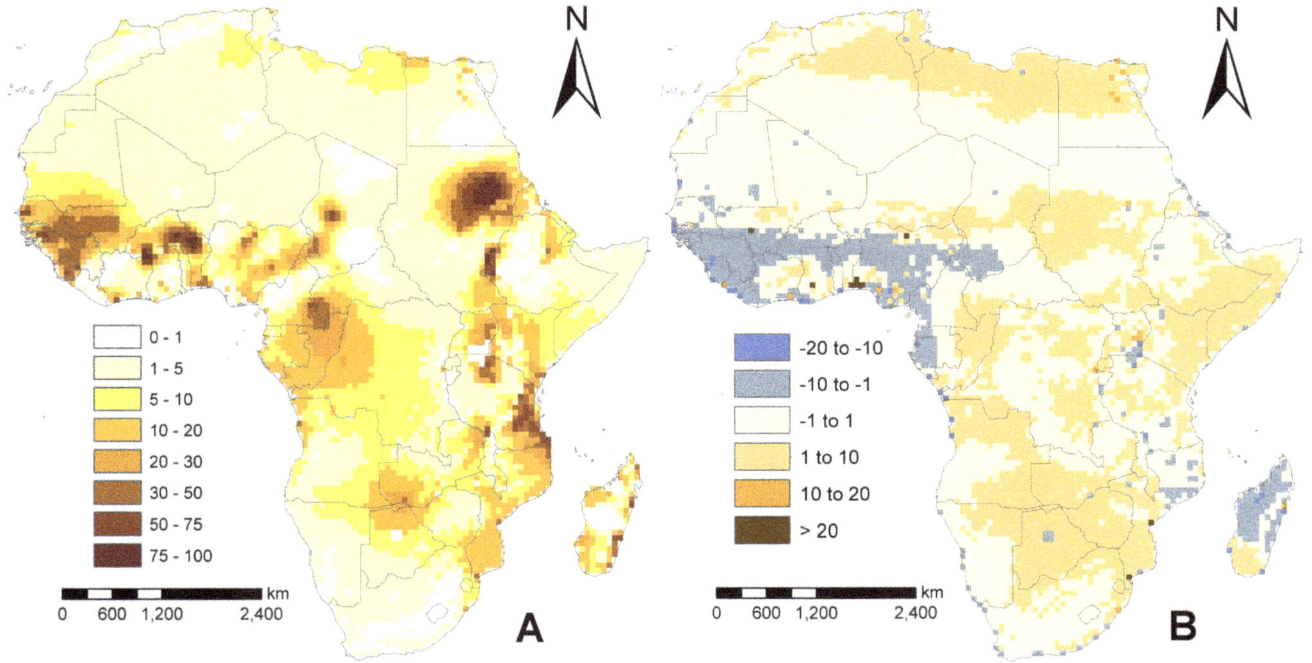

Figure 5. A) Predicted LF prevalence in 2050 using the CCMA model and climate scenario A2a, and B) percentage change in LF prevalence between 2000 and 2050. Areas in grey and blue are where prevalence is predicted to decrease, while areas in orange and brown are where it is predicted to increase.

from fitting of the multivariate GLSM imply a strongly negative correlation with LF prevalence. This is consistent with studies that have investigated LF prevalence in relation to altitude [74–75] where it is suggested that at higher altitudes, the corresponding lower temperatures (owning to the lapse rate, in which temperature is negatively associated with increasing altitude), would reduce LF transmission due to negative effects on mosquito survival rate and the rate at which the parasite develops within the vector [53,74]. By contrast, the form of the relationship estimated in the corresponding univariate analysis suggests that altitude may have a more complex non-linear effect on LF prevalence with both an initial positive followed by a negative effect with increasing values (**Figure 2a**), which is clearly more difficult to explain on experimental or ecological grounds. Similarly, the functional form for the mean annual temperature relationship uncovered in the univariate analysis (**Figure 2e**), suggests that LF prevalence may increase non-linearly and non-monotonically with mean annual temperature, whereas in the multivariate case, the relationship is shown to be a monotonic nonlinear one (**Figure 3e**). The latter function is more biological and ecologically plausible given that experimental findings suggest that larval development and mosquito survivorship are maximised at around 22–30°C as [53,76]. The failure of our functional form to imitate the convexity implied by the experimental results exactly could indicate that mean annual temperature might either not be a useful predictor of LF community prevalence or that our data range is insufficient to conclusively quantify the relationship (given that the upper limit of this variable is restricted at 30 degrees in our study (**Figure 2e**)). The contrast between univariate versus multivariate model estimations of ecological function forms is perhaps most apparent for the NDVI variable (Figure 2b versus **Figure 3b**), which when quantified using a univariate model shows a puzzling decreasing effect on prevalence with increasing values. While this effect may reflect a regulatory effect of forested ecosystems in moderating disease transmission [77], such ecosystem effects might be better captured by the non-linear increase and decrease function shown by the NDVI-prevalence function quantified using the multivariate model. Thus, the multivariate form (**Figure 3b**) suggests that LF prevalence would be highest for NDVI values between 130 and 175, which correspond to shrub or grasslands, or forests, whereas below 130, which refers to barren areas of rock or sand, prevalence, expectedly, is predicted or shown to be low. By contrast, the function indicates that at the highest values of NDVI, prevalence would begin to decrease possibly reflecting the anticipated regulatory effects of forested areas acting to decrease parasite transmission via reductions in vector populations and/or human-vector contacts [77].

Similar complex associations with prevalence are also observed for precipitation values. The association with precipitation levels in the wettest month shows that LF prevalence may increase with precipitation up to a limit of around 1200 mm per month for both models, and above this value, begin to decline. A positive association between LF prevalence and precipitation might be expected because LF vectors require a certain amount of precipitation so that suitable sites to lay eggs are formed. However, our result is the first to support and provide a functional form for the suggestion that when there are heavy downpours these laying sites can get washed away [56], leading to lower overall transmission of infection in areas that experience the highest amounts of precipitation.

In contrast to the purely environmental/climate variables investigated, LF prevalence was found to be associated strongly only with human population density in this study (**Table 2**). While the uncovering of an association between prevalence and host population density is unsurprising given that human to vector density ratios are important drivers of the transmission of vector-borne infections [58], it is clear again that the form estimated by the univariate logistic regression model showing that the association may be negative for LF in Africa may be misleading (**Figure 2c**), as a positive association as predicted by the multivariate GLSM is likely to be the more expected response on theoretical grounds [58]. The statistical evidence for a stronger influence for this variable in comparison with environmental variables *per se* (**Table 2**), however, suggests that this factor may play the biggest role in determining LF prevalence in a location; *ie.* that given a suitable environment for transmission as defined by climatic variables, variations in population density, and more importantly human activities, would play the major role in underlying LF transmission intensity and prevalence levels. This conclusion, and the finding above for NDVI, support suggestions that gaining a better understanding of the association between the environment and LF, and indeed other vector-borne infections, requires a more detailed examination of other non-climatic variables that may confound such relationships and therefore underlie real world parasite transmission, including in particular the socio-ecological dimensions of parasite transmission and the effects of land use or habitat alteration by humans [78].

The multivariate Bayesian GLSM model developed here indicates that LF prevalence distribution in Africa may be highly spatially variable. The map depicted in **Figure 4a** indicates that LF prevalence is high in West Africa, West Central region, along the coast of East Africa, and Sudan, with significantly lower infection prevalence predicted for central Africa and in the northern and southern portions of the continent, except for Egypt. The highest LF prevalence is shown to occur in West Africa, with prevalences higher >23% predicted for Guinea, Guinea-Bissau, Burkina Faso, Senegal and Sierra Leone, while moderately high prevalences are predicted for other West African countries and for the endemic countries of East Africa (**Table 4**). The model further predicts that significant LF infection prevalence (ranging from 6.1% to 12.3%) may occur for several countries, such as Democratic Republic of Congo, Central African Republic and Angola, for which no survey data were available to us for analysis. Similarly, although we have masked the results, the model also predicted infection to occur in several known non-endemic countries, such as South Africa and the other northern Africa countries. While this might indicate the potential for infection in these regions given the covariate distribution in these locations, it is clear that the large errors also predicted for these regions (**Figure 4b**) indicate either absence (as supported by documentation of lack of infection in the latter two regions) or a need to conduct new community surveys in those areas (such as the Democratic Republic of Congo, Central African Republic and Angola) without such clear documentation.

Although the errors behind the generation of the map produced in this work need to be borne in mind, the continuous Africa LF prevalence risk map when used in conjunction with a continuous population density map has allowed a more reliable calculation of the number of people infected with LF in Africa. In total, the overlaying of these maps indicate that some 61.55 million people may have been infected with LF in Africa prior to the advent of large-scale intervention programs, implying that mean LF prevalence was around 7.8% for this continent given a 2000 population of 783.5 million. The last attempt to estimate the burden of LF in Africa [35] estimated that approximately 40 million people were infected. The 1990 population used in their estimate was 726 million, which gave a mean LF prevalence of 5.5% in Africa at that time. However, that figure was based on the assumption of homogenous distribution (unlike the capture of within-country variation by the present continuous map) of prevalence across individual countries, and the disregarding of countries for which no data were available. Thus, the higher LF burden estimated in this study could be a reflection of a better treatment of within country heterogeneities in both infection and population density distributions, and/or the impact of higher overall current population size. Among individual counties, we indicate that those with the largest infected populations in 2000 are found to be in Nigeria (11.01 million), followed by Sudan (7.04 million) and Tanzania (5.40 million), with sizeable infected populations also occurring in Burkina Faso, Uganda, Zaire and Senegal (all close to or >2.5 million) (**Table 4**). This heterogeneity in country-level prevalences clearly imply that efforts to achieve the elimination of LF in Africa will vary markedly between countries, with perhaps more intensive efforts required in the case of those countries exhibiting the highest baseline prevalences [79].

The first result of import from our work on future prevalence change relates to the widely accepted notion that the main effect of climate change on the extent and the severity of vector-borne diseases is to shift rather than increase the geographical range of the diseases [55]. Our findings do not fully agree with this assertion as we predict both changes in severity (**Figure 5**) across Africa, particularly in West Africa, and a potential range shift to the north and south of the continent. Our future climate predictions, however, assume that the biological relationships governing transmission are unchanged between now and 2050. Thus, adaptation by *Wuchereria bancrofti* or the different mosquito vectors to climate change are not taken into account in these first estimates [80], although we note, as per Schmalhausen's law [81], that this is likely to be a major problem only at the extremes of the range of variables that may negatively affect transmission (**Figure 3**). Areas of such extreme changes in the future are also relatively small meaning that vector or parasite biological/physiological adaptation to future climatic changes are unlikely to have a large impact on the results presented here. The sensitivity of our results to different climate scenarios was also explored by running the final model using data from two climate models and two climate scenarios. We found only small differences between the different scenarios/models – the A2a climate scenario lead to a slightly higher prevalence and hence estimates of population infected (211m compared with 184m for the B2a scenario), suggesting that the results are robust to the climate model or development scenario used.

It is clear, however, that given the significance of the contribution of population density in our final model, and the fact that between 2000 and 2050, the population is expected to more than double (http://esa.un.org/unpp/), the primary driver of increased LF burden in Africa will be rapid population growth. This conclusion again underscores the importance of considering the role of non-climatic factors that may modify future climatic influences when assessing possible effects of climate change on infectious diseases [82–85]. It also provides an indication into the types of adaptations and policies that may be required to counter the effects of climate change on a particular disease. For example, our results imply that development-related strategies to strengthen LF control programs and possibly reduce population growth in the short term can increase the capacity of endemic African countries to cope with the projected increase in LF over the medium to long run [83,86]. In the same vein, it might be possible to use estimates of pre-intervention, current and future disease prevalences to investigate whether investing in disease control and wider economic development (which would also bring about knock on effects in reducing population growth) or investing in greenhouse gas emission reduction *per se* would be the better choice in mitigating against the impact of future climate change on parasitic infections [83,85]. This perspective implies that the current global initiative to control LF may also be considered as an adaptation to a major potential health impact of future climate change, a conclusion which provides an additional rationale for its enaction and successful global implementation.

One of the major limitations of studies aimed at identifying, quantifying and predicting the impacts of environmental/climatic risk factors on parasitic diseases clearly lies in the difficulty of locating and matching the temporal and spatial scales of observational versus climate prediction data [87]. This study is no exception. For example, our baseline prevalence prediction map is based on LF survey data collected between 1940 and 2009, while the WorldClim climate predictor data used in model development are based on averages from 1950 to 2000. Although our infection data are thus reasonably conformal to the WorldClim data, it is apparent that the analysis does not consider the possibility of temporal changes occurring in both infection and climate during the above periods. However, given that systematic interventions against LF began only after 2000 in selected countries [88], and that temperature-related changes began to increase above recent natural fluctuation bounds only just before or around the year 2000 in Africa [70], we suggest that such changes are unlikely to significantly bias the present results. The issue of spatial scale, on the other hand, is a more difficult problem to address, as micro-spatial factors below the 12×12 km model resolution, including different age distributions in study locations, proximity to a water source, survey methodological differences, socio-economic effects, and amount of migration, among others, could also have had an impact in influencing the level of variation in LF prevalence observed at the village level in our study. Similarly, the change from the 12 km model development to a coarser 60 km prediction scale implies that the model fitted at the 12 km resolution is also suitable at the coarser scale, an assumption which may or may not be true. These considerations indicate that the present larger scale spatial predictions may provide only an index of the micro-spatial scale heterogeneity underlying the data.

A second limitation of our study is that the models also have not taken into consideration that the spatial dependence structure in the data may well be different in different regions of Africa. Differences in vector species composition which may influence LF transmission are also likely to vary between different regions making taking a stationary modelling approach questionable. Bayesian non-stationary geostatistical models for quantifying parasitic risk have begun to be developed recently [69,89–90], and we expect that the application of such advances will enable development of more reliable LF, and indeed other parasitic disease, maps in the near future.

Nonetheless, this study has provided the first LF prevalence map for a major endemic region developed using Bayesian geostatistical techniques. This approach has allowed us to estimate the prevalence of LF across Africa at both surveyed and unsurveyed locations based on environmental covariates, spatial dependence between data locations, and a better treatment of sampling error. It has also allowed us to identify and analyse the impacts, and functional forms, of key environmental risk factors for LF, which in turn has provided a hint of the complex multidimensional pathways through which these factors may act, possibly with other non-climatic factors, to influence the transmission of this parasitic disease in Africa. We have used the map to estimate the number of people infected with LF in different countries, and make rough predictions of the impact of changes in climate and population growth on the future growth of the disease on the continent in the absence of control. We hope that by identifying the key spatial risk factors for infection, and those areas with high current and future prevalence or with the potential for emergence of LF infection, control programmes will be able to use the present results to make better informed decisions as to where to focus efforts, assess and develop mitigating stategies for the future, and better quantify the magnitude of efforts required to reduce LF prevalence to either zero transmission or below a suitable infection threshold in any given region of Africa [57,91].

Supporting Information

Information S1 Contains two tables and two figures. Table S1 – Details of studies and data used in the Bayesian geostatistical analysis. Table S2 – Convergence statistics for the mcmc model fit. Figure S1 – Matrix of scatterplots between all pairs of variables. Figure S2– Trace plots of the beta coefficients and the spatial parameters, ϕ and σ^2.

Author Contributions

Conceived and designed the experiments: EM HS. Performed the experiments: HS EM. Analyzed the data: HS EM. Contributed reagents/materials/analysis tools: EM HS. Wrote the paper: HS EM.

References

1. Zhou XN, Yang GJ, Yang K, Wang XH, Hong QB, et al. (2008) Potential impact of climate change on schistosomiasis transmission in China. American Journal of Tropical Medicine and Hygiene 78: 188–194.
2. Brooker S, Michael E (2000) The potential of geographical information systems and remote sensing in the epidemiology and control of human helminth infections. Advances in Parasitology: Academic Press. 245–288.
3. Srividya A, Michael E, Palaniyandi M, Pani SP, Das PK (2002) A geostatistical analysis of the geographic distribution of filariasis infection prevalence in Southern India. American Journal of Tropical Medicine and Hygiene 67: 480–489.
4. Clements AC, Lwambo NJ, Blair L, Nyandindi U, Kaatano G, et al. (2006) Bayesian spatial analysis and disease mapping: tools to enhance planning and implementation of a schistosomiasis control programme in Tanzania. Tropical Medicine and International Health 11: 490–503.
5. Clements ACA, Brooker S, Nyandindi U, Fenwick A, Blair L (2008) Bayesian spatial analysis of a national urinary schistosomiasis questionnaire to assist geographic targeting of schistosomiasis control in Tanzania, East Africa. International Journal for Parasitology 38: 401–415.
6. Raso G, Matthys B, N'Goran EK, Tanner M, Vounatsou P, et al. (2005) Spatial risk prediction and mapping of Schistosoma mansoni infections among schoolchildren living in western Côte d'Ivoire. Parasitology 131: 97–108.
7. Beck-Worner C, Raso G, Vounatsou P, N'Goran EK, Rigo G, et al. (2007) Bayesian spatial risk prediction of Schistosoma mansoni infection in western Cote d'Ivoire using a remotely-sensed digital elevation model. American Journal of Tropical Medicine and Hygiene 76: 956–963.
8. Guerra CA, Gikandi PW, Tatem AJ, Noor AM, Smith DL, et al. (2008) The limits and intensity of Plasmodium falciparum transmission: implications for malaria control and elimination worldwide. PLoS Medicine 5: e38.
9. Wardrop NA, Atkinson PM, Gething PW, Fevre EM, Picozzi K, et al. (2010) Bayesian geostatistical analysis and prediction of Rhodesian human African trypanosomiasis. PLoS Neglected Tropical Diseases 4: e914.
10. Kitron U (2000) Risk maps: transmission and burden of vector-borne diseases. Parasitology Today 16: 324–325.
11. Rogers DJ, Randolph SE, Snow RW, Hay SI (2002) Satellite imagery in the study and forecast of malaria Nature 415: 710–715.
12. Hales S, de Welt N, Maindonald J, Woodward A (2002) Potential effect of population and climate changes on global distribution of dengue fever: an empirical model. Lancet 360: 830–834.
13. Hay SI, Omumbo JA, Craig MH, Snow RW (2000) Earth observation, geographical information systems and *Plasmodium falciparum* malaria in sub-Saharan Africa. Advances in Parasitology 47: 174–215.
14. WHO (1998) Research on Rapid Geographical Assessment of Bancroftian Filariasis. TDR/TDF/COMDT/98.2.
15. Richards FO (1993) Use of geographic information systems in control programs for onchocerciasis in Guatemala. Bulletin of the Pan American Health Organization 27: 52–55.
16. Robinson TP (1998) Geographic Information Systems and the selection of priority areas for control of tsetse-transmitted trypanosomiasis in Africa. Parasitology Today 14: 457–461.
17. Mott KE, Nuttall I, Desjeux P, Cattand P (1995) New geographical approaches to control of some parasitic zoonoses. Bulletin of the World Health Organization 73: 247–257.
18. Kazembe L, Kleinschmidt I, Holtz T, Sharp B (2006) Spatial analysis and mapping of malaria risk in Malawi using point-referenced prevalence of infection data. International Journal of Health Geographics 5: 41.
19. Kitron U (1998) Landscape ecology and epidemiology of vector-borne diseases: tools for spatial analysis. Journal of Medical Entomology 35: 435–445.
20. Brooker S (2007) Spatial epidemiology of human schistosomiasis in Africa: risk models, transmission dynamics and control. Transactions of the Royal Society of Tropical Medicine and Hygiene 101: 1–8.
21. Brooker S, Michael E (2000) The potential of geographical information systems and remote sensing in the epidemiology and control of human helminth infections. Advances in Parasitology 47: 246–288.
22. Thompson DF, Malone JB, Harb M, Faris R, Huh OK, et al. (1996) Bancroftian filariasis distribution and diurnal temperature differences in the southern Nile delta. Emerging Infectious Diseases 2: 234–235.
23. Townson H, Nathan MB, Zaim M, Guillet P, Manga L, et al. (2005) Exploiting the potential of vector control for disease prevention. Bulletin of the World Health Organization 83: 942–947.
24. Ottesen EA, Duke BO, Karam M, Behbehani K (1997) Strategies and tools for the control/elimination of lymphatic filariasis. Bulletin of the World Health Organization 75: 491–503.
25. CDC (1993) Recommendations of the International Task Force for Disease Eradication. Morbidity and Mortality Weekly Report 42: 1–38.
26. Ottesen EA (2000) The global programme to eliminate lymphatic filariasis. Tropical Medical and International Health 5: 591–594.
27. Ottesen EA (2006) Lymphatic filariasis: Treatment, control and elimination. Advances in Parasitology 61: 395–441.
28. Brinkmann UK (1976) Epidemiological investigations of Bancroftian filariasis in the Coastal Zone of Liberia Tropenmedizin und Parasitologie 28: 71–76.
29. Juminer B, Diallo S, Diagne S (1971) Le foyer de filariose lymphatique du secteur de Sandiara (Senegal). 1. Evaluation de l'endemicite. Achives de l'Institut Pasteur de Tunis 48: 231–246.
30. Lamontellerie M (1972) Resultats d'enquetes sur les filarioses dans l'Ouest de la Huate-Volta (Cerle de Banfora). Annales de Parasitologie Humaine et Comparee (Paris) 47: 743–838.
31. Sowilem MM, Bahgat IM, el-Kady GA, el-Sawaf BM (2006) Spectral and landscape characterization of filarious and non-filarious villages in Egypt. Journal of the Egyptian Society for Parasitology 36: 373–388.
32. Wijers DJB (1977) Bancroftian filariasis in Kenya. I. Prevalence survey among adult males in the Coast Province. Annals of Tropical Medicine and Parasitology 71: 313–331.
33. Gyapong JO, Kyelem D, Kleinschmidt I, Agbo K, Ahouandogbo F, et al. (2002) The use of spatial analysis in mapping the distribution of bancroftian filariasis in four West African countries. Annals of tropical medicine and parasitology 96: 695–705.
34. Meyrowitsch DW, Nguyen DT, Hoang TH, Nguyen TD, Michael E (1998) A review of the present status of lymphatic filariasis in Vietnam. Acta Tropica 70: 335–347.
35. Michael E, Bundy DAP (1997) Global mapping of lymphatic filariasis. Parasitology Today 13: 472–476.
36. Sabesan S, Palaniyandi M, Michael E, Das PK (2000) Mapping lymphatic filariasis at the district-level in India. Annals of Tropical Medicine and Parasitology 94: 591–606.
37. Beau de Rochars MV, Milord MD, St Jean Y, Desormeaux AM, Dorvil JJ, et al. (2004) Geographic distribution of lymphatic filariasis in Haiti. American Journal of Tropical Medicine and Hygiene 71: 598–601.
38. Lindsay SW, Thomas CJ (2000) Mapping and estimating the population at risk from lymphatic filariasis in Africa. Transactions of the Royal Society of Tropical Medicine and Hygiene 94: 37–45.
39. Diggle PJ, Tawn JA, Moyeed RA (1998) Model-based geostatistics. Journal of the Royal Statistical Society: Series C (Applied Statistics) 47: 299–350.
40. Paterson S, Lello J (2003) Mixed models: getting the best use of parasitological data. Trends in Parasitology 19: 370–375.
41. Dormann CF, McPherson JM, Araújo MB, Bivand R, Bolliger J, et al. (2007) Methods to account for spatial autocorrelation in the analysis of species distributional data: a review. Ecography 30: 609–628.
42. Christensen OF, Ribeiro Jr PJ (2002) geoRglm: A package for generalised linear spatial models. R News 2: 26–28.

43. Belitz C, Brezger A, Kneib T, Lang S (2009) BayesX - Software for Bayesian inference in structured additive regression models. Version 201 Available: http://www.stat.uni-muenchen.de/bayesx/bayesx.html. Accessed 2011 Jun 05.
44. Lunn DJ, Thomas A, Best N, Spiegelhalter D (2000) WinBUGS - A Bayesian modelling framework: Concepts, structure, and extensibility. Statistics and Computing 10: 325–337.
45. Breslow NE, Clayton DG (1993) Approximate Inference in Generalized Linear Mixed Models. Journal of the American Statistical Association 88: 9–25.
46. Zhang H (2002) On Estimation and Prediction for Spatial Generalized Linear Mixed Models. Biometrics 58: 129–136.
47. Buckland ST, Elston DA (1993) Empirical Models for the Spatial Distribution of Wildlife. Journal of Applied Ecology 30: 478–495.
48. Wang XH, Zhou XN, Vounatsou P, Chen Z, Utzinger J, et al. (2008) Bayesian Spatio-Temporal Modeling of Schistosoma japonicum Prevalence Data in the Absence of a Diagnostic 'Gold' Standard. PLoS Negl Trop Dis 2: e250.
49. Staubach C, Schmid V, Knorr-Held L, Ziller M (2002) A Bayesian model for spatial wildlife disease prevalence data. Preventive Veterinary Medicine 56: 75–87.
50. Kleinschmidt I, Bagayoko M, Clarke GPY, Craig M, Le Sueur D (2000) A spatial statistical approach to malaria mapping. International Journal of Epidemiology 29: 355–361.
51. Michael E, Bundy DA, Grenfell BT (1996) Re-assessing the global prevalence and distribution of lymphatic filariasis. Parasitology 112: 409–428.
52. Hassan AN, Dister S, Beck L (1998) Spatial analysis of lymphatic filariasis distribution in the Nile Delta in relation to some environmental variables using geographic information system technology. J Egypt Soc Parasitol 28: 119–131.
53. Lardeux F, Cheffort J (1997) Temperature thresholds and statistical modelling of larval Wuchereria bancrofti (Filariidea:Onchocercidae) developmental rates. Parasitology 114 123–134.
54. Patz JA, Campbell-Lendrum D, Holloway T, Foley JA (2005) Impact of regional climate change on human health. Nature 438: 310–317.
55. Lafferty KD (2009) The Ecology of Climate Change and Infectious Diseases. Ecology 90: 888–900.
56. McMichael AJ (2003) Global climate change: will it affect vector-borne infectious diseases? Internal medicine journal 33: 554–555.
57. Michael E, Malecela MN, Zervos M, Kazura JW (2008) Global eradication of lymphatic filariasis: the value of chronic disease control in parasite elimination programmes. PLoS One 3: e2936.
58. Anderson RM, May RM (1992) Infectious Diseases of Humans. Dynamics and Control. Oxford: Oxford University Press.
59. Mills JN, Gage KL, Khan AS (2010) Potential influence of climate change on vector-borne and zoonotic diseases: a review and proposed research plan. Environmental Health Perspectives.
60. Hijmans R, J., Cameron S, E., Parra J, L., Jones P, G., Jarvis A (2005) Very high resolution interpolated climate surfaces for global land areas. International Journal of Climatology 25: 1965–1978.
61. Diggle PJ, Ribeiro PJ, Christensen OF (2003) An introduction to model-based geostatistics. In: Moller J, editor. Spatial statistics and computational methods Lecture notes in statistics. New York: Springer. 43–86.
62. Ribeiro PJ, Christensen OF, Diggle PJ (2003) Geostatistical software - geoR and geoRglm. DSC 2003 Working Papers.
63. Spiegelhalter TD, Best N, Lunn D (2007) WinBUGS User Manual Version 1.4.3. [Online] Available: http://www.mrc-bsu.cam.ac.uk/bugs/winbugs/contents.shtml. Accessed 2011 June 5.
64. Gelman A, Rubin DB (1992) Inference from iterative simulation using multiple sequences. Statistical Science 7: 457–511.
65. Austin PC, Tu JV (2004) Bootstrap Methods for Developing Predictive Models. The American Statistician 58: 131–137.
66. Craig M, Sharp B, Mabaso M, Kleinschmidt I (2007) Developing a spatial-statistical model and map of historical malaria prevalence in Botswana using a staged variable selection procedure. International Journal of Health Geographics 6: 44.
67. Akaike H (1974) A new look at the statistical model identification. IEEE Transactions on Automatic Control 19: 716–723.
68. Spiegelhalter DJ, Best NG, Carlin BP, van der Linde A (2002) Bayesian measures of model complexity and fit. Journal of the Royal Statistical Society Series B (Statistical Methodology) 64: 583–639.
69. Silue KD, Raso G, Yapi A, Vounatsou P, Tanner M, et al. (2008) Spatially-explicit risk profiling of Plasmodium falciparum infections at a small scale: a geostatistical modelling approach. Malaria Journal 7: 111.
70. IPCC editor (2007) Climate Change 2007: The Physical Basis. Contribution of Working Group I to the Fourth Assessment Report of the Intergovernmental Panel on Climate Change. CambridgeUK: Cambridge University Press. 996 p.
71. Babyak MA (2004) What you see may not be what you get: a brief, nontechnical introduction to overfitting in regression-type models. Psychosometric Medicine 66: 411–421.
72. Susser M (1991) What is a Cause and How Do We Know One? A Grammar for Pragmatic Epidemiology. American Journal of Epidemiology 133: 635–648.
73. Galea S, Riddle M, Kaplan GA (2010) Causal thinking and complex system approaches in epidemiology. International Journal of Epidemiology 39: 97–106.
74. Ngwira B, Tambala P, Perez AM, Bowie C, Molyneux D (2007) The geographical distribution of lymphatic filariasis infection in Malawi. Filaria Journal 6: 12.
75. Onapa AW, Simonsen PE, Baehr I, Pedersen EM (2005) Rapid assessment of the geographical distribution of lymphatic filariasis in Uganda, by screening of schoolchildren for circulating filarial antigens. Annals of Tropical Medicine and Parasitology 99: 141–153.
76. Martens WJM, Jetten TH, Focks DA (1997) Sensitivity of Malaria, Schistosomiasis and Dengue to global warming. Climatic Change 35: 145–156.
77. Patz JA, Confalonieri UEC, Amerasinghe FP, Chua KB, Daszak P, et al. (2005) Ecosystem regulation of infectious diseases. In: Hassan R, Scholes R, Ash N, editors. Ecosystems and Human Well-Being: Curent State and Trends. Washington, D.C.: Island Press, Millennium Ecosystem Assessment Series. 391–415.
78. Ellis BR, Wilcox BA (2009) The ecological dimensions of vector-borne disease research and control. Cadernos de Saúde Pública 25: S155–S167.
79. Michael E, Malecela-Lazaro MN, Simonsen PE, Pedersen EM, Barker G, et al. (2004) Mathematical modelling and the control of lymphatic filariasis. Lancet Infectious Diseases 4: 223–234.
80. Kearney M, Porter W (2009) Mechanistic niche modelling: combining physiological and spatial data to predict species' ranges. Ecology Letters 12: 334–350.
81. Awerbuch T, Kiszewski AE, Levins R (2002) Surprise, nonlinearity and complex behaviour In: Martens P, McMichael AJ, editors. Environmental Change, Climate and Health. Cambridge: Cambridge University Press. pp. 96–119.
82. Béguin A, Hales S, Rocklöv J, Åström C, Louis VR, et al. (2011) The opposing effects of climate change and socio-economic development on the global distribution of malaria. Global Environmental Change 21: 1209–1214.
83. Tol RSJ, Ebi KL, Yohe GW (2007) Infectious disease, development, and climate change: a scenario analysis. Environment and Development Economics 12: 687–706.
84. Bosello F, Roson R, Tol RSJ (2006) Economy-wide estimates of the implications of climate change: human health. Ecological Economics 58: 579–591.
85. Tol RSJ, Dowlatabadi H (2001) Vector-borne diseases, development & climate change. Integrated Assessment 2: 173–181.
86. Campbell-Lendrum D, Woodruff R (2006) Comparative risk assessment of the burden of disease from climate change. Environmental Health Perspectives 114: 1935–1941.
87. Beven K (2007) Towards integrated environmental models of everywhere: uncertainty, data and modelling as a learning process. Hydrology and Earth System Sciences Discussions 11: 460–467.
88. WHO (2004a) Lymphatic filariasis elimination in the African region: progress report. WHO, Congo Brazaville.
89. Gosoniu L, Vounatsou P, Sogoba N, Smith T (2006) Bayesian modelling of geostatistical malaria risk data. Geospat Health 1: 127–139.
90. Clements ACA, Bosqué-Oliva E, Sacko M, Landouré A, Dembélé R, et al. (2009) A comparative study of the spatial distribution of Schistosomiasis in Mali in 1984–1989 and 2004–2006. PLoS Neglected Tropical Diseases 3: e431.
91. Gambhir M, Michael E (2008) Complex ecological dynamics and eradicability of the vector borne macroparasitic disease, lymphatic filariasis. PLoS One 3: e2874.

8

Climate Change Sensitivity Index for Pacific Salmon Habitat in Southeast Alaska

Colin S. Shanley*, David M. Albert

The Nature Conservancy, Juneau, Alaska, United States of America

Abstract

Global climate change may become one of the most pressing challenges to Pacific Salmon conservation and management for southeast Alaska in the 21st Century. Predicted hydrologic change associated with climate change will likely challenge the ability of specific stocks to adapt to new flow regimes and resulting shifts in spawning and rearing habitats. Current research suggests egg-to-fry survival may be one of the most important freshwater limiting factors in Pacific Salmon's northern range due to more frequent flooding events predicted to scour eggs from mobile spawning substrates. A watershed-scale hydroclimatic sensitivity index was developed to map this hypothesis with an historical stream gauge station dataset and monthly multiple regression-based discharge models. The relative change from present to future watershed conditions predicted for the spawning and incubation period (September to March) was quantified using an ensemble global climate model average (ECHAM5, HadCM3, and CGCM3.1) and three global greenhouse gas emission scenarios (B1, A1B, and A2) projected to the year 2080. The models showed the region's diverse physiography and climatology resulted in a relatively predictable pattern of change: northern mainland and steeper, snow-fed mountainous watersheds exhibited the greatest increases in discharge, an earlier spring melt, and a transition into rain-fed hydrologic patterns. Predicted streamflow increases for all watersheds ranged from approximately 1-fold to 3-fold for the spawning and incubation period, with increased peak flows in the spring and fall. The hydroclimatic sensitivity index was then combined with an index of currently mapped salmon habitat and species diversity to develop a research and conservation priority matrix, highlighting potentially vulnerable to resilient high-value watersheds. The resulting matrix and observed trends are put forth as a framework to prioritize long-term monitoring plans, mitigation experiments, and finer-scale climate impact and adaptation studies.

Editor: Inés Álvarez, University of Vigo, Spain

Funding: This project was funded by the Alaska Sustainable Salmon Fund (www.akssf.org; project #44550) and the Gordon and Betty Moore Foundation (www.moore.org). The funders had no role in study design, data collection and analysis, decision to publish, or preparation of the manuscript.

Competing Interests: The authors have declared that no competing interests exist.

* Email: cshanley@tnc.org

Introduction

Pacific Salmon (*Oncorhynchus* spp.) are a key cultural [1], ecological [2], and economic [3] driver in southeast Alaska with global socio-ecological value [4,5]. In a recent study by the U.S. Forest Service, 96% of Alaskans said that salmon are essential to the Alaskan way of life [6]. More than 50 species of animals feed on spawning salmon each year, and 1 in 10 jobs in southeast Alaska is supported by salmon [6]. Yearly salmon productivity has varied considerably by species and watershed—the regional productivity has remained relatively resilient [3], although the influence of hatchery enhancements on wild salmon stocks is of concern [7–9]. Sustained production and harvest opportunities are likely the result of a portfolio effect [10] whereby thousands of watersheds and tens-of-thousands of streams provide a diversity of freshwater habitats, promoting phenotypic diversity, which buffers regional salmon returns against variability in ocean [11], near-shore [12], and freshwater conditions [13]. Looking to the future, predicted hydrologic change associated with climate change may very well present the biggest challenge to Pacific Salmon conservation and management in the 21st Century [13–20].

Southeast Alaska is predicted to experience the largest change in winter days above freezing in all of North America due to climate change [21]. While boreal and arctic Alaska is expected to see the largest changes in absolute temperature [22], a small increase in temperature in southeast Alaska could have transformative ecological effects (e.g., [23–27]). The mean winter temperature is currently near freezing at $-4°C$ [28]; therefore, with a relatively small increase in temperature, many watersheds and portions of sub-basins will no longer receive winter precipitation as snow, and will transition into rain-fed systems. Many watersheds will likely see decreased total snowpack in headwaters that have provided water storage and moderated flows for salmon streams throughout the summer, and maintained cooler summer stream temperatures [17,25,26].

Watersheds of southeast Alaska can be generally categorized into three hydrologic types: (1) rain-fed, (2) snow-fed, and (3) glacial [29]. Each hydrologic type is expected to exhibit a spectrum of change associated with climate change [17,18,29] and provide unique challenges to resident salmon population adaptation [16,30]. In general, rain-fed systems will likely see increased winter flows, reduced summer flows, and higher stream temperatures all year [16,17,19,20]. Snow-fed systems will likely see more variable discharge patterns with increased rain-on-snow events that can cause flooding, an earlier spring melt, and some may transition into rain-fed hydrologic types due to loss of adequate headwater snowpack [17,29,31]. Glacial systems will likely see increased discharge year-round (until some systems lose glaciers altogether), generally colder summer stream temperatures while glaciers are still present, and warmer winter water temperatures [24,32–35].

Pacific Salmon are affected by hydroclimatic factors at every stage of their lifecycle. Starting with egg incubation in freshwater habitats, water temperatures have the greatest effect on development rates [36]. In addition, winter flow extremes can scour salmon eggs from mobile spawning substrates (i.e., smaller gravels) and cause increased sedimentation that eliminates or degrades habitat by reducing oxygenation [19]. Emergent fry and juvenile salmonids require adequate stream flows to maintain high quality, protected off-channel habitats in both summer and winter to support growth and survival [36]. Smolt migration requires adequate stream flows and appropriate temperature cues to reach estuarine rearing habitats in synchrony with high food availability in the near-shore marine environment [37,38]. Development and spawning success during the adult life-phase is in-turn dependent on global ocean circulation patterns, feed availability, and adequate stream flows and suitable temperatures for return migration and spawning [36].

A review of the current literature suggests egg-to-fry lifestage survival under projected climate change may be one of the most important limiting factors for salmon productivity and conservation in southeast Alaska freshwater habitats [15,18,19,27,29]. Recent studies in Washington State have detailed information on air-to-stream water temperature relationships [19], extreme flow events, and survival estimates [15]; population modeling studies thus far suggest egg-to-fry survival may be a key limiting factor under a range of climate change scenarios for the Pacific Northwest [15]—in addition to a suite of temperature related factors that may be less of a concern to salmon in northern parts of their range [17]. However, summer habitat conditions warrant further research in southeast Alaska [17,38]. Washington State also has generally larger river systems that are more affected by climate model predictions for drier, hotter summers for interior parts of the state [19]. Southeast Alaska's mostly small and steep watersheds may behave more like the Olympic Peninsula watersheds where the scouring of eggs due to winter flooding events is currently thought to be more of concern than summer habitat conditions (e.g., [27]). Warmer winter water temperatures in southeast Alaska may accelerate salmon development rates, increase off-channel rearing habitats, and improve productivity in some watersheds (e.g., moderately glacial systems) in northern parts of their range [17,36,38]; this has been evidenced by increased salmon catch records during the warm-phase of the Pacific Decadal Oscillation in Alaska (PDO; [26]). However, depending on specific watershed physiographic characteristics (e.g., mean elevation, slope, and floodplain viability), stream channel geomorphology, genetic diversity, and capacity for phenotypic plasticity [16,20,30,39], directionally changing flow regimes will likely transition into less favorable habitats due to more frequent scouring events in segments of some watersheds [15].

Two primary questions were asked in this paper: (1) effects of future projection trends in temperature and precipitation on seasonal discharge patterns across southeast Alaska, and (2) vulnerability or resilience of watersheds to hydrologic change in relation to the current distribution of high-value salmon habitat. The project was conducted in five phases: (1) development of a comprehensive historical stream gauge station database for the region; (2) development of a transboundary geospatial database of watershed physiographic and climatic characteristics for historical and projected temperature and precipitation; (3) building and testing multiple regression-based monthly discharge models using AIC model selection; (4) mapping the regional discharge model results for projected change during the spawning and incubation period (September to March), and; (5) combining a climate change sensitivity index with an index of current salmon habitat and species diversity to develop a research and conservation priority matrix.

Methods

Historical discharge patterns

A database of historical stream gauge stations was developed for southeast Alaska using USGS mean monthly gauge station records and USGS gauge station catchment polygons. For gauge stations where a catchment polygon was not available, watersheds were delineated using the Shuttle Radar Topography Mission (SRTM) digital elevation model (DEM; 25 m) and the ArcGIS 10.1 Watershed delineation tool (ESRI, Redlands, CA). An inventory of all gauge stations ever recorded via the USGS National Water Inventory System was conducted; only those gauge stations without human alteration (e.g., dams, hydro plants, etc.) were obtained for analysis. When more than one gauge station existed in a watershed, the station with the longest period of record was selected. In order to achieve a spatial distribution necessary for landscape modeling, ≥5 years of monthly discharge data was used as a cut-off for modeling purposes. Furthermore, data post-1976 Pacific Decadal Oscillation (PDO) was also used as a cut-off so as not to confound the analysis with PDO cycles [13,40]. Monthly means (cubic feet per second) for each gauge station across the period of record available were calculated for model development.

Watershed physiography and climatology

A transboundary geospatial database of physical watershed characteristics, and historical and projected climatologies were developed to model present and future patterns of stream discharge for all watersheds. The SRTM DEM was used for the basis of analysis in all U.S. watersheds, and U.S. portions of transboundary rivers. The best available DEMs were used for Canadian reaches (Table 1). The climate modeling software ClimateWNA 4.62 [41] was used at a 1 km spatial resolution to map gridded estimates of monthly temperature and precipitation from the PRISM climate model [42–44]. The three top performing global climate models for the region (ECHAM5, HadCM3, and CGCM3.1; [22,45]) from the Intergovernmental Panel on Climate Change (IPCC) Fourth Assessment were averaged into a single ensemble model for analysis. This ensemble model was used to project temperature and precipitation for the year 2080 using three global greenhouse gas emission scenarios to identify regional trends [41,46]: B1 (low growth), A1B (moderate growth), A2 (high growth). The National Hydrography Dataset (NHD) was used to delineate lake coverage within the U.S., and the best available datasets were used for Canada (Table 1). Glacier coverage for all watersheds was delineated with the Alaska Department of Natural Resources 1:2,000,000 glacier coverage. The transboundary watershed polygons ($n = 1784$) developed by the U.S. Forest Service were used to create a regional watershed database. Mean values were calculated for temperature, precipitation, and elevation. Percent coverage was calculated for lakes and glaciers. Values for the historical gauge station catchment polygons and the complete regional watershed polygons were calculated separately.

Table 1. Climatic and physiographic geospatial data sources used to develop multiple regression-based monthly discharge models for southeast, Alaska, USA.

Variable	Description
Basin Area	USGS gauge station catchment polygons were used for the historical gauge station analysis. The USFS transboundary watershed layer, derived from primarily USGS HUC10 polygons, was used for the regional analysis.
Precipitation	ClimateWNA version 4.62 [41] 1 km downscaling of PRISM climate model [42] with monthly means generated from the period 1961–1990.
Temperature	ClimateWNA version 4.62 [41] 1 km downscaling of PRISM climate model [42] with monthly means generated from the period 1961–1990.
Elevation	Alaska: Shuttle Radar Topography Mission DEM (25 m), British Columbia: Terrain Resource Information Management Program DEM (25 m), Yukon: Department of Environment DEM (70 m). These DEMs were combined and resampled to 70 m.
Lakes	Alaska: National Hydrography Dataset, British Columbia: British Columbia Watershed Altus, Yukon: North American Water Polygons by ESRI. These polygons were combined into one lake coverage.
Glaciers	Alaska Department of Natural Resources 1:2,000,000 glacier coverage.

Discharge models and sensitivity index

A hierarchy of multiple regression models aimed at explaining monthly discharge using six potential explanatory variables were tested (Table 2). These models were tested using multiple regression and AIC model selection criterion [47]. Seven *a priori* models were developed, starting with the simplest model where basin area is the only explanatory variable, to more complex models incorporating climatology and physiographic setting. The same models were tested for each month for consistency and comparability. The most parsimonious, best-fit models with the lowest ΔAIC scores and highest weights were selected for an accuracy evaluation and the final hydroclimatic sensitivity index.

The eight most regionally relevant discharge model variables used in Wiley and Curran [48] for the state of Alaska and conterminous basins in Canada were considered for modeling: basin area, main channel length, mean channel slope, mean basin elevation, % lakes, % forest, % glaciers, mean precipitation and mean temperature, as well as an estimate of precipitation as snow [44]. A Pearson's pairwise correlation analysis was conducted on the data matrix for each month to identify any collinearities ($|r| \geq 0.7$). The resulting dataset used for analysis was paired down to six key variables: basin area, mean basin elevation, % lakes, % glaciers, mean monthly precipitation, and mean monthly temperature. The physiographic variables (i.e., elevation, lakes, and glaciers) and monthly discharge values were normalized with log transformations [49].

The best-fit multiple regression equations for each month were first run on the historical gauge station database. A yearly hydrograph was plotted for a cross-section of rain-fed, snow-fed, and glacial systems for observed, predicted, and future projections for visual interpretation. The model accuracy was evaluated by comparing the observed monthly means with the predicted monthly means in a percent error matrix. The best-fit monthly regression equations were then run on the regional database to calculate monthly means for present and future conditions across all watersheds.

A hydroclimatic sensitivity index was calculated for all watersheds by averaging the percent change in predicted discharge (A1B emission scenario) from September to March when salmon eggs are in the gravel, and broken into a relative rank index by standard deviations from the mean. A salmon habitat and species diversity index was calculated using the 2012 Alaska Department of Fish and Game (ADF&G) Anadromous Waters Catalog (AWC). The AWC mapped presence of salmon species by stream reach was converted to the kilometers of salmon stream in each watershed, weighted by the number of species (≤6 including Pacific salmon and steelhead, *O. mykiss*), and scaled by the percent of total salmon streams in the region. The final priority matrix was the hydroclimatic sensitivity index and the salmon habitat and species diversity index split into four simple risk-value categories with median value cut-points.

Results

The historical gauge station database (n = 41) showed a relatively well distributed spatial pattern with station locations spanning the latitudinal gradient of southeast Alaska from Ketchikan (55° latitude) in the south to Yakutat (59° latitude) in the north (Figure 1, Table S1). The gauge stations catchments were also distributed among islands (n = 18), mainland (n = 23), transboundary/interior (n = 3), and glacial systems (n = 13).

The AIC model selection process showed a characteristic seasonal pattern in variable selection. The simplest three variable models were in April and October with basin area, temperature and precipitation as the most parsimonious, best-fit model (Table 2). Elevation was selected as an additional variable for the early summer months of May, June, July, and again in the fall for the month of November (Table 2). Glaciers were added as a variable in August and September, and then again in December through March (Table 2). The most complex 6-variable model with the addition of lakes was selected in late winter months, February and March (Table 2). The regression coefficients matched seasonal patterns in discharge. Precipitation had a positive association in all models (Table 3). Temperature had a positive association in all months except for mid-summer, June, July and August (Table 3). Elevation had a positive association in summer months, May to September, and a negative association in winter months, November to March (Table 3). Glaciers had a positive association in summer months, August to September, and negative association in winter and spring months, November to March (Table 3). Lakes had a positive coefficient in all models in which it was included (Table 3). All models had high correlation values (Adjusted $R^2 = 0.964$–0.980) with the best models in summer months (Table 3).

Climate model projection trends for gauge station catchments in year 2080 show an increase in temperature, precipitation, and discharge, with the exception of July and August discharge decreases (Figure 2, Table S2, Table S3, Table S4, and Table S5). The mean monthly temperature for gauge station catchments is projected to increase from 3.4°C to an emission scenario range of 5.7–7.1°C, with the largest increases in late fall through spring (Figure 2). The mean monthly precipitation increased from 259.7

Table 2. Differences in AIC scores (ΔAIC), weights (AIC*w*), and number of model parameters (*k*) used to develop monthly multiple regression-based discharge models for southeast, Alaska, USA.

Models		**Month**					
		January		February		March	
Predictors	*k*	Δ AIC	AIC*w*	Δ AIC	AIC*w*	Δ AIC	AIC*w*
Area	1	69	0.000	79	0.000	63	0.000
Area + Precip	2	55	0.000	71	0.000	53	0.000
Area + Precip + Temp	3	6	0.024	13	0.001	9	0.006
Area + Precip + Temp + Elev	4	2	0.214	8	0.010	6	0.021
Area + Precip + Temp + Elev + Lakes	5	3	0.106	2	0.219	3	0.136
Area + Precip + Temp + Elev + Glac	6	0	0.477	3	0.159	1	0.338
Area + Precip + Temp + Elev + Glac + Lakes	7	2	0.180	0	0.611	0	0.499
		April		May		June	
Predictors	*k*	Δ AIC	AIC*w*	Δ AIC	AIC*w*	Δ AIC	AIC*w*
Area	1	62	0.000	39	0.000	56	0.000
Area + Precip	2	44	0.000	23	0.000	45	0.000
Area + Precip + Temp	3	0	0.529	24	0.000	36	0.000
Area + Precip + Temp + Elev	4	2	0.207	0	0.484	0	0.501
Area + Precip + Temp + Elev + Lakes	5	4	0.076	2	0.184	2	0.229
Area + Precip + Temp + Elev + Glac	6	3	0.136	1	0.243	2	0.185
Area + Precip + Temp + Elev + Glac + Lakes	7	5	0.052	3	0.089	4	0.086
		July		August		September	
Predictors	*k*	Δ AIC	AIC*w*	Δ AIC	AIC*w*	Δ AIC	AIC*w*
Area	1	60	0.000	53	0.000	36	0.000
Area + Precip	2	43	0.000	29	0.000	8	0.010
Area + Precip + Temp	3	27	0.000	13	0.001	10	0.004
Area + Precip + Temp + Elev	4	0	0.483	0	0.372	3	0.105
Area + Precip + Temp + Elev + Lakes	5	2	0.220	7	0.013	5	0.040
Area + Precip + Temp + Elev + Glac	6	2	0.193	0	0.343	0	0.467
Area + Precip + Temp + Elev + Glac + Lakes	7	3	0.104	1	0.272	0	0.375
		October		November		December	
Predictors	*k*	Δ AIC	AIC*w*	Δ AIC	AIC*w*	Δ AIC	AIC*w*
Area	1	47	0.000	60	0.000	60	0.000
Area + Precip	2	22	0.000	44	0.000	45	0.000
Area + Precip + Temp	3	0	0.406	0	0.325	8	0.009
Area + Precip + Temp + Elev	4	1	0.309	0	0.407	3	0.096
Area + Precip + Temp + Elev + Lakes	5	2	0.124	0	0.323	2	0.153
Area + Precip + Temp + Elev + Glac	6	3	0.115	0	0.330	0	0.431
Area + Precip + Temp + Elev + Glac + Lakes	7	4	0.046	2	0.186	1	0.311

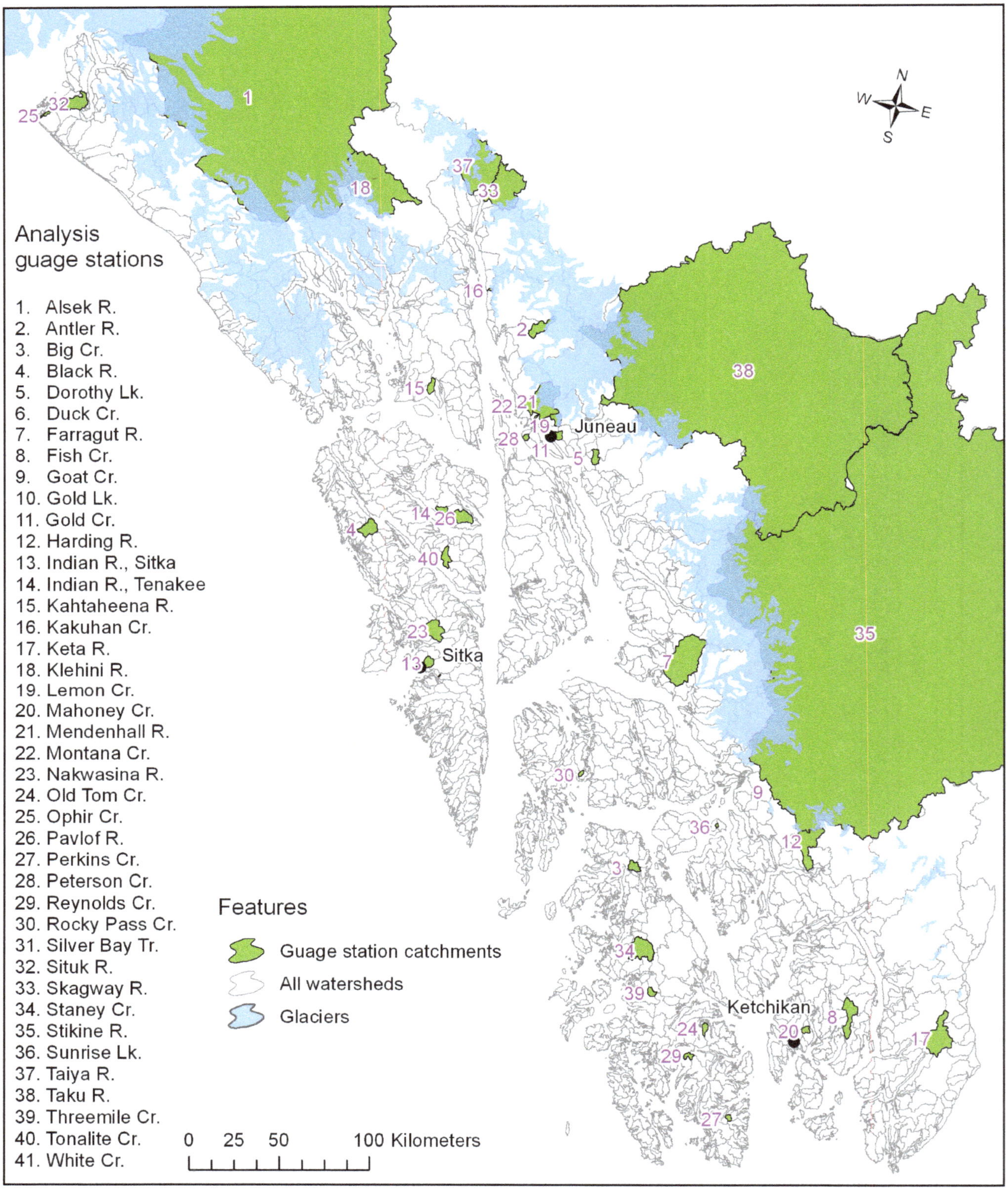

Figure 1. The watersheds of southeast, Alaska, USA, with the 41 gauge station catchments used for the development of regional multiple regression-based monthly discharge models.

Table 3. Coefficients (β), standard error (SE), and adjusted R^2 values for the most parsimonious multiple regression-based monthly discharge models for southeast, Alaska, USA.

	Constant	Area	Precip	Temp	Elevation	Glaciers	Lakes	Adj. R^2
January	−5.410**	0.986**	0.002**	0.059**	−0.167**	−0.911*		0.967
	(0.269)	(0.032)	(0.0002)	(0.009)	(0.077)	(0.508)		
February	−6.327**	1.097**	0.002**	0.093**	−0.192**	−0.981*	2.765*	0.973
	(0.280)	(0.033)	(0.0003)	(0.011)	(0.073)	(0.522)	(1.361)	
March	−6.192**	1.036**	0.002**	0.094**	−0.154*	−1.146*	2.318	0.964
	(0.336)	(0.037)	(0.0004)	(0.015)	(0.079)	(0.571)	(1.501)	
April	−7.090**	1.054**	0.002**	0.134**				0.973
	(0.298)	(0.032)	(0.0003)	(0.015)				
May	−7.339**	0.979**	0.003**	0.068**	0.409**			0.980
	(0.432)	(0.030)	(0.0004)	(0.020)	(0.072)			
June	−6.923	0.946**	0.003**	−0.011	0.609**			0.980
	(0.576)	(0.031)	(0.0007)	(0.026)	(0.083)			
July	−6.210**	0.965**	0.002**	−0.068**	0.535**			0.978
	(0.668)	(0.033)	0.0007	(0.029)	(0.088)			
August	−6.299**	0.989**	0.002**	−0.039	0.322**	0.780		0.979
	(0.595)	(0.033)	(0.0005)	(0.025)	(0.081)	(0.615)		
September	−6.806**	0.991**	0.002**	0.043**	0.219**	1.129**		0.979
	(0.430)	(0.030)	0.0002	(0.018)	(0.073)	(0.530)		
October	−6.430**	1.029**	0.001**	0.070**				0.975
	(0.311)	(0.032)	(0.0001)	(0.013)				
November	−5.935**	1.031**	0.001**	0.068**	−0.106			0.972
	(0.291)	(0.032)	(0.0002)	(0.010)	(0.071)			
December	−5.492**	0.989**	0.001**	0.056**	−0.158**	−1.025**		0.967
	(0.295)	(0.033)	(0.0002)	(0.010)	(0.075)	(0.480)		

Standard errors are reported in parentheses.
*$p<0.10$,
**$p<0.05$.

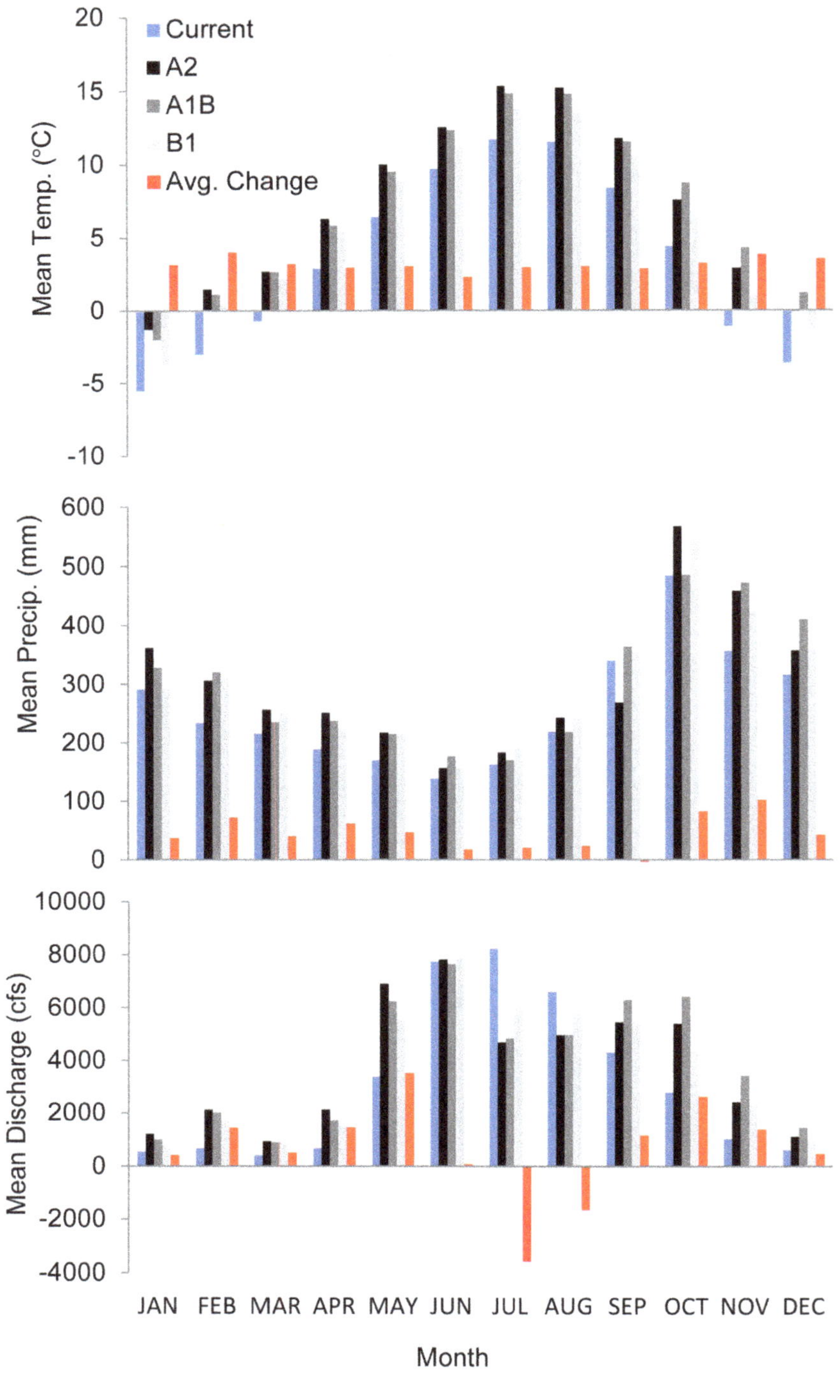

Figure 2. The historical mean (1961–1990) temperature (°C), precipitation (mm), and modeled discharge (cubic feet per second) for the 41 analysis gauge station catchments in southeast, Alaska, USA. These are shown next to projected changes for the year 2080 using an ensemble global climate model average (ECHAM5, HadCM3, and CGCM3.1) and three global greenhouse gas emission scenarios (B1, A1B, and A2) using multiple regression-based monthly discharge models.

to 297.0–302.4 mm (14.4–16.4% change), with the largest increases in late fall (Figure 2). The mean monthly modeled discharge increased from 3088.3 to 3599.6–3905.0 cfs (16.6–26.4% change), with the greatest increases in early fall and spring (Figure 2).

The yearly hydrograph plots for a cross-section of rain-fed, snow-fed, and glacial systems showed characteristic seasonal discharge patterns (Figure 3). Threemile Creek near Klawock, on southern Prince of Wales Island, is a small rain-fed system that generally followed seasonal precipitation patterns with peak flows in the fall with little water storage (Figure 3). The projected

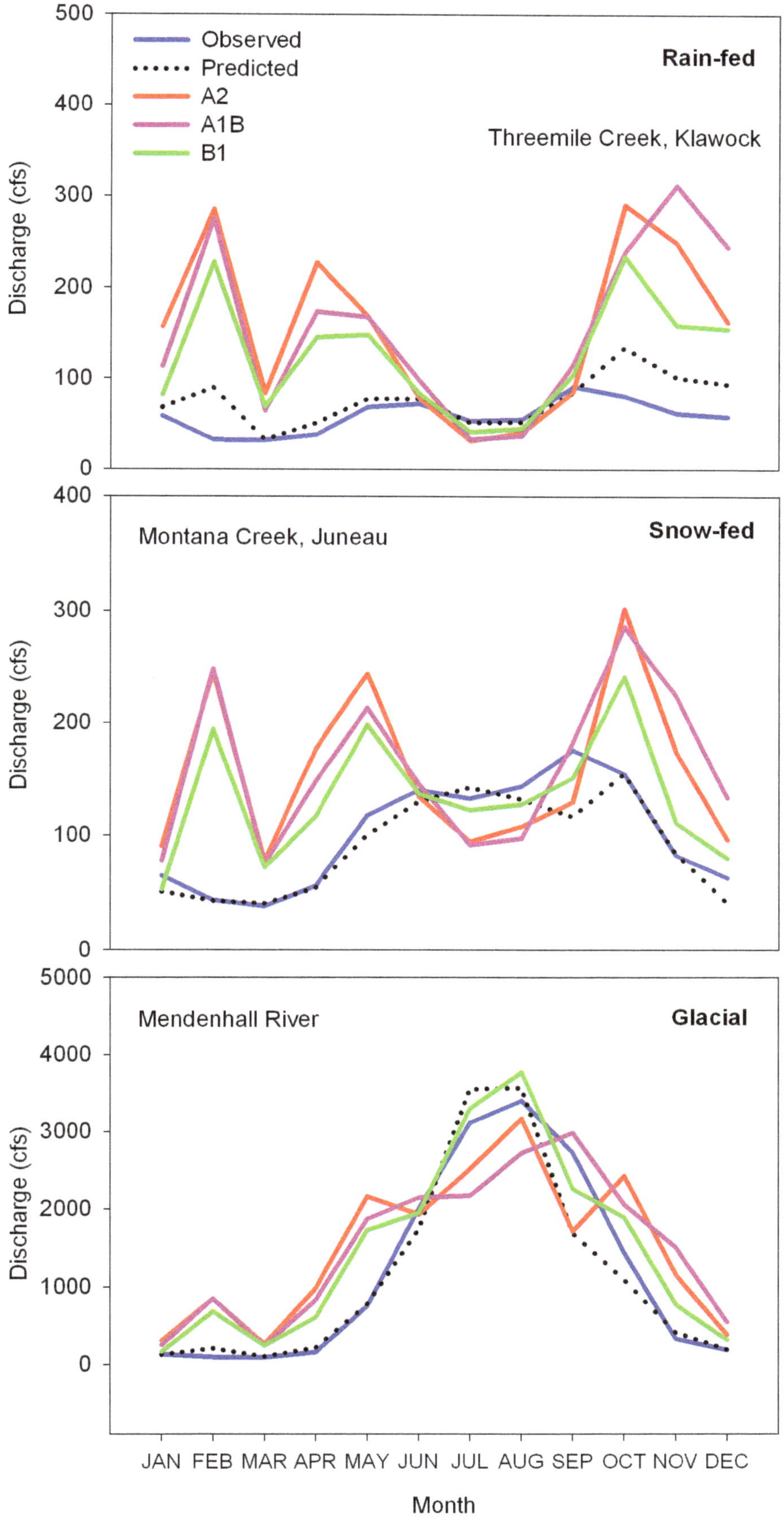
Observed
Predicted
A2
A1B
B1
Rain-fed
Threemile Creek, Klawock
Discharge (cfs)
500
400
300
200
100
0
Montana Creek, Juneau
Snow-fed
Discharge (cfs)
400
300
200
100
0
Mendenhall River
Glacial
Discharge (cfs)
5000
4000
3000
2000
1000
0
JAN FEB MAR APR MAY JUN JUL AUG SEP OCT NOV DEC
Month

Figure 3. Yearly hydrographs of monthly means for a cross section of rain-fed, snow-fed, and glacial watersheds in southeast, Alaska, USA. Recorded observations, modeled predictions, and future projections for the year 2080 are plotted for comparison using multiple regression-based monthly discharge models and an ensemble global climate model average (ECHAM5, HadCM3, and CGCM3.1) run for three global greenhouse gas emission scenarios (B1, A1B, and A2).

hydrograph for Threemile Creek showed increased peak flows in the spring and fall, and similar or lower discharge in summer months (Figure 3). Montana Creek near Juneau is a medium-sized snow-fed system with discharge patterns that showed a spring melt and fall peak (Figure 3). The projected hydrograph for Montana Creek showed increased spring and fall peak flows, and lower discharge in summer months (Figure 3). The Mendenhall River near Juneau is a large glacial system with characteristic peak flows in mid-summer with a bell-shaped yearly hydrograph (Figure 3). The projected hydrograph for the Mendenhall River showed generally increased discharge year-round with a more spread out summer peak (Figure 3).

The historical gauge station model evaluation showed that monthly discharge was on average over predicted (Table S6). The mean monthly discharge absolute error for all models was 36.4%. The most accurate models were in summer months with a mean absolute error of 28.5%. The mean absolute error during the analysis period of September to March was 41.9%.

The regional hydroclimatic sensitivity index showed northern mainland and steeper, snow-fed mountainous watersheds with current winter temperatures close to freezing exhibiting the greatest change (Figure 4). The highest ranking group (>1.5 SD; 184–280%) were distributed along the northern mainland, including some glacial systems, and others were scattered throughout the region. Next, there was a group of watersheds with high spatial variability and moderate ranking (0.50–1.5 SD; 156–184%) with the highest concentration in the southeast corner of the region. Watersheds with average percent change (−0.50–0.50; 128–156%) occurred along the mainland, and included many glacial systems and mountainous areas of the major island groups. Fair to low ranking sensitivity (−0.50–<−1.5 SD; 93–128%) watersheds were located throughout the region in low elevation areas, and included several of the larger transboundary watersheds. The combined salmon habitat and species diversity hydroclimatic sensitivity index showed highest priority in systems that exhibited a combination of high salmon habitat and species diversity with changing hydrology (Figure 5).

Discussion

The most striking result of the analysis was the transition in mean winter temperatures across the freezing threshold for many watersheds, and how it translated into changes in regional discharge patterns that are important for the reproduction and survival of Pacific Salmon in southeast Alaska. Combined with predicted increases in precipitation across all months, mean monthly discharge was forecasted to increase by approximately 1-fold to 3-fold during September to March when salmon eggs are in the gravel and exposed to more frequent scouring events. Plots of yearly hydrographs showed substantially increased peak flows in rain-fed and snow-fed systems in early spring and late fall across emission scenarios. The hydroclimatic sensitivity index showed the interaction between projected climate trends and the region's diverse physiography—a relatively predictable pattern that could be used as a preliminary framework to develop targeted long-term monitoring and potential mitigation strategies. The combined salmon habitat and species diversity sensitivity index showed clusters of high-value watersheds that could be prioritized for: (1) conservation of available genetic and life history diversity (e.g., run timing; [30]); (2) evaluation of local, reach-scale geomorphic sediment mobility and susceptibility to scour; (3) restoration to improve natural variability and general ecological resilience [15,50]; (4) monitoring harvest pressures and escapement goals in light of new environmental factors, and; (5) developing finer-scale climate impact and salmonid adaptation studies (e.g., [19,20,27,34,38,51]).

The ability of the discharge models to accurately reflect seasonal flow patterns among watersheds and watershed types is a positive indicator for the models' ability to predict regional trends in ungauged watersheds across the landscape. An evaluation of model selection, coefficient direction, and yearly hydrographs agreed with the current general understanding of regional hydrologic patterns [13,24,32,52]. The models general over-prediction of discharge is common among discharge models [48]. The percent error scores (both positive and negative) are likely attributed to a combination of sources: resolution of the climate data; older and more remote faulty gauge station records, especially during winter months when stations are more prone to freezing issues, and; water storage ecological processes (e.g., wetlands) simply not captured in the generalized linear regional model. This regional modeling effort is an important step forward in the development of future predictive models for southeast Alaska. Uncertainty associated with the climate modeling and potential effects could be further reduced with higher spatial and temporal resolution datasets that can capture ecological processes such as geomorphologic change and generate flood frequency statistics. Future forecasting studies would also benefit from a broader stream gauge network that represents the spectrum of potentially vulnerable to resilient salmon producing watersheds, with an emphasis on monitoring extreme events such as the frequency of flooding, coupled with experimental studies on egg-to-fry survival and salmonid adaptive capacity.

Nested within the hydroclimatic sensitivity framework, river channel types will change with flow rates and different species of salmon will likely be affected differently based on their body size, freshwater life-history and residence time, which warrants further research. The body size of salmon species directly relates to the burial depth of eggs and therefore exposure to scouring events in mobile spawning substrates [53,54]. The spawning location in watersheds directly relates to the rate of flow and sedimentation exposure [19]. The highest flow and sedimentation rates are generally expected in the central parts of the mainstem and significant tributaries, as opposed to the headwaters or floodplains [19]. Steelhead (*O. mykiss*) spawn in the spring within the highest reaches of the watershed [36] and could, for example, be less prone to winter souring events. Coho salmon (*O. kisutch*), sockeye salmon (*O. nerka*), and Chinook salmon (*O. tshawytscha*) spawn in the mid-portions of the watersheds, have longer residence times [36], and may therefore be the most exposed species to flooding events; however, their relative body sizes will help egg burial depths and exposure to scour [53]. Pink salmon (*O. gorbuscha*) and chum salmon (*O. keta*) spawn and rear in the lowest reaches of watersheds and migrate a few weeks after they emerge from the gravel (April–May) [36], and may therefore be less vulnerable to scouring where floodplains are intact. However, potential sea level

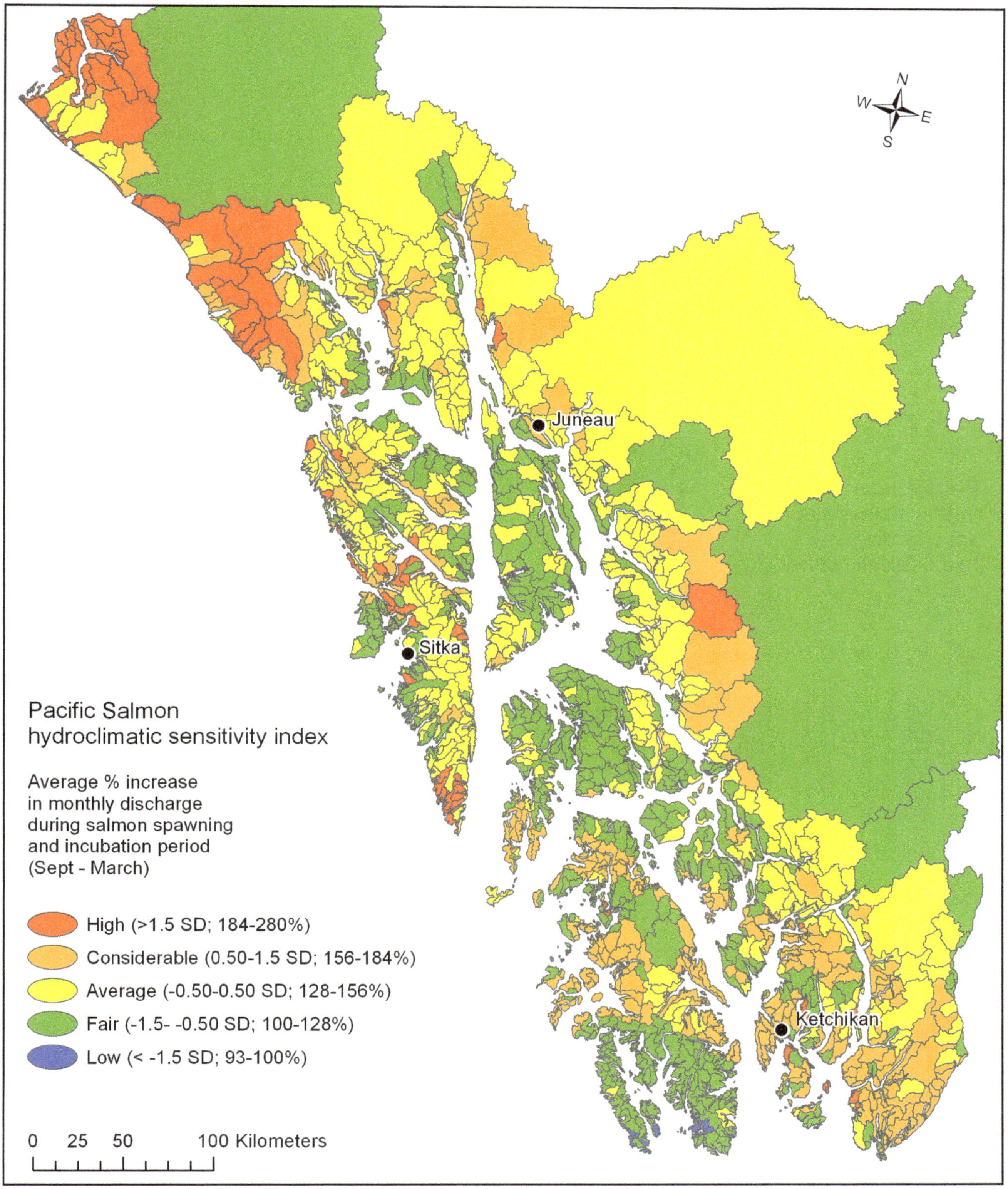

Figure 4. The watersheds of southeast, Alaska, USA (n = 1784) showing the Pacific Salmon hydroclimatic sensitivity index for predicted hydrologic change. This was derived using multiple regression-based monthly discharge models, an ensemble global climate model average (ECHAM5, HadCM3, and CGCM3.1) for temperature and precipitation, and the A1B global greenhouse gas emission scenario projections for the year 2080 to illustrate regional trends.

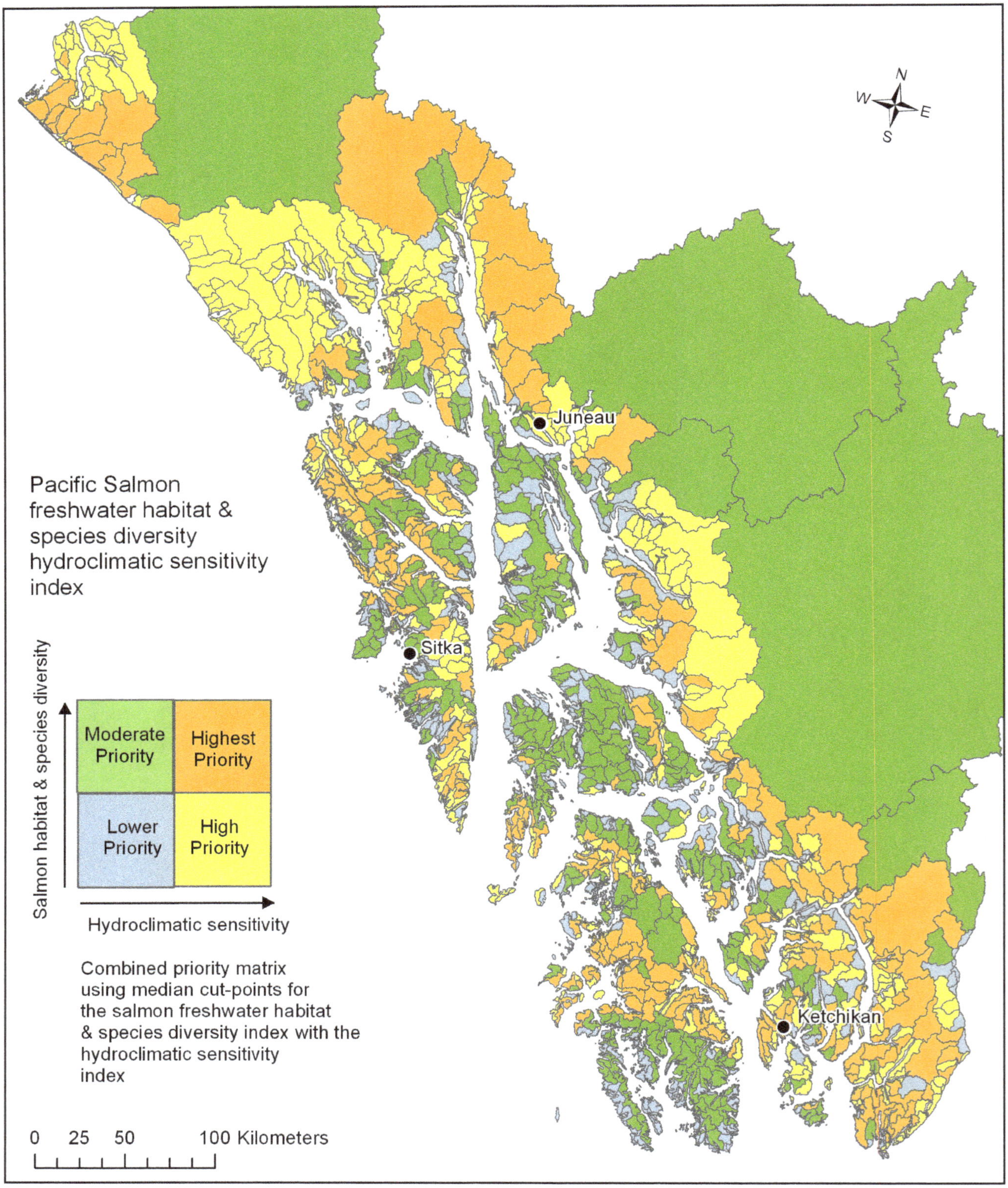

Figure 5. The watersheds of southeast, Alaska, USA (n = 1784) showing the Pacific Salmon freshwater habitat and species diversity index combined with the hydroclimatic sensitivity index developed in this study.

rise is a factor for low-elevation pink salmon and chum salmon spawning habitat in southern southeast Alaska [17,55].

Hydrologic projections for Washington State predict a complete loss of snow-fed hydrologic systems by the year 2080 under the same A1B emission scenario, with only a few watersheds in the North Cascade mountains retaining transitional rain/snow characteristics [18,56]. The snow-fed watersheds are predicted to transition into rain-fed hydrologic patterns with low flows in the summer and more intense flooding events during the winter months [18]. Increases in temperature appear to have a stronger influence on flows than changes in precipitation in Washington State [19]; this also appears to be the case with the largest changes in projected flows for southeast Alaska occurring during months when many watersheds cross the freezing threshold. Thus far, simulation studies suggest low-elevation floodplain and wetland connectivity restoration efforts to slow increased flow rates and improve summer water storage are the most effective climate change mitigation strategies in population viability models for Washington State [15]. These ecosystem engineering techniques could be further tested in historically impacted watersheds in southeast Alaska [57].

The results of this analysis should be treated as a hypothesis of potential change, and a framework for finer-scale experimental studies that investigate the long-term effects of changing hydrologic regimes, inter-annual variability, extreme events, and salmonid adaptive capacity in southeast Alaska. As global climate models, hydrologic forecasting and downscaling techniques improve, the absolute values for projected temperature, precipitation, and stream discharge will change and model predictions will improve. The results of this study provide a compelling case for how climate change trends could interact with the regions diverse physiography in a relatively predictable pattern where research and mitigation experiments should be prioritized for effective salmon conservation investments.

Supporting Information

Table S1 Recorded mean monthly discharge (cubic feet per second) for 41 southeast, Alaska, USA gauge stations with ≥5 year period of record.

Table S2 Multiple regression-based monthly discharge model predictions for 41 southeast, Alaska, USA gauge station catchments.

Table S3 Multiple regression-based monthly discharge models for projected future discharge (cubic feet per second) for 41 gauged catchments in southeast, Alaska, USA using an ensemble global climate model average (ECHAM5, HadCM3, and CGCM3.1) and the B1 global greenhouse gas emission scenario for the year 2080.

Table S4 Multiple regression-based monthly discharge models for projected future discharge (cubic feet per second) for 41 gauged catchments in southeast, Alaska, USA using an ensemble global climate model average (ECHAM5, HadCM3, and CGCM3.1) and the A1B global greenhouse gas emission scenario for the year 2080.

Table S5 Multiple regression-based monthly discharge models for projected future discharge (cubic feet per second) for 41 gauged catchments in southeast, Alaska, USA using an ensemble global climate model average (ECHAM5, HadCM3, CGCM3.1) and the A2 global greenhouse gas emission scenario for the year 2080.

Table S6 Multiple regression-based monthly discharge model percent error ((predicted-observed)/observed) for 41 southeast, Alaska, USA gauged catchments.

Acknowledgments

C.L. Woll, M.V. McPhee, and two anonymous reviewers made helpful comments on the developing manuscript.

Author Contributions

Conceived and designed the experiments: CS DA. Performed the experiments: CS. Analyzed the data: CS. Contributed reagents/materials/analysis tools: CS. Wrote the paper: CS DA.

References

1. Bottom DL, Jones KK, Simenstad CA, Smith CL (2009) Reconnecting social and ecological resilience in salmon ecosystems. Ecology and Society 14: 5.
2. Gende SM, Edwards RT, Willson MF, Wipfli MS (2002) Pacific salmon in aquatic and terrestrial ecosystems. Bioscience 52: 917–928.
3. Clarke JH, McGregor A, Mecum RD, Krasnowski P, Carroll A (2006) The commercial salmon fishery in Alaska. Alaska Fishery Research Bulletin 12: 1–146.
4. Gustafson RG, Waples RS, Myers JM, Weitkamp LA, Bryant GJ, et al. (2006) Pacific salmon extinctions: quantifying lost and remaining diversity. Conservation Biology 21: 1009–1020.
5. Pinsky ML, Springmeyer DB, Goslin MN, Augerot X (2009) Range-wide selection of catchments for Pacific Salmon conservation. Conservation Biology 23: 680–691.
6. USFS (2012) Salmon in the Tongass National Forest. R10-PR-028. Juneau, Alaska: USDA Forest Service.
7. Araki H, Cooper B, Blouin MS (2007) Genetic effects of captive breeding cause a rapid, cumulative fitness decline in the wild. Science 318: 100–103.
8. Buhle ER, Holsman KK, Scheuerell MD, Albaugh A (2009) Using an unplanned experiment to evaluate the effects of hatcheries and environmental variation on threatened populations of wild salmon. Biological Conservation 142: 2449–2455.
9. Ruggerone GT, Peterman RM, Dorner B, Myers KW (2010) Magnitude and trends in abundance of hatchery and wild Pink salmon, Chum salmon, and Sockeye salmon in the North Pacific Ocean. Marine and Coastal Fisheries: Dynamics, Management, and Ecosystem Science 2: 306–328.
10. Schindler DE, Hilborn R, Chasco B, Boatright CP, Quinn TP, et al. (2010) Population diversity and the portfolio effect in an exploited species. Nature 465: 609–612.
11. Mantua NJ, Hare SR, Zhang Y, Wallace JM, Francis RC (1997) A Pacific interdecadal climate oscillation with impacts on salmon production. Bulletin of the American Meteorological Society 78: 1069–1079.
12. Koski KV (2009) The fate of coho salmon nomads: The story of an estuarine-rearing strategy promoting resilience. Ecology and Society 14: 4. http://www.ecologyandsociety.org/vol14/iss11/art14/.
13. Neal EG, Walter MT, Coffeen C (2002) Linking the pacific decadal oscillation to seasonal stream discharge patterns in southeast Alaska. Journal of Hydrology 263: 188–197.
14. Adkison MD, Finney BP (2003) The long-term outlook for salmon returns to Alaska. Alaska Fishery Research Bulletin 10: 83–94.
15. Battin J, Wiley MW, Ruckelshaus MH, Palmer RN, Korb E, et al. (2007) Projected impacts of climate change on salmon habitat restoration. PNAS 104: 6720–6725.
16. Beechie T, Buhle E, Ruckelshaus M, Fullerton A, Holsinger L (2006) Hydrologic regime and the conservation of salmon life history diversity. Biological Conservation 130: 560–572.
17. Bryant MD (2009) Global climate change and potential effects on Pacific salmonids in freshwater ecosystems of southeast Alaska. Climatic Change 95: 169–193.
18. Elsner MM, Cuo L, Voisin N, Deems JS, Hamlet AF, et al. (2010) Implications of 21st century climate change for the hydrology of Washington State. Climatic Change: DOI: 10.1007/s10584-10010-19855-10580.

19. Mantua N, Tohver I, Hamlet A (2010) Climate change impacts on streamflow extremes and summertime stream temperature and their possible consequences for freshwater salmon habitat in Washington State. Climatic Change: DOI 10.1007/s10584-10010-19845-10582.
20. Taylor SG (2008) Climate warming causes phenological shift in Pink Salmon, Oncorhynchus gorbuscha, behavior at Auke Creek, Alaska. Global Change Biology 14: 229–235.
21. Meehl GA, Tebaldi C, Nychka D (2004) Changes in frost days in simulations of twentyfirst century climate. Climate Dynamics 23: 495–511.
22. Walsh JE, Chapman WL, Romanovsky V, Christensen JH, Stendel M (2008) Global climate model performance over Alaska and Greenland. Journal of Climate 21: 6156–6174.
23. Hennon PE, D'Amore DV, Schaberg PG, Wittwer DT, Shanley CS (2012) Shifting climate, altered niche, and a dynamic conservation strategy for Yellow-cedar in the North Pacific coastal rainforest. Bioscience 62: 147–158.
24. Hood E, Berner L (2009) Effects of changing glacial coverage on the physical and biogeochemical properties of coastal streams in southeastern Alaska. Journal of Geophysical Research 114: G03001, doi:03010.01029/02009JG000971.
25. Hodgson S, Quinn TP (2002) The timing of adult sockeye salmon migration into freshwater: adaptations by populations to prevailing thermal regimes. Canadian Journal of Zoology 80: 542–555.
26. Mote PW, Parson EA, Hamlet AF, Keeton WS, Lettenmaier D, et al. (2003) Preparing for climatic change: The water, salmon, and forests of the Pacific Northwest. Climatic Change 61: 45–88.
27. Milner AM, Robertson AL, McDermott MJ, Klarr MJ, Brown LE (2013) Major flood disturbance alters river ecosystem evolution. Nature Climate Change 3: 137–141.
28. Simpson JJ, Hufford GL, Daly C, Berg JS, Fleming MD (2005) Comparing maps of mean monthly surface temperature and precipitation for Alaska and adjacent areas of Canada produced by two different methods. Arctic 58: 137–161.
29. Edwards RT, D'Amore DV, Norberg E, Biles F (2013) Riparian ecology, climate change, and management in North Pacific Coastal Rainforests. In: G. H. Orians and J. Schoen, editors. North Pacific temperate rainforests: ecology and conservation. Seattle, WA: University of Washington Press. pp. 43–72.
30. Halupka KC, Willson M, Bryant MD, Everest FH, Gharrett AJ (2003) Conserving of population diversity of Pacific salmon in southeast Alaska. North American Journal of Fisheries Management 23: 1057–1086.
31. Stewart IT, Cayan DR, Dettinger MD (2004) Changes in snowmelt runoff timing in western North America under a 'business as usual' climate change scenario. Climatic Change 62: 217–232.
32. Neal EG, Hood E, Smikrud K (2010) Contribution of glacier runoff to freshwater discharge into the Gulf of Alaska. Geophysical Research Letters 37: L06404, doi:06410.01029/02010GL042385.
33. Nolin AW, Phillippe J, Jefferson A, Lewis SL (2010) Present-day and future contributions of glacier runoff to summertime flows in a Pacific Northwest watershed: Implications for water resources. Water Resources Research 46: W12509 DOI:12510.11029/12009WR008968.
34. Fellman JB, Nagorski S, Pyare S, Vermilyea AW, Scott D, et al. (2013) Stream temperature response to variable glacier cover in coastal watersheds of southeast Alaska. Hydrologic Processes doi: 10.1002/hyp.9742
35. Milner AM, Brown LE, Hannah DM (2009) Hydroecological response of river systems to shrinking glaciers. Hydrologic Process 23: 62–77.
36. Quinn TP (2005) The behavior and ecology of Pacific Salmon and trout. Seattle, WA.: University of Washington Press. 378 p.
37. Wipfli MS, Baxter CV (2010) Linking ecosystems, food webs, and fish production: subsidies in salmonid watersheds. Fisheries 35: 373–387.
38. Kovach RP, Joyce JE, Echave JD, Lindberg MS, Tallmon DA (2013) Earlier migration timing and decreasing phenotypic variation for multiple salmonid species. PLOS ONE 8: e53807.
39. Kovach RP, Gharrett AJ, Tallmon DA (2012) Genetic change for earlier migration timing and loss of biocomplexity in a pink salmon population. Proceeding of the Royal Society Biology 279: 3861–3869.
40. Schwarz TC (2010) Analysis of select stream discharge models in southeast Alaska. Masters Thesis, Cornell University, Ithaca, New York.
41. Wang T, Hamann A, Spittlehouse DL, Murdock TQ (2012) ClimateWNA–High-resolution spatial climate data for western North America. Journal of Applied Meteorology 51: 16–29.
42. Daly C, Gibson WP, Taylor GH, Johnson GL, Pasteris P (2002) A knowledge-based approach to the statistical mapping of climate. Climate Research 22: 99–113.
43. Hamann A, Wang GA (2005) Models of climatic normals for genecology and climate change studies in British Columbia. Agriculture and Forest Meteorology 128: 211–221.
44. Wang T, Hamann A, Spittlehouse D, Aitken SN (2006) Development of scale-free climate data for western Canada for use in resource management. International Journal of Climatology 26: 383–397.
45. Radic V, Clarke GKC (2011) Evaluation of IPCC models' performance in simulating late-twentieth-century climatologies and weather patterns over North America. Journal of Climate 24: 5257–5274.
46. Fussel H (2009) An updated assessment of the risks from climate change based on research published since the IPCC fourth assessment report. Climatic Change 97: 469–482.
47. Anderson DR (2008) Model based inference in the life sciences: a primer on evidence. New York, New York: Springer. 184 p.
48. Wiley JB, Curran JH (2003) Estimating annual high-flow statistics and monthly and seasonal low-flow statistics for ungauged sites on streams in Alaska and conterminous basins in Canada. Water-Resources Investigations Report 03-4114. Anchorage, Alaska: U.S. Geological Survey.
49. McGarigal K, Cushman S, Stafford S (2000) Multivariate statistics for wildlife ecology and research. New York: Springer.
50. Bisson PA, Dunham JB, Reeves GH (2009) Freshwater ecosystems and resilience of Pacific salmon: Habitat management based on natural variability. Ecology and Society 14: 45. http://www.ecologyandsociety.org/vol14/iss41/art45/.
51. Dorava JM, Milner AM (2000) Role of lake regulation on glacier-fed rivers in enhancing salmon productivity: the Cook Inlet watershed, south-central Alaska, USA. Hydrologic Processes 14: 3149–3159.
52. Schoch GM, Albert DM, Shanley CS (2013) An estuarine habitat classification for a complex fjordal island archipelago. Estuaries and Coasts: doi 10.1007/s12237-12013-19622-12233.
53. Montgomery DR, Beamer EM, Pess GR, Quinn TP (1999) Channel type and salmonid spawning distribution and abundance. Canadian Journal or Fisheries and Aquatic Sciences 56: 377–387.
54. Jonsson B, Finstad AG, Jonnson N (2012) Winter temperature and food quality affect age at maturity: an experimental test with Atlantic salmon (*Salmo salar*). Canadian Journal or Fisheries and Aquatic Sciences 69: 1817–1826.
55. Larsen CF, Motyka RJ, Freymueller JT, Echelmeyer KA, Ivins ER (2005) Rapid viscoelastic uplift in southeast Alaska caused by post-Little Ice Age glacial retreat. Earth and Planetary Science Letters 237: 548–560.
56. Mote PW, Salathe EP (2010) Future climate in the Pacific Northwest. Climatic Change DOI 10.1007/s10584-010-9848-z.
57. Albert DM, Schoen JW (2013) Use of historical logging patterns to identify disproportionately logged ecosystems within temperate rainforests of southeastern Alaska. Conservation Biology 27: 774–784.

9

Future Bloom and Blossom Frost Risk for *Malus domestica* Considering Climate Model and Impact Model Uncertainties

Holger Hoffmann*¤, Thomas Rath

Biosystems Engineering, Institute for Biological Production Systems, Leibniz Universität Hannover, Hannover, Germany

Abstract

The future bloom and risk of blossom frosts for *Malus domestica* were projected using regional climate realizations and phenological (= impact) models. As climate impact projections are susceptible to uncertainties of climate and impact models and model concatenation, the significant horizon of the climate impact signal was analyzed by applying 7 impact models, including two new developments, on 13 climate realizations of the IPCC emission scenario A1B. Advancement of phenophases and a decrease in blossom frost risk for Lower Saxony (Germany) for early and late ripeners was determined by six out of seven phenological models. Single model/single grid point time series of bloom showed significant trends by 2021–2050 compared to 1971–2000, whereas the joint signal of all climate and impact models did not stabilize until 2043. Regarding blossom frost risk, joint projection variability exceeded the projected signal. Thus, blossom frost risk cannot be stated to be lower by the end of the 21st century despite a negative trend. As a consequence it is however unlikely to increase. Uncertainty of temperature, blooming date and blossom frost risk projection reached a minimum at 2078–2087. The projected phenophases advanced by 5.5 d K^{-1}, showing partial compensation of delayed fulfillment of the winter chill requirement and faster completion of the following forcing phase in spring. Finally, phenological model performance was improved by considering the length of day.

Editor: Vanesa Magar, Plymouth University, United Kingdom

Funding: The study was supported by the Ministry for Science and Culture of Lower Saxony within the network KLIFF - climate impact and adaptation research in Lower Saxony. The funders had no role in study design, data collection and analysis, decision to publish, or preparation of the manuscript.

Competing Interests: The authors have declared that no competing interests exist.

* E-mail: hhoffmann@uni-bonn.de

¤ Current address: Institute of Crop Science and Resource Conservation (INRES), University of Bonn, Bonn, Germany

Introduction

Apple production and its economic efficiency are clearly influenced by blossom frosts [1]. In addition, global warming could increase the risk due to greater changes in the date of flowering than in the last spring freeze or increasing variability in both. A generally higher risk of frost after bud burst for warmer winters was further stated as due to faster completion of the chilling requirement [2]. Past observations of late frosts and blossom frosts around the world have indicated a decreasing [3,4] up to increasing risk [4–8] for fruit trees. However, findings cannot be generalized as they vary regionally. For instance, observed damages due to late frost increased in Northern Japan while other regions of Japan exhibited different tendencies [4]. An analysis of meteorological and phenological records of the Rhineland fruit-growing region in the West of Germany revealed, that risk of apple yield loss due to frosts in April remained unchanged during the period 1958 to 2007 [9–11]. This is consistent with studies showing an advance during the past of about 2.2 d/decade for both the last spring freeze (≤0°C, Central Europe, 1951–1997) [12] and for apple flowering (BBCH 60 [13], Germany, 1961–2000) [14].

Regardless of its development during the past, future blossom frost risk development remains uncertain as published estimates diverge (Table 1). Discrepancies are mainly due to differences in selected regions and varieties, as well as to the fact, that blossom frost risk computation requires estimates for flowering dates in addition to consistent climate time series which reproduce temperature thresholds (e.g. 0°C) accurately. For this purpose climate model temperature time series are used as input for empirical phenological models accounting for chilling and/or forcing phases in winter and spring respectively [15]. While most climate scenarios describe an enhanced warming beyond 2040 [16], the following risk estimates are given. For the apple cultivar *Golden delicious* a "decreasing trend … of little significance" was found (Trentino, Italy), concluding that blossom frost risk "will not differ greatly from its present level" [17]. Similarly, for Finland the risk is expected to generally "stay at the current level or to decrease" for the period 2011–2040 compared to 1971–2000, excepting the southern inland which exhibits increases [18]. Increases in frost damage to apple blossom (*Malus pumila* Mill. cv. Cox's Orange Pippin) were estimated for Britain [19] and an increase in the frequency of apple blossom frost damage was projected for Saxony (East Germany) by applying a simple thermal model to predict flowering, beginning on each 1 January [20]. Using the same approach, no increase in the mean apple blossom frost risk for Lower Saxony (Saxony and Lower Saxony are non-

Table 1. Published projections of future apple blossom frost risk.

Region	Increase (+) Decrease (−) No change (°)	Model	Statistics on time series[a]	Ref.
Trentino, Italy	−, °[b]	Modified Utah	yes	[17]
Finland	−, °,+[b]	Thermal Time	no	[18]
Britain	+	Thermal Time-Chilling	no	[19]
Saxony, Germany	+	Thermal Time	no	[20]
Lower Saxony, Germany	−, °,+[b]	Thermal Time	yes	[21]

[a]Tests on blossom frost risk.
[b]depending on subregion.

adjacent states) was found [21], despite temporarily/regionally increasing blossom frost risk.

These differences in estimates can be attributed to two deficits:

1) The modeling properties of the mentioned model [20,21] are very limited for climate impact studies, as it solely calculates the onset of a phenophase based on accumulation of a heat requirement (forcing), hence assuming that dormancy has already been satisfied by a fixed starting date (see [22] for more details). Since future fulfillment of dormancy cannot be guaranteed, models including chilling phases seem to be more suitable for future climate impact simulations [23]. With their help, a possible impact of climate change on the fulfillment of dormancy [6] can be assessed. However, most of these models rely only on air temperature, ignoring possible influences of other climatic variables. Nevertheless improvement was found after including light conditions in the form of day length [24,25], despite ongoing discussions about the influence of light conditions on tree phenological phases [26].
2) Published estimates of future blossom frost risk (Table 1) are based on single climate realizations and out of five studies, only two presented statistics for future blossom frost risk [17,21]. However, assessing climate impact on the basis of models involves error concatenation resulting from the following chain of information. The future climatic impact is studied with the help of simulated climate time series, generated by global circulation models (GCM) and regionalized or downscaled by regional climate models (RCM). For this purpose these climate models are forced with greenhouse gas emissions scenarios of an evolving world (IPCC scenarios, SRES emission scenarios, [16,27]). In order to estimate climate projection uncertainty, ensembles of GCM-RCM combinations or several realizations of one GCM-RCM combination (runs) are usually produced. These climate time series are used after down-scaling to drive impact models in order to assess the climatic impact in such different fields as coastal protection, water management, environmental research, food supply, urban planning and land use. Since models cannot reproduce every environmental aspect in real accuracy and resolution, systematic deviations of simulated and observed climate time series as well as of simulated and observed climate impact have to be taken into account. Depending on model sensitivity and question at hand, these biases can be removed by bias correction (e.g. 1-dimensional [28]; 2-dimensional [29]). Hence the chain of information for climate impact is: Scenario - emission - GCM - RCM - climate run - (bias correction) - impact model. Further chain members (e.g. prevention, adaptation strategies) or influences (e.g. feedbacks, interpolation, statistics) are possible. Since each member of this chain exists in different versions, numerous computations have to be conducted in order to cover the whole set of information available. Therefore most impact studies focus on "likely" scenarios [30], often not considering the full range of possibilities. This leads to the effect of possibly biased but significant trends of single or similar time series.

Taking these deficits into account, the objective of this work is to present a robust estimate of future blossom frost risk, taking the climate-model-impact-model uncertainty into account, including two new developed extensions of one sequential and one parallel chilling-forcing model considering light conditions.

Methods

General Procedure and Regional Focus

Thirteen simulated time series of air temperature from varying regional climate models were used to drive seven phenological models for the projection of apple bloom in Lower Saxony, Germany, whereas blossom frost risk was obtained by evaluating the temperature following bloom. Changes of these variables over time and compared to a reference period are referred to as "signal" in the following. The behavior of signal and variance across climate and impact models was analyzed subsequently, extracting the fractional uncertainty (inverse of signal-to-noise ratio). From this the meaningful horizon of projection was obtained, being basically the year at which the investigated signal exceeds the variation of the signal. This climatological approach [31,32] originally divides time series into their internal variability, scenario and model uncertainty. Advancing this approach beyond climatology, the present work estimates the extension of uncertainty from the climate signal to the climatic impact by dividing time series into their internal variability, climate model and impact model uncertainty of one scenario.

In order to project apple bloom, phenological models were calibrated with measurements of daily air temperature and observations of phenophases. Subsequent projection of future apple bloom was carried out with bias-corrected climate projections from physical-dynamical regional climate models (Table 2). Calibrated models were validated for accuracy in prediction of bloom by cross-validation as well as testing for different locations. Blossom frost risk estimates were validated first by calculating the accuracy of the phenological model (comparing measured blossom frost risk with blossom frost risk simulated with measured temperature) and secondly through calculating the influence of

Table 2. Overview of employed data.

Data	Specification	Climate model runs	Resolution (spatial, temporal)	Period	Ref.
observed	early ripeners, BBCH 60		0.116°, d	1991–2012	[a]
flowering	early ripeners, BBCH 65		0.116°, d	1991–2012	[a]
(DOY)	late ripeners, BBCH 60		0.116°, d	1991–2012	[a]
	late ripeners, BBCH 65		0.116°, d	1991–2012	[a]
measured T (°C)[b]	115 stations		0.126°, d	variable	[c]
simulated		1. EH5-REMO5.7, C20 1/A1B 1[d]	0.088°, h	1951–2100	[58]
T (°C)[b]		2. EH5-REMO5.8, C20 1/A1B 2[e]	0.088°, h	1961–2100	[59]
		3. EH5-REMO2008, C20 3/A1B 3[f]	0.088°, h	1950–2100	[f]
		4. EH5-CLM2.4.11 D2 C20 1/A1B 1	0.165°, 3 h	1961–2100	[60]
		5. EH5-CLM2.4.11 D2 C20 2/A1B 2	0.165°, 3 h	1961–2100	[61]
		6. C4IRCA3_A1B_HadCM3Q16	0.223°, d	1951–2099	[62]
		7. CNRM-RM5.1_SCN_ARPEGE	0.232°, d	1951–2100	[62]
		8. DMI-HIRHAM5_BCM_A1B	0.223°, d	1961–2099	[62]
		9. DMI-HIRHAM5_A1B_ARPEGE	0.223°, d	1951–2100	[62]
		10. DMI-HIRHAM5_A1B_ECHAM5	0.223°, d	1951–2099	[62]
		11. ICTP-REGCM3_A1B_ECHAM5_r3	0.232°, d	1951–2100	[62]
		12. KNMI-RACMO2_A1B_ECHAM5_r3	0.223°, d	1951–2100	[62]
		13. MPI-M-REMO_SCN_ECHAM5	0.223°, d	1951–2100	[62]

[a]German Meteorological Service. Phenological observation program. URL: http://www.dwd.de (April 20, 2013).
[b]air temperature at 2 m elevation.
[c]German Meteorological Service. Station network. URL: http://www.dwd.de (April 20, 2013).
[d]"UBA"-Run, experiments 6215/6221.
[e]"BFG"-Run, experiments 29001/29002.
[f]experiments 1518/1518, Max Planck Institute for Meteorology, Hamburg, Germany.

the time series on blossom frost risk projection accuracy (comparing simulated blossom frost risk from measured temperature with that from simulated temperature).

Climatic Data and Models

Data sources. Measured as well as simulated air temperature time series for Lower Saxony, Germany, (Table 2, Figure 1) were processed and applied as follows. Simulated temperature of regional climate model projections of the IPCC-emission A1B [27] was obtained from the Max Planck Institute for Meteorology, Hamburg, Germany, (in the following climate runs 1–5) and from the ENSEMBLES project (in the following climate runs 6–13).

Temporal interpolation. Temporal interpolation of measured daily temperature time series was used to obtain hourly time series, following a stepwise procedure of spline interpolation [21]. Resulting hourly temperature time series showed a year-round mean error of -0.031 K h^{-1} and mean absolute error (MAE) of 0.448 C h^{-1} as well as an error of 0.587 hours of frost ($\leq 0°C$) per month of April, compared to measured hourly time series at 56 sites. Time series of the climate model CLM (3 h resolution) were brought to hourly resolution by applying cubic spline interpolation.

Spatial interpolation. Spatial interpolation through ordinary kriging [33] was used to bring measured as well as simulated data to common and regular grids (0.1°·0.1° as well as 0.2°·0.2°) for the area 51° to 54° latitude north and 6.5° to 12° longitude east. While measured data was interpolated directly, simulated hourly temperatures (climate runs 1–5) were previously aggregated by taking the mean of each hour of nine neighboring model grid points (area approximately 30 km·30 km for REMO). By doing so for every model grid point and hence obtaining a spatial floating mean, the original model resolution was maintained. Simulated daily mean and minimum temperature time series were not aggregated due to the coarser spatial resolution.

Bias correction. Since several climate models underestimate the occurrence of frosts, simulated temperature series were bias-corrected for each month by distribution-based quantile mapping [28], using non-parametric transfer functions obtained by applying a Gaussian kernel with bandwidth h = 0.1 [34]. The period of comparison from which transfer functions were derived for bias correction was 54.4±7.3 years for climate runs 1 and 3, 49.8±4.9 years for climate runs 2, 4 and 5 as well as 57.9±4.4 years for climate runs 6–13 (mean ± standard deviation). Hence, the influence of the multidecadal variability was assumed to be negligible. Information on bias correction dynamics with climate runs 6–13 (Table 2) have been published [35].

Projection of temperature. In the following, temperature time series are presented as anomaly from the 1971–2000 mean as indicated by $\Delta T_{y1,y2,s}$ with the centers of the respective periods $y1$ and $y2$ and grid points s (see Methods S1 for equation).

Projection of last spring freeze. The last spring freeze was defined as the last day before July 31st, exhibiting a minimum air temperature $\leq 0°C$, and taken directly for every year from temperature time series.

Phenological Data and Models

Data sources. In order to simulate apple bloom phenophases, time series (Table 2, Figure 1) from the German National

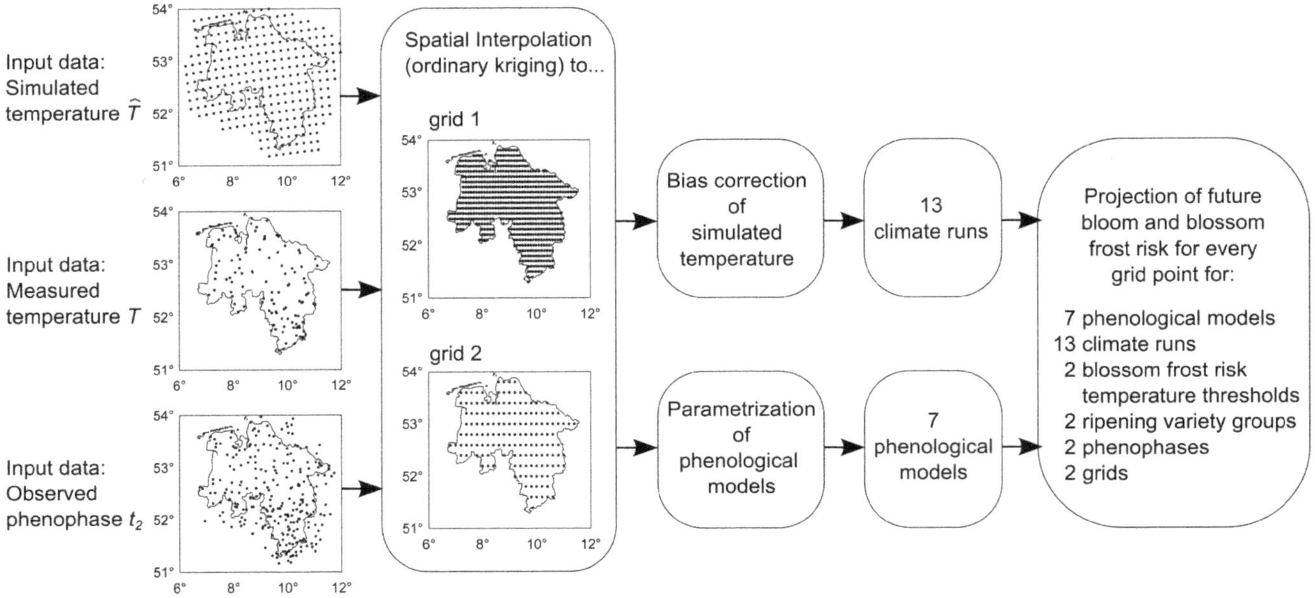

Figure 1. Scheme of used input data and projection. Note that for simulated temperature the grid of the regional climate model CLM is shown exemplarily.

Meteorological Service (htp://www.dwd.de) of observed beginning of flowering (first flowers open) as well as onset of full bloom (50% of flowers open), defined as phenophases 60 and 65 on the BBCH-scale [13], were processed and used to calibrate phenological models for early and late ripening varieties as follows.

Spatial interpolation. Phenological time series were spatially interpolated as described above for measured temperature time series.

Basic phenological models. In principle, all applied phenological models (Table 3, 4, Methods S1) assume that the time of bloom is related to so-called sums of chilling and heat units (*Sc*, *Sf*) accumulated during winter (chilling phase) and spring (forcing phase), (see Table 4 for denominations). It is assumed, that *Sf* is related to *Sc* [36,37]. The basic models (Table 3, models 1–4) have been described in the literature [17–21] and their equations are given in Methods S1.

Extended phenological models. Models including an additional day-length-parameter for the calculation of the forcing phase were included in the ensemble (Table 3, models 5–7), as a higher performance of model no. 5 has been reported. Models 6–7 are new model variations of the sequential and parallel chilling-forcing models [23], which were extended for a factor for the length of day *D*, assuming that bloom is influenced by radiation only during the forcing phase. For both, the rate of forcing *Rf* was calculated as follows:

$$Rf(T_i) = \begin{cases} 0 & \text{if} \quad T_i \leq Tbf \\ \dfrac{28.4}{1+e^{(-0.185(T_i - Tbf - 18.4))}} \cdot \left(\dfrac{D}{10}\right)^c & \text{else} \end{cases} \quad \text{with} \tag{1}$$

Rf : Rate of forcing $[-]$
T_i : Daily mean air temperature at day i $[^\circ\mathrm{C}]$
Tbf : Base temperature $[^\circ\mathrm{C}]$
D : Length of the day $[\mathrm{h}]$
c : Calibration parameter $[-]$

Table 3. Phenological models.

No.	Type	Daylength	*Tbf*	*Tbc*	$Sf(t_2)$	$Sc(t_1)$	t_1	*a*	b	c	Ref.
1	Thermal time	−	+	−	+	−	+[a]	−	−	−	[20]
2	Sequential chilling-forcing	−	+	+	+	+	−	+	+	−	[23]
3	Parallel chilling forcing	−	+	+	+	+	−	+	+	−	[23]
4	Modified Utah	−	+	+	+	+	−	−	−	−	[17,43]
5	Thermal time	+	+	−	+	−	+	−	−	+	[25]
6	Sequential chilling-forcing	+	+	+	+	+	−	+	+	+	−
7	Parallel chilling forcing	+	+	+	+	+	−	+	+	+	−

[a]For model 1, t1 was set to January 1.

Table 4. Denomination of variables and parameters.

Notation	Description	Unit
T	Air temperature	°C
Tbc, Tbf	Base temperature for chilling, forcing	°C
t	Time	hour [h], day [d] or year [a]
t_0	Start of the chilling period (dormancy)	day of the year [DOY]
t_1	Chilling requirement completed, start of forcing	day of the year [DOY]
t_2	Forcing completed (BBCH 60, BBCH 65)	day of the year [DOY]
Sc, Sf	State of chilling, state of forcing	–
Rc, Rf	Rate of chilling, rate of forcing	–
D	Daylength	h
a, b, c	Calibration parameters	–
i, s, z	Index variables	–
θ	Blossom frost risk	–
β	Temperature threshold for blossom frost	°C
λ	Parameter for calculation of mean and confidence level	–

Parameter estimation and model validation. Models were parametrized for each grid point by fitting the models to observed bloom (BBCH-scale [13], stages 60 and 65 for early and late ripening varieties of *Malus domestica*) and measured daily air temperature (Table 2). Fitting was performed through bound-constrained simulated annealing, minimizing the root mean square error (RMSE) between observed and simulated day of the year (DOY) of bloom. Simulated annealing for parameter estimation of phenological models has been described in detail [38] and was performed in the present study by using the Global Optimization Toolbox (The Mathworks Inc., Natick, Massachusetts) on a computing cluster system (http://www.rrzn.uni-hannover.de/clustersystem.html). For this, *Tbc* and *Tbf* were searched between 0°C and 10°C, as this is believed to be the effective range of temperature on the development of apple trees [23]. The models were validated internally (same location) as well as externally (different location) by calculating the prediction root mean square error (PRMSE) determined by full-cross validation ("leave-one-out") and by applying the model with optimized parameters to six different and randomly chosen locations in the range of 20 to 28.3 km distance.

All models accounting for *Sc* were initiated with $t_0 = 1$ August. The simple thermal-time model (1) was started with fixed $t_1 = 1$ (January 1st, model 1), whereas the extended thermal-time model (5) was started on August 1st (DOY 213, 214) in order to optimize t_1. Models 1 and 5 do not account for a chilling phase and hence implicitly assume that chilling is already completed at t_1.

Projection of Bloom

Bias-corrected air temperature time series of 13 climate realizations (Table 2, Figure 1) were used as input for seven phenological models for 792 locations in Lower Saxony on a 0.1°·0.1° grid (climate runs 1–5) and for 274 locations on a 0.2°·0.2° grid (climate runs 6–13, Table 2, Figure 1) to project future apple bloom. Projections were conducted for all grid points whereas presented results were restricted to the area of Lower Saxony (Figure 1) in order to avoid boundary effects due to interpolation. Comparison of results from all 13 projections took place on the grid of lower resolution. All simulations were conducted with early as well as late ripening varieties and for two phenological stages (BBCH 60, 65). The change in blooming date $\Delta t_2 y1,y2,s$ with the centers of the respective periods $y1$ and $y2$ and grid points s was calculated as the difference in the 30-year-mean for each grid point. Years with unfulfilled chilling were recorded by counting years without bloom or bloom projected for DOY > 200 as fraction of occurrences in a 30-year-mean. Please see Methods S1 for equations.

Projection of Blossom Frost Risk

Subsequently, years with occurrences of frosts (daily minimum temperature ≤0°C) and possibly blossom damaging situations (daily minimum temperature ≤2°C) during the time from simulated bloom (BBCH 60, BBCH 65) to the 31st of July of each year were counted separately. The additional threshold of 2°C was chosen in order to account for spatial discrepancies of observed bloom and measured temperature as well as for possible radiation frosts with tissue temperatures falling below air temperature [19], measured at standard meteorological conditions. Blossom frost risk was defined as the ratio of number of years with temperatures lower or equal to a predefined threshold occurring after a specific phenophase in 30 years:

$$\theta_{y,s} = \frac{1}{30} \cdot \sum_{i=-14}^{15} \mu_{i,s} \text{ with}$$

$$\mu_{i,s} = \begin{cases} 1 & \text{if} \quad min(\{T_{y+i,t_2 y,s,s} \dots T_{y+i,\omega,s}\}) \leq \beta \\ 0 & \text{else} \end{cases}$$

$\theta_{y,s}$: blossom frost risk of year y at grid point s, $[-]$

$T_{y,d,s}$: array of daily minimum temperature of year y, day d and grid point s $[^\circ C]$

β : temperature threshold, either 0 or 2 $[^\circ C]$ (2)

ω : 212 or 213 (leap year) for 31.7., [DOY]

$t_2 y,s$: onset of phenophase, e.g. begin of bloom of year y at grid point s

y : year of calculation, e.g. 1980

i : index

s : grid point

The change in blossom frost risk $\Delta\theta$ was calculated from 30-year-means of each grid point:

$\Delta\theta_{y1,y2,s} = \theta_{y2,s} - \theta_{y1,s}$ with

$\Delta\theta_{y1,y2,s}$: projected change in blossom frost risk from year $y1$ to year $y2$ of every grid point s in Lower Saxony, $[-]$ (3)

$y1, y2$: year of calculation (past, future)

s : grid point

Probability mass functions were calculated in order to estimate the distribution of changes in blossom frost risk till the end of the 21st century (2070–2099 minus 1971–2000). The values of these probability mass functions were estimated non-parametrically with the help of kernel density estimation, applying a Gaussian kernel. Please see Methods S1 for equations.

Partitioning of Uncertainty of Temperature, Bloom and Blossom Frost Risk

In order to estimate the meaningful projection horizon (= 'Time of emergence', [39]) of the results obtained as described above, the fractional variance of the system was calculated and the total variance of the projection was partitioned. For this purpose the methodology of Hawkins and Sutton [31] was applied to the presented projections for the day of bloom t_2. Instead of looking at different climate models and scenarios, the present work analyzes the internal variability, the uncertainty from climate realizations of one IPCC-scenario (A1B) and the variance resulting from the impact models. Impact models were weighted by their error as described for climate models [31]. The following calculations were carried out with 10 year mean moving average time series of the area mean of Lower Saxony (mean of all grid points s, please see Methods S1 for equations). In brief, the total variance for bloom was calculated as described below. Projection uncertainty of temperature and blossom frost risk was calculated as described for bloom (temperature analysis only for internal and climate realization variability).

$$\begin{aligned} &B_{total}(y) = B_1 + B_2(y) + B_3(y) \text{ with} \\ &B_{total} : \text{Total variance of projected bloom, } [d^2] \\ &B_1 : \text{Internal variability (residual variance), } [d^2] \\ &B_2 : \text{Uncertainty of climate realizations} \\ &\quad \text{(variance across climate runs), } [d^2] \\ &B_3 : \text{Uncertainty of impact models (variance} \\ &\quad \text{across phenological models), } [d^2] \\ &y : \text{year of calculation, e.g. 1980} \end{aligned} \tag{4}$$

The contribution of B_1,B_2 and B_3 to the total variance can be expressed as fraction of the total variance:

$$\begin{aligned} &H_z = \frac{B_z \cdot 100}{B_{total}} \\ &H : \text{Fraction of the total variance, } [\%] \\ &z : 1, 2, 3 \end{aligned} \tag{5}$$

The mean change in blooming dates of all projections (climate impact signal) over the reference period was obtained as:

$$\begin{aligned} &G(y) = \frac{1}{n}\sum_{s,z} W_s x_{s,z,y} \text{ with} \\ &W : \text{model weight, } [-] \\ &x : \text{change of phenophase, } \Delta t_2, \\ &\quad \text{compared to 1971} - \text{2000 [d]} \\ &s : \text{impact model } (2-7) \\ &z : \text{climate realization } (1-13) \\ &n : \text{number or climate realizations, } [-] \\ &y : \text{year of calculation, e.g. 1980} \end{aligned} \tag{6}$$

Models were weighted (eq. 6) with weights W inversely proportional to their model error (see [31]), giving models with lower errors comparatively more importance. From G and B_{total} the fractional uncertainty F, which is the inverse of the signal-to-noise ratio, was calculated as follows:

$$\begin{aligned} &F(y) = \frac{\lambda\sqrt{B_{total}(y)}}{G(y)} \text{ with} \\ &\lambda : \text{parameter for calculation of confidence} \\ &\quad \text{levels } 50\%\ (\lambda = 0.67),\ 68\%\ (\lambda = 1) \text{ and} \\ &\quad 90\%\ (\lambda = 1.65) \end{aligned} \tag{7}$$

Statistics of Single Time Series

Continuous time series of calculated completion of dormancy, blooming date and last spring freeze were analyzed using a Mann-Kendall-test [40], whereas trends in blossom frost risk were analyzed with a test by Cox & Lewis [41].

Results

Validation of Methods

The presented methodology was evaluated at the levels climate, quality of phenological model in order to simulate phenophases as well as blossom frost risk. A bias correction had no influence on the mean temperature pattern, whereas the accuracy of simulated frost distribution was drastically improved (Table 5), see also [35]). While climate model time series underestimated frosts in April, this was corrected through the bias correction.

Models could be fitted to reproduce bloom with 3.2 to 5.7 d mean accuracy (RMSE), whereas testing models with fitted parameters (see Methods S1) for different locations revealed an external PRMSE of 3.9 to 8.0 d (Table 6). While the thermal time model (1) exhibited the highest mean error (1.8 d higher than mean of other models), the thermal time model with extension for day length exhibited the lowest mean error (2.0 d lower than mean of other models). On average models (1–3) were improved by 2.0 d when accounting for day length (models 5–7), whereas performance did not differ greatly between BBCH-stages 60 and 65 nor between early and late ripening varieties.

Blossom frost projection accuracy was verified at different levels, since direct comparison of measured blossom frost with blossom frost from simulated time series is not possible in a direct manner

Table 5. Stepwise error of simulation chain segments. SE: Simulation error, ABS: absolute level from measured data.

Parameter		T bias corrected	Frost occurrences per month of April		Bloom[a]	Blossom[b] frost risk θ
			[h]	[d]	[d, DOY]	[–]
Frost	ABS	–	25	4	–	–
Frost	SE[c]	no	7	3	–	–
Frost	SE[c]	yes	<1	<1	–	–
Bloom	ABS	–	–	–	117–126	0.163
Bloom	SE[c]	no	–	–	–	–
Bloom	SE[c]	yes	–	–	4–8	–
Blossom frost	ABS	–	–	–	–	0.163
Blossom frost	SE[c]	no	–	–	–	–
Blossom frost	SE from phenol. models[cd]	yes	–	–	–	0.001–0.034
Blossom frost	SE from time series[ce]	yes	–	–	–	0.021–0.075

[a]min-to-max range across all ripening groups and phenophases.
[b]min-to-max range across all ripening groups, phenophases and phenological models.
[c]Mean absolute error (MAE), average over all grid points.
[d]Error from comparison of measured blossom frost risk with blossom frost risk simulated with measured temperature (1991–2012).
[e]Error from comparison of blossom frost risk simulated with measured temperature with blossom frost risk simulated.
with simulated temperature (1951–2012).

for short periods (<30 a). Therefore the influences of phenological models and of time series on blossom frost incidents were extracted separately. Applying the phenological models to measured climate data of the calibration period 1991–2012 reproduced blossom frost incidences from measured temperature and measured bloom (Figure 2, Table 5). Subsequently the influence of the time series on blossom frost projection accuracy was tested by applying the validated phenological models on measured and on simulated-bias corrected time series (1951–2012). Despite bias correction, projection with simulated-bias corrected time series showed a mean absolute error (MAE) of blossom frost risk of up to 7.5 percentage points (Table 5). However, mean influences of impact model and time series on blossom frost risk projection accuracy were 1.4 and 3.6 percentage points respectively (mean MAE). Finally blossom frost risk was biased by +0.9 and −3.6 percentage points by impact model and time series, respectively, still resulting in an overall underestimation of blossom frost.

Dormancy and Bloom

In the mean, observed bloom from 1991 to 2012 changed by -3.3 d K^{-1} ($R^2 = 0.87$) while air temperature increased by 0.037 K a^{-1}. Phenological models, which were calibrated with these data, gave the following results when applied to simulated temperatures. All chilling-forcing models consistently showed a delay for the release of dormancy t_1 (Figure 3) with major changes not occurring before 2030, following the temperature warming patterns of both simulated climate data sets. However, t_1 showed a larger spread across ENSEMBLES runs than for ECH5-REMO/CLM simulations, while the number of years with unfulfilled chilling requirement increased in both cases (Figure 4). Unlike t_1, projection of the onset of the phenological phases for t_2 (BBCH 60, 65) revealed an advancement. While models 2–7 follow a relatively homogeneous pattern, model 1 projects a faster advance. These main patterns also become visible on a regional scale (Figure 5,6). However, changes in the day of bloom vary regionally depending

Table 6. Prediction Root Mean Squared Error PRMSE of phenological models [d].

Model	early ripeners		late ripeners		mean
	BBCH 60	BBCH 65	BBCH 60	BBCH 65	
1	7.97	7.26	7.28	7.27	7.45
2	6.67	5.95	6.24	6.03	6.22
3	7.10	6.30	6.54	6.25	6.55
4	6.81	6.83	6.54	6.67	6.71
5	4.14	4.12	3.91	4.34	4.13
6	4.96	5.08	4.88	5.10	5.00
7	5.13	5.19	4.89	5.29	5.13
mean	6.11	5.82	5.75	5.85	5.88

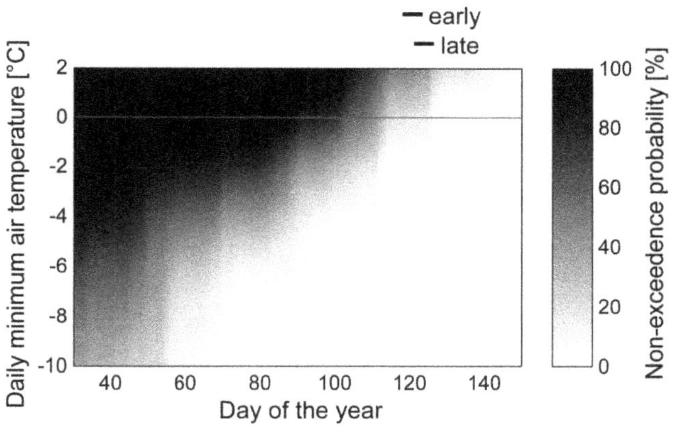

Figure 2. Present temperature incidence of Lower Saxony (1991–2010). Bars indicate mean flowering period (BBCH 60–65) of early and late ripening varieties.

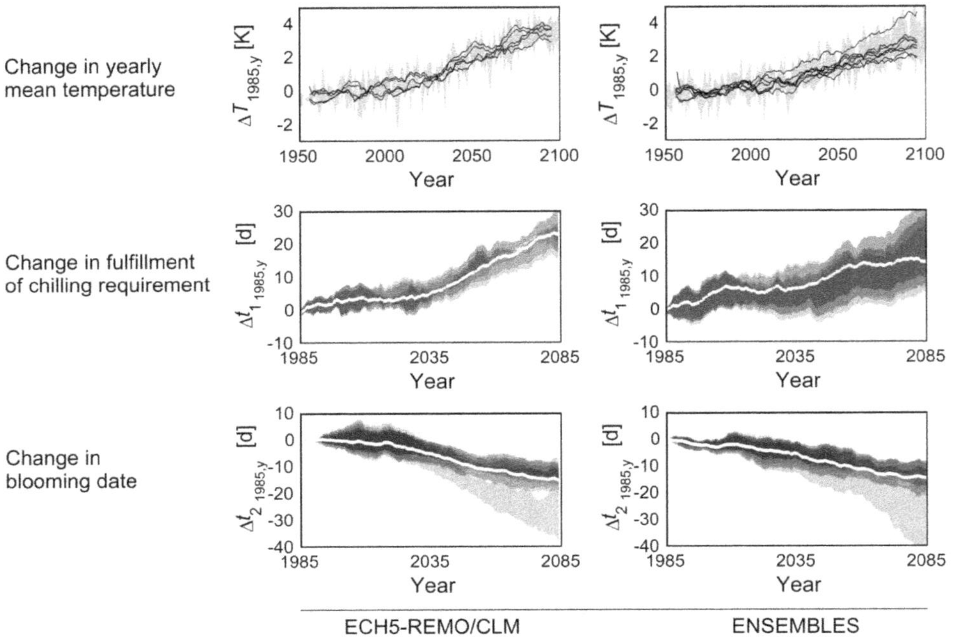

Figure 3. Projected changes in air temperature, fulfillment of chilling requirement and onset of flowering. Projected with 5 (ECH5-REMO/CLM) and 8 (ENSEMBLES) climate runs and five (Δt_1) and seven (Δt_2) phenological models for Lower Saxony (area mean), relative to the 1971–2000 mean. ΔT: single year-mean, min-to-max range of climate runs (shaded area), 10 year moving average of each run (solid lines, see Methods S1 for equation). Δt_1, Δt_2: BBCH 65, early ripeners, 30-year-moving-average, all impact model mean (solid white line), single model range (shaded areas). The range of each phenological model (min-to-max) obtained from climate runs is plotted with 20% transparency (darker areas illustrate coinciding results).

on the model. Regarding the timescale, all models project a shift in the day of bloom of −5.4±3.0 d by 2035 compared to 1971–2000 (area mean, all varieties and stages), whereas results for 2084 differ. While model 1 shows the strongest change (−26.7±8.2 d), models 2–7 project a mean shift of approx. −12.9±3.3 days. The latter again differ in their regional variation. Although the classic sequential and parallel chilling forcing models (2–3) show a similar mean shift of bloom as their versions extended for daylength (models 5–7; −13.5 d and −11.2 d respectively), the former exhibit higher variation (±3.6 d and ±2.2 d respectively). A similar variation was also found for model 4 (±3.3 d).

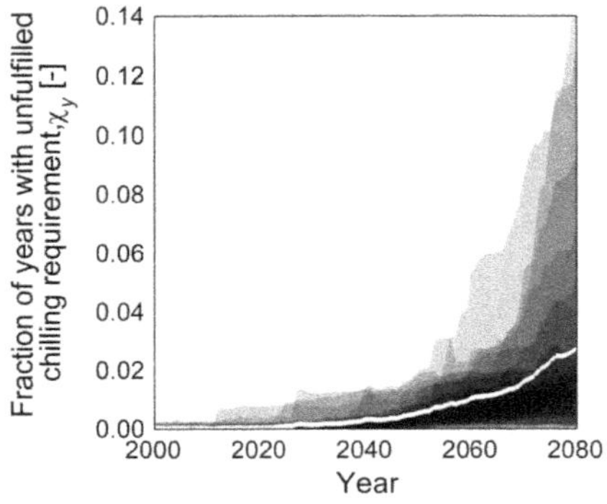

Figure 4. Proportion of years with unfulfilled chilling requirement. Areas: min-to-max range across seven phenological models for each climate run (area mean of Lower Saxony, 30-year moving average); white line: Mean of impact models and climate runs.

Projected Last Spring Freeze and Blossom Frost Risk

According to the scenario and climate runs considered, the last spring freeze (≤0°C) will shift by −10.0±4.2 days and −27.3±7.4 days by 2035 and 2084 respectively, with regard to the reference period 1971–2000 (Figure 7). Hence these 30-year-mean trends indicate an increasing discrepancy of the day of bloom and the last spring freeze. Correspondingly the mean occurrences of blossom frost (θ) are projected to decrease in the long run (Figure 5,6). Nevertheless model 1, which showed the fastest advancement of bloom, projects a mean increase of blossom frost risk by 3.4 percentage points whereas models 2–7 project a mean change by −4.1± 3.3 percentage points, ranging from −2.6 percentage points for late ripeners (BBCH 65) to −6.0 percentage points for early ripeners (BBCH 60). In the mean, runs of EH5-REMO/CLM and ENSEMBLES runs produced similar estimates for changes in blossom frost risk (−2.7±4.4 percentage points and −3.2±4.5 percentage points respectively). However, all models also exhibited regional and temporary increases in blossom frost occurrences. The resulting probability mass function values (*pmf*) are shown in Figure 8, displaying also the contrary result of model 1. A larger spread and stronger decrease was observed for the probability of temperatures of ≤2°C after onset of phenophases.

Projection Uncertainty

Phenophases followed temperature patterns closely, with early and late ripening varieties advancing at 5.6 and 5.4 d K^{-1} respectively and BBCH 60 and BBCH 65 advancing at 5.6 and 5.4 d K^{-1} respectively, resulting in a mean change of −5.5 d K^{-1} (Figure 9). Higher correlations were found between changes in begin of flowering date and mean temperatures between February and April (−6.1 d K^{-1}, $R^2 = 0.93$). However no correlation was

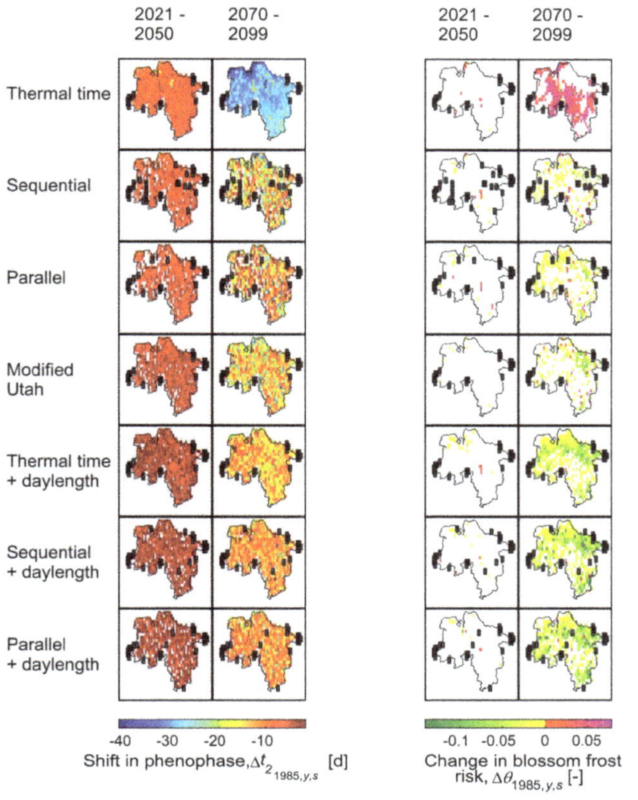

Figure 5. Changes in bloom and blossom frost risk as projected by different phenological models and climate runs 1–5. Early ripeners, BBCH 65, temperature threshold $\beta = 0°C$, reference period 1971–2000, resolution 0.1°. White fields denote non-significant results, black fields denote missing/insufficient data. 1–99% percentile range. $y = 1985$ and 2084, $s =$ grid point.

found between changes in the respective variances of temperature and flowering dates, with exception of the simple thermal time model (model 1, data not shown).

The projection uncertainty increased with increasing lead time (Figure 10, top) and for the period investigated, the accuracy of the projection of t_2 in the short run is mainly dependent on the projected climate and internal variability. With increasing horizon of projection, the climate signal (temperature) becomes stable while impact/phenological model results diverge. Consistently fractions of climate and internal variability of the total variance decreased with increasing lead time (Figure 10, bottom). Finally, the projection accuracy at the end of projection horizon depended equally on the climate and impact/phenological model variance.

The resulting fractional uncertainty F decreased over time. Comparing the sources of uncertainty, the fractional uncertainty of temperature time series decreased faster than of blooming date and blossom frost risk time series. Accordingly, the lowest level of fractional uncertainty at any of the confidence levels investigated was also reached by temperature. While the 90% percentile for temperature and bloom reached 1 in 2019 and 2042–2044 respectively, the uncertainty of blossom frost risk passed 1 only by the 68% percentile (±1 standard deviation) by 2077 (Figure 11). From this point on, the projected change (signal) exceeded the variance of the projection (noise). A minimum of the fractional uncertainty was found for 2078 (temperature), 2083–2084 (bloom) and 2085–2088 (blossom frost risk), after which it was projected to increase. This result was similar for early as well as late ripening varieties and for both BBCH stages.

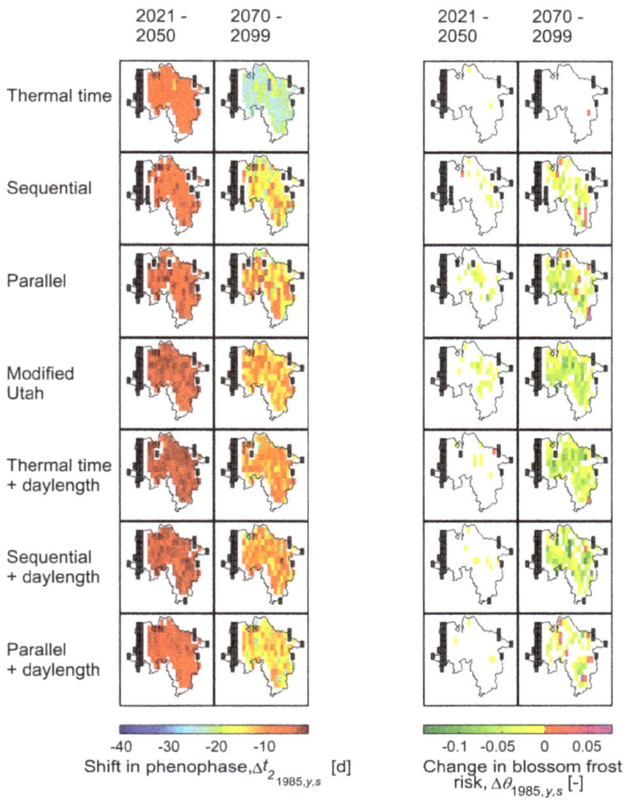

Figure 6. Changes in bloom and blossom frost risk as projected by different phenological models and climate runs 6–13. Early ripeners, BBCH 65, temperature threshold $\beta = 0°C$, reference period 1971–2000, resolution 0.2°. White fields denote non-significant results, black fields denote missing/insufficient data. 1–99% percentile range. $y = 1985$ and 2084, $s =$ grid point.

Discussion

Phenological Models

Projections with pure forcing models [20,21] are subject to changes in dormancy completion [23] and varying warming of the seasons. The application of such a model in the present study produced similar results of increasing risk as in the mentioned literature, but different to the main outcome of the present

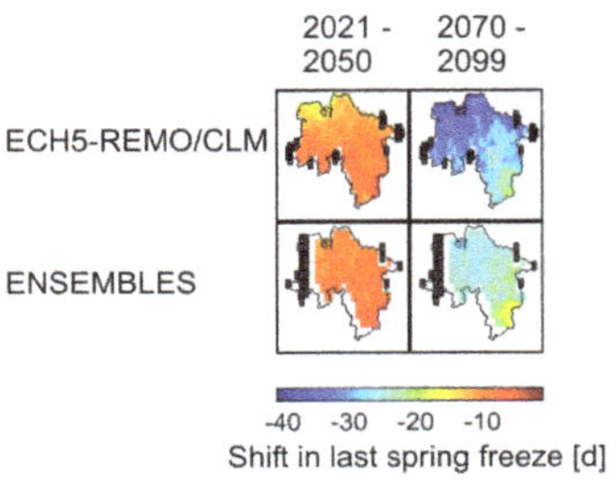

Figure 7. Changes in last spring freeze. Reference period: 1971–2000. White fields denote non-significant results, black fields denote missing/insufficient data. 1–99% percentile range.

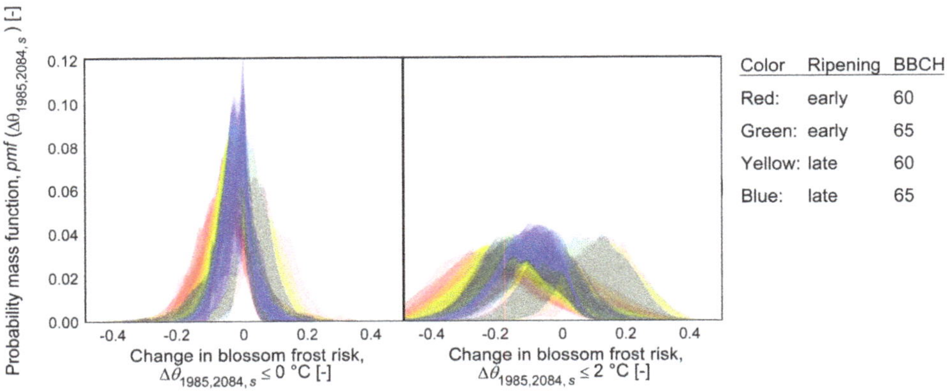

Figure 8. Distribution of projected changes in blossom frost risk by the end of the 21st century (2070–2099 minus 1971–2000) for early and late ripening varieties, phenophases BBCH 60 and 65 and 7 phenological models: Temperature thresholds ≤ 0°C and ≤ 2°C; inter-quartile range across 13 climate runs; phenological models are presented by same colors. Calculated from all grid points s (see Methods S1 for equation).

ensemble study. For this reason, sequential or parallel chilling-forcing models have been recommended [23], as well as models including nearly time-invariant factors as day length [25]. The mean error of all models presented (5.9 d) was in the range of published model performances [15,20,21,23,25,42,43]. This error must be seen in context to the observed flowering duration (BBCH 60 to BBCH 67), which ranged during the calibration period from 6 to 27 d (1 to 99% percentile range). As large errors in simulated flowering dates can erroneously increase the blossom frost risk, the influence of the RMSE on the simulated blossom frost risk was tested (not shown), but no significant influence was found in the range of the calibrated models errors. Having further a negligible bias, the models were rated as suitable for blossom frost risk projections from this point of view. Furthermore, in the present work models were improved by including day length, thus confirming previous findings [25]. Also other models including

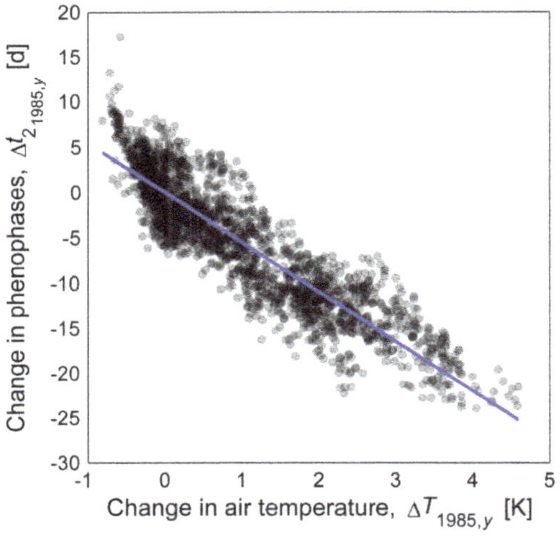

Figure 9. Simulated relation between projected absolute changes in decadal mean air temperature and changes in the day of bloom compared to the 1971–2000 mean. Depicted values are related to 139 years ($y = \{1956 .. 2094\}$, see Methods S1 for equation) and 13 climate realizations for the area mean of 2 phenophases and 2 variety groups. Slope of regression (solid line) = −5.4842, offset = 0.0385, $R^2 = 0.81$.

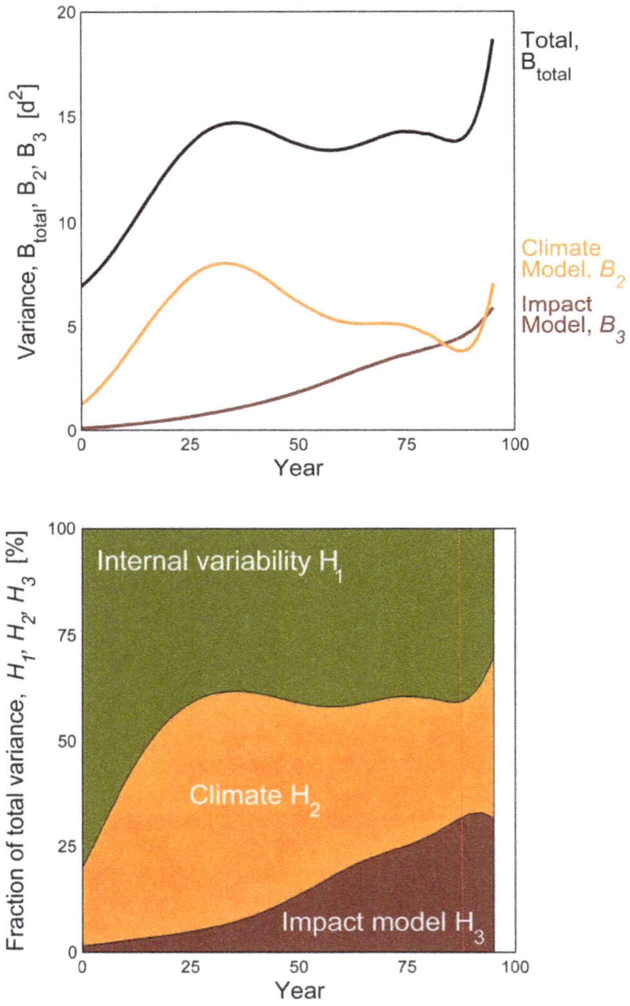

Figure 10. Uncertainty in the projection of apple bloom (t_2). Drawn from phenological impact models 2–7 and 13 climate projections. Mean uncertainty of phenophases (BBCH 60, 65) and ripening groups (early, late).

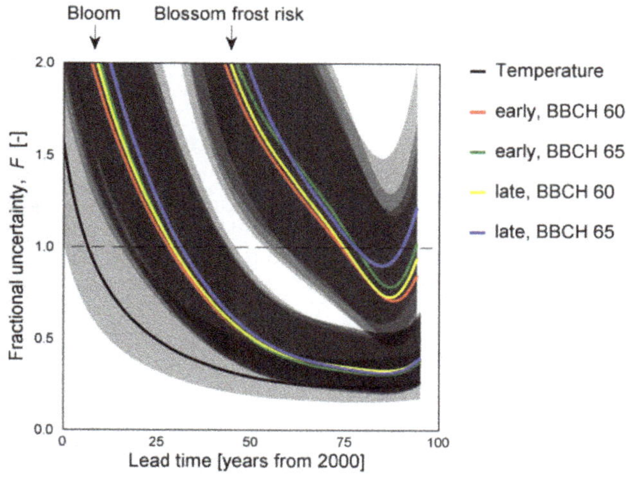

Figure 11. Uncertainty pattern of projected temperature (T), apple bloom (t_2) and apple blossom frost risk (θ). 68.3% percentile (solid lines) and 50-to-90% percentile ranges (gray areas) from 13 climate projections and phenological impact models 2-7 (bloom, blossom frost risk).

exponential terms were applied in blossom frost risk estimation [17,43], relying solely on temperature as input. As they increase the "resistance" for each computation of a day of the year for flowering, exponential models eliminate one deficit of pure temperature sum models which is a calculated flowering date beyond summer in exceptional years, leading to high errors (given that dormancy is completed). In addition, the error of models including a parameter for day length might be lower due to a higher number of parameters. This statistical effect should be separated from the physiological meaning of the parameter. As the role of the length of day in flowering physiology of apple is still under debate [26], these model properties cannot be isolated for the present study, but should be regarded in the future. Finally, while presented combination of sequential or parallel models with an exponential term for day length improves model robustness, these models are also more complex.

Influence of Climate Change on the Onset of Phenophases

The observed effects of delayed completion of the chilling requirement and earlier flowering due to faster completion of heat requirement are well known[6,15,42,44–46]. Thereby the extension of the growing season [47,48] and the advancement of flowering dates during the past due to climate change have been studied largely for several tree species[44,49–51] including apple flowering phenology [9,14,42], allowing the assumption of a general trend. Accordingly "very similar" reactions of apple and cherry blossoming (BBCH 60) as well as winter rye stem elongation (BBCH 31) to early spring conditions were observed [14]. However, the observed mean change of onset of flowering (BBCH 60) of -3.3 d K^{-1} during the short calibration period of phenological models (1991–2012) were lower than those reported from other studies for the entire second half of the 20th century. These published estimates range from -7 to -8 d K^{-1} of year-mean temperatures (values calculated from [9,42]) for late ripeners up to -5 d K^{-1} of mean temperatures from February to April [14] for early ripeners. Still these discrepancies should result from geographic and orographic differences from the present to the mentioned publications: Analyzing the present model projections for the same periods as in the mentioned literature (1958–2007, 1976–2002, 1969–1998) fairly reproduced these dependencies with -7.5 up to -8.6 d K^{-1} for late, and -6.5 d K^{-1} (February-April temperatures). Consistently, also the projected findings for changes in the onset of apple flowering of -5.4 to -5.6 d K^{-1} (all varieties and stages and years) and -6 d K^{-1} (BBCH 60, February-April temperatures) are in a comparable range. From this can be concluded, that apple flowering phenophases have a clear and comparable reaction to changes in temperature despite differences in region and varieties and that this impact can be tracked by one-dimensional phenological models in combination with climate ensembles.

Furthermore, despite a continuous advancement of flowering dates, an opposing effect of delayed release of dormancy and enhanced spring warming was observed. While warmer winters result in reduced chilling, they can be compensated to a certain extent by warmer springs [52]. For apple bloom this has been reported for the past [42]. However, reduced chilling will eventually slow down the advancement of flowering dates as postulated [42,52] and as deduced from the relative changes for t_1 and t_2 in the present study for the 2nd half of the 21st century. In addition, eventually years with unfulfilled dormancy will occur. Such events have not been observed in Germany during the past century [6], but are discussed for the future [6,45,46]. A rough estimate for the probability of years with unfulfilled chilling requirement of up to 15% can be found for the largest producing area in Lower Saxony (Niederelbe) [53]. While this estimate coincides with the here presented range, the mean fraction of years with unfulfilled chilling requirement is lower (3.7%). Following the authors, it must be stated, that these projections are subject to large uncertainties and require further investigation.

Spring Freeze and Blossom Frost Risk

Last spring freeze follows the warming pattern with changes of increasing speed towards the second half of the 21st century. The projected shifts for the period 1985–2035 (30-year-means) of -2.0 d/decade are in the range of those changes reported for the second half of the 20th century for Central Europe (-2.2 d/decade [12]). Following the future warming pattern in simulations, last spring freeze is likely to change about -3.5 d/decade (2035–2084).

Blossom frost risk possibly decreases in the long term. This result can be obtained roughly by putting together the relative advancement of projected bloom and last spring freezes, as well as in more detail through the present computation with single models. Starting with a blossom frost risk of up to 16%, simulations showed a decline in blossom frost occurrence to about half by the end of the 21st century. Nevertheless, blossom frost is unlikely to disappear and staying at a comparable level as present until the middle of the century. As blossom frost risk strongly depends on the region, period, variety and BBCH stages, publications are hardly comparable. While the present observations and computations for the past are in the range of other studies [9,19,20], projected results differ. The often stated hypothesis of an increase in blossom frost risk due to advanced bloom in combination with increased variance in the last spring freeze date [19] does not hold true for the present study, as spring freezes declined comparably faster than flowering dates.

Projection Uncertainty

Climate impact projection to a near future is often highly uncertain since the internal variability of the system at hand is larger than the expected changes at point of time. As these changes increase with time and relatively to the total variance of the projection, more confidence in the projection signal is gained. Future climate is commonly assessed in ensemble run projections, including RCMs [54] and bias-corrected simulations [35]. Sampling, climate model, radiative and boundary uncertainties have been investigated for climate models, varying for RCMs across field, region and season [54]. While such climate ensembles are also increasingly used to drive impact models [55], the impact models error adds to the signal strength. Uncertainty of climate projections increases with increasing simulation members, as clearly shown by the different patterns of fractional uncertainty of temperature and bloom as well as blossom frost risk. Thereby projection uncertainty of surface temperature depended only on the different climate models, whereas bloom depended on climate and impact models and blossom frost risk additionally depended on the interaction of projected bloom and temperature.

In the present approach times of emergence of 34 years and 57 to 59 years were estimated for temperature and blooming date respectively (compared to the mean 1971–2000), considering one SRES scenario (A1B). This is in the range of the estimated time of emergence for regional surface temperatures of SRES scenarios A2, A1B and B1 from GCMs [39]. While the approach relies heavily on the chosen climate ensemble and impact models, larger variance can be expected with increasing spatial (or temporal) resolution. Therefore the estimated lead time for the minimum of uncertainty of ~100 years (2078–2088) is consistent with ~30 to 80 years established for temperature [31]. However, the present works investigated a range of climate and impact models of one scenario, while the cited publications investigated three scenarios for climate models. Hence further projections of future bloom are required in order to remove this lack of comparability. Nonetheless, looking at the cooler scenario B1 and neglecting the similar scenario A2 for central Europe, a larger spread in the day of bloom and hence in the estimated blossom frost risk can be expected, increasing the time of emergence of the climate impact signal. Transferring the estimated time of emergence to other climate impact studies from different research fields by assuming similar variability across models would imply, that a large fraction of these studies operates at the very edge of statistical significance. For example, from a review on 14 publications on future risks through wheat diseases [56], 8 include statements and 2 are solely based on statements for a time horizon ≤2030. From the present findings, the statistical meaning of these studies must be carefully put into context.

Two effects arise: On the one hand, using a location parameter (e.g. mean or median) of a climate ensemble as input for impact models may produce significant future changes while ignoring climate projection uncertainty. On the other hand, using single impact models and/or fixed impact model parameters can give only mean tendencies, similarly ignoring parameter ranges in climate impact. The presented results show these effects, as single impact models with climate ensemble mean as input show consistently significant trends of advancing bloom and, with one exception, of decreasing blossom frost risk. Regarding the total uncertainty of climate and impact models, this may hold true for bloom beyond the estimated projection horizon. However, projected changes in blossom frost risk are low compared to the variability across models. While this is a particularly pronounced problem of extreme events such as blossom frost, it has severe consequences. From the present results, despite a tendency of decreasing blossom frost risk, it must only be concluded that future blossom frost risk is very unlikely to increase.

Limitations

The present work does not consider the severity and distribution of frosts. Hence it must be taken into account, that other plant reactions than those investigated and resulting from frost distributions may dominate in the future. As actual blossom frost damages were not evaluated, the presented results depict the blossom frost risk tendency. Although blossom frost damage severity increases with decreasing temperature [5], temperatures cannot be translated directly into economic losses, as frost protection (e.g. sprinkler) takes place in practice. Furthermore employed models accounted for day length, but did not use actual surface radiation from climate models. Hence possible effects due to changes in light conditions (e.g. phenological effects) and effects due to severe radiation (radiation frosts) are not represented to full extent. Additionally, the influence of the day length on apple flowering physiology remains uncertain. Despite low availability of consistently bias corrected climate time series of high temporal resolution [29], future approaches should consider this. Finally, future changes in varieties were not taken into account albeit varieties might respond differently to blossom frost [57].

Conclusions

Regarding the aspects of phenological model structure, simulation uncertainty as well as blossom frost risk, the following conclusions must be drawn from the present findings. Despite a lack of physiological explanation, phenological model performance is improved by including the length of the day. However, projection results from single time series must be put into context to the uncertainty of the modeling chain, considering the significant projection horizon. The latter depends on the investigated variable and was determined for the present simulation of bloom at 2042–2044. Differently, a minimum of uncertainty was estimated for temperature, bloom and blossom frost risk for the range 2078–2088. Finally the resulting regional blossom frost risk cannot be expected to increase in the long term, as compensatory effects of delayed fulfillment of chilling requirement and faster completion of the forcing phase in spring take place.

Acknowledgments

We kindly thank the RRZN cluster system team at the Leibniz Universität Hannover, Germany, for their support in the production of this work.

Author Contributions

Conceived and designed the experiments: HH. Performed the experiments: HH. Analyzed the data: HH. Contributed reagents/materials/analysis tools: HH. Wrote the paper: HH TR. Review: TR.

References

1. Rodrigo J (2000) Spring frosts in deciduous fruit trees – morphological damage and flower hardiness. Sci Hortic-Amsterdam 85: 155–173.
2. Farajzadeh M, Rahimi M, Kamali G, Mavrommatis T (2009) Modelling apple tree bud burst and frost risk in iran. Meteorol Appl 17: 45–52.
3. Sunley R, Atkinson C, Jones H (2006) Chill unit models and recent changes in the occurrence of winter chill and spring frost in the united kingdom. J Hortic Sci Biotech 81: 949–958.

4. Sugiura T (2010) Characteristics of responses of fruit trees to climate changes in japan. Acta Hortic 872: 85–88.
5. Asakura T, Sugiura H, Sakamoto D, Sugiura T, Gemma H (2011) Frost risk evaluation in apple by modelling phenological changes in critical temperatures. Acta Hortic 919: 65–70.
6. Luedeling E, Blanke M, Gebauer J (2009) Auswirkungen des Klimawandels auf die Verfügbarkeit von Kältewirkungen (Chilling) für Obstgehölze in Deutschland. Erwerbs-Obstbau 51: 81–94.
7. Chitu E, Topor E, Paltineanu C, Dumitru M, Sumedrea D, et al. (2011) Phenological and climatic modelling of the late frost damage in apricot orchards under the changing climatic conditions of south-eastern romania. Acta Hortic 919: 57–64.
8. Fan X, Wang W, Yang X, Wu Y (2010) Responses of apple tree's phonology in east and west sides of liupanshan mountain to climate change. Chinese J Ecol 29: 50–54.
9. Blanke M, Kunz A (2009) Einfluss rezenter Klimaveränderungen auf die Phänologie bei Kernobst am Standort Klein-Altendorf - anhand 50-jähriger Aufzeichnungen. Erwerbs-Obstbau 51: 101–114.
10. Blanke M, Kunz A (2011) Effects of climate change on pome fruit phenology and precipitation. Acta Hortic 922: 381–386.
11. Kunz A, Blanke M (2011) Effects of global climate change on apple 'golden delicious' phenology – based on 50 years of meteorological and phenological data in klein-altendorf. Acta Hortic 903: 1121–1126.
12. Scheifinger H, Menzel A, Koch E, Peter C (2003) Trends in spring time frost events and phenological dates in central europe. Theor Appl Climatol 74: 41–51.
13. Meier U, Graf H, Hack H, Hess M, Kennel W, et al. (1994) Phänologische Entwicklungsstadien des Kernobstes (*Malus domestica Borkh.* und *Pyrus communis* L.), des Steinobstes (Prunus-Arten), der Johannisbeere (Ribes-Arten) und der Erdbeere (*Fragaria x ananassa Duch.*). Nachrichtenbl Deut Pflanzenschutzd 46: 141–153.
14. Chmielewski F, Müller A, Bruns E (2004) Climate changes and trends in phenology of fruit trees and field crops in Germany, 1961–2000. Agric For Meteorol 121: 69–78.
15. Luedeling E (2012) Climate change impacts on winter chill for temperate fruit and nut production: A review. Sc Hortic-Amsterdam 144: 218–229.
16. Solomon S, Quin D, Manning M, Chen Z, Marquis M, et al. (2007) Contribution of Working Group I to the Fourth Assessment Report of the Intergovernmental Panel on Climate Change. Cambridge University Press, Cambridge, United Kingdom and New York, NY, USA. Available: http://www.ipcc.ch/. Accessed 20 April 2013.
17. Eccel E, Rea R, Caffarra A, Crisci A (2009) Risk of spring frost to apple production under future climate scenarios: the role of phenological acclimation. Int J Biometeorol 53: 273–286.
18. Kaukoranta T, Tahvonen R, Ylämäki A (2010) Climatic potential and risks for apple growing by 2040. Agr Food Sci 19: 144–159.
19. Cannell G, Smith R (1986) Climatic warming, spring budburst and frost damage on trees. J Appl Ecol 23: 177–191.
20. Chmielewski F, Müller A, Küchler W (2005) Climate changes and frost hazard for fruit trees. Annalen der Meteorologie 41 2: 488–491.
21. Hoffmann H, Langner F, Rath T (2012) Simulating the influence of climatic warming on future spring frost risk in northern german fruit production. Acta Hortic 957: 289–296.
22. Chmielewski F, Müller A (2005) Possible impacts of climate change on natural vegetation in Saxony (Germany). Int J Biometeorol 50: 96–104.
23. Chmielewski F, Blümel K, Henniges Y, Blanke M, Weber R, et al. (2011) Phenological models for the beginning of apple blossom in Germany. Met Z 20: 487–496.
24. Hökkinen R, Linkosalo T, Hari P (1998) Effects of dormancy and environmental factors on timing of bud burst in *Betula pendula*. Tree Physiol 18: 707–712.
25. Blümel K, Chmielewski F (2012) Shortcomings of classical phenological forcing models and a way to overcome them. Agric For Meteorol 164: 10–19.
26. Körner C, Basler D (2010) Warming, photoperiods and tree phenology. Science 329: 277–278.
27. Nakicenovic N, Alcamo J, Davis G, de Vries B, Fenhann J, et al. (2000) Special report on emission scenarios. Cambridge University Press, Cambridge, United Kingdom and New York, NY, USA. Available: http://www.ipcc.ch/. Accessed 20 April 2013.
28. Piani C, Haerter J, Coppola E (2010) Statistical bias correction for daily precipitation in regional climate models over europe. Theor Appl Climatol 99: 187–192.
29. Hoffmann H, Rath T (2012) Meteorologically consistent bias correction of simulated climate time series for agricultural models. Theor Appl Climatol 110: 129–141.
30. Hoffmann H, Rath T (2012) High resolved simulation of climate change impact on greenhouse energy consumption in Germany. Eur J Hortic Sci 77: 241–248.
31. Hawkins E, Sutton R (2009) The potential to narrow uncertainty in regional climate predictions. B Am Meteorol Soc 90: 1095–1107.
32. Yip S, Ferro CAT, Stephenson DB, Hawkins E (2011) A simple, coherent framework for partitioning uncertainty in climate predictions. J Climate 24: 4634–4643.
33. Oliver M, Webster R (1990) Kriging: a method of interpolation for geographical information systems. Int J Geogr Inf Syst 4: 313–332.
34. Bowman A, Azzalini A (1997) Applied Smoothing Techniques for Data Analysis. New York: Oxford University Press.
35. Dosio A, Paruolo P, Rojas R (2012) Bias correction of the ensembles high resolution climate change projections for use by impact models: Analysis of the climate change signal. J Geophys Res Atmos 117.
36. Landsberg J (1974) Apple fruit bud development and growth; analysis and an empirical model. Ann Bot-London 28: 1013–1023.
37. Murray M, Cannell G, Smith R (1989) Date of budburst of fifteen tree species in britain following climatic warming. J Appl Ecol 26: 693–700.
38. Chuine I, Cour P, Rousseau D (1998) Fitting models predicting dates of flowering of temperate-zone trees using simulated annealing. Plant Cell Environ 21: 455–466.
39. Hawkins E, Sutton R (2012) Time of emergence of climate signals. Geophys res lett 39: 1–7.
40. Mann H (1945) Nonparametric tests against trend. Econometrica 13: 245–259.
41. Cox D, Lewis P (1966) The Statistical Analysis of Series of Events. London: Methuen & Co. Ltd.
42. Legave J, Farrera I, Almeras T, Calleja M (2008) Selecting models of apple flowering time and understanding how global warming has had an impact on this trait. J Hortic Sci Biotech 83: 76–84.
43. Rea R, Eccel E (2006) Phenological models for blooming of apple in a mountaineous region. Int J Biometeorol 51: 1–16.
44. Schwartz M, Ahas R, Aasa A (2006) Onset of spring starting earlier across the northern hemisphere. Glob Change Biol 12: 343–351.
45. Luedeling E, Zhang M, Luedeling V, Girvetz E (2009) Sensitivity of winter chill models for fruit and nut trees to climatic changes expected in california's central valley. Agric Ecosys Environ 133: 23–31.
46. Luedeling E, Zhang M, Girvetz EH (2009) Climatic changes lead to declining winter chill for fruit and nut trees in california during 1950–2099. PLoS ONE 4: 1–9.
47. Chmielewski F, Rötzer T (2001) Response of tree phenology to climate change across europe. Agric For Meteorol 108: 101–112.
48. Tooke F, Battey N (2010) Temperate flowering phenology. J Exp Bot 61: 2853–2862.
49. Menzel A, Sparks T, Estrella N, Koch E, Aasas O, et al. (2006) European phenological response to climate change matches the warming pattern. Glob Change Biol 12: 1969–1976.
50. Ibáñez I, Primack R, Miller-Rushing A, Ellwood E, Higuchi H, et al. (2011) Forecasting phenology under global warming. Phil Trans R Soc 365: 3247–3260.
51. Jie B, Quansheng G, Junhu D (2011) The response of first flowering dates to abrupt climate change in beijing. Adv atmos sci 28: 564–572.
52. Harrington C, Gould P, StClair J (2010) Modeling the effects of winter environment on dormancy release of douglas-fir. Forest Ecol Manag 259: 798–808.
53. Chmielewski F, Görgens M, Kemfert C (2009) KliO: Klimawandel und Obstbau in Deutschland. Abschlussbericht. Humboldt University of Berlin, Institute of Crop Sciences, Subdivision of Agricultural Meteorology. Available: http://www.agrar.hu-berlin.de/fakultaet/departments/dntw/agrarmet/forschung/fp/AB-HU.pdf. Accessed: 20 Apr 2013.
54. Dqu M, Rowell DP, Lthi D, Giorgi F, Christensen JH, et al. (2007) An intercomparison of regional climate simulations for europe: Assessing uncertainties in model projections. Climatic Change 81: 53–70.
55. Rojas R, Feyen L, Bianchi A, Dosio A (2012) Assessment of future flood hazard in europe using a large ensemble of bias-corrected regional climate simulations. J Geophys Res Atmos 117.
56. Juroszek P, von Tiedemann A (2012) Climate change and potential future risks through wheat diseases: a review. Eur J Plant Pathol : 1–13.
57. Rugienius R, Siksnianas T, Gelvonauskiene D, Staniene G, Sasnauskas A, et al. (2009) Evaluation of genetic resources of fruit crops as donors of cold and disease resistance in lithuania. Acta Hortic 825: 117–124.
58. Jacob D (2005) REMO A1B scenario run, UBA project, 0.088 degree resolution, run no. 006211, 1h data. cera-db "'REMO- UBA -A1B -1-R006211-1H'". World Data Center for Climate. Available: http://cera-www.dkrz.de/WDCC/ui/Compact.jsp?acronym = REMO_UBA_A1B_1_R006211_1H. Accessed: 20 Apr 2013.
59. Jacob D, Nilson E, Tomassini L, Bülow K (2009) REMO A1B scenario run, BFG project, 0.088 degree resolution, 1h values. cera-db "remo- bfg- a1b-1h". World Data Center for Climate. Available: http://cera-www.dkrz.de/WDCC/ui/Compact.jsp?acronym = REMO_BFG_A1B_1H. Accessed: 20 Apr 2013.
60. Keuler K, Lautenschlager M, Wunram C, Keup-Thiel E, Schubert M, et al. (2009) Climate simulation with CLM, scenario A1B run no.1, data stream 2: European region MPI-M/MaD. World Data Center for Climate. DOI:10.1594/WDCC/CLM_A1B_1_D2. Available: http://dx.doi.org/10.1594/WDCC/CLM_A1B_1_D2. Accessed: 20 Apr 2013.
61. Keuler K, Lautenschlager M, Wunram C, Keup-Thiel E, Schubert M, et al. (2009) Climate simulation with CLM, scenario A1B run no.2, data stream 2: European region MPI-M/MaD. World Data Center for Climate. DOI:10.1594/WDCC/CLM_A1B_2_D2. Available: http://dx.doi.org/10.1594/WDCC/CLM_A1B_2_D2. Accessed: 20 Apr 2013.
62. van der Linden P, Mitchell J (2009) Ensembles: Climate change and its impacts: Summary of research and results from the ensembles project, technical report. Met Off Hadley Cent., Exeter, U.K. Available: http://ensembles-eu.metoffice.com/. Accessed 20 April 2013.

10

Climate Change May Boost the Invasion of the Asian Needle Ant

Cleo Bertelsmeier[1]*, Benoît Guénard[2], Franck Courchamp[1]

1 Ecologie, Systématique & Evolution, Univ. Paris Sud, Orsay, France, **2** Biodiversity and Biocomplexity Unit, Okinawa Institute of Science and Technology, Okinawa, Japan

Abstract

Following its introduction from Asia to the USA, the Asian needle ant (*Pachycondyla chinensis*) is rapidly spreading into a wide range of habitats with great negative ecological affects. In addition, the species is a concern for human health because of its powerful, sometimes deadly, sting. Here, we assessed the potential of *P. chinensis* to spread further and to invade entirely new regions. We used species distribution models to assess suitable areas under current climatic conditions and in 2020, 2050 and 2080. With a consensus model, combining five different modelling techniques, three Global Circulation (climatic) Models and two CO_2 emission scenarios, we generated world maps with suitable climatic conditions. Our models suggest that the species currently has a far greater potential distribution than its current exotic range, including large parts of the world landmass, including Northeast America, Southeast Asia and Southeast America. Climate change is predicted to greatly exacerbate the risk of *P. chinensis* invasion by increasing the suitable landmass by 64.9% worldwide, with large increases in Europe (+210.1%), Oceania (+75.1%), North America (+74.9%) and Asia (+62.7%). The results of our study suggest *P. chinensis* deserves increased attention, especially in the light of on-going climate change.

Editor: Deborah M. Gordon, Stanford University, United States of America

Funding: This paper was supported by Région Ile-de-France (03-2010/GV-DIM ASTREA) and ANR (2009 PEXT 010 01) grants. The funders had no role in study design, data collection and analysis, decision to publish, or preparation of the manuscript.

Competing Interests: The authors have declared that no competing interests exist.

* E-mail: cleo.bertelsmeier@u-psud.fr

Introduction

Among the over 12,000 described species of ants [1], more than 200 species have established populations outside their native range [2]. The rate of new species introductions continues to increase due to the ever growing human-mediated transportation via international trade and tourism [3]. Only a small subset of introduced ant species eventually becomes invasive, but these species can have a large impacts [4–6]. They can cause significant biodiversity losses, in particular as extremely efficient predators and competitors [7]. For example, most native ant species may be eliminated from the invaded habitat and a variety of other taxa, ranging from soil microbes to small mammals, can be negatively affected [6,8]. In addition, invasive ants can disturb ecological networks, such as seed dispersal mutualism, thereby impairing ecosystem functioning [7]. Finally, they often damage agroecosystems and are a nuisance to humans by infesting estates, leading to high economic costs [9].

Ants are known to be very sensitive to changes in temperature and humidity, because it affects their survival [10], foraging activity [11] and foraging networks [12] and dominance hierarchies [13]. It is generally accepted that with climate change, many invasive ant species will progressively colonize higher latitudes and altitudes, where the currently too cold climatic conditions are expected to become more suitable [14,15]. In this regard, several studies have used species distribution models [16–19] or physiological experiments [11,20,21] to investigate the relationship between temperature and humidity and ant distribution. Climate is one of the most important factors influencing the distribution of ants [22–24] and climatic suitability has been shown to be even the most important factor responsible for the current global distribution of the invasive Argentine ant, *Linepithema humile* [25]. Climate can therefore serve as an important proxy to estimate the potential distribution of invasive ants worldwide. It is generally recognized that climate change is going to be a major determinant of species physiology, phenology and range shifts during this century [26]. However, few studies have gone beyond the estimation of current potential invasive range and forecast also the future potential distributions of invasive species in general (but see [27–29]). Furthermore, such pioneer studies estimating the future potential ranges of invasive ants have concentrated on the few species of *Linepithema humile* [25,30] *Solenopsis invicta* [31] and *Pheidole megacephala* [16], leaving a great knowledge gap for most other major invasive ant species.

Risk assessments conducted prior to the arrival of an invasive species are a vital component of biosecurity preparedness, because most species, including invasive ants, are extremely difficult to eradicate once they become established [32]. Providing a spatial model of relative climatic suitability is an important component of risk assessments to prioritize surveillance efforts for invasive species with a high likelihood of establishing in a particular region. The greater the extent of an invasion, the higher the environmental impacts of management attempts and the difficulty of achieving successful eradication [33]. It is known that reactive programs have generally a higher cost than proactive programs [34]. In this context, the Asian needle ant, *Pachycondyla chinensis*, is of utmost interest. Despite its introduction from Asia to the eastern part of North America in the first part of the 20th century, the invasion of this fast-spreading species was only detected recently in a wide range of habitats in North America, including mature temperate forests, where it causes a strong decline in native ant abundance [35]. In addition, the species has been shown to disrupt an ant-seed dispersal mutualism by displacing a native keystone ant species [36]. The species' negative impact on native seed dispersers has been compared to the impact of Argentine ant [36], which is among the "100 of the worst invasive species" list of the IUCN [37] and has enormous impacts on biodiversity [38]. Additionally, *P. chinensis* is a growing concern for public health due to its powerful, and sometimes deadly, sting [35].

Consequently, there is a strong need to develop predictive models of the potential distribution of this highly invasive species, both currently and in the future with predictions of climate change. Here we use species distribution models to: (1) to quantify the current potential distribution worldwide and within six broad geographic regions; and (2) quantify the change in potential distribution with global climate change at the global and regional levels.

Materials and Methods

Species Distribution Data

Species distribution models search for a non-random association between environmental predictors and species occurrence data to make spatial predictions of potential distribution. Because our models should include the full set of climatic conditions under which the target species can exist, we included occurrence points (presence only data) from both invaded and native habitats [39]. In total, we used 283 occurrence points, 219 from North America (invaded range) and 64 from Asia (native range) (Fig. S1). The exact distribution of *P. chinensis*' native range is problematic as this species belongs to a large and taxonomically unresolved complex of species [40]. To maximize data integrity, the data used for modeling were limited to specimens collected in its native range and identified by one of the authors (BG) and specimens strictly identified as *P. chinensis* in literature [40,41]. In its introduced range, where *P. chinensis* identification is not problematic, localities were extracted from literature, museum records and personal collecting (BG).

For models requiring absence data, 10,000 pseudo-absence (background) points were generated randomly from all around the world to provide background data. This is a classic procedure because confirmed absence data is difficult to obtain for most species and requires great sampling efforts [42]. True absence data might improve the model accuracy because some pseudo-absence points may be drawn from regions where the species is actually present, but has not been recorded. However, it is not possible to base our projections on true absences due to lacking large-scale absence data of the species. In addition, in the case of invasive species, even a true absence point may indicate a suitable location that the species has not yet been introduced to due to a lack of opportunity. Therefore, we believe that pseudo-absence data can serve as a reasonable proxy.

Climatic Predictors

Climatic predictor data was sourced from the Worldclim dataset. The 19 Worldclim variables represent annual trends (e.g. mean annual temperature, annual precipitation) and extreme limiting environmental factors (e.g. temperature of the coldest and warmest months, precipitation of the wettest or driest quarter) and are known to influence species distributions [45,46]. All Worldclim variables are 30-year averages of monthly temperature and rainfall values from 1960–1990 [43], which is characteristic of the climate that the species experienced when the occurrence point was collected or the species established in this locality. We modelled the species niche based on 4 of the 19 bioclimatic variables that were not collinear (pair-wise $r_{Pearson} < 0.75$). The selected variables were (in the order of their relative contribution to the Maxent model): Precipitation of the driest month, isothermality, precipitation of the warmest quarter and maximum temperature of the warmest month. These variables are believed to directly influence ant distributions because many features of ant biology are sensitive to small differences in temperature [16] or humidity, for example foraging [11], oviposition rates [44], survival [10], the structure of foraging networks [12].

Future climatic data were sourced from the 4th IPCC assessment report [47]. The direct output of Global Climate Models is provided in the form of very large (500 km) grid cells because of the heavy calculations needed for the simulation of geophysical processes. To get a better resolution required for species distribution modelling, climate centres use statistical models to infer climatic variation at a more local scale, "downscaling" the data by using the WorldClim data for 'current' conditions for calibration. Therefore the projections at different time horizons can be compared. In order to consider a range of possible future climates, we used downscaled climate data from three Global Circulation Models (GCMs), provided by different climate centres, each based on different geophysical assumptions: the CCCMA-GCM2 model; the CSIRO-MK2 model; and the HCCPR-HADCM3 model [47]. We also used two extreme Special Report on CO_2 Emission Scenarios (SRES): the optimistic B2a and pessimistic A2a scenarios. In total, we used six future climatic scenarios (3 GCM $\times$ 2 SRES). Data for the future climatic projections were climate data averaged across a decade, centred on the focal year (e.g. 2020) [47].

Worldclim data is the standard source of climate data for species distribution models. However it is poor at interpolating climate in topographically complex regions such as mountains or coastal regions [48]. But the focus of our study is projections at the global scale with a spatial resolution of 10 arcmin (approx. 18.5 $\times$ 18.5 km pixel), where complex coastlines are not visible. Predictions based on coarser resolutions are more likely to be controlled by climatic predictors, whereas fine-scale, patchy distributions at a smaller scale are more likely to be determined by micro-topographic variations or habitat fragmentation [49].

Species Distribution Modelling

In order to make spatial predictions of potential distribution, we used species distribution models (SDMs), which explain the species' current distribution based on a set of climatic predictor variables. It has been shown that model outputs are sensitive to the algorithms, climatic data from different global climate models and different human development scenarios [50]. One way to deal with these uncertainties in species distribution modelling is to conduct a consensus forecast which can be defined as combining multiple simulations across a range of possible initial conditions and different classes of models [51].

To generate the consensus forecasts we used five machine learning methods, which are a set of algorithms that learn the mapping function or classification rule inductively from the input data [52]. The first two models were based on one- and two- class Support Vector Machines (SVMs) [39,31]. Two-class SVMs (SVM2) seek to find a hyperplane that maximally separates the two target classes. Recently, one-class SVMs have also been developed [42] that distinguish one specific category from all other categories. Third, we used Artificial Neural Networks (ANN), which extract linear combinations of the input variables as derived features (synthetic variables), and model the output as a nonlinear function of these derived features [42,54]. Fourth, we used Classification Trees (CT), which partition the response variable into increasingly pure binary subsets with splits and stop criteria [41,42,53]. Finally we used the Maximum Entropy Method (Maxent) which estimates a probability distribution of a species being present by seeking the most widespread distribution, given a set of constraints [18,57,58]. For a more detailed description of these algorithms see [16,19]. All models were run using the ModEco Platform with default parameters [59].

A clear limitation of modelling is that outputs are dependent on the specifically chosen input settings, in this instance the algorithms, global climate models and scenarios of human development. To minimise potential resulting variation, we conducted consensus forecasts [51] using the outputs of the five different modelling techniques detailed above with each of three climate models (GCMs) and two CO_2 emission scenarios (SRES). The purpose of consensus forecasts is to separate the signal from the "noise" associated with the errors and uncertainties of individual models, by superposing the maps based on individual model outputs. Areas where these individual maps overlap are defined as areas of "consensual prediction" [51]. This is different from averaging the individual projections, as the area predicted by the consensus forecast can be smaller than any individual forecast if there is little spatial agreement (*i.e.,* overlap) between individual forecasts. Simple averaging across individual forecasts is considered unlikely to match reality [51].

The contribution of the individual models (i.e. the spatial prediction of "suitable range") was weighted according to their AUC (section on model validation) in order to enhance contribution of models with higher model performance values (see [18]). Only binary projections (present or absent) have been combined to generate the consensus model because continuous outputs can have different meanings for different models and cannot be simply added together [59]. The combination of the individual forecasts then yields a projection (the consensus model), where the value of pixels vary between 0 and 1 and can be interpreted as a probability of the species occurring in each pixel [51].

The consensus model was generated using all 30 individual projections, each based on a different combination of CO_2 scenario × GCM × modelling technique, yielding a consensus projection for 4 time horizons (current, 2020, 2050 and 2080). The future climatic projections that we used as a basis of our models are in fact averaged climate data across a decade, centred on the focal year (e.g. 2020) [47].

Model Validation

Model robustness was evaluated using the AUC of the ROC curve, which is a nonparametric threshold-independent measure of accuracy commonly used to evaluate species distribution models (e.g., [18,60]). We used the AUC because it does not depend on the selected classification threshold, and it readily indicates if a model discriminates correctly between presence and absence points [18,60]. AUC values range from 0 to 1, where a value of 0.5 can be interpreted as a random prediction. AUC between 0.5 and 0.7 are considered low (poor model performance), 0.7–0.9 moderate and >0.9 high ([42] and references therein). For model evaluation, the data needs to be split into a train and a test group. Here, we used 10-fold cross-validation, whereby the data was split into 10 equal parts, with 9/10 of the observations used to build the models and the remaining 1/10 used to estimate performance. Validation was repeated ten times and the estimated performance measures were averaged [42,61].

Assessing Climatically Suitable Areas

Studies with *a priori* objectives may use a range of different threshold values [62] to determine habitat suitability. As this was not the case here, we applied a limit whereby pixels with a probability of presence exceeding 0.5 were classified as "suitable" area, as is frequently done for binary classification for species distribution modelling [42,63]. Users of our models may want to minimize the chance of either over- or under-prediction of potential distribution (omission or commission errors) and to apply a different threshold. For example, for management decisions it could be better to apply a more "prudent" (lower) threshold that lowers the probability of omission errors. To allow these user-specific applications of our models, we provide maps with a continuous output with a probability of presence between 0 and 1 (with 0.1 intervals).

In addition, we created a difference map (future suitability map – current suitability map), which showed relative differences that are independent of any classification threshold and indicated areas where the climatic suitability improved or decreased. Second, we generated a "shift" map where we mapped the net gains, losses and stable ranges under current and future climatic conditions. Third, we calculated two indices that have been recently proposed as complementary measures to evaluate the extent of spatial shift [64]: the spatial congruence index (2a/2a+b+c), based on the Sorensen-Dice dissimilarity measure, and the stable range (a/(a+b)), which is a measure of spatial shift vs stability, whereby a = area suitable currently and in the future, b = area suitable currently only, and c = area suitable in the future only. Spatial analyses were carried out using DIVA-GIS [65] and Arcgis v. 9.3.

Results

The AUC values indicated excellent model performance for all five algorithms in predicting the species' distribution based on the consensus model (AUC values: SVM1 = 0.968, SVM2 = 0.991, Maxent = 0.998, ANN = 0.991, CT = 0.997).

Current Climatic Conditions

Maps of the consensus model under current and future climatic conditions by 2020, 2050 and 2080 indicated large and increasing suitable areas for *P. chinensis* (Fig. 1a–d). Under current climatic conditions, 3.33% of global landmass was predicted to be suitable for *P. chinensis.* The suitable range was unequally distributed among biogeographic regions, with the highest relative amount of suitable landmass found in North America (45%), followed by Asia (38%), South America (11%), Europe (3%) and Oceania (2%) (Fig. 2a). The relative proportion of suitable landmass was also highest in North America (Fig. 2b).

Climate Change Impacts

The suitable range for *P. chinensis* increased dramatically with projected climate change. In 2020 the potential range increased by 15.6%, in 2050 by 29.3% and in 2080 this increase reached +64.9% of the currently suitable landmass (Fig. 3).

By 2080, changes in suitable landmass differed greatly among biogeographic regions, with large increases in Europe (+210.1%, *i.e.,* +363,117 km^2), Oceania (+75.1%, *i.e.,* +94,332 km^2), North America (+74.9%, *i.e.,* +1,972,781 km^2) and Asia (+62.7%, *i.e.,* +1,403,693 km^2) and a decrease in Africa (−22.9%, *i.e.,* −2,042 km^2).

Spatial Shifts of Suitable Conditions

The net changes in suitable landmass were almost exclusively due to gains in potential distribution (+4,350,682 km^2, Fig. 4a).

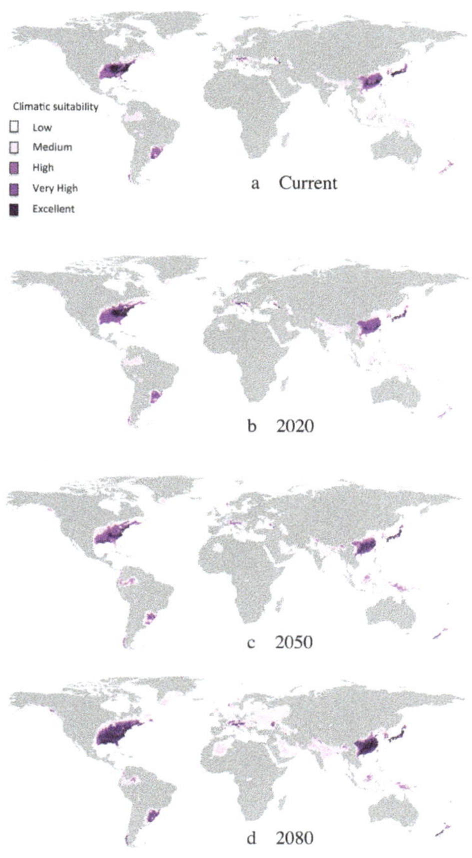

Figure 1. Maps of potential distribution. Climatic suitability ranges from "low" (light purple) to "excellent" (dark purple). (**a**) Current climatic conditions (**b**) Consensus model of 6 future climatic scenarios (3 GCM×2 SRES) for 2020, (**c**) 2050 and (**d**) 2080.

Very small areas of suitable landmass was lost (i.e., suitable under current but not under future climatic conditions: $-88{,}300$ km^2, Fig. 4a). These increases were from range expansions at the edge of the suitable range, but also entirely of new areas that became suitable (e.g. Europe, Northern Brazil or Indonesia, Fig. 4a). The stable range index was 0.569, meaning that only 56.9% of *P. chinensis*' current potential range will remain suitable in the future. Spatial congruence was 0.722, which provides a measure of the stability (stable range) vs shift (losses and gains), indicating that 72.2% of all current and future suitable areas can be considered as "stable range" over time, whereas 27.8% will be either suitable currently or in the future, but not under both climatic scenarios.

The relative differences between the current and future areas of suitability, indicated that *P. chinensis* was likely to experience higher relative climatic suitability in all biogeographic regions, including large parts of north Africa, Arabia, India, South East Asia, north east America and eastern Europe (Fig. 4b). Few areas, such as the eastern USA showed a slight decrease of suitability (areas in light blue, Fig. 4b).

Discussion

The Asian needle ant is native to Asia and has already invaded the South Eastern part of the USA [35]. Our models suggest that the species has a far greater invasive potential and is capable of invading large parts of the global landmass on several continents, particularly north east America, South East Asia and south east America. In addition the consensus models suggest that climate change will exacerbate the risk of invasion by *P. chinensis* globally. Strikingly, the global suitable landmass is predicted to increase by 64.9% with climate change. At the regional scale, most of *P. chinensis*' potential distribution was found in its Asian native range and in North America where the species is now considered invasive [35]. With climate change, the amount of suitable area for *P. chinensis* is predicted to greatly increase based on more suitable conditions, by 74.9% in North America. In addition, our models also predict a potential expansion into new biogeographic regions that should become suitable by 2080, in particular Europe, South America or Indonesia and an expansion in Asia relative to its current known distribution range.

In addition to direct climatic suitability, climate change could enhance the invasion likelihood by disadvantaging other competing ants in the invaded areas. Ant community structures are known to be temperature-dependent [13,66], and therefore a change in temperature might lead to new invaders dominating. For example, *P. chinensis* has recently been found to establish in sites dominated by the invasive Argentine ant, *L. humile*, resulting in a dominance swap and even localised extinction of *L. humile* [65]. Because *P. chinensis* does not seem to be behaviourally dominant, this dominance shift has been attributed to differences in the climatic preferences of the two species, with *P. chinensis* establishing nests and expanding is population earlier in the season when temperatures were lower before *L. humile* populations could expand [65].

Our consensus models for the potential distribution had a high accuracy (good to excellent AUC) and were designed to include a broad range of climate change scenarios, with an optimistic B2a and a pessimistic A2a CO_2 emission scenario and three global circulation climatic models [47]. We additionally reduced uncertainty due to single modelling methods by building models with five different algorithms that contributed to the final consensus forecast [51]. Nevertheless, inherent uncertainty in the spatial projection remains because of the underlying assumptions shared by all species distributions models [49], [46] in that they assume that the species is in equilibrium with its environment and therefore its current distribution reflects the ideal climatic conditions for the species, which can be used to model its potential distribution. That means that a model of the potential distribution of an invasive species with climate change has to make two extrapolations: 1) in space (invasion of a different place); and 2) in time (with future climate change). However, niche shifts during invasions are possible and have already been observed [67–69] and species may display phenotypic plasticity or show evolutionary adaptations, such as has been shown for the Asian ladybeetle which has the same invasion pattern as *P. chinensis*, dispersing from Asia to the east coast of North America [70]. Therefore, projections for invasions under climate change come necessarily with some uncertainty and should be only viewed as an attempt to evaluate future trends and invasion risks, and not as a precise prediction at a small scale. For example, a new occurrence point of *P. chinensis* has been recently recorded in Washington D.C. by the School of Ants project [71], where the species does not find 'excellent' climatic conditions according to our projections. One of the inherent problems of species distribution models is that the species may be able to occur within areas predicted to be of relatively low suitability, if the microclimatic conditions favour the species or if it is associated to human infrastructure. Despite these limitations, species distribution

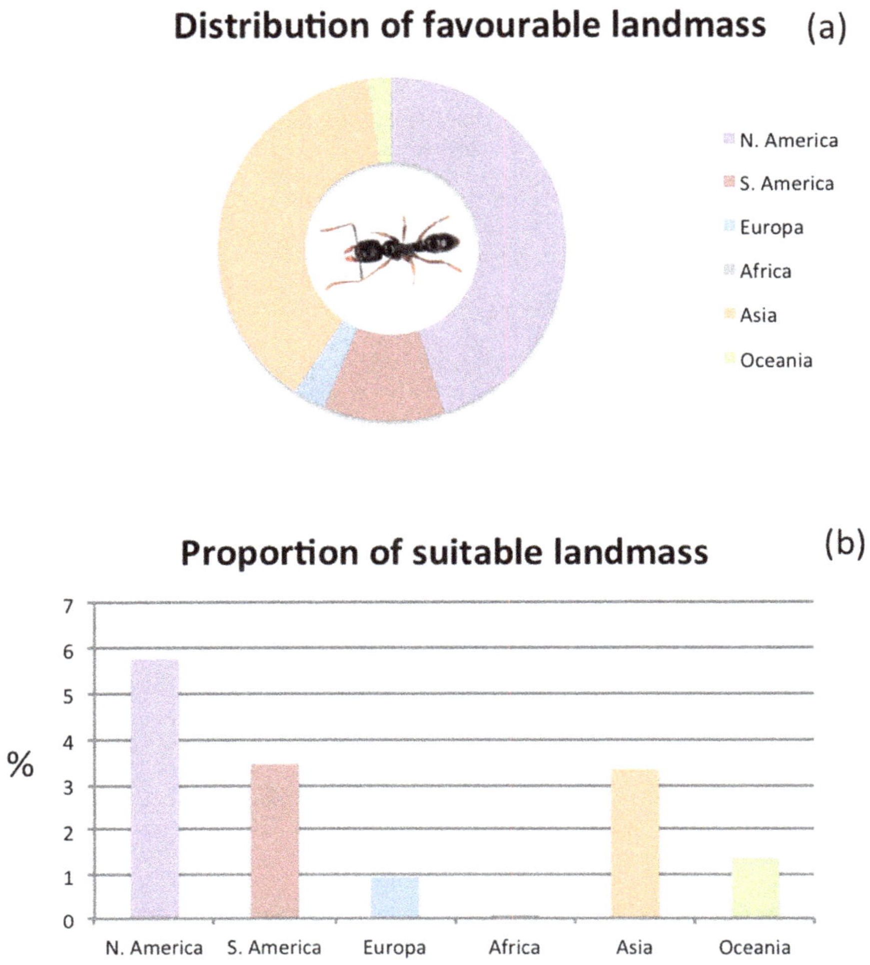

Figure 2. a+b Distribution and proportion of suitable landmass under current climatic conditions among six biogeographic regions.

models are generally considered to deliver useful approximations [72,73].

A further factor to consider is that numerous native species are predicted to suffer the effects from climate change [26]. This may

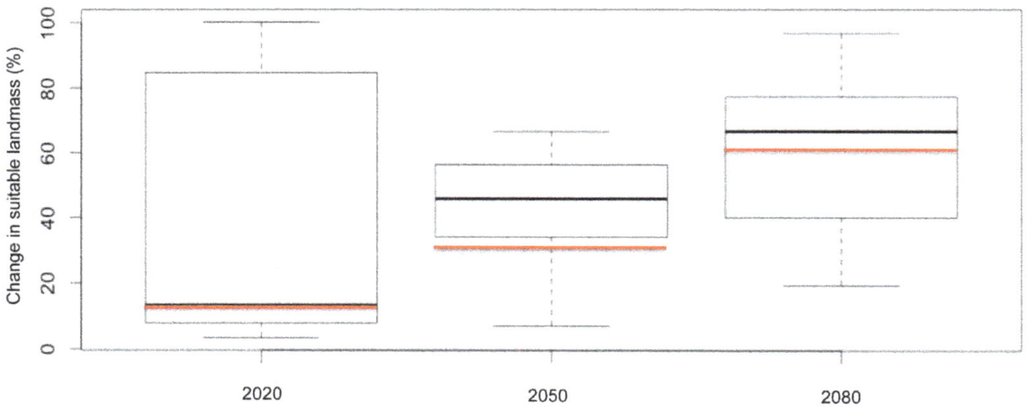

Figure 3. Change in suitable landmass over time relative to the currently suitable landmass. The boxplot represents variation of projections across six future climatic scenarios, each based on a different combination of Global Climate Model×CO_2 emission scenario, per time horizon (± s.d). The red line indicates the value of the consensus model.

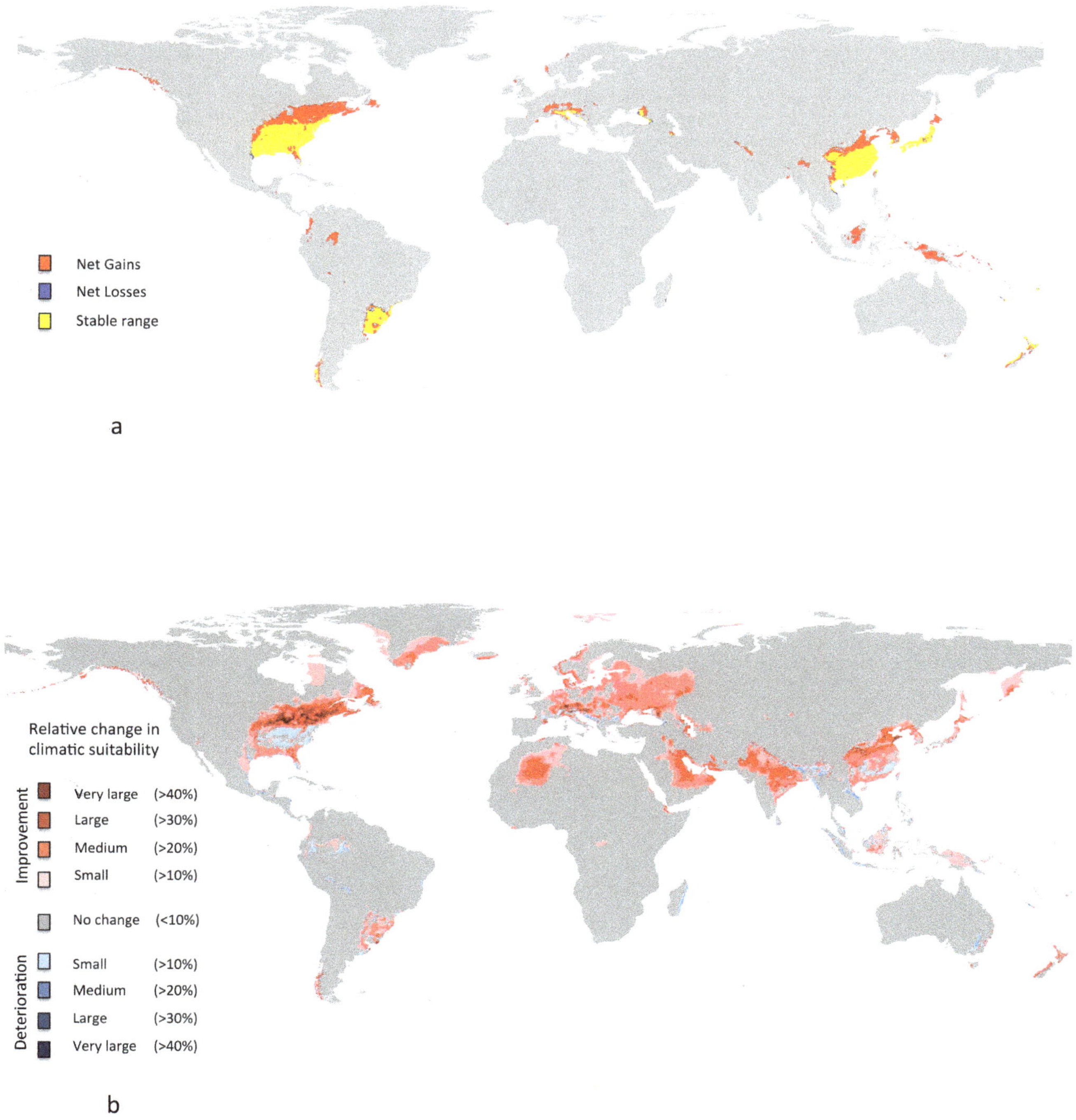

Figure 4. Differences in suitable climatic conditions between the current climate and the consensus projection in 2080. (a) Net gains (red), losses (blue) and stable range (yellow), (b) Relative differences in suitability : red colors indicate improved climatic suitability (light to dark red gradient indicates the relative change in suitability), blue colors indicate deteriorated climatic suitability for *P. chinensis*.

increase the vulnerability of the community by decreasing the biotic resistance to new invaders [74,75]. This could even further exacerbate the invasion risk of *P. chinensis*.

Our study solely focused on the role of climate change on the potential distribution of a newly invasive ant species because climate has been shown to be very important in determining the distribution of ants at a global scale [22–24] as well as being the most important factor influencing the global invasion of the Argentine ant [25]. Therefore, climate is probably a crucial factor limiting the distribution of other invasive ants, such as *P. chinensis*. For future studies, predictions of invasive potential will be advanced by investigations into the influence of other abiotic factors and drivers of species displacement. At a finer scale for example, topographic or terrain variables, such as elevation, geomorphology or hydrology can be important [42]. For example, appropriate soil moisture levels can be an important requirement for nest location of ants. Also important is the effect of disturbance regimes, because invasive ants are frequently associated with disturbed habitats [76,77]. However, it should be noted that *P. chinensis* has been found to invade intact forests [36].

Our results support the view that biological invasions could increase due to climate change [60,61,62,10,11] and show it can do so dramatically. In this way two of the most important threats to global biodiversity (invasive species and climate change) may interact synergistically. An important observation is that the potential distribution of *P. chinensis* exists at a wide range of latitudes, and thus this species' potential range did not simply shift to higher latitudes. Consequently, invasion risk was exacerbated globally with entirely new areas covering large amounts of landmass becoming suitable. Given the important ecological impacts of *P. chinensis* [35] and its ability to potentially even displace one of the most aggressive and dominant species of invasive ants, the *L. humile* [81], clearly it is important that surveillance efforts of this species are increased to prevent further spread and aid early detection. Eradication of well established large invasive ant populations can be extremely challenging [32] if not impossible, but early detection of small incipient populations can enable managers to carry out early responses and achieve eradication [34]. The use of species distribution models to inform risk assessments is increasingly becoming standard, but a spatial model should be always viewed in the light of the many uncertainties associated with the approach [58]. Species distribution models can serve as a guide to prioritize surveillance efforts in certain regions. Ideally, this approach should be complemented with interception data at ports of entry. The results of our study suggest *P. chinensis* deserves increased attention in new regions of the world (e.g., Europe and South America) as a rising exotic species of significance.

Acknowledgments

We thank Ben Hoffmann, Melissa Burt, and two anonymous referees for their comments.

Author Contributions

Conceived and designed the experiments: CB FC BG. Performed the experiments: CB. Analyzed the data: CB FC. Contributed reagents/materials/analysis tools: CB GB FC. Wrote the paper: CB FC GB.

References

1. Bolton B, Alpert G, Ward PS, Naskrecki P (2007) Bolton's catalogue of the ants of the world. Harvard University Press, Cambridge, Mass.
2. Suarez A V, McGlynn T, Tsutsui ND (2010) Biogeographic and Taxonomic Patterns of Introduced Ants. In: Lach L, Parr CL, Abbott KL, editors. Ant Ecology. Oxford: Oxford University Press.
3. Vitousek PM, Dantonio CM, Loope LL, Rejmanek M, Westbrooks R (1997) Introduced species: A significant component of human-caused global change. New Zealand Journal of Ecology 21: 1–16.
4. Holway D, Lach L, Suarez A V, Tsutsui ND, Case TJ (2002) The causes and consequences of ant invasions. Annual Review of Ecology and Systematics 33: 181–233.
5. Rabitsch W (2011) The hitchhiker's guide to alien ant invasions. BioControl 56: 551–572.
6. Lach L, Hooper-Bui LM (2010) Consequences of Ant Invasions. In: Lach L, Parr CL, Abbott KL, editors. Ant Ecology. Oxford: Oxford University Press. 261–286.
7. Ness JH, Bronstein JL, Andersen AN, Holland JN (2004) Ant body size predicts dispersal distance of ant-adapted seeds: Implications of small-ant invasions. Ecology 85: 1244–1250.
8. Lessard JP, Fordyce JA, Gotelli NJ, Sanders NJ (2009) Invasive ants alter the phylogenetic structure of ant communities. Ecology 90: 2664–2669.
9. Pimentel D, Zuniga R, Morrison D (2005) Update on the environmental and economic costs associated with alien-invasive species in the United States. Ecological Economics 52: 273–288.
10. Walters AC, Mackay DA (2004) Comparisons of upper thermal tolerances between the invasive Argentine ant (Hymenoptera: Formicidae) and two native Australian ant species. Annals of the Entomological Society of America 97: 971–975.
11. Brightwell R, Labadie P, Silverman J (2010) Northward Expansion of the Invasive *Linepithema humile* (Hymenoptera: Formicidae) in the Eastern United States is Constrained by Winter Soil Temperatures. Environmental Entomology 39: 1659–1665.
12. Heller NE, Gordon DM (2006) Seasonal spatial dynamics and causes of nest movement in colonies of the invasive Argentine ant (*Linepithema humile*). Ecological Entomology 31: 499–510.
13. Cerda X, Retana J, Cros S (1997) Thermal disruption of transitive hierarchies in Mediterranean ant communities. Journal of Animal Ecology: 363–374.
14. Walther G, Roques A, Hulme P, Sykes M, Pysek P, et al. (2009) Alien species in a warmer world: risks and opportunities. Trends in ecology & evolution 24: 686–693.
15. Hellmann JJ, Byers JE, Bierwagen BG, Dukes JS (2008) Five potential consequences of climate change for invasive species. Conservation Biology 22: 534–543.
16. Bertelsmeier C, Luque GM, Courchamp F (2013) Global warming may freeze the invasion of big-headed ants. Biological Invasions 15: 1561–1572.
17. Steiner FM, Schlick-Steiner BC, VanDerWal J, Reuther KD, Christian E, et al. (2008) Combined modelling of distribution and niche in invasion biology: a case study of two invasive *Tetramorium* ant species. Diversity and Distributions 14: 538–545.
18. Roura-Pascual N, Brotons L, Peterson a T, Thuiller W (2009) Consensual predictions of potential distributional areas for invasive species: a case study of Argentine ants in the Iberian Peninsula. Biological Invasions 11: 1017–1031.
19. Bertelsmeier C, Luque GM, Courchamp F (2013) Increase in quantity and quality of suitable areas for invasive species as climate changes. Conservation Biology in press. doi:10.1111/cobi.12093.
20. Abril S, Oliveras J, Gomez C (2010) Effect of temperature on the development and survival of the Argentine ant, *Linepithema humile*. Journal of Insect Science 10: 97.
21. Menke SB, Fisher RN, Jetz W, Holway DA (2007) Biotic and abiotic controls of Argentine ant invasion success at local and landscape scales. Ecology 88: 3164–3173.
22. Dunn RR, Agosti D, Andersen AN, Arnan X, Bruhl CA, et al. (2009) Climatic drivers of hemispheric asymmetry in global patterns of ant species richness. Ecology letters 12: 324–333.
23. Sanders NJ, Lessard JP, Fitzpatrick MC, Dunn RR (2007) Temperature, but not productivity or geometry, predicts elevational diversity gradients in ants across spatial grains. Global Ecology and Biogeography 16: 640–649.
24. Jenkins CNC, Sanders NNJ, Andersen AN, Arnan X, Brühl CA, et al. (2011) Global diversity in light of climate change: the case of ants. Diversity and Distributions: 1–11.
25. Roura-Pascual N, Hui C, Ikeda T, Leday G, Richardson DM, et al. (2011) Relative roles of climatic suitability and anthropogenic influence in determining the pattern of spread in a global invader. Proceedings of the National Academy of Sciences of the United States of America 108: 220–225. 3.
26. Bellard C, Bertelsmeier C, Leadley P, Thuiller W, Courchamp F (2012) Impacts of climate change on the future of biodiversity. Ecology Letters 15: 365–377.
27. O'Donnell J, Gallagher RV., Wilson PD, Downey PO, Hughes L, et al. (2011) Invasion hotspots for non-native plants in Australia under current and future climates. Global Change Biology 18: 617–629.
28. Ficetola GF, Maiorano L, Falcucci A, Dendoncker N, Boitani L, et al. (2010) Knowing the past to predict the future: land-use change and the distribution of invasive bullfrogs. Global Change Biology 16: 528–537.
29. Bradley BA, Blumenthal DM, Wilcove DS, Ziska LH (2010) Predicting plant invasions in an era of global change. Trends in Ecology & Evolution 25: 310–318.
30. Roura-Pascual N, Suarez A V, Gomez C, Pons P, Touyama Y, et al. (2004) Geographical potential of Argentine ants (*Linepithema humile* Mayr) in the face of global climate change. Proceedings of the Royal Society of London Series B-Biological Sciences 271: 2527–2534.
31. Peterson AT, Nakazawa Y (2008) Environmental data sets matter in ecological niche modelling: an example with *Solenopsis invicta* and *Solenopsis richteri*. Global Ecology and Biogeography 17: 135–144. A.
32. Hoffmann BD, Abbott KL, Davis PD (2010) Invasive Ant Management. In: Lach L, Parr CL, Abbott KL, editors. Ant Ecology. Oxford: Oxford University Press. 287–304.
33. Simberloff D (2003) How much information on population biology is needed to manage introduced species? Conservation Biology 17: 83–92.

34. Kaiser B, Burnett K (2010) Spatial economic analysis of early detection and rapid response strategies for an invasive species. Resour. Energy Econ 32: 566–585.
35. Guénard BB, Dunn RR (2010) A New (Old), Invasive Ant in the Hardwood Forests of Eastern North America and Its Potentially Widespread Impacts. Plos One 5: e11614.
36. Rodriguez-Cabal M a., Stuble KL, Guénard B, Dunn RR, Sanders NJ (2011) Disruption of ant-seed dispersal mutualisms by the invasive Asian needle ant (*Pachycondyla chinensis*). Biological Invasions 14: 557–565.
37. Lowe S, Browne M, Boudjelas S, De Poorter M (2000) 100 of the World's Worst Invasive Alien Species - A selection from the Global Invasive Species Database.
38. IUCN SSC Invasive Species Specialist Group (2012) Global Invasive Species Database. available from http://www.issg.org/database: accessed 24 January 2012.
39. Beaumont LJ, Gallagher R V, Thuiller W, Downey PO, Leishman MR, et al. (2009) Different climatic envelopes among invasive populations may lead to underestimations of current and future biological invasions. Diversity and Distributions 15: 409–420.
40. Yashiro T, Matsuura K, Guénard B, Terayama M, Dunn R (2010) On the evolution of the species complex *Pachycondyla chinensis* (Hymenoptera: Formicidae: Ponerinae) including the origin of its invasive form and description of a new species. Zootaxa 2685: 39–50.
41. Emery C (1895) Viaggio di Leonardo Fea in Birmania e regioni vicine. LXIII. Formiche di Birmania del Tenasserim e dei Monti Carin raccolte da L. Fea. Parte II. Annali del Museo Civico di Storia Naturale 34: 450–483.
42. Franklin J (2009) Mapping Species Distributions - Spatial Inference and Prediction. Cambridge: Cambridge University Press.
43. Hijmans RJ, Cameron SE, Parra JL, Jones PG, Jarvis A (2005) Very high resolution interpolated climate surfaces for global land areas. International Journal of Climatology 25: 1965–1978.
44. Abril S, Oliveras J, Gómez C, Gomez C (2008) Effect of temperature on the oviposition rate of Argentine ant queens (*Linepithema humile* Mayr) under monogynous and polygynous experimental conditions. Journal of Insect Physiology 54: 265–272.
45. Root BA, Price JT, Hall K (2003) Fingerprints of global warming on wild animals and plants. Nature 421: 47–60.
46. Austin M (2007) Species distribution models and ecological theory: A critical assessment and some possible new approaches. Ecological Modelling 200: 1–19.
47. GIEC (2007) Climate Change 2007: Synthesis Report. An Assessment of the Intergovernmental Panel on Climate Change.
48. Daly C, Halbleib M, Smith J, Gibson W, Doggett M, et al. (2008) Physiographically sensitive mapping of climatological temperature and precipitation across the conterminous United States. International Journal of Climatology 28: 2031–2064.
49. Guisan A, Thuiller W (2005) Predicting species distribution: offering more than simple habitat models. Ecology Letters 8: 993–1009.
50. Buisson L, Thuiller W, Casajus N, Lek S, Grenouillet G (2010) Uncertainty in ensemble forecasting of species distribution. Global Change Biology 16: 1145–1157.
51. Araújo MB, New M (2007) Ensemble forecasting of species distributions. Trends in Ecology & Evolution 22: 42–47.
52. Elith J, Graham C, Anderson R, Dudik M (2006) Novel methods improve prediction of species distributions from occurrence data. Ecography 29: 129–151.
53. Cristianini N, Schölkopf B (2002) Support Vector Machines and Kernel Methods, The New Generation of Learning Machines. AI Magazine 23: 31–41.
54. Maravelias C, Haralabous J, Papaconstantinou C (2003) Predicting demersal fish species distributions in the Mediterranean Sea using artificial neural networks. Marine Ecology Progress Series 255: 240–258.
55. De'ath G, Fabricius KE (2000) Classification and regression trees: a powerful yet simple technique for ecological data analysis. Ecology 81: 3178–3192.
56. Broennimann O, Thuiller W, Hughs G, Midgely GF, Alkemade JMR, et al. (2006) Do geographic distribution, niche property and life form explain plants' vulnerability to global change? Global Change Biology 12: 1079–1093.
57. Phillips SJ, Anderson RP, Schapire RE (2006) Maximum entropy modeling of species geographic distributions. Ecological Modelling 190: 231–259.
58. Jimenez-Valverde A, Peterson AT, Soberon J, Overton JM, Aragon P, et al. (2011) Use of niche models in invasive species risk assessments. Biological Invasions 13: 2785–2797.
59. Guo QH, Liu Y (2010) ModEco: an integrated software package for ecological niche modeling. Ecography 33: 637–642.
60. Pearce J, Ferrier S (2000) Evaluating the predictive performance of habitat models developed using logistic regression. Ecological Modelling 133: 225–245.
61. Fielding AH, Bell JF (1997) A review of methods for the assessment of prediction errors in conservation presence/absence models. Environmental Conservation 24: 38–49.
62. Nenzén HK, Araújo MB (2011) Choice of threshold alters projections of species range shifts under climate change. Ecological Modelling 222: 3346–3354.
63. Klamt M, Thompson R, Davis J (2011) Early response of the platypus to climate warming. Global Change Biology 17: 3011–3018.
64. Franklin J, Davis FW, Ikegami M, Syphard AD, Flint LE, et al. (2012) Modeling plant species distributions under future climates: how fine-scale do climate projections need to be? Global Change Biology 19: 473–483.
65. Hijmans RJ, Cruz M, Rojas E (2001) Computer tools for spatial analysis of plant genetic resources data: 1. DIVA-GIS. Genetic Resources Newsletter 127: 15–19.
66. Lessard J-PP, Dunn RR, Sanders NJ (2009) Temperature-mediated coexistence in temperate forest ant communities. Insectes Sociaux 56: 149–156.
67. Broennimann O, Treier UA, Muller-Scharer H, Thuiller W, Peterson AT, et al. (2007) Evidence of climatic niche shift during biological invasion. Ecology Letters 10: 701–709.
68. Gallagher R V, Beaumont LJ, Hughes L, Leishman MR (2010) Evidence for climatic niche and biome shifts between native and novel ranges in plant species introduced to Australia. Journal of Ecology 98: 790–799.
69. Pearman PB, Guisan A, Broennimann O, Randin CF (2008) Niche dynamics in space and time. Trends in Ecology & Evolution 23: 149–158.
70. Sloggett JJ (2012) Harmonia axyridis invasions: Deducing evolutionary causes and consequences. Entomological Science 15: 261–173.
71. Lucky A, Dunn R (2013) School of Ants. Available: http://www.schoolofants.org. Accessed on 30 July 2013.
72. Warren DL (2012) In defense of "niche modeling." Trends in ecology & evolution 27: 497–500.
73. Araujo MB, Peterson AT (2012) Uses and misuses of bioclimatic envelope modelling. Ecology 93: 1527–1539.
74. Jiménez M a, Jaksic FM, Armesto JJ, Gaxiola A, Meserve PL, et al. (2011) Extreme climatic events change the dynamics and invasibility of semi-arid annual plant communities. Ecology letters 14: 1227–1235.
75. Aronson RB, Thatje S, Clarke A, Peck LS, Blake DB, et al. (2007) Climate change and invasibility of the antarctic benthos. Annual Review of Ecology and Systematics 38: 129–154.
76. Fitzgerald K, Gordon DM (2012) Effects of vegetation cover, presence of a native ant species, and human disturbance on colonization by Argentine ants. Conservation Biology 26: 525–538.
77. King JR, Tschinkel WR (2006) Experimental evidence that the introduced fire ant, *Solenopsis invicta*, does not competitively suppress co-occurring ants in a disturbed habitat. Journal of Animal Ecology 75: 1370–1378.
78. Brook BW, Sodhi NS, Bradshaw CJA (2008) Synergies among extinction drivers under global change. Trends in Ecology & Evolution 23: 453–460.
79. Dukes JS, Mooney HA (1999) Does global change increase the success of biological invaders? Trends in Ecology & Evolution 14: 135–139.
80. Sala OE, Sala O.E., Chapin F.S., Armesto J.J., Berlow E., et al. (2000) Global Biodiversity Scenarios for the Year 2100. Science 287: 1770–1774.
81. Rice ES, Silverman J (2013) Propagule Pressure and Climate Contribute to the Displacement of *Linepithema humile* by *Pachycondyla chinensis*. PLoS ONE 8: e56281.

11

Incorporating Cold-Air Pooling into Downscaled Climate Models Increases Potential Refugia for Snow-Dependent Species within the Sierra Nevada Ecoregion, CA

Jennifer A. Curtis[1]*, Lorraine E. Flint[2], Alan L. Flint[3], Jessica D. Lundquist[4], Brian Hudgens[5], Erin E. Boydston[6], Julie K. Young[7]

1 U. S. Geological Survey, California Water Science Center, Eureka, California, United States of America, 2 U. S. Geological Survey, California Water Science Center, Sacramento, California, United States of America, 3 U. S. Geological Survey, California Water Science Center, Placer Hall, California, United States of America, 4 University of Washington, Department of Civil and Environmental Engineering, Seattle, Washington, United States of America, 5 Institute for Wildlife Studies, Arcata, California, United States of America, 6 U. S. Geological Survey, Western Ecological Research Center, Thousand Oaks, California, United States of America, 7 U. S. Department of Agriculture, Wildlife Services, National Wildlife Research Center and Utah State University, Wildland Resources Department, Logan, Utah, United States of America

Abstract

We present a unique water-balance approach for modeling snowpack under historic, current and future climates throughout the Sierra Nevada Ecoregion. Our methodology uses a finer scale (270 m) than previous regional studies and incorporates cold-air pooling, an atmospheric process that sustains cooler temperatures in topographic depressions thereby mitigating snowmelt. Our results are intended to support management and conservation of snow-dependent species, which requires characterization of suitable habitat under current and future climates. We use the wolverine (*Gulo gulo*) as an example species and investigate potential habitat based on the depth and extent of spring snowpack within four National Park units with proposed wolverine reintroduction programs. Our estimates of change in spring snowpack conditions under current and future climates are consistent with recent studies that generally predict declining snowpack. However, model development at a finer scale and incorporation of cold-air pooling increased the persistence of April 1st snowpack. More specifically, incorporation of cold-air pooling into future climate projections increased April 1st snowpack by 6.5% when spatially averaged over the study region and the trajectory of declining April 1st snowpack reverses at mid-elevations where snow pack losses are mitigated by topographic shading and cold-air pooling. Under future climates with sustained or increased precipitation, our results indicate a high likelihood for the persistence of late spring snowpack at elevations above approximately 2,800 m and identify potential climate refugia sites for snow-dependent species at mid-elevations, where significant topographic shading and cold-air pooling potential exist.

Editor: Juan A. Añel, University of Oxford, United Kingdom

Funding: Funding provided by Park Oriented Biologic Support, a program funded jointly by the National Park Service and United States Geological Survey. The funders had no role in study design, data collection and analysis, decision to publish, or preparation of the manuscript.

Competing Interests: The authors have declared that no competing interests exist.

* Email: jacurtis@usgs.gov

Introduction

Previous studies document a general decline in snowpack over the last half century throughout the western United States [1,2]. The loss of snowpack coincides with regional warming, and within California the warming trend is expected to continue with expected increases in mean annual temperatures of 2.2 to 8.3°C over the next 100 years [3]. Documented effects of regional warming in California's alpine regions include: earlier onset of spring snowmelt [4–6], reduced summer base flows [7,8], declines in snowpack volumes at mid-elevations [9], and migration of the rain to snow transition line to higher elevations [10].

In the midst of regional declines in snowpack previous studies documented a net increase in snowpack at elevations above 2,500 m within the southern Sierra Nevada [1,11,12]. This reversal from the regional decline in snowpack may be related to increases in atmospheric moisture due to the presence of warmer air masses capable of holding more water [2]. Increases in atmospheric moisture combined with adiabatic cooling can lead to increased snow accumulation at the highest elevations where winter and early spring temperatures remain below the temperature threshold for the transition from snow to rain, even under predicted regional warming.

Due to its complex glaciated terrain the Sierra Nevada Ecoregion (Figure 1) is prone to cold-air pooling (CAP), a nocturnal atmospheric process that sustains cooler air temperatures in CAP–prone areas. On calm clear nights air in contact with the ground cools due to radiative energy loss. Being denser than the free atmosphere at the same elevation, this colder-denser air becomes decoupled and sinks into topographic depressions. CAP occurs where cooled air collects on the landscape. CAP generally occurs in concavities and areas cut off from the free atmosphere, typically along flat valley bottoms in mountainous terrain with valley constrictions [13–21]. Exposed ridges and topographic convexities are not prone to CAP.

Due to the decoupling and sinking of colder denser air, CAP-prone areas may respond differently to predicted climate change

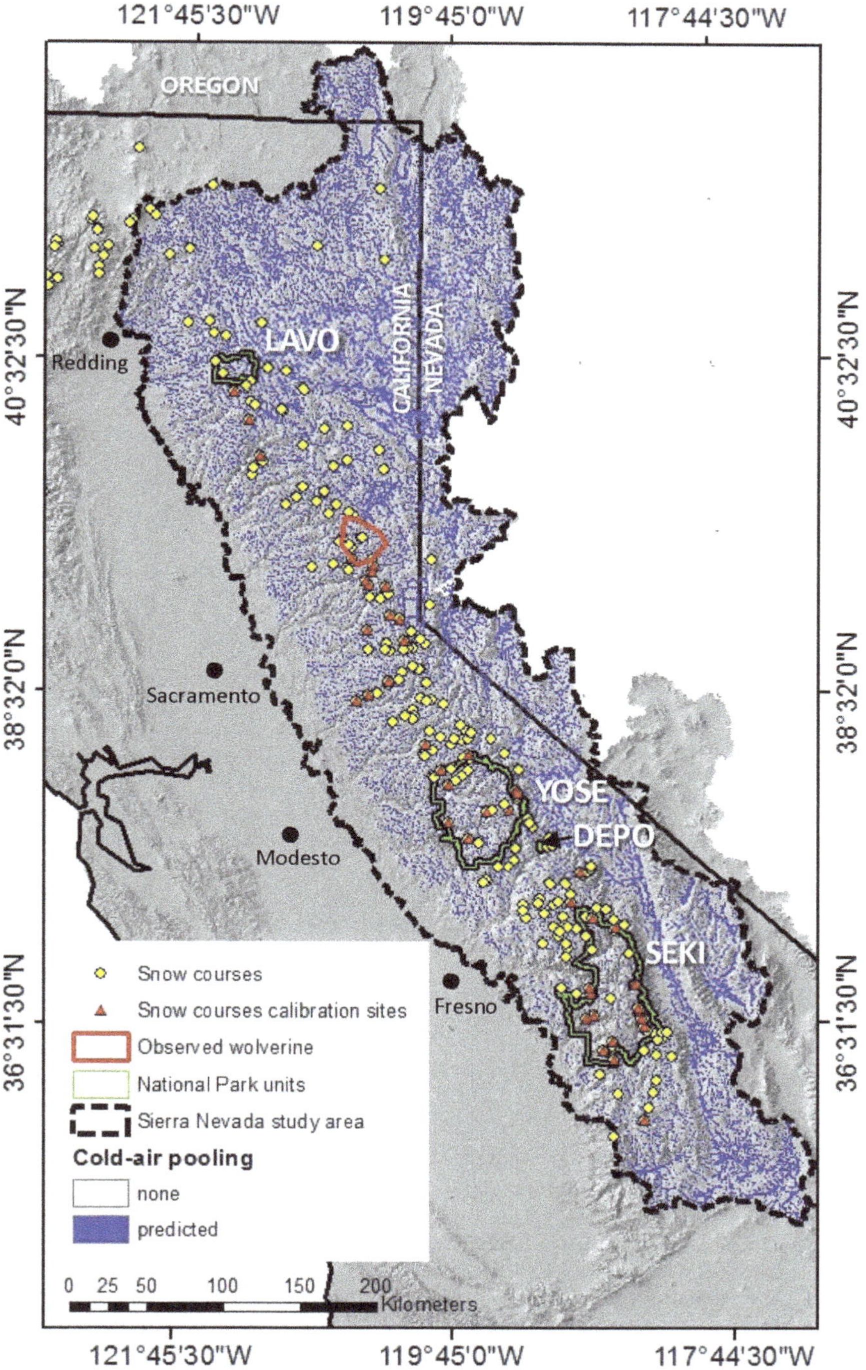

Figure 1. Study area map showing cold-air pooling potential. Map of Sierra Nevada Ecoregion showing the location of snow courses, snow model calibration sites, range of resident wolverine sitings, National Park units (Lassen Volcanic National Park, LAVO; Yosemite National Park, YOSE; Sequoia-Kings National Parks, SEKI; and Devils Postpile National Monument, DEPO), and cold-air pooling potential.

than surrounding terrain. For example, CAP-prone snow covered sites in Idaho, Montana, and Wyoming, warm less rapidly than non-CAP sites located in exposed locations [22]. Furthermore, CAP may not only diminish the impacts of regional warming in high mountain valleys [23,24] but, along with coincident topographic shading provided by surrounding terrain, CAP may actually delay snowmelt despite increases in air temperatures [6].

In this study we investigate the influence of CAP on air temperatures and snowmelt and assess whether incorporating CAP into future climate scenarios leads to snowpack persistence.

The trajectory of change in snowpack under future climates will undoubtedly impact snow-dependent ecology. Consequently, habitat-scale projections of future snowpack conditions are needed to provide information to support conservation and management decisions. In this study we use the wolverine (*Gulo gulo*) as an example species. Because the wolverine is an obligate carnivore with a northern circumpolar distribution [25] it's range limits are strongly correlated with snow covered areas [26,27]. Current research indicates that spring snowpack is a critical resource for providing suitable denning sites to support successful wolverine reproduction [28].

Historically, the Sierra Nevada wolverine population represented the southernmost extent of the species' range, but wolverines were believed to be extirpated from this region by the 1920s [26,29–32]. In 2008, a lone male wolverine (Figure 1), most likely an immigrant from Idaho, was photo-documented [33]. If a population of wolverines is re-established in the Sierra Nevada Ecoregion, it would provide population redundancy to this threatened species, which currently only occupies a small area of the northern Cascades and Rocky Mountains in the conterminous United States [26,34]. Determining the feasibility of wolverine reintroduction to the Sierra Nevada Ecoregion requires investigation of potential suitable habitat that includes adequate spring snowpack under current and future climates.

We present a unique approach for developing spatially distributed estimates of snowpack under future climates. Currently, global climate models (GCMs) are developed at coarse spatial scales and future climate projections cannot represent fine-scale processes. It is well recognized that fine-scale modeling incorporates the effects of topographic heterogeneity and results in improved precipitation estimates [35,36]. A fine-scale is further required to capture processes that control accumulation, melt, and sublimation of snow [37]. However, this is the first study that incorporates a fine-scale process such as CAP into projections of snowpack under future climates. Our methodology includes the definition and application of a temperature correction factor such that the role of CAP, in mitigating snowmelt and sustaining the depth and extent of snow covered areas, is incorporated into future projections of snowpack. We then present an example ecological application by investigating late spring snowpack conditions under historic, current and future climates and the presence of snow covered habitat in four National Park units with proposed wolverine reintroduction programs.

Study Area

The study area encompasses 131,650 km^2 of the Sierra Nevada Ecoregion (Figure 1). We further highlight results for four National Park units with proposed wolverine reintroduction programs. The park units include Yosemite (YOSE), Sequoia-Kings Canyon (SEKI), and Devils Postpile (DEPO) within the Sierra Nevada National Park Network and Lassen (LAVO) National Park from the Klamath Network. Elevations within the study area range from 40 to 4,415 m, with the highest elevations occurring in the southern Sierra Nevada. Under the current climate, mean annual precipitation ranges from 50 cm on the western edge of the study area to more than 150 cm at the highest elevations along the Sierra Nevada crest [38]. Precipitation is generated primarily by Pacific frontal systems that supply approximately 85% of annual precipitation between November and April.

Regions above 3,000 m were glaciated repeatedly throughout the Quaternary geologic period [39,40]. Topographic shading and CAP allowed small glaciers in sheltered high mountain cirques to persist at elevations well below the regional permanent snowline defined at 4,500 m [41]. The persistence of permanent snowpack throughout the Sierra Nevada Ecoregion over geologic time scales and at relatively lower elevations highlights a distinct ability for producing and sustaining snowpack and snow covered habitat.

The glaciated terrain of the Sierra Nevada is uniquely suited for producing CAP due to characteristic valley morphologies. Fluvial and glacial valley profiles are typically concave, but glacial valley profiles are over-deepened with higher concavities and steeper headwater profiles [42,43] and typically have wide flat valley bottoms and lower gradients at the glacial terminus [44,45]. CAP is further influenced by valley constrictions that occur where wider U-shaped glacial valleys transition into narrower V-shaped fluvial valleys. Hanging valleys, located at the confluence of tributary and trunk streams, and stepped topography, formed due to an increase in cross sectional area to accommodate the added tributary stream discharge, are additional topographic complexities [41] that may increase the potential for CAP in glaciated terrain.

Methods

We used a published regional water balance model (BCM) developed at a 270 m grid scale [46], to estimate historic (1951 to 1980), current (1981 to 2010), and future (2011 to 2100) snowpack conditions throughout the Sierra Nevada Ecoregion. The BCM employs a deterministic water-balance calculation and utilizes SNOW-17, a temperature index model [47,48], to estimate monthly snow accumulation and snow melt. The 270 m grid required spatial downscaling of available air temperature and precipitation data. The downscaling approach [49] preserves and enhances topographic effects.

Initially the adequacy of the snowmelt calculation was evaluated based on the timing of streamflow in high elevation subbasins [46]. The snow model calibration was refined in this study by comparing model results to snow covered area (SCA) at the basin scale and to measured snow water equivalents (SWE) at 45 spatially distributed calibration sites. We selected 1998 to 2009 as the calibration period and used the root-mean squared error (RMSE) and the sum of differences (SUMDIFF) to compare model results with measured snowpack data.

The calibration process was iterative such that we started with literature values for the adjustable snow parameters [50], which were changed accordingly to improve the goodness-of-fit between measured and modeled SWE. We visually assessed spatial correlations with the presence of permanent snowpack and the spatial distribution of snow covered areas. The calibration required the snow model to produce persistent snowpack (SWE>0) at locations with permanent snowpack (http://glaciers.research.pdx.edu/) and reasonable distributions of snow cover area, which we assessed by comparing the spatial extent of modeled snow cover area with published estimates [51]. Modeled snow covered area was compared to the gap-filled daily fractional snow cover area [51], estimated using MODIS satellite images (http://zero.eng.ucmerced.edu/snow/csnwis/), for dates reflecting winter accumulation (February) and spring snowmelt (May) conditions from 2000–2003.

Modeled monthly SWE was compared to measured monthly SWE at 45 snow course calibration locations (data available from the California Data Exchange at http://cdec.water.ca.gov/). There are 360 snow courses within California, and we selected 45 representative stations within the study region (Figure 1) that were spatially distributed over a range of elevations (1,325 to 3,460 m). Each station had records adjusted by DWR for snow density differences and records that spanned December to May for the duration of the calibration period (1998 to 2009). We did not use available snow pillow data, which is measured at discreet locations, but rather calibrated the snow model solely using snow course data collected along transects, which better represent spatially averaged conditions.

A spatial representation of CAP potential for the study region (Figure 1) was estimated using an automated algorithm [23] that evaluates a digital elevation model (DEM) and identifies flat valley bottoms and concave areas where cold air pools are likely to form. The algorithm requires an estimate of the average peak-to-peak distance, defined as the distance between two peaks divided by a valley. We used a peak-to-peak distance of 1,500 m for the entire study region. Because the Sierra Nevada Ecoregion covers a large geographic extent, a correction for diverging valleys where cold air drains rather than pools in a given location [23], was not applied.

Gridded air temperature data from the PRISM Climate Group (http://prism.oregonstate.edu) capture the CAP process in the historic record by including a vertical layer weighting function [35]. Since CAP is a nocturnal process that reduces minimum air temperatures and cold-air pools are dissipated by mid-day, maximum temperatures are not impacted. We evaluated the extent to which the PRISM data represents CAP by comparing minimum monthly air temperatures at 243 snow course stations located in CAP and non-CAP areas spatially distributed throughout the study region (Figure 1). A temperature correction factor was then defined as the mean difference between monthly minimum air temperatures estimated in CAP-prone and non-CAP areas during the primary snow accumulation and melt season (December to May). The correction factor represents a spatial and temporal mean estimated using 243 snow course locations distributed throughout the study region over a range of representative elevations (1,265 to 3,490 m) and temporally averaged from 1896 to 2009.

Because GCM projections do not include the effects of CAP we adjusted the downscaled monthly air temperature data in CAP-prone areas by subtracting the temperature correction factor to incorporate the CAP process and then generated monthly gridded snow maps for future climatic conditions. We necessarily assumed that the magnitude and frequency of CAP under future climates will remain unchanged and applied the same temperature correction to all future climate datasets.

Changes in snow-cover habitat were evaluated using the wolverine as an example species. We defined the presence of suitable wolverine habitat based on an April 1st threshold depth of snowpack. April 1st represents the midway point between February 15th and May 15th which is the average period during which female wolverines typically give birth, raise and then graduate kits such that they are no longer dependent on their snowpack dens [28,52]. April 1st snowpack is therefore a suitable surrogate for determining available wolverine breeding and denning habitat. A conservative threshold for April 1st snow depth was defined at 1 meter, which equates to an April 1st SWE of 400 mm based on a snow density of 2.5 g/cm^3 (http://cdec.water.ca.gov/snow/misc/density.html).

Four projections, defined in the Fourth Assessment Report (AR4) of the United Nation's Intergovernmental Panel on Climate Change [53], were used in this study. The four projections were selected to portray a range of future climate conditions. Although updated projections were recently released in the Fifth Assessment Report [54], the fine-scale resolution necessary for this study required that we use available downscaled AR4 scenarios [49].

We selected two GCMs and two emissions scenarios capable of simulating the distribution of monthly temperatures and the strong seasonal cycle of precipitation that exists in the region [3]. Cayan and others [55] describe the four projections in detail for California. The two GCMs are the Parallel Climate Model (PCM) developed by the National Center for Atmospheric Research and the Department of Energy [56,57] and the Geophysical Fluid Dynamics Laboratory CM2.1 model (GFDL) developed by National Oceanic and Atmospheric Administration [58,59]. The greenhouse gas emissions scenarios include A2, which represents "business as usual" with no reduction in emissions, and B1 which represents "mitigated emissions." We refer to the four future climate scenarios as "GFDL-A2," "GFDL-B1," "PCM-A2," "PCM-B1." The GFDL scenarios represent "hot-dry" future conditions and the PCM scenarios represent "warm-wet" conditions.

Results

Model Performance

Comparison of measured data with initial model runs indicated that we were generally underestimating snowpack throughout the study region. Increases in snow accumulation and decreases in snowmelt were achieved by adjusting snow parameters iteratively to improve the goodness-of-fit between measured and modeled SWE. We started with maximum and minimum melt factors of 3.6°C and 2.2°C and a minimum temperature threshold of 4.5°C for the transition from rain to snow. Final melt factor values of 1.8°C and 1.3°C and a minimum temperature threshold of 6°C generated more snow throughout the study region, allowed permanent snowpack to occur in regions where it naturally occurs, and improved the goodness-of-fit between measured and modeled SWE. Because the model operated on a monthly time-step, these values differ greatly from those required when SNOW-17 is run at finer temporal resolutions.

We assessed the spatial accuracy of modeled SWE by visual comparison with permanent snowpack zones and satellite imagery depicting actual snow covered area. The spatial extents of modeled SWE consistently intersected locations with permanent snowpack (Figure 2). Similarly, modeled SWE consistently intersected estimates of snow covered area [51] determined using MODIS satellite imagery (Figure 3).

Our final regression analysis of measured versus modeled SWE at the 45 calibration sites resulted in an r^2 of 0.416, RMSE of 328 mm, and SUMDIFF of −8.44E+04 mm indicating modeled SWE values reasonably approximate measured SWE. Separate regression analyses specifically assessed the influence of elevation (Table 1) and latitude (Table 2) on the relation between measured and modeled SWE values. When the results are categorized into three elevation classes, modeled SWE is most accurate at higher elevations above 2,500 m (RMSE = 238 mm) and least accurate at mid-elevations between 2,130 to 2,500 m (RMSE = 432 mm) where the rain to snow transition typically occurs (Table 1). When the results are separated by latitude, the RMSE indicates modeled SWE is more accurate in the northern region (RMSE = 277 mm) in comparison to the southern region (RMSE = 412 mm) but the SUMDIFF indicates the magnitude of the differences was higher in the northern region in comparison to the southern region (Table 2). These are expected results as the density of meteorological stations is higher, and thus PRISM data better represent the climate, in the Northern Sierra where the topography is less complex.

Our analysis of monthly air temperatures from 1896 to 2009 indicates average monthly minimum air temperatures (AVG-TMIN) in CAP-prone locations are 1.6°C lower than AVG-TMIN in non-CAP locations during the primary snow accumulation and melt season from December to May (Table 3). Based on AVG-TMIN differences, the effects of CAP appear to be greatest at the beginning of the snow accumulation season from December to February, when the AVG-TMIN difference is ~1.8°C. CAP declines moderately throughout the spring and the AVG-TMIN difference by May is ~1.3°C. Based on these data, a spatially and

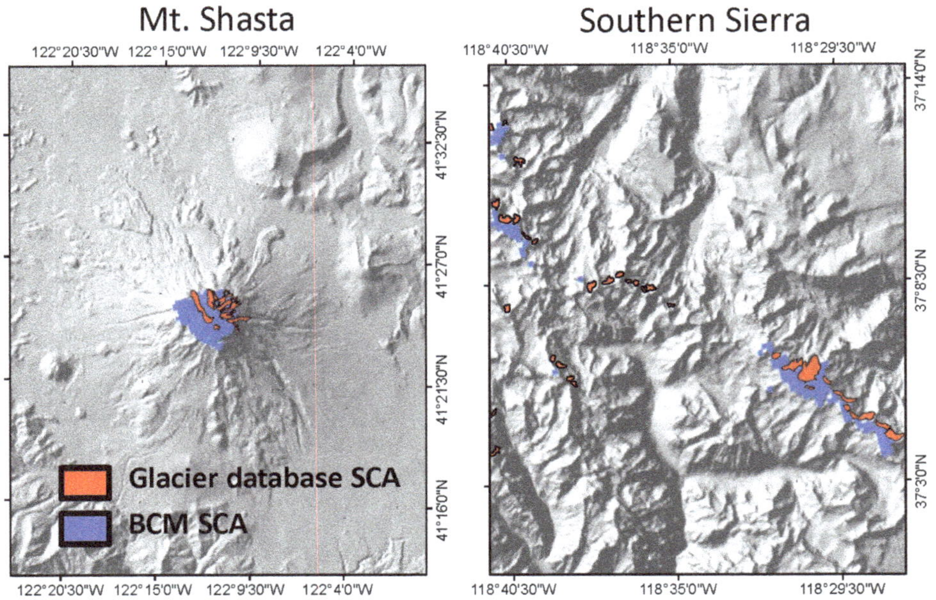

Figure 2. Visual comparison between modeled snow covered area and permanent snowpack locations. Comparison of snow covered area (SCA) simulated using the Basin Characterization Model (BCM) for September 2009 and areas of persistent snowpack for Mount Shasta, located in the northern study region, and along the southern Sierra Nevada crestline within the headwaters of the San Joaquin and Kings River watersheds.

temporally averaged monthly air temperature correction factor of 1.6°C was used to simulate CAP under future climate scenarios. The temperature correction was applied to air temperature datasets for the four future climate scenarios to incorporate the effects of CAP.

Incorporating CAP into future climate projections increased SWE for all snow dominated months (January to July). The

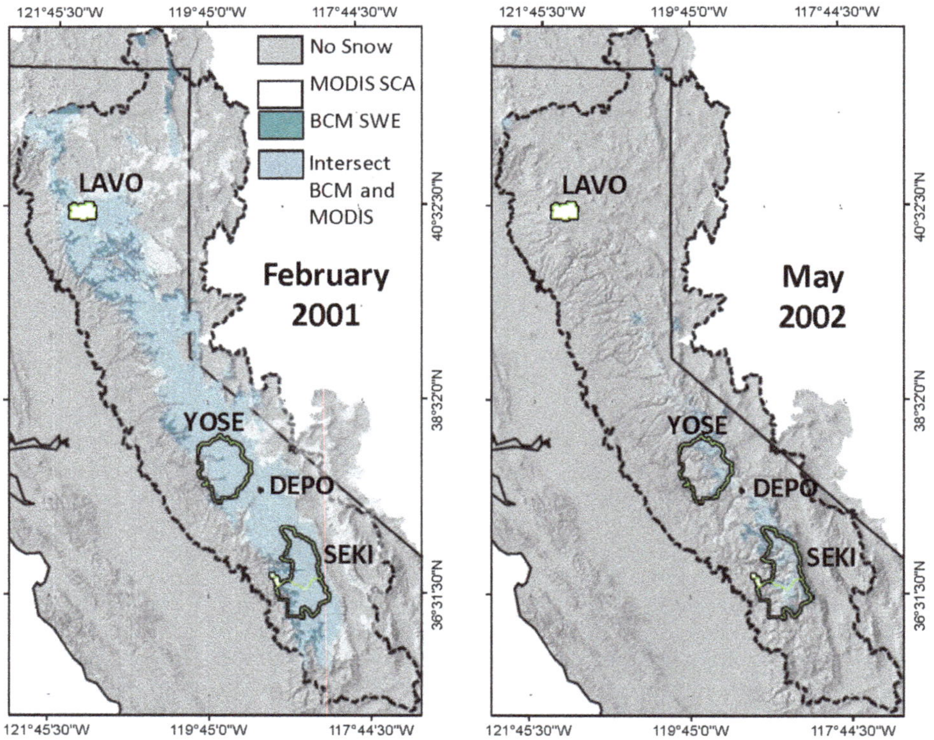

Figure 3. Visual comparison between modeled snow covered area and satellite imagery. Comparison of snow water equivalent (SWE) simulated with the Basin Characterization Model (BCM) and MODIS snow covered area (SCA) for February 2001 and May 2002.

Table 1. Elevation effects on goodness-of-fit parameters for modeled snow water equivalents during the calibration period from 1998 to 2009.

	Root Mean Squared Error (mm)	Sum of Differences (mm)
Average	328	−8.44E+04
<2130 m	361	9.51E+03
2130 to 2500 m	432	−6.64E+04
>2500 m	238	−2.75E+04

increase in April 1st SWE, averaged over the entire study region, was 6.5% and ranged from 5.5% to 11.1% for the pessimistic GFDL-A2 mid-21st century projection (Figure 4). We compared the 6.5% simulated effect of CAP on April 1st SWE to measured snowpack data collected at the 45 calibration sites from 1998 to 2009. The average April 1st SWE was 690 mm at the non-CAP locations and 742 mm at the CAP-prone locations. The 52 mm or 8% increase in April 1st SWE is comparable to the 6.5% simulated increase. An example of modeled SWE without CAP (Figure 5a) and modeled SWE with simulated CAP (Figure 5b) demonstrates the effect of CAP on the depth, spatial extent, and persistence of snowpack into early summer (June). The difference map (Figure 5c) highlights potential climatic refugia identified by incorporating CAP into future climate projections. Even for the most pessimistic mid-century scenario (GFDL-A2) simulation of CAP increased early summer SWE in CAP-prone areas by 100 to 400 mm.

Estimating April 1st Snowpack

We used the model results to quantify changes in spring snowpack under historic (1951 to 1980), current (1981 to 2010), and future (2011 to 2100) climates. Future climate results are separated into three 30-year projections: early-century (2011 to 2040), mid-century (2041 to 2070), and late-century (2071 to 2100). Additional results are focused on estimates of April 1st SWE, a surrogate variable suitable for characterizing wolverine breeding habitat and we highlight changes in April 1st SWE for four National Park units with proposed wolverine relocation programs [34].

We first present a spatial representation of the change in estimated April 1st SWE by comparing historic and current climates (Figure 6). From a regional perspective, declining April 1st SWE in the northern portion of the study area is balanced somewhat by increases in the southern Sierra Nevada. Declines in April 1st SWE throughout the northern Sierra generally range from −50 to −400 mm whereas smaller magnitude increases of 50 to 100 mm were estimated for higher elevations along the southern Sierra Nevada crestline. The highest magnitude increases in April 1st SWE were estimated for Lassen Peak located in the southwest corner of Lassen National Park (Figure 6). Generally, the net increase in April 1st SWE under current climate conditions shown in Figure 6 occurred only at elevations above 2,800 m, a slightly higher elevation than reported in previous studies based on observational data [1,12].

Figure 7 illustrates mean April 1st SWE, spatially averaged across each National Park boundary, for historic, current, and future climate periods. Lassen National Park is located in the northern Sierra, where we estimated widespread declines in April 1st SWE between historic and current climates (Figure 6); however Lassen Peak represents a relatively small high-elevation area with significant gains in April 1st SWE under the current climate leading to an overall slight increase in April 1st SWE between historic and current climates (Figure 7). The remaining three park units (Yosemite, Sequoia-Kings Canyon, and Devils Postpile) are located in the southern Sierra. We estimated declines in April 1st SWE on lower elevation eastern slopes in the southern Sierra Nevada that were of a similar magnitude (−50 to −400 mm) to declines estimated in the northern region of the study area, but declines in the southern Sierra covered a smaller spatial extent. There are also significant areas with no net change in April 1st SWE and extensive high elevation areas along the Sierra Nevada crest with increases in April 1st SWE under current climatic conditions. This is especially evident in Devil's Postpile, which is predominately located within a CAP zone, where future scenarios predict an increase in projected April 1st SWE, relative to historic April 1st SWE, until the mid-century (Figure 7).

Although projections of April 1st SWE under future climates indicate Lassen may experience the largest declines among the National Park units, projections of April 1st SWE remain higher for Lassen in comparison to the other park units for all four scenarios (Figure 7). Percent declines in Lassen's April 1st SWE for the late-century projection range from 33 to 69%, but abundant spring snowpack will likely persist in Lassen due to high elevation topography (Table 4). Percent declines in April 1st SWE for the late-century projection for Yosemite and Devils Postpile were comparable to estimates for Lassen and range from 34 to 60% and 14 to 76% respectively; whereas projected percent declines for Sequoia-Kings Canyon were significantly less ranging from 13 to

Table 2. Latitude effects on goodness-of-fit parameters for modeled snow water equivalents during the calibration period from 1998 to 2009.

	Root Mean Squared Error (mm)	Sum of Differences (mm)
Average	328	−8.44E+04
Northern Region >/=38.5°	277	−6.71E+04
Southern Region <38.5°	412	−1.73E+04

Table 3. Comparison of mean minimum air temperatures from 1896 to 2009 for snow courses with and without the potential to produce cold-air pooling (CAP).

	Minimum Air Temperature (°C)						
	Dec	Jan	Feb	Mar	Apr	May	Average
Non-CAP Sites (N = 106)	−6.05	−6.86	−6.94	−5.75	−3.77	0.25	−4.85
CAP-prone Sites (N = 137)	−7.85	−8.73	−8.75	−7.34	−4.97	−1.07	−6.45
Temperature difference between Non-CAP and CAP sites	1.80	1.87	1.81	1.59	1.20	1.32	1.60

See Figure 1 for station locations.

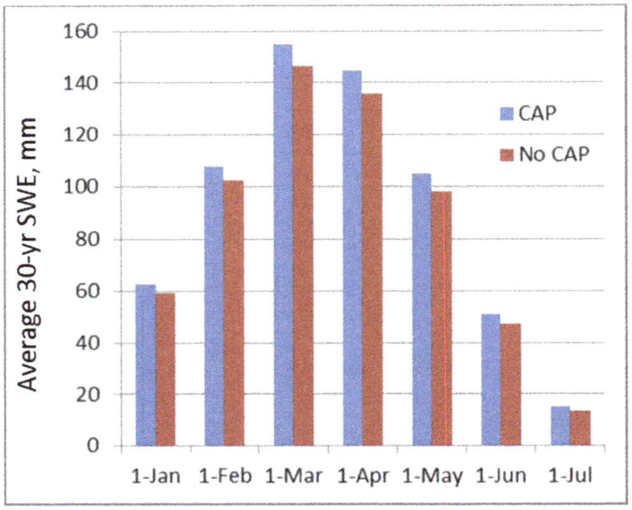

Figure 4. Influence of cold-air pooling on simulations of regional snow water equivalents. Monthly snow water equivalent (SWE) spatially averaged over the entire Sierra Nevada Ecoregion for the early-century period (2011–2040) using GFDL-A2, showing increases in SWE related to the use of a temperature correction factor (−1.6°C) to adjust minimum air temperatures and simulate cold-air pooling (CAP).

36% (Table 4). Based on the late-century worst case projection (GFDL-A2) all National Park units may experience a 36 to 76% decline in April 1st SWE, and even under the best case scenario (PCM-B1) the parks may experience a 13 to 34% decline (Table 4).

We further investigated changes in April 1st SWE between the current, mid-century and late-century periods for the A2 scenario over a range of representative elevations (Figure 8). By mid-century most of the change in April 1st SWE occurs at mid-elevations from approximately 1,800 m to 3,000 m. By late-century the change in April 1st SWE moves upslope and is greatest between about 2,500 and 3,500 m. The mean elevation of each park unit is also shown in Figure 8 further indicating that Sequoia-Kings Canyon is projected to experience the greatest declines in April 1st SWE by the late-century, while projections for Lassen indicate the least amount of change.

Impacts of changing snowpack conditions on snow covered habitat within the four National Park units were evaluated using the wolverine as an example species. Figure 9 shows a spatial representation of April 1st SWE for each park unit under current climate conditions and for two late-century projections. The threshold April 1st snowpack condition that represents the presence of suitable breeding and denning habitat is defined at 400 mm of SWE. Under historic and current climates the spatially averaged April 1st snowpack is above the 400 mm threshold for Lassen, Yosemite, and Sequoia-Kings Canyon but not for Devils Postpile; however Devils Postpile is generally surrounded by higher elevations with sufficient snowpack. The relatively drier GFDL-A2 projection represents a worst case scenario and generally indicates larger magnitude declines in snowpack by the late-century in comparison to the wetter PCM-A2 projection (Figure 9) for all four National Park units. The worst case scenario (GFDL-A2) indicates average April 1st snowpack will not meet the 400 mm threshold at any of the park units by the end of the 21st century but substantial areas within and around the parks will be adjacent to potential higher elevation wolverine habitat that could be used for breeding and denning. The best case scenario (PCM-

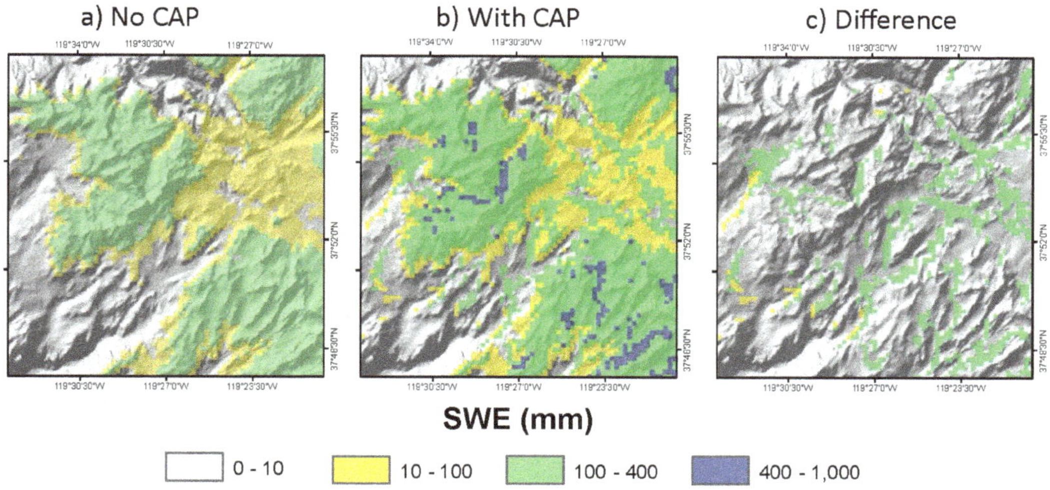

Figure 5. Influence of cold air pooling on persistence of snowpack at a finer scale. Comparison of June snow pack conditions shown for the headwaters of the Tuolumne River estimated using the worst case (GFDL-A2) future climate projection. Snowpack was simulated a) without an adjustment for cold-air pooling (CAP) and b) with a temperature correction factor (−1.6°C) to adjust air temperatures and simulate cold-air pooling (CAP). Panel c) shows the difference in the spatial extent and depth of SWE achieved by incorporating CAP into the future climate projection.

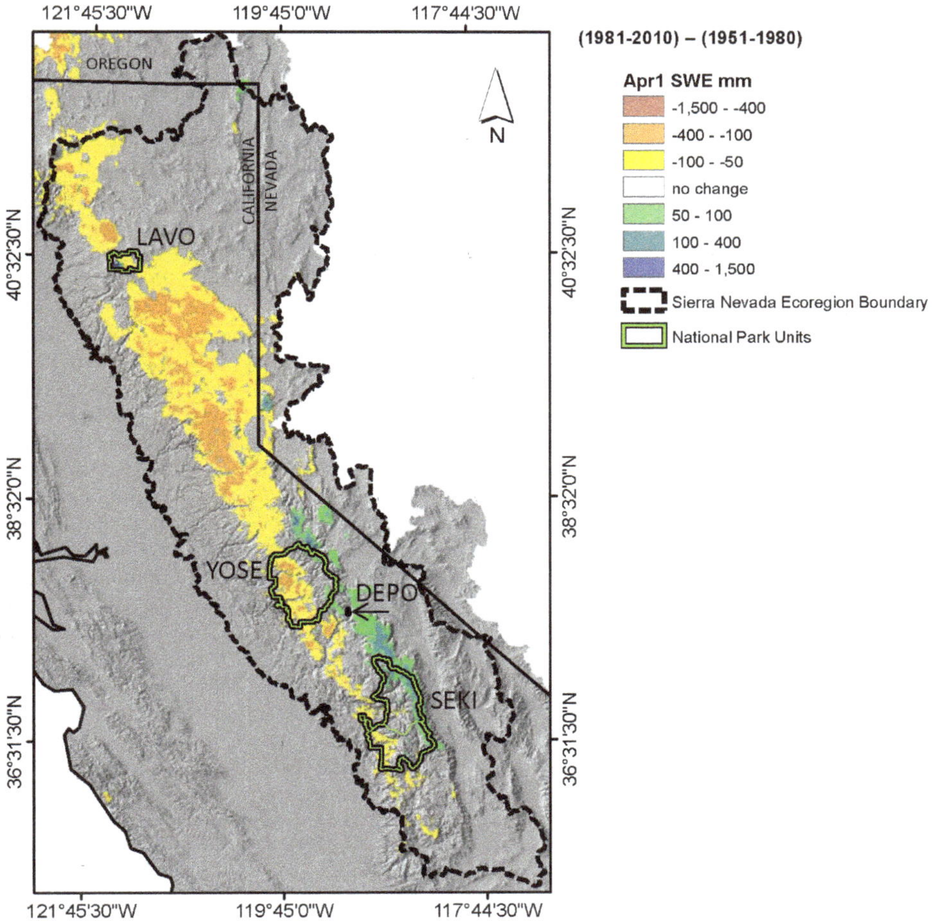

Figure 6. Historic changes in April 1st snow water equivalents. A spatially distributed estimate of the change in April 1st snow water equivalent (SWE) from historic (1951–1980) to current (1981–2010) climatic conditions.

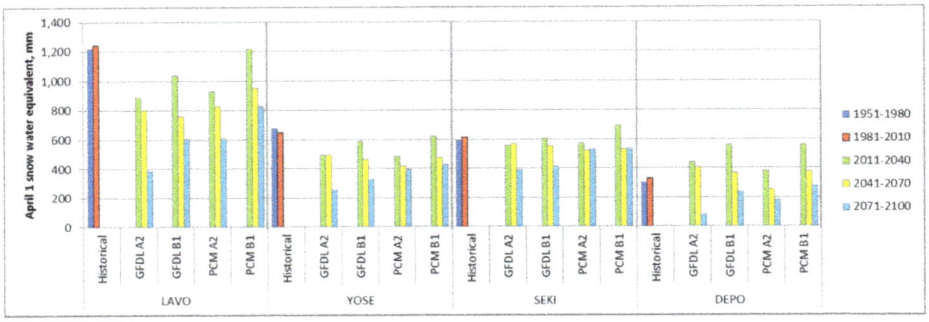

Figure 7. Changes in April 1st snowpack under various climatic conditions in four National Park units. Simulated April 1st snow water equivalent (SWE) spatially averaged across the boundaries of four National Park units for historic (1951–1980), current (1981–2010), and future (2011 to 2100) climatic conditions. The data, organized from left to right, show a worst case scenario (GFDL-A2), two moderate case scenarios (GFDL-B1 and PCM-A2) and a best case scenario (PCM-B1).

B1) indicates that the 400 mm April 1st SWE threshold will be met in Lassen, Yosemite, and Sequoia-Kings Canyon by the end of the 21st century. The "drier-mitigated emissions" scenario (GFDL-B1) indicates the April 1st threshold will be met in Lassen and Sequoia-Kings Canyon. Under the "wetter-business as usual" scenario (PCM-A2) we estimate the wolverine habitat threshold will be met in Lassen, Yosemite, and Sequoia-Kings Canyon.

Under the two "business as usual" scenarios projected late-century snow covered areas in Lassen, above the 400 mm SWE threshold, represent about 30% of the total park area under the drier GFDL-A2 scenario whereas about 80% of Lassen's snow covered areas exceed the threshold under the wetter PCM-A2 scenario. Notably, both "business as usual" scenarios indicate high elevation regions in Lassen will continue to maintain deep (> 800 mm SWE) April 1st snowpack. In Yosemite and Sequoia-Kings Canyon snow cover areas above the 400 mm threshold represent about 60% of the total park area in the wetter PCM-A2 scenario, but only about 20% for the drier GFDL-A2 scenario. Projections for Devils Postpile indicate no snow covered areas above the 400 mm SWE threshold within the park boundary under either late-century "business as usual" scenario but surrounding hillslopes provide snow covered areas with suitable SWE under the wetter scenario, while the zone with suitable SWE decreased and moved upslope under the drier scenario (Figure 9).

Discussion

Results from the calibrated snow model presented here provide a spatial representation of snowpack for a vast regional area over a 150-year time period. The spatially and temporally averaged RMSE for the calibrated model is 328 mm, indicating estimated SWE values are typically within ~0.3 m of measured data. The SUMDIFF estimates indicate SWE is generally underestimated throughout the study area, resulting in conservative projections of SWE and potential wolverine breeding and denning habitat. The modeled SWE values include uncertainties associated with the PRISM approach to spatially distribute measured climatic data, variability related to the spatially continuous model parameterization, and potential errors associated with measured snowpack at snow course calibration sites. Future improvements to the snow model could be realized by utilizing site-specific calibrations or latitudinal variations in parameterization as recommended by Raleigh and Lundquist [58].

Previous work showed that April 1st SWE is better correlated to temperature than to precipitation in the northern Sierra Nevada [2] indicating regional warming will significantly impact the persistence of spring snowpack in the northern regions of the study area. Our results for current and future climates confirm this assertion. Conversely, the same study showed no correlation of April 1st SWE with temperature at lower elevations but some correlation at higher elevations within the Southern Sierra Nevada [2]. Because April 1st SWE is better correlated to precipitation in the Southern Sierra Nevada an increase in the accumulation of snowpack may occur at cooler high elevation sites where deeper snowpack has a higher potential to persist into late spring and early summer even under warmer climates. Conversely, regional warming without an increase in precipitation may result in large decreases in snowpack even at the coolest and highest elevations. If predictions of increased precipitation under future climates occur, our results indicate a high likelihood for the persistence of late spring snowpack at elevations above approximately 2,800 m. Our analysis of CAP also identifies a zone of potential climate refugia for snow-dependent species at mid-elevations where significant topographic shading and CAP potential exist.

Our estimates of changes in spring snowpack conditions under future climates are consistent with recent studies that generally predict significant declines. When future projections are compared to current conditions [11,56,61], earlier studies predict the highest reduction in snowpack volumes at low to mid-elevations (1,000 to 3,000 m). Figure 8 shows a comparison of the change in April 1st snowpack, between current and worst case mid-century and late-century projections, for elevation bands throughout the study region. Our results indicate large magnitude changes in April 1st SWE at the highest elevations, but the trajectory of change reverses at mid-elevations (2,000 to 3,500 m), where the National Park units are located, indicating mid-elevations may represent a zone of climatic refugia with an increased resilience leading to the persistence of April 1st snowpack. Lassen, located at the lowest mean elevation of 2,090 m, is projected to experience the largest change in April 1st SWE, whereas Sequoia-Kings Canyon, the park unit located at the highest mean elevation of 2,850 m, is projected to experience the least change in April 1st SWE (Figure 7 and Table 4).

We compared our results to two recent studies that characterize potential changes in available wolverine habitat under future climates [62,63]. Notably, the two previous studies selected less conservative spring snowpack depths to characterize potential wolverine habitat and were developed at coarser spatial scales. The first study used the Community Climate System model [62], developed at a much coarser grid of 250 km, and used a 20 cm snow depth threshold that produced pessimistic results indicating large declines in available wolverine habitat under future climates.

Table 4. Percent change in April 1st snow water equivalent from current climatic conditions (1981–2010) to early, mid and late 21st century climatic projections simulated for four National Park units located within the Sierra Nevada Ecoregion.

	LAVO				YOSE				DEPO				SEKI			
% change from 1981–2010	GFDL A2	GFDL B1	PCM A2	PCM B1	GFDL A2	GFDL B1	PCM A2	PCM B1	GFDL A2	GFDL B1	PCM A2	PCM B1	GFDL A2	GFDL B1	PCM A2	PCM B1
2011–2040	−29	−17	−25	−2	−23	−9	−25	−4	35	71	16	71	−10	−2	−7	12
2041–2070	−36	−39	−34	−23	−24	−28	−35	−26	24	13	−22	15	−8	−10	−15	−14
2071–2100	−69	−51	−51	−33	−60	−49	−39	−34	−76	−28	−44	−14	−36	−36	−14	−13

The second study used an ensemble climate model [63], developed at a 12 km grid, and used a 13 cm snow depth threshold that produced more optimistic results. We believe characterization of snow-dependent habitat requires finer-scale analysis that incorporates topographic heterogeneities and CAP. We used a 270 m grid and incorporated a temperature correction factor into future climate scenarios in CAP-prone areas. Because we were specifically interested in characterizing wolverine breeding habitat for potential species reintroduction [34] we selected a more conservative spring snowpack threshold at 1 meter (equivalent to 400 mm SWE) which is associated with successful denning and reproduction [28].

Our results identify potential climatic refugia for snow dependent species throughout the Sierra Nevada Ecoregion. Three National Park units (Lassen, Yosemite, and Sequoia-Kings Canyon) all have areas that lie above 2,800 m, an equilibrium elevation defined in this study, where snow accumulation is occurring under current climatic conditions. Although Devils Postpile is located in a valley prone to CAP, our results indicate that this park unit is located at an elevation that has not accumulated large historic snowpack volumes and generally does meet our threshold of late spring snowpack (400 mm SWE) under current or future climates. Since home ranges of female wolverines are typically much larger (200 to 300 km^2) than the area of Devils Postpile (2.8 km^2), the presence of snowpack in the surrounding forest lands (Figure 9) indicates that wolverines could use the park as part of their home range although use of the park for denning would be rare.

Implementation of habitat-scale modeling and incorporation of CAP into the future climate projections were dominant factors contributing to optimistic projections of wolverine breeding and denning habitat throughout the Sierra Nevada Ecoregion. Utilization of a 270 m grid preserved topographic heterogeneity such that valley bottoms prone to CAP, high elevation peak and ridges that provide topographic shading, and north facing slopes where late spring snowpack typically persists remained distinct and not subdued by the spatially averaging that occurs at coarser grid scales.

The temperature correction applied to simulate CAP under future climate projections increased simulated April 1st SWE by 6.5% over the regional model domain and identified fine-scale climatic refugia where snowpack is predicted to be maintained into the late spring and early summer. CAP zones may also represent climatic refugia for other snow-dependent species including predator species such as the pine marten (*Martes americana*), which hunt small mammals that rely on winter snow cover for survival [64], and species such as ermine (*Mustella ermine*), long-tailed weasels (*Mustella frenata*) and snowshoe hare (*Lepus americanus*), that rely on seasonally variable coat color for camouflage [65]. Thus, water balance models used to inform habitat studies must utilize appropriate scales to represent topographic complexities and must begin to incorporate fine-scale processes such as CAP.

Conclusions

Results from this study provide data necessary to assess management and conservation strategies for snow-dependent species throughout the Sierra Nevada Ecoregion. The snow model explicitly incorporates topographic heterogeneity and CAP to estimate temporal and spatial variability of snowpack and the presence of snow covered habitat. Our estimates of changes in spring snowpack conditions under future climates are consistent with recent regional studies that generally predict significant

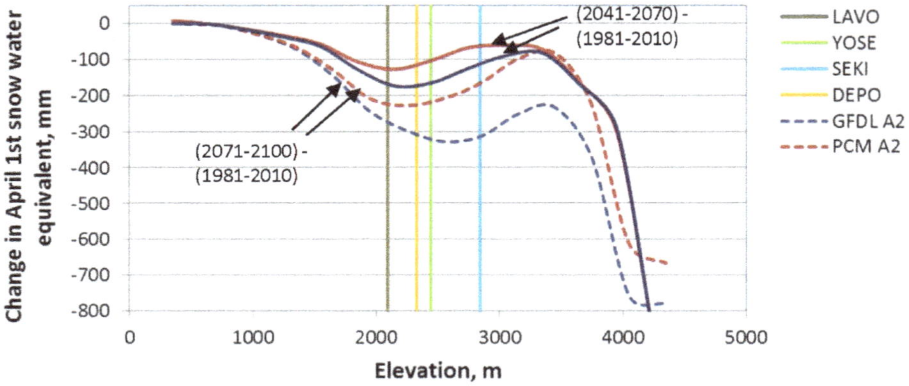

Figure 8. Influence of elevation on April 1st snowpack throughout the Sierra Nevada Ecoregion. Change in April 1st snow water equivalent (SWE) from current to mid-21st century (solid lines) and late 21st century (dashed lines) for elevation bands throughout Sierra Nevada Ecoregion. The location of National Park units are represented by their respective average elevations and future climate scenarios represent "business as usual" carbon emissions for warmer-wetter (PCM-A2) and warmer-drier conditions (GFDL-A2).

declines. Notably, model development at a habitat-scale and incorporation of CAP resulted in significant mitigation of snow loss at mid-elevations, which may represent a zone of climatic refugia with a higher resilience for sustaining patches of snow suitable for wolverine breeding and denning and other species dependent on snow cover. We could not directly test whether CAP improves future snowpack projections, but our results show that incorporation of CAP produced more optimistic spring snowpack projections in CAP-prone areas supporting our central thesis that

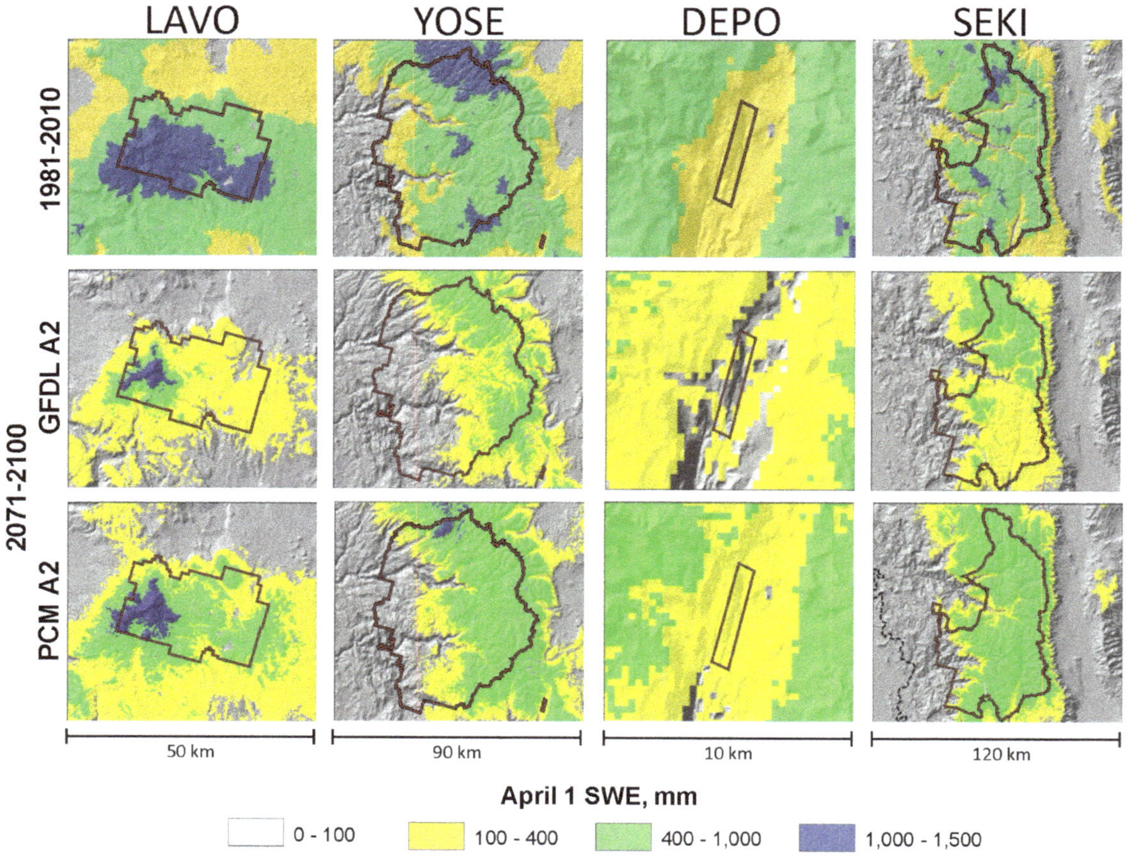

Figure 9. Simulated April 1st snowpack under current and late-century climates for four National Park units. Detail of four National Park units in the Sierra Nevada Ecoregion showing April 1st snow water equivalent (SWE) for current (1981–2010) and late 21st century (2071–2100) climatic conditions. Late-century scenarios represent "business as usual" carbon emissions for warmer-wetter (PCM-A2) and warmer-drier future climates (GFDL-A2).

CAP creates climatic refugia for species dependent upon late spring snowpack.

We acknowledge that this is a relatively simple approach and recognize that the occurrence of CAP under future climates requires the presence of clear skies and high pressure systems, which are governed by large scale atmospheric circulation that could change under future conditions. Current GCMs do a poor job of predicting fine-scale changes in atmospheric circulation; however we expect that the next generation of GCMs will be able to better represent changes in atmospheric circulation and the presence of high pressure systems, thereby enabling more rigorous predictions of the magnitude and frequency of smaller scale atmospheric processes such as CAP.

Regional warming under the current climate has resulted in a higher percentage of precipitation falling as rain rather than snow throughout Western North America. This phenomenon will likely continue under future climates and even a modest temperature increase may significantly alter snow accumulation and melt processes. However, the highest elevations in the southern Sierra Nevada accumulated snow under the current climate. If predictions of increased precipitation under future climates are realized, our results indicate a high likelihood for the persistence of late spring snowpack at elevations above approximately 2,800 m and identify climate refugia sites for snow-dependent species at mid-elevations where significant topographic shading and cold-air pooling potential exist.

Author Contributions

Conceived and designed the experiments: JAC LEF ALF JDL BH EEB JKY. Performed the experiments: JAC LEF ALF. Analyzed the data: JAC LEF ALF. Wrote the paper: JAC LEF ALF JDL BH EEB JKY.

References

1. Mote PW, Hamlet AF, Clark MP, Lettenmaier DP (2005) Declining mountain snowpack in western North America. Bulletin of the American Meteorological Society 86: 39–49.
2. Mote PW (2006) Climate-driven variability and trends in mountain snowpack in western North America. Journal of Climate 19: 6209–6220.
3. Cayan DR, Maurer EP, Dettinger MD, Tyree M, Hayhoe K, et al. (2006) Climate scenarios for California. Public Interest Energy Research, California Energy Commission, San Francisco, CA, 22p.
4. Cayan DR, Kammerdiener SA, Dettinger MD, Caprio JM, Peterson DH (2001) Changes in the onset of spring in the western United States. Bull Amer Meteor Soc 82: 399–415.
5. Stewart IT, Cayan DR, Dettinger MD (2005) Changes towards earlier streamflow timing across Western North America. Journal of Climate 18: 1136–1155.
6. Lundquist JD, Flint AL (2006) 2004 onset of snowmelt and streamflow, how shading and the solar equinox may affect spring runoff timing in a warmer world. Journal of Hydroclimatology 7: 1199–1217.
7. Dettinger MD, Cayan DR, Meyer M, Jeton AE (2004) Simulated hydrologic responses to climate variations and change in the Merced, Carson, and American River basins, Sierra Nevada, California,1900–2099. Climatic Change 62: 283–317.
8. Hidalgo H, Dettinger MD, Cayan DR (2008) Downscaling with constructed analogues: Daily precipitation and temperature fields over the United States. California Energy Commission Report No. CEC-500-2007-123. 62p.
9. Knowles N, Cayan DR (2002) Potential effects of global warming on the Sacramento/San Joaquin watershed and the San Francisco estuary. Geophysical Research Letters 29.
10. Hunsaker CT, Whitaker TW, Bales RC (2012) Snowmelt Runoff and Water Yield Along Elevation and Temperature Gradients in California's Southern Sierra Nevada. Journal of the American Water Resources Association 1–12. DOI:10.1111/j.1752-1688.2012.00641.x.
11. Howat IM, Tulaczyk S (2005) Trends in spring snowpack in California over a half-century of climate warming in California, USA. Annals of Glaciology 40: 151–156.
12. Andrews ED (2012) Hydrology of the Sierra Nevada Network national parks: Status and trends. Natural Resource Report NPS/SIEN/NRR–2012/500. National Park Service, Fort Collins, Colorado. 196p.
13. Marvin CF (1914) Air drainage explained. Monthly Weather Review 42: 583–585. DOI:10.1175/1520-0493(1914)42<583:ADE>2.0.CO;2.
14. Barr S, Orgill MM (1989) Influence of external meteorology on nocturnal valley drainage wind. Journal of Applied Meteorology 28: 497–517.
15. Neff WD, King CW (1989) The accumulation and pooling of drainage flows in a large basin. Journal of Applied Meteorology 28: 518–529. DOI:10.1175/1520-0450(1989)028<0518:TAAPOD>2.0.CO;2.
16. Blennow K (1998) Modeling minimum air temperature in partially and clear felled forests. Agric For Meteorol 91: 223–235. DOI:10.1016/S0168-1923(98)00069-0.
17. Gustavsson T, Karlsson M, Bogren J, Lindqvist S (1998) Development of temperature patterns on clear nights. Journal of Applied Meteorology 37: 559–571. DOI:10.1175/1520-0450(1998)037<0559:DOTPDC>2.0.CO;2.
18. Whiteman CD, Bian X, Zhong S (1999) Wintertime evolution of the temperature inversion in the Colorado Plateau Basin. Journal of Applied Meteorology 38: 1103–1117. DOI:10.1175/1520-450(1999)038<1103:WEOTTI>2.0.CO;2.
19. Lindkvist L, Gustavsson T, Borgren J (2000) A frost assessment method for mountainous areas. Agricultural and Forest Meteorology 102: 51–67. DOI:10.1016/S0168-1923(99)00087-8.
20. Halley V, Eriksson M, Nunez M (2003) Frost prevention and prediction of temperatures and cooling rates using GIS. Aust Geogr Stud 41: 287–302. DOI:10.1046/j.1467-8470.2003.00235.x.
21. Chung U, Seo HH, Hwang KH, Hwang BS, Choi JJ, et al. (2006) Minimum temperature mapping over complex terrain by estimating cold air accumulation potential. Agricultural and Forest Meteorology 137: 15–24.
22. Pepin N, Daly C, Lundquist JD (2011) The Influence of Surface/Free-Air Decoupling on Temperature Trend Patterns in the Western U.S. Journal of Geophysical Research - Atmospheres 116: D10109. DOI:10.1029/2010JD014769.
23. Lundquist JD, Pepin NC, Rochford C (2008) Automated algorithm for mapping regions of cold-air pooling in complex terrain. Journal of Geophysical Research 113: D22107. DOI:10.1029/2008JD009879.
24. Daly C, Conklin DR, Unsworth MH (2009) Local atmospheric decoupling in complex topography alters climate change impacts. International Journal of Climatology, 30: 1857–1864. DOI:0.1002/joc.2007.
25. Wilson DE (1982) Wolverine *Gulo gulo*. In: Chapman JA and Feldhamer GA editors. Wild Mammals of North America: The Johns Hopkins University Press, Baltimore, Maryland. 644–652.
26. Aubry KB, McKelvey KS, Copeland JP (2007) Distribution and broad scale habitat relations of the wolverine in the contiguous United States. Journal of Wildlife Management 71: 2147–2158.
27. Copeland JP, McKelvey KS, Aubry KB, Landa A, Persson J, et al. (2010) The bioclimatic envelope of the wolverine (*Gulo gulo*): do climatic constraints limit its geographic distribution? Canadian Journal of Zoology 88: 233–246.
28. Magoun AJ, Copeland JP (1998) Characteristics of wolverine reproductive den sites. Journal of Wildlife Management 62: 1313–1320.
29. Grinnell J, Dixon JS, Lisdale JM (1937) Fur-bearing mammals of California: Their natural history, systematic status and relations to man. Berkeley: University of California Press.777p.
30. Banci V (1994) The Scientific Basis for Conserving Forest Carnivores: American Marten, Fisher, Lynx and Wolverine in the Western United States. In: Ruggiero LF, Aubry KB, Buskirk SW, Lyon LJ and Zielinski WJ, editors. General Technical Report RM-254: USDA Forest Service, Rocky Mountain Forest and Range Experiment Station, Fort Collins, CO. 99–127.
31. Zielinski WJ, Truex RL, Schlexer FV, Campbell LA, Carroll C (2005) Historical and contemporary distributions of carnivores in forests of the Sierra Nevada, California, USA. Journal of Biogeography 32: 1385–1407. DOI:10.1111/j.1365-2699.2004.01234.x.
32. Schwartz MK, Aubry KB, McKelvey KS, Pilgrim KL, Copeland JP, et al.(2007) Inferring geographic isolation of wolverines in California using historical DNA. Journal of Wildlife Management 71: 2170–2179.
33. Moriarty KM, Zielinski WJ, Gonzales AG, Dawson TE, Boatner KM (2009) Wolverine confirmation in California after nearly a century: native or long-distance immigrant? Northwest Science 83: 154–162.
34. Garcelon DK, Rall R, Hudgens BR, Young JK, Brown R, et al. (2009) Feasibility Assessment and Implementation Plan for Population Augmentation of Wolverines in California. Arcata, CA: Institute for Wildlife Studies. 120p.
35. Ohmura A, Wild M, Bengtsson L (1996) A Possible Change in Mass Balance of Greenland and Antarctic Ice Sheets in the Coming Century. Journal of Climate 9: 2124–2135.
36. Ackerly DD, Loarie, SR Cornwell, WK (2010) The geography of climate change: implications for conservation biogeography. Diversity and Distributions 16: 476–487. DOI:10.1111/j.1472-4642.2010.00654.x.
37. Garen DC, Marks D (2005) Spatially distributed energy balance snowmelt modeling in a mountainous river basin: estimation of meteorological inputs and verification of model results. Journal of Hydrology 315: 126–153.
38. Daly C, Halbleib M, Smith JI, Gibson WP, Doggett MK, et al. (2008) Physiographically-sensitive mapping of temperature and precipitation across the conterminous United States. International Journal of Climatology 28: 2031–2064.

39. Warhaftig C, Birman JH (1965) The Quaternary of the Pacific mountain system in California. In: Wright HE and Frey DG editors. The Quaternary of the United States: Princeton, Princeton University Press. 299–340.
40. Gillespie AR, Zehfuss PH (2004) Glaciations of the Sierra Nevada, California, USA. In: Ehlers J and Gibbard PL, editors. Quaternary Glaciations and Chronology Part II: North America, Developments in Quaternary Science Vol 2b: Amsterdam, Elsevier. 51–62.
41. Flint RF (1957) Glacial and Pleistocene geology. New York: John Wiley & Sons, 553p.
42. Sugden DE, John BS (1976) Glaciers and landscape: A geomorphological Approach. Wiley, New York, NY, 376p.
43. MacGregor KR, Anderson RS, Anderson SP, Waddington ED (2000) Numerical simulations of glacial-valley longitudinal profile evolution. Geology 28: 1031–1034.
44. Penck AC (1905) Glacial features in the surface of the Alps. Journal of Geology 13: 1–19.
45. Anderson RS, Molnar P, Kessler MA (2006) Features of glacial valley profiles simply explained. Journal of Geophysical Research 111: F01004. DOI:10.1029/2005JF000344.
46. Flint LE, Flint AL, Thorne JH, Boynton R (2013) Fine-scale hydrologic modeling for regional landscape applications: the California Basin Characterization model development and performance. Ecological Processes 2: 25. doi:10.1186/2192-1709-2-25.
47. Anderson EA (1973) National Weather Service River Forecast System – Snow Accumulation and Ablation Model, NOAA Technical Memorandum NWS HYDRO-17. U.S. Department of Commerce, Silver Spring, MD. 217p.
48. Anderson EA (1976) A Point Energy and Mass Balance Model of a Snow Cover. NOAA Technical Report 19. U.S. Department of Commerce, Silver Spring, MD. 150p.
49. Flint LE, Flint AL (2012) Simulation of climate change in San Francisco Bay Basins, California: Case studies in the Russian River Valley and Santa Cruz Mountains. U.S. Geological Survey Scientific Investigations Report 2012–5132, 55p.
50. Shamir E, Georgakakos KP (2006) Distributed snow accumulation and ablation modeling in the American River basin. Advances in Water Resources 29-4: 558–570.
51. Rice R, Bales RC, Painter TH, Dozier J (2011). Snow water equivalent along elevation gradients in the Merced and Tuolumne River basins of the Sierra Nevada. Water Resources Research 47: W08515. DOI:10.1029/2010WR009278.
52. Inman RM, Magoun AJ, Persson J, Pedersen DN, Mattisson J, et al. (2007) Reproductive chronology of wolverines. In: Greater Yellowstone Wolverine Program Cumulative Report, Wildlife Conservation Society North America Program, Montana, USA, 55–63.
53. Intergovern-mental Panel on Climate Change (2007) Climate change 2007: The Physical Science Basis. In: Solomon S, Qin D, Manning M, Chen Z, Marquis M, Averyt KB, Tignor M, Miller HL, editors. Contribution of Working Group I to the Fourth Assessment Report of the Intergovernmental Panel on Climate Change. Cambridge University Press, Cambridge, United Kingdom and New York, NY, USA, 996.
54. Intergovern-mental Panel on Climate Change (2013) Climate Change 2013: The Physical Science Basis. In: Stocker TF, Qin D, Plattner GK, Tignor M, Allen SK, Boschung J, Nauels A, Xia Y, Bex V, Midgley PM, editors. Contribution of Working Group I to the Fifth Assessment Report of the Intergovern-mental Panel on Climate Change Cambridge University Press, Cambridge, United Kingdom and New York, NY, USA, 1535.
55. Cayan DR, Maurer EP, Dettinger MD, Tyree M, Hayhoe K (2008) Climate change scenarios for the California region. Climatic Change 87: S21–S42.
56. Washington WM, Weatherly JW, Meehl GA, Semtner AJ, Bettge TW, et al. (2000) Parallel climate model (PCM) control and transient simulations. Climate Dynamics 16-10/11: 755–774.
57. Meehl GA, Washington WM, Wigley TML, Arblaster JM, Dai A (2003) Solar and greenhouse gas forcing and climate response in the twentieth century. Journal of Climate 16-3: 426–444.
58. Stouffer RJ, Broccoli AJ, Delworth TL, Dixon KW, Gudgel R, et al. (2006) GFDL's CM2 global coupled climate models. Part IV: Idealized climate response. Journal of Climate 19-5: 723–740.
59. Delworth TL, Broccoli AJ, Rosanti A, Stouffer RJ, Balaji V, et al. (2006) GFDL's CM2 Global Coupled Climate Models. Part I: Formulation and Simulation Characteristics. J. Climate 19: 643–674. DOI:http://dx.doi.org/10.1175/JCLI3629.1.
60. Raleigh MS, Lundquist JD (2012) Comparing and combining SWE estimates from the SNOW-17 model using PRISM and SWE reconstruction. Water Resources Research 48: W01506. DOI:10.1029/2011WR010542.
61. Young CA, Escobar-Arias MI, Fernandes M, Joyce B, Kiparsky M, et al. (2009) Modeling the Hydrology of Climate Change in California's Sierra Nevada for Subwatershed Scale Adaptation. Journal of the American Water Resources Association 45-6: 1409–1423. DOI:10.1111 / j.1752-1688.2009.00375.x.
62. Peacock S (2011) Projected 21st century climate change for wolverine habitats within the contiguous United States, Environmental Research Letters 6, 9p.
63. McKelvey KS, Copeland JP, Schwartz MK, Littell JS, Aubry KB, et al. (2011) Climate change predicted to shift wolverine distributions, connectivity, and dispersal corridors. Ecological Applications 21: 2882–2897.
64. Powell RA, Buskirk SW, Zielinski WJ (2003) Fisher and marten. In: Feldhamer, GA, Thompson BC, Chapman JA, editors. Wild mammals of North American 2nd edn. Johns Hopkins University Press, Baltimore, MD. 635–649.
65. Mills LS, Zimova M, Oyler J, Running S, Abatzoglou JT, et al. (2013) Camouflage mismatch in seasonal coat color due to decreased snow duration. Proceedings of the National Academy of Sciences 110–18: 7360–7365.

12

Effects of Landscape-Scale Environmental Variation on Greater Sage-Grouse Chick Survival

Michael R. Guttery[1]*, David K. Dahlgren[2], Terry A. Messmer[3], John W. Connelly[4], Kerry P. Reese[5], Pat A. Terletzky[3], Nathan Burkepile[6], David N. Koons[7]

1 Department of Forest and Wildlife Ecology, University of Wisconsin, Madison, Wisconsin, United States of America, **2** Kansas Department of Wildlife and Parks, Hays, Kansas, United States of America, **3** Department of Wildland Resources, Utah State University, Logan, Utah, United States of America, **4** Idaho Department of Fish and Game, Blackfoot, Idaho, United States of America, **5** Department of Fish and Wildlife Sciences, University of Idaho, Moscow, Idaho, United States of America, **6** Northland Fish and Game, Whangarei, New Zealand, **7** Department of Wildland Resources and the Ecology Center, Utah State University, Logan, Utah, United States of America

Abstract

Effective long-term wildlife conservation planning for a species must be guided by information about population vital rates at multiple scales. Greater sage-grouse (*Centrocercus urophasianus*) populations declined substantially during the twentieth century, largely as a result of habitat loss and fragmentation. In addition to the importance of conserving large tracts of suitable habitat, successful conservation of this species will require detailed information about factors affecting vital rates at both the population and range-wide scales. Research has shown that sage-grouse population growth rates are particularly sensitive to hen and chick survival rates. While considerable information on hen survival exists, there is limited information about chick survival at the population level, and currently there are no published reports of factors affecting chick survival across large spatial and temporal scales. We analyzed greater sage-grouse chick survival rates from 2 geographically distinct populations across 9 years. The effects of 3 groups of related landscape-scale covariates (climate, drought, and phenology of vegetation greenness) were evaluated. Models with phenological change in greenness (NDVI) performed poorly, possibly due to highly variable production of forbs and grasses being masked by sagebrush canopy. The top drought model resulted in substantial improvement in model fit relative to the base model and indicated that chick survival was negatively associated with winter drought. Our overall top model included effects of chick age, hen age, minimum temperature in May, and precipitation in July. Our results provide important insights into the possible effects of climate variability on sage-grouse chick survival.

Editor: Mark S. Boyce, University of Alberta, Canada

Funding: This research was supported by the Idaho Department of Fish and Game Federal Aide in Wildlife Restoration Project W-160-R, the United States Natural Resources Conservation Service, U.S. Bureau of Land Management, Utah State University Quinney Professorship for Wildlife Conflict Management, Jack H. Berryman Institute of Wildlife Damage Management, Utah State University Extension Service. The funders had no role in study design, data collection and analysis, decision to publish, or preparation of the manuscript.

Competing Interests: The authors have declared that no competing interests exist.

* E-mail: micrgutt@gmail.com

Introduction

Selective pressures result in the evolution of a life history conducive to species persistence under the environmental conditions encountered throughout the species' evolutionary history. Environmental conditions are not static, but rather experience climatic, geological, and successional changes through time. While such changes continue to occur naturally, anthropogenic disturbances have critically altered many of these processes, resulting in environments changing at rates that exceed the ability of some species to adapt [1]. The impact of rapidly changing environments may be particularly severe for species with limited dispersal opportunities (i.e., those existing in highly fragmented habitats; [2]). Efforts to conserve such species must focus on identifying the key demographic rates that are limiting population growth and the environmental factors that affect these rates [3].

During the 20th century, greater sage-grouse (*Centrocercus urophasianus*; hereafter sage-grouse) populations experienced precipitous declines as a result of anthropogenic habitat destruction, degradation, conversion, and fragmentation [4,5]. In response to declining populations and increasing threats to remaining habitat, the Canadian Committee on the Status of Endangered Wildlife in Canada declared sage-grouse to be an endangered species in 1998 [6]. The United States Fish and Wildlife Service (USFWS) designated the sage-grouse as a candidate for protection under the Endangered Species Act in 2010 [7].

Sage-grouse are endemic to sagebrush (*Artemisia* sp.) dominated habitats of western North America, which have historically been very stable given that sagebrush is a long-lived and persistent plant. As such, sage-grouse evolved to use sagebrush for food and cover throughout the majority of their annual cycle. However, sage-grouse chicks do not consume sagebrush during their early development but instead require forbs and their associated arthropod communities. These components of the sagebrush ecosystem are highly dependent upon precipitation levels and therefore may exhibit high interannual variability. Thus, sage-grouse evolved a life history characterized by high annual adult survival but relatively low and variable reproductive rates compared to most other tetronids [8,9].

Recently, researchers have applied life cycle models to gain a better understanding of factors affecting greater sage-grouse at the population [10] and range-wide scales [9]. Although both studies found sage-grouse population growth rates to be most sensitive to variability in adult female survival, they also found chick survival to have the second largest impact on population growth. While numerous studies have evaluated factors which influence survival rates of adult female sage-grouse [11,12], little is known about factors affecting chick survival. Generally, demographic rates to which the population growth rate is highly sensitive have low temporal variability [13,14]. Thus, chick survival should exhibit greater inter-annual variability and could therefore contribute more to spatio-temporal variability in population growth rate [15] even though sage-grouse populations are more sensitive to hen survival.

Previously published studies of factors affecting sage-grouse chick survival [16,17,18] have focused on micro-scale habitat factors such as percent coverage and height of forbs and grasses and availability of arthropods at chick location sites. These studies follow logically from previous research on sage-grouse brood habitat selection [19,20,21,22] and chick diets [23,24,25,26]. Collectively, these studies clearly demonstrate that broods typically select relatively mesic habitats with abundant forbs and arthropods and that these choices are related to chick survival. However, existing studies have not investigated the impacts of large-scale environmental processes (drought, temperature, etc.) on sage-grouse chick survival.

Landscape-scale environmental factors such as habitat condition, drought, and climate may be correlated with chick survival. Normalized Difference Vegetation Index (NDVI) is a commonly used index of plant production and habitat quality [27,28,29,30], with higher values of the index corresponding to increased levels of "greenness". Despite being less sensitive to plant phenology in sagebrush steppe ecosystems [29] and potential biases due to image quality, NDVI has been shown to be positively related to sage-grouse recruitment and population growth [27]. Drought and climatic variables can work independently and in concert to affect habitat parameters and can be reflected in NDVI values. For example, measures of drought, precipitation, and temperature can be correlated to winter snow pack which is known to be a major driver of vegetation dynamics throughout much of the mountainous regions of western North America [31]. However, climatic variables may affect sage-grouse chick survival in ways other than through their influence on habitat quality. Young grouse may be susceptible to exposure mortality during periods of extreme temperatures [32]. Additionally, numerous studies have documented increased nest and chick predation rates following precipitation events (i.e., moisture facilitated predation hypothesis; [33,34,35]). This effect is typically attributed to increased scent production resulting from increased bacterial growth when skin and feathers are wet [36]. Although the assumptions underlying the moisture facilitated predation hypothesis have not been thoroughly evaluated in the context of the hypothesis, the processes of moisture facilitating microbial activity and increased microbial activity resulting in increased scent production have been well documented in other fields of study [37,38,39,40].

The objective of our study was to model the effects of landscape scale biotic (habitat greenness) and abiotic (climate and drought) factors on sage-grouse chick survival. We demonstrate the utility of data that can be readily obtained for virtually any geographic region or temporal period via web-based resources for predicting sage-grouse chick survival.

Materials and Methods

Study areas

Data were collected as part of 2 larger studies conducted in Idaho and Utah (Fig. 1). The Idaho study was conducted from 1999–2002 in sagebrush-grassland habitats of the Upper Snake River Plain in southeastern Idaho (44°13′N, 112°38′W). This area was characterized by relatively low topographic relief with elevation across the site ranging from 1300–2500 m. Approximately 50% of the area was privately owned, with the remainder being public lands administered by the U.S. Bureau of Land Management (BLM). Annual precipitation varied by elevation with low elevation areas receiving 17.5 to 30.0 cm of precipitation. Most sage-grouse habitat at lower elevations was dominated by Wyoming big sagebrush (*A. tridentata wyomingensis*). At higher elevations precipitation ranged from 30.5 to 45.5 cm annually and the habitat was dominated by mountain big sagebrush (*A. t. vaseyena*). Livestock grazing and cropland agriculture were the dominant land uses across the area [41].

The Utah study area was located on Parker Mountain in south-central Utah (38°17′N, 111°51′W). Research on sage-grouse chick survival was conducted on this site from 2005 through 2009. The area encompassed 107,478 ha and was administered by the Utah School and Institutional Trust Lands Administration (40.8%), United States Forest Service (20.2%), BLM (33.9%), and private ownership (5.1%). Parker Mountain is a sagebrush-dominated plateau at the southern edge of the sage-grouse range. It is one of the few areas remaining in Utah with relatively stable numbers of sage-grouse, and it includes some of the largest contiguous tracts of sagebrush in the state [42]. Grazing by domestic livestock is the predominant land use practice across the study site. The area receives between 40 and 51 cm of precipitation annually, which generally exhibits a bi-modal pattern, occurring either as rain during the seasonal monsoonal period from late summer and early autumn or as snow during winter.

Field methods

We captured female sage-grouse on and around leks using spotlights, binoculars and long handled nets [43,44] during early spring (March–April). Captured hens were classified as being either second-year (SY) or after second-year (ASY) birds based on wing characteristics as described by Beck et al. [45]. Birds were fitted with 15–19 g necklace style radio-transmitters (Advanced Telemetry Systems, Isanti, MN, USA; Holohil Systems, Carp, Ontario, Canada) and released at the capture location.

Marked hens were monitored during April and May to determine if they initiated a nest. Nesting was confirmed visually, but hens were never intentionally flushed from their nest due to the tendency of female sage-grouse to abandon their nest if disturbed [46,47]. Nesting hens were visually monitored every 2–3 days to determine nest fate. Nests were monitored daily as the anticipated hatch date approached.

We captured chicks by using telemetry equipment to locate radio-marked hens. During capture events, the brood hen was flushed and chicks were captured by hand and placed in an insulated container to help maintain body temperature. We captured most broods within 48 hours of hatching with all broods being captured within 1 week of hatching. Captured chicks were weighed to the nearest gram and marked with a ≤1.5 g backpack-style radio-transmitter (Advanced Telemetry Systems, Insanti, MN in 1999–2001 and 2005, Holohil Systems, Carp, Ontario, Canada in 2006–2008, and American Wildlife Enterprises, Monticello, FL in 2009) attached with 2 sutures [48]. For the Idaho study site, 2–3 chicks per brood were selected at random to receive radio-

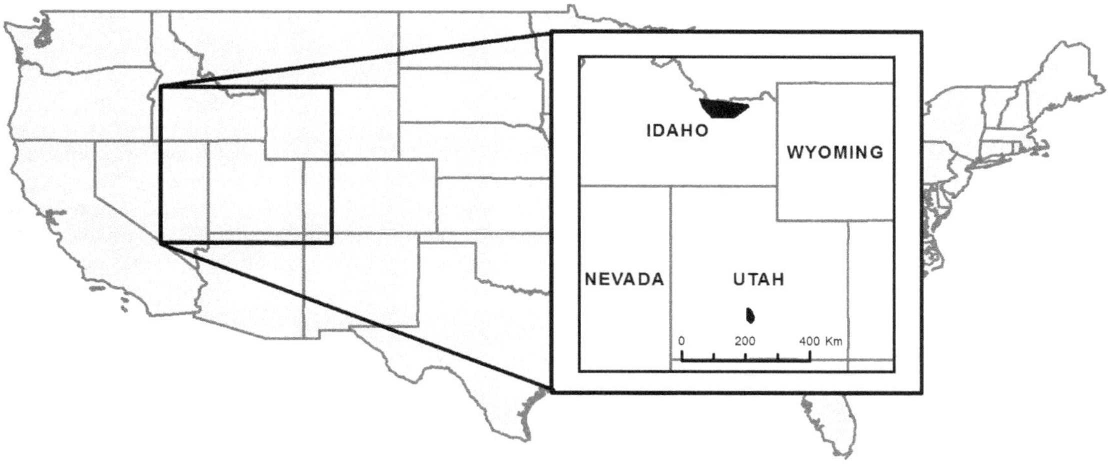

Figure 1. Map of study areas in Idaho and Utah.

transmitters. At the Utah study area, we marked all captured chicks except in 2006 when 3 chicks from each brood were randomly selected to receive transmitters. Chicks found dead in the immediate vicinity of the capture site were considered to have died as a result of handling and were excluded from subsequent analyses. Broods were typically checked within 12 hours of being marked and all chicks classified as capture mortalities were found intact within a few meters of the release site, indicating that their death was directly attributable to the capture event. Our decision to exclude chicks classified as capture mortalities from our analysis may have inflated survival estimates if some of these mortalities were in fact not related to capture. However, we do not believe this was a common occurrence if it occurred at all.

Marked chicks were located every 1–2 days until they reached 42 days of age. Monitoring intervals did occasionally exceed 2 days due to inclement weather events or difficulties locating broods following large movements. Extensive efforts were made to find any chicks missing from a brood. We occasionally recovered chick transmitters with no chick remains or signs of predation. These recoveries were classified as mortalities although it is possible that transmitters may have been lost for reasons other than chick death. Alternatively, we could have right-censored these specific events. While this would have been a valid option, we chose to treat the events as mortalities to ensure that our survival estimates were conservative. Due to the difficulty of distinguishing predation from scavenging, we did not assign specific causes of mortality.

All necessary permits were obtained for the described field studies. Permission to capture and mark sage-grouse in Idaho was obtained from the Idaho Department of Fish and Game and from the Utah Division of Wildlife Resources for the Utah study site. Grouse capture and transmitter attachment procedures were approved under the Utah State University Institutional Animal Care and Use Committee (IACUC) permit #945R and #942 and University of Idaho IACUC permit #2000-7.

Covariate data

We compiled year and site specific covariate data pertaining to drought, landscape greenness, and climate. We included seasonal (preceding winter and current summer) and monthly (May–July) Palmer Drought Severity Index (PDSI) and Palmer Z-Index (PZI) values. For climate and drought covariates, we defined winter as the period from 1 November to 30 April because precipitation would likely fall as snow on both study sites during these months. Summer was defined as the period from 1 May through 31 July. We did not include August because very few broods were monitored beyond July.

While both the PDSI and PZI indices are measures of drought and their values interpreted similarly (negative values correspond to drought conditions while positive values indicate wet conditions), the PDSI is most appropriate for measuring conditions across long time periods (several months) while the PZI is designed to measure conditions across shorter time periods (several weeks to a few months, [49]). Although drought is often thought about in terms of the presence or absence of precipitation, PDSI and PZI also account for site specific rates of evapotranspiration, soil moisture recharge, runoff, and moisture loss [50]. Additionally, both drought indices are calculated relative to the long-term average drought conditions at a specific site. As such, values of each index are standardized to have a common interpretation across locations [50]. All drought data were downloaded from the National Oceanic and Atmospheric Administration's National Climate Data Center (http://www.ncdc.noaa.gov/temp-and-precip/time-series/index.php).

Climate variables of interest included total precipitation, minimum temperature, and maximum temperature for the same seasonal and monthly periods described above. Unlike drought covariates, climate covariates were not adjusted to account for other physical processes or long-term site-specific averages. Because complete and representative weather station data were not available for both study sites, we used the Parameter-elevation Regressions on Independent Slopes Model (PRISM; http://www.prism.oregonstate.edu/) to estimate climatic data for both sites. PRISM is a knowledge-based climate analysis system capable of generating gridded predictions of climate data from known point climate data and a digital elevation map [51]. We used ArcMap10 to generate minimum convex polygons around all chick locations at both sites to define our study sites. We then extracted climate variable data from the corresponding PRISM layer.

Phenological change in landscape greenness was measured using NDVI for each study area. We generated NDVI values using Landsat 4–5 satellite images obtained from the United States Geological Survey EarthExplorer website (http://earthexplorer.usgs.gov/). We selected images captured between 1 May and 31 August with minimal cloud coverage. Due to variability in image quality (i.e., cloud cover) and capture date, we were not able to use images taken on identical dates across years. Images were processed using the ERDAS Imagine remote sensing image analysis software (Intergraph, Madison, AL, USA) to apply radiometric corrections that eliminated background noise while retaining temporal variance in vegetative reflectance [52]. We used ERDAS Imagine to calculate NDVI on a pixel-by-pixel basis for each image based on the ratio of red to near-infared reflectance [53]. For each site, we fit our observed NDVI values to a linear model:

$$NDVI = Year + Date + Date^2 + Year * Date + Year * Date^2 + \varepsilon,$$

to estimate daily NDVI values where $\varepsilon \approx N(0,\sigma^2)$. This model provided a good fit to the data ($R^2 > 0.80$, $F_{11} > 4.50$, $P < 0.001$). We used predicted values to estimate mean and maximum NDVI values for May, June, July, and summer (as defined above). We also estimated the mean NDVI value at the date of hatching, 15 days before hatch, date of each survival observation, and 5 and 10 days prior to the date of each observation. Finally, because variability in NDVI was low due to sagebrush obscuring the phenological progression of forbs and grasses, we adjusted all NDVI values by subtracting out the year and site specific NDVI value on 1 May. This linear transformation effectively removed baseline site and year variation, thereby allowing our analysis to focus more directly on the effect of within-year plant phenology at a site.

Analysis

Missing chicks whose fate could not be determined were removed (i.e., right-censored) from the data set at the time of their last confirmed detection. Failure to locate chicks may have been the result of transmitter failure, the chick being removed from the study site by a predator, or long distance movements that exceeded the range of the transmitters. On a few rare occasions, chicks were found alive several weeks after going missing. The flexibility of our model allowed us to reintroduce these chicks back into the data set once rediscovered. Alternatively, we could have assumed that missing chicks were either dead or alive but our approach likely provides the most realistic estimate of chick survival because only chicks with known fate were allowed to influence daily survival rates [17].

We modeled sage-grouse chick daily survival rates from hatch to 42-days of age using the known-fate maximum likelihood estimator developed by Manly and Schmutz [54] and extended by Fondell et al. [55]. This model assumes a piecewise survival function such that the survival rate from age t to age t+1 is:

$$\Phi_t = \exp(-\alpha_i)$$

where $\alpha_i \geq 0$, for $t_{i-1} \leq t < t_i$, with $t_0 = 0$, and i = 1,2,…,p. Therefore the daily survival rate (DSR) for ages 0 to t_1 days is assumed to be exp(-α_1), the DSR for ages t_1 to t_2 days is assumed to be exp(-α_2), and so forth, with p survival intervals. If N_a chicks are observed in a brood at age a then the number of survivors at age $b>a$, N_b, has a binomial distribution with mean:

$$E(N_b|N_a) = N_a\Phi_a\Phi_{a+1}\cdots\Phi_{b-1}.$$

To account for extra-binomial variance, the variance term is:

$$V_1(N_b|N_a) = D * V(N_b|N_a),$$

Where D is a constant and $V(N_b|N_a)$ is the binomal variance given by:

$$V(N_b|N_a) = N_a\Phi_a\Phi_{a+1}\cdots\Phi_{b-1}(1-\Phi_a\Phi_{a+1}\cdots\Phi_{b-1}).$$

This variance formulation assumes that most extra-binomial variation is the result of lack of independence in the fates of chicks within broods. Given this formulation and assumptions, the log-likelihood function for the observed number n_b at the end of a survival period is derived from the normal density function and takes the form:

$$L(\alpha_1,\alpha_2,\cdots,\alpha_p,D) = \sum\left[-0.5\log_e\{2\pi V_1(N_b|N_a)\} - \frac{0.5\{n_b - E(N_b|N_a)\}^2}{V_1(N_b|N_a)}\right],$$

where the summation is over all the instances in the data set where a brood size is observed at time a and then observed next at time b [54].

This generalized linear model is appropriate because it allows for variable observation intervals, changes in brood size due to missing chicks, and accounts for lack of independence in fates among chicks within a brood by using a quasi-likelihood approach [17,54,55]. Values of D near 1 indicate minimal dependence in the fates of brood mates whereas larger values correspond to decreasing independence among brood mates [54]. Covariates were modeled using a logit-link. Maximum likelihood estimates for all parameters were estimated using the 'OPTIM' function in R 2.14.1 [56].

To examine processes affecting chick survival in our populations, we first developed models that included alternative parameterizations of chick age. For example, we created models with categorical age classes wherein the categories were based on biological development of chicks, such as pre- versus post-flight ages or early ages when the diet consists primarily of insects versus later ages when forbs become important. We also considered linear and quadratic models of age treated as a continuous variable. Competing models of the various chick age parameterizations were ranked using the quasi-likelihood version of Akaike's Information Criterion adjusted for sample size ($QAIC_c$: [57,58]). Models with $\Delta QAIC_c \leq 2$ were considered to be equally supported by the data, and when this occurred we applied the principle of parsimony and based our inference on the model with the fewest parameters [59]. Upon identifying the best parameterization of chick age, we next considered the addition of hen age (SY or ASY) and hatch date effects, as both have been shown to be important predictors of sage-grouse chick survival [17,60]. Year and site effects were not modeled explicitly because all covariates of interest were site and year specific (i.e., site and year effects were modeled implicitly). The validity of the approach was assessed by adding year and site effects to our final top model and monitoring the change in $QAIC_c$.

We then developed candidate model sets for each of the 3 covariate groups. Covariates within each group tended to be correlated. To insure the interpretability of parameter estimates (i.e., to avoid multicollinearity), covariates with a Pearson correlation coefficient (ρ) greater than 0.50 were not included in the same model. To determine which of the 3 groups of covariates had the greatest impact on chick survival, we did not include covariates from different groups in the same model. These restrictions limited the complexity of models we considered. Upon identifying the top model for each of the 3 covariate groups, we obtained 95% bootstrap confidence intervals for model parameters using 5,000 samples with replacement from our dataset [61]. All continuous covariates were Z-standardized prior to analyses. We calculated the proportional reduction in deviance [62] for each model relative to the null model:

$$D_I = 1 - (dev_I / dev_N)$$

where D_I is the Zheng-score for the model of interest, dev_I is the deviance for the model of interest, and dev_N is the deviance for the null model (unless otherwise noted, an intercept-only model) and deviance was calculated as -2*quasi-log-likelihood. The Zheng-score is a goodness-of-fit measure for generalized linear models of longitudinal data and can be interpreted similarly to a standard coefficient of determination, R^2, in a linear model [62]. We then further assessed model fit by calculating the ratio of the Zheng-score for the model of interest relative to the spatially and temporally saturated model [63]:

$$R_I = D_I / D_{FS}$$

where D_I is the Zheng-score for the model of interest and D_{FS} is the Zheng-score for the fully spatial and temporally saturated model. Values of R close to zero indicate little improvement in model fit over the null model, whereas values of R that approach one indicate model fit similar to the fully saturated model.

Results

Chick statistics

Most hens had a single brood during the course of our multi-year study; however, 24 of the 142 hens had broods during more than one year of the study. Peak hatch date ranged from 25 May to 7 June at the Utah study area and from 19 May to 30 May for the Idaho area. During the 9 years of study we attached radio transmitters to 518 chicks from 142 broods, resulting in 11,188 chick exposure days (Table 1). Chick age at the time of capture ranged from 1 to 8 days. A total of 18 chicks were determined to have died as a result of capture, and were excluded from analyses.

Table 1. Capture statistics for greater sage-grouse chicks marked in Idaho (1999–2002) and Utah (2005–2009).

Year	Broods[1]	Chicks[2]	Hen Ages[3]	Marked[4]
1999	13	30	SY = 3, ASY = 10	2.31
2000	15	42	SY = 4, ASY = 11	2.80
2001	14	40	SY = 1, ASY = 13	2.86
2002	24	71	SY = 5, ASY = 19	2.96
2005	21	89	SY = 11, ASY = 10	4.24
2006	21	61	SY = 0, ASY = 21	2.90
2007	12	55	SY = 4, ASY = 8	4.58
2008	11	66	SY = 2, ASY = 9	6.00
2009	11	64	SY = 1, ASY = 10	5.82
Total	142	518	SY = 31, ASY = 111	3.65

[1]Number of broods captured.
[2]Total number of chicks marked with radio-transmitters.
[3]SY = second year hen (hatched the previous year), ASY = after second year hen (hatched ≥2 years earlier).
[4]Average number of chicks marked per brood.

We censored an additional 159 missing chicks from the dataset after the last date of telemetry observation.

Base null model

Our best intra-annual model of chick survival included linear and quadratic effects of chick age (Table S1). This model clearly out-performed all other intra-annual temporal models in terms of QAICc value. This model was then used as the base model for evaluation of the main effects of hen age and hatch date. Comparison of QAICc scores for these models (Table S2) shows that the model including only hen age and the additive effects of hen age and hatch date were both competitive (Δ QAICc<2). Because the model containing only the effect of hen age was more parsimonious, we chose to retain this model as the base null model for comparison of climate, drought, and greenness phenology covariates [64].

NDVI models

All NDVI covariates considered were highly correlated (all $\rho > 0.62$). Thus, we did not construct models containing multiple NDVI covariates. All 13 single NDVI-effect models produced positive beta estimates (Table S3). Five models were equally supported by the data (ΔQAICc<2.0, Table S3). However, none of the 13 models, including the top 5 models, resulted in a meaningful increase in model fit (as measured by the R-score) relative to the base model. Additionally, the 95% confidence intervals for the effect of the average NDVI in July relative to May 1 for a given site and year (the model with lowest $QAIC_c$) were symmetrical around zero, indicating a weak and imprecise effect (Table S4).

Drought models

As with our NDVI covariates, all drought covariates were highly correlated (all $\rho > 0.77$) so only single-effect models were considered. Of the 10 drought models considered, the model including the effect of the PZI for the preceding winter performed best (Table S5). The addition of winter PZI to the base model resulted in an approximate 40% increase in the R-score, indicating a substantial improvement in model fit. Further, the 95% confidence interval for winter PZI indicated a significant positive effect of the covariate on chick survival (Fig. 2, Table S6).

Climate models

Several combinations of climatic covariates had correlation coefficients below our critical value. Additive effects of multiple climatic covariates were modeled if the correlation between all covariates was less than 0.5 and the model was deemed to be ecologically meaningful. These conditions resulted in the construction of 18 models (Table S7). The top climate model (minimum temperature in May+total precipitation in July) fit the data well (R-score = 0.766) and was the overall top model (Table 2). Both climatic effects in the top model were negatively associated with chick survival (Figs. 3–4). Despite the model fitting the data well, only the effect of July precipitation was significantly different from zero (Table S8). To ensure that the effects in our top model were robust and were not confounded by underlying effects of site and/or year, we added site and year effects to our top climate model (Table S9). Models containing year effects did not converge and models including additive and interactive site effects were not supported by the data (based on $\Delta QAIC_c$), indicating that our results are robust across the 2 study sites. Allowing daily survival to change with chick age and holding all other covariate values at the sample mean, predicted values from our top model yielded a 42-

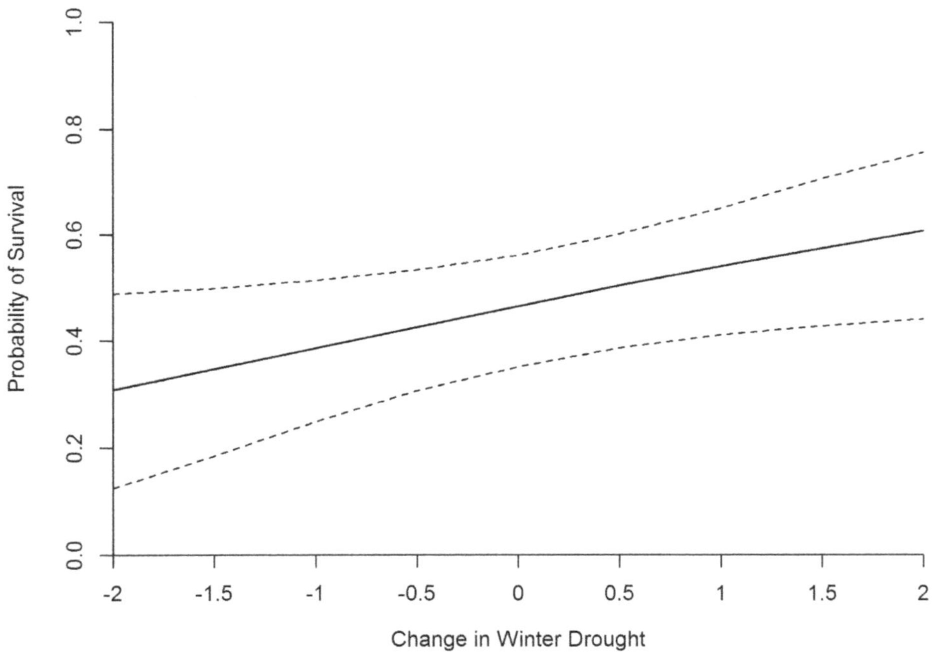

Figure 2. Effects of changes in winter drought severity (PZI) on the probability of greater sage-grouse chick survival to 42 days of age. Dashed lines indicate 95% confidence intervals. Negative values correspond to increasingly severe drought conditions. A change of 0.0 is equal to the mean Winter Palmer Z-Index score observed during the extent of this study. Palmer (1965) stated that a drought score of -2 was indicative of moderate drought.

day survival probability of 0.475 (95% CI = 0.375 to 0.566). Estimates of D from the 3 top models all indicate that dependence in the fates among brood mates was low (1.6149 to 1.7085) but non-negligible (Tables S4, S6, S8).

Discussion

Studies of avian survival are often short-term and conducted on a single study area. While such studies provide important

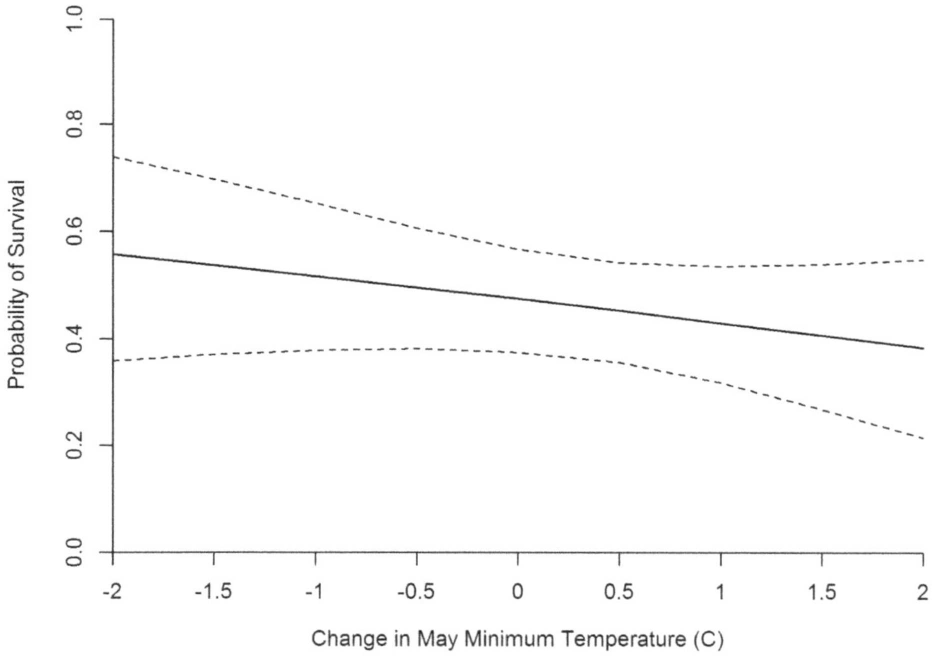

Figure 3. Effects of May minimum temperature on the probability of greater sage-grouse chick survival to 42 days of age. Dashed lines indicate 95% confidence intervals. A change of 0.0 is equal to the mean May minimum temperature observed during the extent of this study.

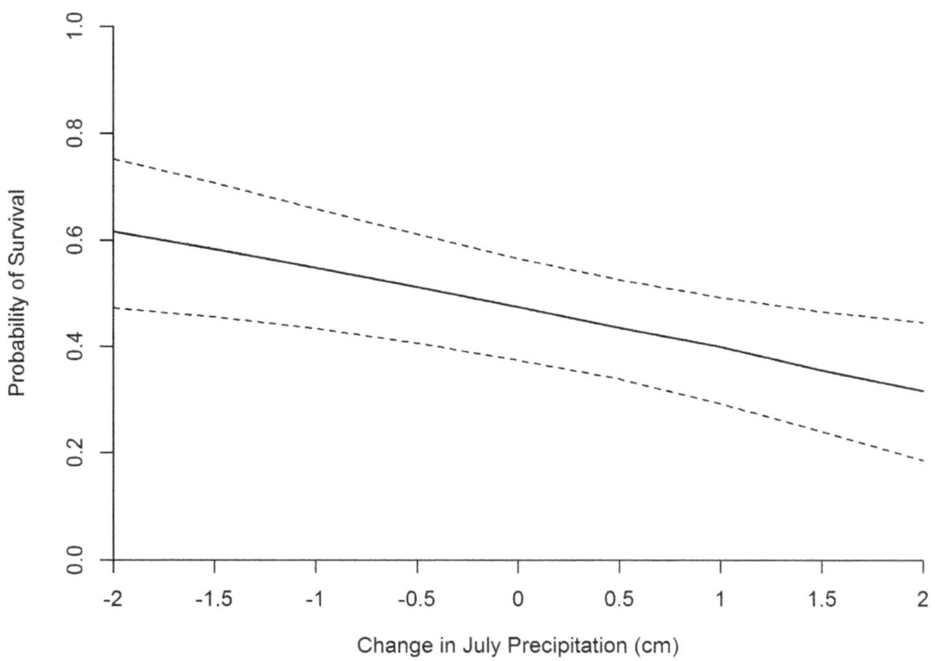

Figure 4. Effects of July precipitation on the probability of greater sage-grouse chick survival to 42 days of age. Dashed lines indicate 95% confidence intervals. A change of 0.0 is equal to the mean July precipitation observed during the extent of this study.

information, for many species there is a lack of knowledge concerning general large-scale factors which influence dynamics across space and time. An understanding of large-scale population drivers is essential for effective wildlife conservation planning and provides a baseline for developing meaningful hypotheses about specific local factors affecting populations at smaller spatial and temporal scales. Our study is the first to attempt to establish this baseline for the survival of greater sage-grouse chicks across multiple populations.

Table 2. Comparison of top chick survival models among the landscape-scale covariate groups.

Model	K	QAICc	ΔQAICc	w_i	R-score
Base+Saturated Model	13	−121.45	0.00	0.999	1.000
Base+May Min Temp+July Precip (−,−)	7	−58.30	63.15	0.000	0.766
Base+Winter PZI (+)	6	58.21	179.66	0.000	0.396
Base+July Mean NDVI (+)	6	178.53	299.98	0.000	0.022
Quadratic Chick Age+Hen Age (Base)	5	183.48	304.93	0.000	0.000

All models contain the base effects of quadratic chick age and hen age. Models were evaluated using the Quasi-Akaike's Information Criterion (QAIC). K = number of parameters. w_i = model weight (i.e. the likelihood of a particular model being the best model). R-score = percent reduction of deviance relative to the base model (Quadratic Chick Age+Hen Age). The saturated model contains effects for site (1 parameter) and each year (7 parameters). Typically 8 parameters would be required to model the effects of 9 years. However, because years did not overlap between the 2 sites we were able to fully specify year effects with only 7 parameters.

Independence of brood mates

Our modeling approach allowed simultaneous incorporation of commonly collected demographic information (hatch date, chick age, hen age) as well as publically accessible landscape level biotic (NDVI) and abiotic (temperature, precipitation, drought) information into survival models implemented in R [56]. Additionally, our approach allowed us to account for the potential lack of independence among chicks from the same brood [55]. Estimates of *D* from the top 3 models (Tables S4, S6, S8) ranged from 1.6149 to 1.7085 and, in all cases, confidence intervals did not include one or the mean number of chicks marked per brood (3.65, Table 1). This finding indicates that, while not highly dependent, fates were not independent among brood mates. This supports our decision to use the Manly and Schmutz [54] survival estimator rather than traditional known-fate survival estimators that assume fates of individuals are independent.

Survival rate

Overall, chick survival during this study was relatively high. Our top model produced an average 42-day survival probability of 0.475 (95% CI = 0.375 to 0.566). This is similar to the 42-day survival rate of 0.50 reported for sage-grouse chicks by Dahlgren et al. [17] and considerably higher than the 28-day survival rate of 0.392 reported by Gregg and Crawford [18]. However, comparison of our observed survival rate to those of Gregg and Crawford [18] are potentially confounded by the use of different transmitter attachment methods (suture attachment versus subcutaneous implant). Gregg and Crawford [18] report a total of 32 chick mortalities attributable to capture compared to only 18 in our study despite a similar total number of chicks being marked in both studies. It is possible that our survival rates may be inflated if some chicks treated as capture mortalities were incorrectly classified as such. However, the low incidence of chicks being classified as capture mortalities makes it unlikely that any misclassifications would significantly influence our findings.

Our analysis supported previous research that has shown both chick age and hen age to be important predictors of sage-grouse chick survival [17,60] (Tables S1 and S2). Interestingly, our models indicate that chicks hatched to second-year hens experience higher survival rates than chicks hatched to older hens. This effect has been previously reported for sage-grouse [17,60] although the mechanism underlying it has not been thoroughly explained. Despite being poorly estimated throughout (Tables S4, S6, S8), we chose to retain chick and hen age covariates in all models to minimize bias in estimates of the effects of interest.

Effects of NDVI

In recent years, NDVI has proven to be a useful tool for understanding various aspects of animal ecology [65]. We found positive relationships between all of our measures of NDVI phenology and chick survival (Table S3). However, none of the NDVI measures resulted in substantial improvements in model fit, as measured by the R-score, relative to the base (chick age+hen age) model, nor were the effects significant. Blomberg et al. [27] similarly found that NDVI was positively associated with sage-grouse recruitment and population growth, but that NDVI provided weak predictive power relative to other predictors such as precipitation.

Given the importance of invertebrates and herbaceous vegetation in the diet of sage-grouse chicks [23,18,25], the poor predictive power of NDVI for sage-grouse chick survival is somewhat surprising because NDVI is a well-established index of net primary production [65,66,67], and invertebrate production is positively related to plant production [68]. We suggest that the extensive coverage of sagebrush across both study sites resulted in phenological measures of NDVI being less sensitive to changes in coverage of forbs and grasses, thereby diminishing the ability of NDVI to measure changes of direct relevance to sage-grouse chicks. Correspondingly, Paruelo and Lauenroth [29] found a generally smaller range of NDVI values in sagebrush-steppe ecosystems than in grasslands where plant phenological changes are likely easier to detect.

Effects of drought

Although local availability and abundance of specific invertebrates and forbs is proximally related to sage-grouse chick survival [17,18], survival is likely under the primary influence of physical factors such as precipitation (amount and timing), temperature, and drought. Accordingly, our analysis indicated that abiotic factors were better predictors of sage-grouse chick survival than phenology of NDVI. Our top drought model (Table S5) indicated the presence of a significant relationship between winter drought and chick survival (Table S6). Since smaller PZI (and PDSI) values correspond to increasingly severe drought conditions, the positive parameter estimate associated with the winter drought effect implies that winter droughts lead to reduced chick survival (Fig. 2). Unfortunately, our data do not allow us to identify the true causal mechanism(s) underlying this relationship. Schwinning et al. [69] found that winter drought, even more so than summer drought, affects plant production during the following summer. Therefore, winter drought may affect sage-grouse chick survival via its influence on brood habitat quality. Additionally, winter drought may influence chick survival by affecting resource provisioning during egg formation. Forb abundance during the pre-nesting period is positively associated with hen nutrition [70], and hen nutrition prior to nest initiation is positively related to reproductive investment [71]. Thus, we suggest that either or both of these effects may be the mechanism behind the relationship between winter drought and chick survival that we observed.

Effects of climate

Blomberg et al. [27] reported relatively stable survival rates for adult sage-grouse but found that recruitment was variable and strongly influenced by annual climatic variation. These findings led the authors to conclude that stability of sage-grouse populations is dependent upon stable annual survival rates and occasional large inputs of new individuals into the population when climatic conditions are amenable to chick and juvenile survival. Our results support this assertion that climatic variables play a primary role in determining sage-grouse reproductive success. Of the 3 groups of predictors of chick survival we considered, models containing climatic effects clearly outperformed all other models (Table S7 versus Tables S5 and S3).

Our top climatic model fit the data well (Table 2, R-score = 0.766). In addition to the effects of chick and hen age, the top model included the minimum temperature in May (MMT) and precipitation in July, both producing negative parameter estimates (Table S8). We initially hypothesized that MMT could have either a positive or negative association with chick survival. Specifically, we predicted that MMT could be positive if higher minimum temperature resulted in fewer chicks dying due to exposure. Alternatively, we predicted that a negative effect of MMT would be attributable to high minimum temperatures leading to early snow melt and thus lower soil moisture and poor habitat quality throughout the brood-rearing period. We conclude that the latter is the case. Although particularly cold temperatures in late May could potentially result in increased exposure mortality, consideration of our peak hatch dates (see Section 3.1) reveals that it is unlikely that many chicks would be hatched early enough to be exposed to extreme low temperatures likely occurring in the first half of May. We also note that the minimum temperature in June is positively associated with chick survival, possibly indicating that exposure mortality does increase as temperature decreases during this timeframe.

Our interpretation of the negative effect of MMT on chick survival does raise concerns about the impact of projected climate on future sage-grouse reproductive success. Significant temperature increases have been documented across western North America in recent decades, and climate models consistently predict that temperatures will continue to increase into the foreseeable future [72]. Observed and projected warming trends have also been connected to observed and projected transitions from winter precipitation falling as rain rather than snow and consequently reduced spring snow pack [72,73,74].

The trend in warming temperatures could impact sage-grouse population dynamics as a result of phenological asynchrony [75], increased spread of exotic species such as red fox (*Vulpes vulpes*; [76]) and cheatgrass (*Bromus tectorum*; [5]), and increased frequency and severity of wildfires [77]. Blomberg et al. [27] concluded that projected climate change could result in reduced recruitment of sage-grouse. Our results support this conclusion. Figure 3 shows model-derived chick survival estimates across a range of MMT. According to our results, a 2°C increase in mean MMT, well within the range projected by most climate models [72], will result in an approximate 10% reduction in sage-grouse chick survival. This effect could be mitigated if sage-grouse are able to adjust their hatch dates to correspond with earlier snow melt and advanced plant phenology. A simple linear regression of our observed median hatch dates on MMT shows a significant correlation

($p = 0.0171$, $R^2 = 0.58$). This demonstrates that sage-grouse may be capable of synchronizing the timing of nesting with MMT for at least the range of MMT observed during our study. However, it is not clear if the level of plasticity in breeding phenology is sufficient to compensate for future climatic changes. Additionally, if warming results in a shift in the form of winter precipitation from snow to rain [73], chick survival may still be negatively affected by poor habitat quality, even if hens are able to adjust nest initiation to correspond with early snowmelt.

We initially hypothesized that July precipitation (JP) would have a positive effect on sage-grouse chick survival due to a moisture associated increase in plant and arthropod forage. However, our analysis showed a significant negative effect of JP (Table S8). While this result may be less intuitive, we conclude that it is real and meaningful. At least 2 mechanisms may underlie the relationship between chick survival and JP. First, chicks may be susceptible to exposure mortality in July. As noted above, the Utah study area is located in a monsoonal zone and receives a substantial proportion of the annual precipitation during late summer (primarily July and August). Monsoonal storms across the Utah study area often build quickly and result in significant temperature reductions followed by rain, hail, or both. By July, chicks are larger in size and are more independent of the brood hen [78]. If chicks are too large to be effectively brooded, severe monsoonal storms may result in chicks becoming soaked by rain and/or losing body temperature due to low temperatures or hail. While the Idaho study area is not in a monsoonal zone, it is possible that occasional severe July storms could produce similar effects. However, by July chicks should be more capable of thermoregulation relative to the early development period in June. Thus, we conclude that if exposure were a major source of chick mortality, models including the effect of June precipitation or the minimum temperature in June would have performed better.

Alternatively, JP may negatively affect sage-grouse chick survival through an interaction between increased moisture/ humidity and predator search efficiency (i.e., moisture facilitated predation hypothesis). Moisture on skin and feathers increases bacterial activity, subsequently increasing scent production [36,37,38,39,40]. Mammalian predators have been shown to respond rapidly to the presence of prey odor [79] and increased scent production may lead to enhanced prey detection rates. A number of studies have found increased nest predation rates following precipitation events for greater sage-grouse and other gallinaceous birds [33,80,81], and the phenomenon of moisture-facilitated predation may apply to chicks and adult birds as well [34,82]. We do not present observed chick predation rates due to concerns about correctly distinguishing between predation and scavenging. However, both of our study sites were inhabited by a suite of potential mammalian predators. Both study sites received predator management to reduce coyote (*Canis latrans*) predation on livestock, but coyotes and other common predators of sage-grouse chicks (red fox, badger [*Taxidea taxus*], weasels [*Mustella* sp.], and rattlesnakes [*Crotalus viridis*]) were present on both sites. In addition to the potential effects of moisture-facilitated predation by olfactory predators, JP may increase predation by avian predators if sage-grouse broods move to areas with less sagebrush cover following precipitation events to expedite drying and/or warming. Although not formally documented, we did observe broods along roadways at a higher frequency following precipitation events than at other times.

Effects of climate change on precipitation are less clear than the effects on temperature [72]. Climate models are inconclusive as to the sign of the effect on precipitation [83], and effects may vary by season [84]. In the absence of a consensus about effects of climate change on summer precipitation, anticipating the effect of changing precipitation on ecological communities and populations is difficult. Our analysis indicates that a 2 cm change in JP (positive or negative) would result in an approximate 15% change in sage-grouse chick survival (Fig. 4).

Sage-grouse are a species of great conservation concern in western North America. Chick survival has been shown to be an important determinant of population growth rates [9,10], yet relatively little is known about climatic or other large-scale environmental factors affecting survival rates. Previous studies have identified specific habitat characteristics that influence survival [17,18]. These studies have led to a proliferation of efforts to manage brood-rearing habitats without thorough consideration of the abiotic factors influencing both habitat quality and chick survival. Our study clearly demonstrates that large-scale abiotic factors such as drought, temperature and precipitation have significant effects on chick survival. These factors are beyond the control of state and federal wildlife management agencies and highlight the importance of considering current and future climatic conditions when developing policy and conservation strategies for this species. However, the effects we observed were measured for populations inhabiting large intact tracts of sagebrush habitat. The availability of adequate amounts of suitable habitat is a prerequisite that must be met for the effects of the abiotic factors we studied to be relevant.

Supporting Information

Table S1 Models for effect of age on greater sage-grouse chick survival.

Table S2 Models for effect of hen age and hatch date on greater sage-grouse chick survival. 'Age' is the top age varying model from Table S1. Signs in parentheses indicate the direction of respective covariate effects excluding chick age.

Table S3 Models for the effects of habitat greenness as measured by the Normalized Difference Vegetation Index (NDVI) on greater sage-grouse chick survival. Signs in parentheses indicate the direction of respective covariate effects excluding chick age. All models (except the intercept only model) contain the base effects of quadratic chick age and hen age. Models were evaluated using the Quasi-Akaike's Information Criterion (QAIC). K = number of parameters. w_i = model weight (i.e. the likelihood of a particular model being the best model). R-score = percent reduction of deviance relative to the base model (Quadratic Chick Age+Hen Age).

Table S4 Parameter estimates with 95% confidence intervals for the top model of the effects of Normalized Difference Vegetation Index (NDVI) on greater sage-grouse chick survival. Confidence intervals were calculated based on 5,000 bootstraps of the original data set.

Table S5 Models for the effects of drought on greater sage-grouse chick survival. Signs in parentheses indicate the direction of respective covariate effects excluding chick age. All models (except the intercept only model) contain the base effects of quadratic chick age and hen age. Models were evaluated using the Quasi-Akaike's Information Criterion (QAIC). K = number of parameters. w_i = model weight (i.e. the likelihood of a particular model being the best model). R-score = percent reduction of deviance relative to the base model (Quadratic Chick Age+Hen Age).

Table S6 Parameter estimates with 95% confidence intervals for the top model of the effect of drought on greater sage-grouse chick survival. Confidence intervals were calculated based on 5,000 bootstraps of the original data set.

Table S7 Models for the effects of climate on greater sage-grouse chick survival. Signs in parentheses indicate the direction of respective covariate effects excluding chick age. All models (except the intercept only model) contain the base effects of quadratic chick age and hen age. Models were evaluated using the Quasi-Akaike's Information Criterion (QAIC). K = number of parameters. w_i = model weight (i.e. the likelihood of a particular model being the best model). R-score = percent reduction of deviance relative to the base model (Quadratic Chick Age+Hen Age).

Table S8 Parameter estimates with 95% confidence intervals for the top model of the effects of climate on greater sage-grouse chick survival. Confidence intervals were calculated based on 5,000 bootstraps of the original data set.

Table S9 Evaluation of the effects of site and year effects on the top model from Table S7. All models contain the base effects of quadratic chick age and hen age. Models were evaluated using the Quasi-Akaike's Information Criterion (QAIC). K = number of parameters. w_i = model weight (i.e. the likelihood of a particular model being the best model).

Acknowledgments

We thank the many technicians who assisted with collecting data for this research. D. Menuz, J. Guttery and 2 anonymous reviewers provided valuable comments on earlier versions of this manuscript.

Author Contributions

Conceived and designed the experiments: MRG DKD TAM JWC KPR NB. Performed the experiments: MRG DKD NB. Analyzed the data: MRG PAT DNK. Wrote the paper: MRG.

References

1. Jump AS, Peñuelas J (2005) Running to stand still: adaptation and the response of plants to rapid climate change. Ecol Letters 8: 1010–1020.
2. Thomas CD, Cameron A, Green RE, Bakkenes M, Beaumont LJ, et al. (2004) Extinction risk from climate change. Nature 427: 145–148.
3. Norris K (2004) Managing threatened species: the ecological toolbox, evolutionary theory and declining-population paradigm. J Applied Ecol 41: 413–426.
4. Garton EO, Connelly JW, Horne JS, Hagen CA, Moser A, et al. (2011) Greater sage population dynamics and probability of persistence. In: Knick, ST, Connelly, JW, eds. Greater sage-grouse: ecology and conservation of a landscape species and its habitats. Studies in Avian Biology, Vol. 38. University of California Press, Berkeley, USA. pp 293–382.
5. Miller RF, Knick ST, Pyke DA, Meinke CW, Hanser SE, et al. (2011) Characteristics of sagebrush habitats and limitations to long-term conservation. In: Knick, ST, Connelly, JW, eds. Greater sage-grouse: ecology and conservation of a landscape species and its habitats. Studies in Avian Biology, Vol. 38. University of California Press, Berkeley, USA. pp 145–184.
6. Harris W, Lungle K, Bristol B, Dickinson D, Eslinger D, et al. (2001) Canadian sage grouse recovery strategy. Saskatchewan Environment and Resource Management and Alberta Natural Resources Services.
7. USFWS (2010) Endangered and Threatened Wildlife and Plants; 12-month findings for petitions to list the greater sage-grouse (*Centrocercus urophasianus*) as threatened or endangered. Federal Register 50 CFR Part 17 [FWS-R6-ES-2010-00189]. http://www.fws.gov/mountain-prairie/species/birds/sagegrouse/FR03052010.pdf
8. Connelly JW, Hagen CA, Schroeder MA (2011) Characteristics and dynamics of greater sage populations. In: Knick, ST, Connelly, JW, eds. Greater sage-grouse: ecology and conservation of a landscape species and its habitats. Studies in Avian Biology, Vol. 38. University of California Press, Berkeley, USA. pp 53–68.
9. Taylor RL, Walker BL, Naugle DE, Mills LS (2012) Managing multiple vital rates to maximize greater sage-grouse population growth. J Wildl Manage 76: 336–347.
10. Dahlgren DK (2009) Greater sage-grouse ecology, chick survival, and population dynamics, Parker Mountain, Utah. Utah State University, Logan: Ph.D. Dissertation.
11. Anthony RG, Willis MJ (2009) Survival rates of female greater sage-grouse in autumn and winter in southeastern Oregon. J Wildl Manage 73: 538–545.
12. Moynahan BJ, Lindberg MS, Thomas JW (2006) Factors contributing to process variance in annual survival of female greater sage-grouse in Montana. Ecol Appl 16: 1529–1538.
13. Pfister CA (1998) Patterns of variance in stage-structured populations: evolutionary predictions and ecological implications. Proc Nat Acad Sci 95: 213–218.
14. Gaillard JM, Yoccoz NG (2003) Temporal variation in survival of mammals: a case of environmental canalization? Ecology 84: 3294–3306.
15. Caswell H (2001) Matrix population models: construction, analysis and interpretation. 2nd edition.Sinauer Associates, Sunderland, Massachusetts, USA.
16. Aldridge CL, Boyce MS (2007) Linking occurrence and fitness to persistence: habitat-based approach for endangered greater sage-grouse. Ecol Appl 17: 508–526.
17. Dahlgren DK, Messmer TA, Koons DN (2010) Achieving better estimates of greater sage-grouse chick survival in Utah. J Wildl Manage 74: 1286–1294.
18. Gregg MA, Crawford JA (2009) Survival of greater sage-grouse chicks and broods in the northern Great Basin. J Wildl Manage 73: 904–913.
19. Drut MS, Crawford JA, Gregg MA (1994) Brood habitat use by sage grouse in Oregon. Great Basin Naturalist 54: 170–176.
20. Klebenow DA (1969) Sage grouse nesting and brood habitat in Idaho. J Wildl Manage 33: 649–662.
21. Sveum CM, Crawford JA, Edge WD (1998) Use and selection of brood-rearing habitat by sage grouse in south central Washington. Great Basin Nat 58: 344–351.
22. Wallestad RO (1971) Summer movements and habitat use by sage grouse broods in central Montana. J Wildl Manage 35: 129–136.
23. Drut MS, Pyle WH, Crawford JA (1994) Diets and food selection of sage grouse chicks in Oregon. J Range Manage 47: 90–93.
24. Johnson GO, Boyce MS (1990) Feeding trials with insects in the diet of sage grouse chicks. J Wildl Manage 54: 89–91.
25. Klebenow DA, Gray GM (1968) Food habits of juvenile sage grouse. J Range Manage 21: 80–83.
26. Peterson JG (1970) The food habits and summer distribution of juvenile sage grouse in central Montana. J Wildl Manage 34: 147–155.
27. Blomberg EJ, Sedinger JS, Atamian MT, Nonne DV (2012) Characteristics of climate and landscape disturbance influence the dynamics of greater sage-grouse populations. Ecosphere 3: 55.
28. Box EO, Holben BN, Kalb V (1989) Accuracy of the AVHRR vegetation index as a predictor of biomass, primary production, and new CO_2 flux. Vegetatio 80: 71–89.
29. Paruelo JM, Lauenroth WK (1995) Regional patterns of normalized difference vegetation index in North American shrublands and grasslands. Ecology 76: 1888–1898.
30. Pettorelli N, Vik JO, Mysterud A, Gaillard JM, Tucker CJ, et al. (2005) Using the satellite-derived NDVI to assess ecological responses to environmental change. TREE 20: 503–510.
31. Walker DA, Halfpenny JC, Walker MD, Wessman CA (1993) Long-term studies of snow-vegetation interactions. BioScience 43: 287–301.
32. Hannon SJ, Martin K (2006) Ecology of juvenile grouse during the transition to adulthood. Journal of Zoology 269: 422–433.
33. Herman-Brunson KM, Jensen KC, Kaczor NW, Swanson CC, Rumble MA, et al. (2009) Nesting ecology of greater sage-grouse *Centrocercus urophasianus* at the eastern edge of their historic distribution. Wildl Biol 15: 237–246.
34. Lehman CP, Flake LD, Rumble MA, Thompson DJ (2008) Merriam's turkey poult survival in the Black Hills, South Dakota. Intermountain J Sci 14: 78–88.
35. Roberts SD, Coffey JM, Porter WF (1995) Survival and reproduction of female wild turkeys in New York. J Wildl Manage 59: 437–447.
36. Syrotuck WG (1972) Scent and the scenting dog. Arner Publishers, Rome, New York, USA.
37. Barros N, Gomez-Orellana I, Feijóo S, Balsa R (1995) The effects of soil moisture on soil microbial activity studied by microcalorimetry. Thermochimica Acta 249: 161–168.
38. Bohn HL, Bohn KH (1999) Moisture in biofilters. Enviro Progress 18: 156–161.

39. Schimel JP, Gulledge JM, Clein-Curley JS, Lindstrom JE, Braddock JF (1999) Moisture effects on microbial activity and community structure in decomposing birch litter in the Alaskan taiga. Soil Bio Biochem 31: 831–838.
40. Zavala MAL, Funamizu N (2005) Effects of moisture content on the composting process in a biotoilet system. Compost Sci Util 13: 208–216.
41. Beck JL, Reese KP, Connelly JW, Lucia MB (2006) Movements and survival of juvenile greater sage-grouse in southeastern Idaho. Wildl Soc Bull 34: 1070–1078.
42. Beck JL, Mitchell DL, Maxfield BD (2003) Changes in the distribution and status of sage-grouse in Utah. Western North American Naturalist 63: 203–214.
43. Giesen KM, Schoenberg TJ, Braun CE (1982) Methods for trapping sage grouse in Colorado. Wildl Soc Bull 10: 224–231.
44. Wakkinen WL, Reese KP, Connelly JW, Fischer RA (1992) An improved spotlighting technique for capturing sage grouse. Wildl Soc Bull 20: 425–426.
45. Beck TDI, Gill RB, Braun CE (1975) Sex and age determination of sage grouse from wing characteristics. Colorado Department of Natural Resources Game Information Leaflet 49, Denver, USA.
46. Baxter RJ, Flinders JT, Mitchell DL (2008) Survival, movement, and reproduction of translocated greater sage-grouse in Strawberry Valley, Utah. J Wildl Manage 72: 179–186.
47. Holloran MJ, Heath BJ, Lyon AG, Slater SJ, Kuipers JL, et al. (2005) Greater sage-grouse nesting habitat selection and success in Wyoming. J Wildl Manage 69: 638–649.
48. Burkepile NA, Connelly JW, Stanley DW, Reese KP (2002) Attachment of radiotransmitters to one-day-old sage grouse chicks. Wildl Soc Bull 30: 93–96.
49. Karl TR (1986) The sensitivity of the Palmer drought severity index and Palmer's Z-index to their calibration coefficients including potential evapotranspiration. J Climate Applied Meteorology25: 77–86.
50. Palmer WC (1965) Meteorological drought. U.S. Weather Bureau Research Paper No.45.
51. Daly C, Taylor GH, Gibson WP, Parzybok TW, Johnson GL, et al. (2000) High-quality spatial climate data sets for the United States and beyond. Trans Amer Soc Ag Eng 43: 1957–1962
52. Schroeder TA, Cohen WB, Song C, Canty MJ, Yang Z (2006) Radiometric correction of multi-temporal Landsat data for characterization of early successional forest patterns in western Oregon. Remote Sensing of Environ 103:16–26.
53. Jensen J R (2005) Introductory digital image processing: a remote sensing perspective, 3rd edition. Pearson Prentice Hall.
54. Manly BFJ, Schmutz JA (2001) Estimation of brood and nest survival: comparative methods in the presence of heterogeneity. J Wildl Manage 65: 258–270.
55. Fondell TF, Miller DA, Grand JB, Anthony RM (2008) Survival of dusky Canada goose goslings in relation to weather and annual nest success. J Wildl Manage 72: 1614–1621.
56. R Development Core Team (2011) R: a language and environment for statistical computing. R Foundation for Statistical Computing, Vienna. http://www.R-project.org/
57. Akaike H (1973) Information theory and an extension of the maximum likelihood principle. In: Petrov BN, Csaki BF, eds. Second International Symposium on Information Theory, Academiai Kiado, Budapest. pp 267–281.
58. Burnham KP, Anderson DR (2002) Model selection and multimodel inference: a practical information-theoretic approach. Springer-Verlag, New York, New York, USA.
59. Hamel S, Côté SD, Festa-Bianchet M (2010) Maternal characteristics and environment affect the costs of reproduction in female mountain goats. Ecology 91: 2034–2043.
60. Guttery MR (2011) Ecology and management of a high elevation southern range greater sage-grouse population: vegetation manipulation, early chick survival, and hunter motivations. Utah State University, Logan: Ph.D. Dissertation.
61. Dixon PM (1993) The bootstrap and the jackknife: describing the precision of ecological indices. In: Scheiner SM, Gurevitch J, eds. Design and analysis of ecological experiments. Chapman and Hall, New York, New York, USA. pp 290–318.
62. Zheng B (2000) Summarizing the goodness of fit of generalized linear models for longitudinal data. Statistics in Medicine 19: 1265–1275.
63. Iles DT, Rockwell RF, Matulonis P, Robertson GJ, Abraham KF, et al. (2013) Predators, alternative prey, and climate influence annual breeding success of a long-lived sea duck. J Animal Ecology 82: 683–693.
64. Arnold TW (2010) Uninformative parameters and model selection using Akaike's Information Criterion. J Wildl Manage 74: 1175–1178.
65. Pettorelli N, Ryan S, Mueller T, Bunnefeld N, Jędrzejewska B, et al. (2011) The normalized difference vegetation index (NDVI): unforseen successes in animal ecology. Climate Research 46: 15–27.
66. Fang J, Piao S, Tang Z, Peng C, Ji W (2001) Interannual variability in net primary production and precipitation. Science 293: 1723a.
67. Field CB, Randerson JT, Malmström CM (1995) Global net primary production: combining ecology and remote sensing. Remote Sensing of Environ 51: 74–88.
68. Wenninger EJ, Inouye RS (2008) Insect community response to plant diversity and productivity in a sagebrush-steppe ecosystem. J Arid Environ 72: 24–33.
69. Schwinning S, Starr BI, Ehleringer JR (2005) Summer and winter drought in a cold desert ecosystem (Colorado Plateau) part II: effects on plant carbon assimilation and growth. J Arid Environ 61: 61–78.
70. Gregg MA, Barnett JK, Crawford JA (2008) Temporal variation in diet and nutrition of preincubating greater sage-grouse. Range Ecol Manage 61: 535–542.
71. Dunbar MR, Gregg MA, Crawford JA, Giordano MR, Tornquist SJ (2005) Normal hematologic and biochemical values for prelaying greater sage-grouse (*Centrocercus urophasianus*) and their influence in chick survival. J Zoo and Wildl Medicine 36: 422–429.
72. Karl TR, Melillo JM, Patterson TC, eds(2009) Global climate change impacts in the United States. Cambridge University Press.
73. Knowles N, Dettinger MD, Cyan DR (2006) Trends in snowfall versus rainfall in the western United States. J Climate 19: 4545–4559.
74. Mote PW, Hamlet AF, Clark MP, Lettenmaier DP (2005) Declining mountain snowpack in western North America. Bull American Meteorological Society 86: 39–49.
75. Parmesan C (2006) Ecological and evolutionary responses to recent climate change. Annual Review of Ecology, Evolution, and Systematics 37: 637–669.
76. Walther G, Post E, Convey P, Menzel A, Parmesan C, et al. (2002) Ecological responses to recent climate change. Nature 416: 389–395.
77. Brown TJ, Hall BL, Westerling AL (2004) The impact of twenty-first century climate change in wildland fire danger in the western United States: an applications perspective. Climatic Change 62: 365–388.
78. Schroeder MA, Young JR, Braun CE (1999) Sage grouse (*Centrocercus urophasianus*). In: Poole A, Gill F, eds. The Birds of North America, No. 425. The Academy of Natural Science, Philadelphia, Pennsylvania. pp 1–28.
79. Hughes NK, Price CJ, Banks PB (2010) Predators are attracted to the olfactory signals of prey. Plos ONE 5: e13114.
80. Lehman CP, Rumble MA, Flake LD, Thompson DJ (2008) Merriam's turkey nest survival and factors affecting nest predation by mammals. J Wildl Manage 72: 1765–1774.
81. Webb SL, Olson CV, Dzialak MR, Harlu SM, Winstead JB, et al. (2012) Landscape features and weather influence nest survival of a ground-nesting bird of conservation concern, the greater sage-grouse, in human-altered environments. Ecol Proc 1: 1–15.
82. Hohensee SD, Wallace MC (2001) Nesting and survival of Rio Grande turkeys in northcentral Texas. Proceedings of the National Wild Turkey Symposium 8: 85–91.
83. Chambers JC, Pellant M (2008) Climate change impacts on northwestern and intermountain United States rangelands. Rangelands 30: 29–33.
84. Mote PW (2006) Climate-driven variability and trends in mountain snowpack in western North America. J Climate 19: 6209–6220.

13

Pathogen-Host Associations and Predicted Range Shifts of Human Monkeypox in Response to Climate Change in Central Africa

Henri A. Thomassen[1,2], Trevon Fuller[1], Salvi Asefi-Najafabady[3,4], Julia A. G. Shiplacoff[1], Prime M. Mulembakani[5], Seth Blumberg[6,7], Sara C. Johnston[8], Neville K. Kisalu[9], Timothée L. Kinkela[5], Joseph N. Fair[10], Nathan D. Wolfe[10,11], Robert L. Shongo[12], Matthew LeBreton[10], Hermann Meyer[13], Linda L. Wright[14], Jean-Jacques Muyembe[15], Wolfgang Buermann[1,16], Emile Okitolonda[5], Lisa E. Hensley[17], James O. Lloyd-Smith[6,7], Thomas B. Smith[1,7], Anne W. Rimoin[9,18]*

1 Center for Tropical Research, University of California Los Angeles, Los Angeles, California, United States of America, **2** Department of Comparative Zoology, University of Tübingen, Tübingen, Germany, **3** School of Life Sciences, Arizona State University, Tempe, Arizona, United States of America, **4** Institute of the Environment and Sustainability, University of California Los Angeles, Los Angeles, California, United States of America, **5** Kinshasa School of Public Health, Kinshasa, Democratic Republic of Congo, **6** Fogarty International Center, National Institutes of Health, Bethesda, Maryland, United States of America, **7** Department of Ecology and Evolutionary Biology, University of California Los Angeles, Los Angeles, California, United States of America, **8** United States Army Medical Research Institute of Infectious Diseases, Fredrick, Maryland, United States of America, **9** Department of Microbiology, Immunology, and Molecular Genetics, University of California Los Angeles, Los Angeles, California, United States of America, **10** Global Viral Forecasting, San Francisco, California, United States of America, **11** Stanford University, Program in Human Biology, Stanford, California, United States of America, **12** Ministry of Health, Kinshasa, Democratic Republic of Congo, **13** Bundeswehr Institute of Microbiology, Munich, Germany, **14** The Eunice Kennedy Shriver National Institute of Child Health and Human Development, Bethesda, Maryland, United States of America, **15** National Institute of Biomedical Research, Kinshasa, Democratic Republic of Congo, **16** Department of Atmospheric and Oceanic Sciences, University of California Los Angeles, Los Angeles, California, United States of America, **17** Medical Countermeasures Initiative, Silver Spring, Maryland, United States of America, **18** Department of Epidemiology, School of Public Health, University of California Los Angeles, Los Angeles, California, United States of America

Abstract

Climate change is predicted to result in changes in the geographic ranges and local prevalence of infectious diseases, either through direct effects on the pathogen, or indirectly through range shifts in vector and reservoir species. To better understand the occurrence of monkeypox virus (MPXV), an emerging Orthopoxvirus in humans, under contemporary and future climate conditions, we used ecological niche modeling techniques in conjunction with climate and remote-sensing variables. We first created spatially explicit probability distributions of its candidate reservoir species in Africa's Congo Basin. Reservoir species distributions were subsequently used to model current and projected future distributions of human monkeypox (MPX). Results indicate that forest clearing and climate are significant driving factors of the transmission of MPX from wildlife to humans under current climate conditions. Models under contemporary climate conditions performed well, as indicated by high values for the area under the receiver operator curve (AUC), and tests on spatially randomly and non-randomly omitted test data. Future projections were made on IPCC 4th Assessment climate change scenarios for 2050 and 2080, ranging from more conservative to more aggressive, and representing the potential variation within which range shifts can be expected to occur. Future projections showed range shifts into regions where MPX has not been recorded previously. Increased suitability for MPX was predicted in eastern Democratic Republic of Congo. Models developed here are useful for identifying areas where environmental conditions may become more suitable for human MPX; targeting candidate reservoir species for future screening efforts; and prioritizing regions for future MPX surveillance efforts.

Editor: Yury E. Khudyakov, Centers for Disease Control and Prevention, United States of America

Funding: This work was made possible by the support of the Faucett Family Foundation. The authors thank the DRC Ministry of Health and local health workers who were responsible for specimen collection and case investigation. Additional support for this study was provided by the National Institutes of Health, the Eunice Kennedy Shriver National Institute of Child Health and Human Development, Bethesda, Maryland, United States of America, by the joint National Science Foundation-National Institutes of Health Ecology of Infectious Diseases Program (grant number EF-0430146), and the RAPIDD program of the Science & Technology Directorate, Department of Homeland Security, by the Fogarty International Center, National Institutes of Health, and by the National Institute of Allergy and Infectious Diseases (grant number EID-1R01AI074059-01). NDW is supported by the NIH Director's Pioneer Award (DP1-OD00370). Global Viral Forecasting is supported by Google.org, the Skoll Foundation, the Henry M. Jackson Foundation for the Advancement of Military Medicine, the US Armed Forces Health Surveillance Center Division of GEIS Operations, and the United States Agency for International Development (USAID) Emerging Pandemic Threats Program, PREDICT project, under the terms of Cooperative Agreement Number GHN-A-OO-09-00010-00. This study is made possible by the generous support of the American people through the United States Agency for International Development (USAID). The contents are the responsibility of the authors and do not necessarily reflect the views of USAID or the United States Government. Further support came from the Department of Homeland Security/National Biodefense Analysis and Countermeasures Center Project #HSQDC-05-X-000-35/P400006 (RSCB-09-00343). Opinions, interpretations, conclusions, and recommendations are those of the authors and are not necessarily endorsed by the U.S. Army. The funders had no role in study design, data collection and analysis, decision to publish, or preparation of the manuscript.

Competing Interests: The authors have declared that no competing interests exist.

* E-mail: arimoin@ucla.edu

Introduction

Climate change is predicted to result in shifts in the incidence and prevalence of infectious diseases (e.g. [1,2]), and may impose health risks on human populations in previously unexposed regions. As a result, the incidence of a variety of viruses is predicted to increase. Examples include arboviruses such as dengue, blue tongue, and African horse sickness viruses; tickborne encephalitis virus in the UK; and Usutu virus in Austria [3,4,5,6,7]. To date, climate-related studies of such diseases have focused on vector-borne pathogens (reviewed in [2]) with little attention paid to viruses with vertebrate reservoirs. The effect of climate on this type of viruses is likely to be complex since the host species themselves have niches limited by a variety of environmental conditions. Yet, a better understanding of the ecology of viruses and the ability to predict future outbreaks will be helpful in identifying potential risk areas for human infection, and in subsequent disease management and control efforts.

An emerging infectious disease of concern in tropical Africa is monkeypox (MPX). MPX virus (MPXV) is a zoonotic Orthopoxvirus that can cause serious smallpox-like illness in humans. It is endemic to forested regions of West and Central Africa, and since the global eradication of smallpox in 1977, MPXV has been considered the most important poxvirus affecting human health [8]. Once considered a rare, sporadic infection in humans, recent studies in the Democratic Republic of the Congo (DRC, previously Zaire), where most cases have been reported, suggest that the incidence of human MPX has markedly increased in the Congo Basin since the 1980s [9]. The post-eradication cessation of routine smallpox vaccination, which is known to provide cross-immunity against infection with MPXV, is likely to have played a role in the observed increase over the past three decades. Other driving factors such as land use and climate change likely contribute to the increase in incidence of MPX, but have so far not been studied.

The rise in incidence of MPX and its emergence in areas where it has not been previously known to occur [10] are important reminders that Africa is particularly vulnerable to emerging infectious diseases [1,11]. First, Africa is the continent projected to be most severely impacted by climate change [12], likely affecting the distributions and prevalence of infectious diseases [1,2]. Predictions for the end of the 21st century are that annual mean temperatures will increase by up to 5°C, precipitation will significantly increase in tropical Africa and decrease further away from the equator, evaporation will increase across most of the continent, and seasonality in precipitation will be significantly affected (multi-model mean predictions from the IPCC 4th Assessment Report [12]). Rainfall in West Africa is linked to sea surface temperatures in the Atlantic Ocean, and increasing temperatures could perturb the system leading to century-long droughts [13]. Second, human populations in Africa have a high disease burden [11], and may suffer from multiple infections and other health stressors such as malnutrition, which may depress immune response and increase susceptibility to new pathogens. In some populations of Sub-Saharan Africa, co-infection with one pathogen has been shown to increase the virulence of another [14]. These populations, which have a high incidence of HIV and other endemic infections, are potentially more sensitive to MPXV. Third, the health infrastructure in place is inadequate as a result of limited financial resources. These issues are further exacerbated by armed conflicts and the complexity of government institutions that have resulted in a large number of internally displaced persons (IDP), estimated to be around two million in 2009 [11,15,16]. Given these challenges, there is an urgent need to examine the effects of climate change on infectious diseases in Africa in order to most efficiently direct limited health resources [11].

Human MPX has been reported throughout most of tropical Africa, but the majority of cases are from the Congo basin (e.g. [23] and references therein). Little is known about the ecology of MPXV and its dependence on environmental conditions. The majority of human MPX infections result from close contact with infected animals that often serve as food sources, but person-to-person transmission occurs (e.g. [17]) and may be on the rise. Human-to-human transmission might spread monkeypox more widely as humans become more mobile, but third and fourth generation cases appear to be very rare, the transmission cycle is likely to cease relatively quickly, and the virus cannot sustain large-scale outbreaks in human populations in the rural settings where it has been observed [17]. . Antibodies to MPXV have been detected serologically in multiple animal species, suggesting that the host range could be quite large. A variety of potential reservoir species have been identified, including rope squirrels (genus *Funisciurus*; [18,19,20,21]) and African dormice (genus *Graphiurus*; [21]). An outbreak in prairie dogs in the United States in 2003 occurred via imported rodents and caused clinical and subclinical cases in humans (e.g. [24]). A recent review of models for zoonotic infections highlighted MPXV as a key priority for modeling research [22].

New developments in spatial modeling and the availability of global climate data and satellite remotely-sensed variables characterizing the environment have provided the tools necessary for a better understanding of diseases' spatial ecology [25,26,27]. Few studies have predicted shifts in reservoir species and the infectious diseases associated with them, such as presented here, yet some recent studies have shown the potential of such an approach [28,29]. Using these approaches, we recently demonstrated the importance of rope squirrels (*Funisciurus*) in predicting the locations of human MPX cases in the Sankuru district, DRC [30]. Here, we expand our study to a larger set of potential reservoir species, and to projections of human MPX occurrence under future climate conditions. We examine the spatial heterogeneity of human MPX across Tropical Africa, focusing on the Congo Basin. Our study focuses on both a small geographical scale combined with a short timeframe, as well as a large geographical scale combined with human MPX and environmental data covering multiple decades. To better understand the potential impacts of future climate change on the distribution of MPXV in Central Africa, we: 1) examine the relative contributions of climate and deforestation over the past decade to MPX occurrence in DRC (small scale, short timeframe); 2) identify the potential reservoir species that best predict human MPX occurrence (large scale, multi-decadal); 3) predict the distribution of MPXV across Tropical Africa under current climate conditions (large scale, multi-decadal); and 4) project the current distribution of MPXV onto future climate layers (large scale, multi-decadal). A schematic representation of aims 2), 3), and 4) is shown in Fig. 1. With the approach presented here, we aim to contribute to the development of a toolbox that is generally applicable to model a broad range of emerging infectious diseases with vertebrate reservoirs.

Results

Assessing the effects of present-day climate and forest clearing on MPX outbreaks on a local scale in DRC

To test whether covariates such as deforestation since 2000 and climate affected the locations of MPX hotspots in Sankuru, DRC,

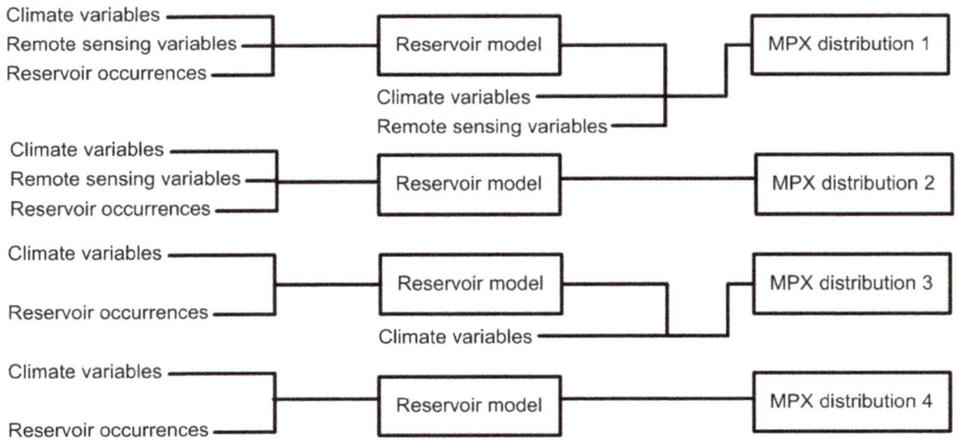

Figure 1. Flowchart of models created using the indicated input data. We modeled human MPX distributions under current conditions and assessed the concordance among models using the following predictor datasets: 1) reservoir species distributions (based on climate and remote sensing variables) and in addition the climate and remote sensing variables (the full model), which may contain additive information on top of the reservoir species distributions that were also based on both types of data; 2) only reservoir species distributions (based on climate and remote sensing variables); 3) reservoir species distributions (based on climate variables) plus climate variables; and 4) only reservoir species distributions (based on climate variables). To project future human MPX distributions under different climate change scenarios, we projected reservoir species distributions onto future climate variables. We then used the results as input for models of human MPX under approaches 3) and 4) above, since future remote sensing variables are not available.

as observed since 2001, we constructed a discrete Poisson model without covariates followed by a model that included ecological covariates and compared the hotspots indicated by the two models. The model without covariates identified clusters of MPX cases based only upon the fraction of the human population infected with MPX in each secteur. The ecological covariates were factors hypothesized to control habitat suitability for MPX reservoirs and included canopy moisture, a measure of drought, the deforestation rate, and the temperature of the coldest and the warmest quarter of the year (see Materials and Methods for details and Text S2 for a description of the drought calculation). If these ecological covariates were important drivers of MPX infections in humans, we would expect the model with covariates to classify different secteurs as MPX hotspots than the model without covariates. According to the model without covariates (Fig. 2A), the prime hotspots of MPX are in three secteurs in northwestern Sankuru: Batetela-Lomela, Okutu, and Batetela-Dibele. These secteurs are smaller than other secteurs in Sankuru ($t=-4.712$, $df=22.957$, $p=4.79\times10^{-5}$) but the fraction of the human population infected with MPX in these secteurs is significantly higher than average for Sankuru ($t=2.7289$, $df=4$, $p=0.0263$). This preliminary model also identified a second hotspot in Ngandu in eastern Sankuru.

Next, we adjusted the preliminary model, which assumed that the ecological covariates were the same throughout Sankuru, to account for differences in rainfall, temperature, and the deforestation rate among the secteurs. Forest clearing emerged as an important driver of MPX transmission risk because including

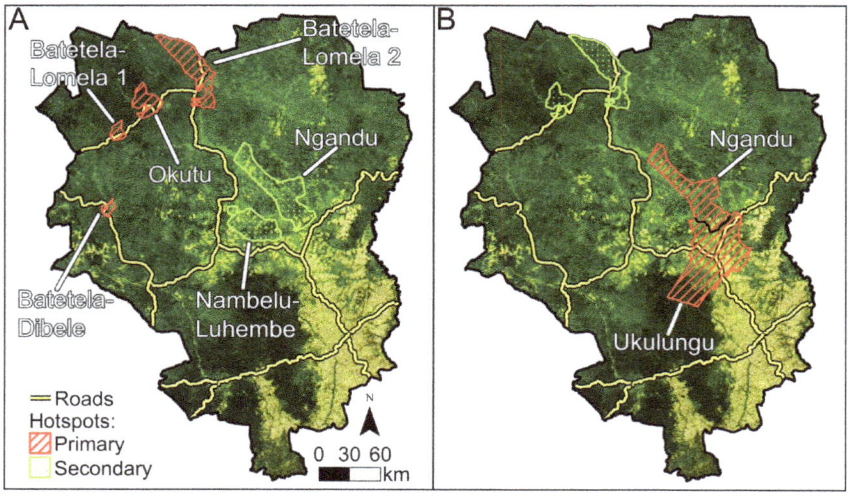

Figure 2. MPX hotspots in Sankuru district DRC. (A) secteurs classified as hotspots by a model that assumes that ecological factors are constant across the district; (B) MPX hotspots according to the model adjusted for spatial heterogeneity in deforestation and climate.

Table 1. Effect of recent forest clearing and climate on hotspots of MPX transmission in Sankuru, DRC.

Covariates	Hotspot	Secteurs	Observed No. MPX Cases	Expected No. MPX Cases	RR	LLR	p
None	P	Batetela-Lomela 1,2,3, Okutu, Batetela-Dibele	33	5.07	7.6	35.998	8.5×10^{-12}
	S	Nabelu-Luhembe, Ngandu1	45	17.7	2.99	16.823	1.3×10^{-6}
Deforestation	P	Ngandu1,2 Watambulu-Sud Bahamball, Ukulungu Watambulu Batetela-Lomela3, Nambelu-Luhembe	121	67.4	3	29.775	4.6×10^{-10}
	S	Batetela-Dibele Batetela-Lomela1	15	0.82	19.61	29.8756	2.7×10^{-9}
Climate*	P	Ukulungu, Ngandu1	34	13.3	2.87	12.4	7×10^{-4}
	S	Batetela-Lomela2,3, Okutu	18	5.85	3.28	8.46	8.2×10^{-3}
Deforestation+Climate	P	Ukulungu, Ngandu1	34	13.68	2.79	11.77	9.2×10^{-4}
	S	Batetela-Lomela3, Okutu	17	4.85	3.73	9.55	2.7×10^{-3}

*The climate variables were temperature of the warmest quarter, temperature of the coldest quarter, canopy moisture, and the correlation between maximum water deficit, which is a measure of drought, and the Atlantic Multidecadal Oscillation (for details, see Table 6). P = primary hotspot, *p* = *p*-value, LLR = log likelihood ratio, RR = relative risk, and S = secondary hotspot.

deforestation in the model reduced the log likelihood from 35.998 to 29.775, indicating that the deforestation-adjusted model provided a better fit to the data ([31] Table 1). Climate was also significant; adjusting the model to account for temperature and rainfall variation within Sankuru reduced the log likelihood to 12.44. Both the model with no covariates and the model adjusted for deforestation and climate identify Batetela-Lomela and Okutu in northwestern Sankuru and Ngandu in northeastern Sankuru as MPX hotspots (Fig. 2). However, according to the model with no covariates, Batetela-Lomela and Okutu are the primary hotspots and Ngandu is a secondary hotspot, whereas in the model adjusted for climate and deforestation, Ngandu becomes the primary hotspot. Furthermore, the climate- and deforestation-adjusted model identifies Ukulungu, a secteur south of Ngandu, as part of the primary MPX hotspot (Fig. 2B).

Regional scale study of MPX under current climate conditions

We used ecological niche modeling to first reconstruct the potential distributions of reservoir species, which were used in subsequent models of human MPX occurrence. Models for reservoir species based on climate variables performed well, as suggested by AUC values >0.85 and test AUC values >0.80 (Table 2), and by model tests using spatial subdivision of the presence points into training and test data sets (Table 3). The only model that did not perform significantly better than random was the one for the fire-footed rope squirrel (*F. pyrropus*) using western occurrence data to train the model. This may be explained by a relative lack of occurrence data in Central Africa (Nigeria, Cameroon), combined with a relatively high density of occurrences in West Africa (Ivory Coast, Ghana). Model tests using a spatial separation of eastern and western sites were qualitatively the same when the "minimum training presence" threshold was employed compared to those based on the "balance" threshold (see Materials and Methods). Temperature seasonality and annual precipitation stood out as important variables in predicting reservoir species' distributions (Fig. S1). The predicted distributions were largely confined to the African tropical forests (Fig. S2), and corresponded well with known distributions based on expert knowledge [33,34]. The modeled distributions were therefore considered useful for subsequent application as predictor variables in modeling human MPX.

Results for models of human MPX under contemporary climate conditions are shown in Table 4. We assessed model accuracy using the Akaike Information Criterion with a correction for small samples sizes (hereafter "AICc"). The model with the minimum AICc was considered to have the best support. The lowest AICc value (2777.3), and thus the least complex model with high performance, was found when using reservoir species distributions as predictor variables, with only linear features allowed to be included in the model. The fact that a model with 'auto features' did not perform significantly better than a model with only linear features may be somewhat surprising given the often complex relationships between species' occurrence and environmental variables. In our case, however, this complexity of biotic and abiotic interactions may be sufficiently captured by the putative reservoir species' distributions, and result in relatively simple, monotonic relationships between reservoir and human MPX (note that features are similar to basis functions, and linear features can thus result in non-linear, monotonic response curves; see also [35]). Human MPX was predicted to be present in large parts of Central and Western Africa (Fig. 3A). In a model using reservoir species distributions that were based on both climate and remote sensing variables, Thomas's rope squirrel (*F. anerythrus*) was the

Table 2. Results of Maxent runs for reservoir species.

Species common name	Species scientific name	N sites	AUC	Test AUC
African brush-tailed porcupine	*Atherurus africanus*	52	0.958	0.943
Long-tailed pangolin	*Manis tetradactyla*	100	0.893	0.810
Tree pangolin	*Manis tricuspis*	100	0.919	0.846
Demidoff's galago	*Galagoides demidoff*	100	0.923	0.855
Greater cane rat	*Thryonomys swinderianus*	65	0.888	0.858
Gambian rat	*Cricetomys gambianus*	127	0.911	0.870
Wolf's monkey	*Cercopithecus wolfi*	100	0.930	0.852
Grey-cheeked mangabey	*Lophocebus albigena*	100	0.923	0.865
Thomas's rope squirrel	*Funisciurus anerythrus*	75	0.951	0.937
Congo rope squirrel	*Funisciurus congicus*	48	0.938	0.930
Fire-footed rope squirrel	*Funisciurus pyrropus*	94	0.956	0.915

most important variable when used on its own (Fig. 4). In the same model, the Congo rope squirrel (*F. congicus*) was the most important when omitted from the model, indicating that it contained unique information not captured by other reservoir species distributions. Variable importance changed when the reservoir species distributions were based on only climate data. In this case, rope squirrels were not the most important species, and instead the pangolin *Manis tetradactyla* was most important, both used on its own, or when omitted from the model. A close second in determining human MPX distribution in this model was Wolf's monkey (*Cercopithecus wolfi*). This difference in variable importance between the two models can likely be explained by the more fine-scale resolution of the remote sensing variables used in the former model as compared to the climate data used in the latter. The fine resolution may more effectively capture small-scale heterogeneity in human MPX occurrence that is not picked up by lower resolution variables. Nevertheless, the broader-scale predictions of human MPX occurrence were highly concordant among the different data sets used (Fig. S3), despite the fact that several reservoir models were more complex due to interactions among predictor variables. We therefore used the least complex model – which used reservoir species based on climate, and allowed for linear features only – in subsequent projections of human MPX occurrence from future reservoir species distributions.

Tests of human MPX model performance

We subdivided our study region into western ($n = 45$) and eastern ($n = 48$) sites. AUC values; balance thresholds, which balances the training omission rate, cumulative threshold, and fractional predicted area; and proportion of total area above the balance threshold for each of the models are shown in Table 5. Tests of model performance presented here are particularly conservative, because pseudo-absence background points were drawn from the entire study region, including the area where MPX positive locations were omitted (i.e. the western or eastern

Table 3. Results of model performance tests for potential reservoir species using spatial subdivisions.

Species name	AUC		% area>threshold		P binomial test	
	W	E	W	E	W>E	E>W
African brush-tailed porcupine (*A. africanus*)	0.988	0.976	0.157	0.130	**<0.001**	**<0.001**
Long-tailed pangolin (*M. tetradactyla*)	0.944	0.925	0.252	0.370	**<0.001**	**<0.001**
Tree pangolin (*M. tricuspis*)	0.944	0.898	0.278	0.594	**<0.001**	**<0.001**
Demidoff's galago (*G. demidoff*)	0.943	0.902	0.274	0.466	**<0.001**	**<0.001**
Greater cane rat (*T. swinderianus*)	0.938	0.919	0.499	0.484	**0.003**	**0.030**
Gambian rat (*C. gambianus*)	0.936	0.969	0.370	0.234	**<0.001**	**<0.001**
Wolf's monkey (*C. wolfi*)	0.949	0.908	0.242	0.479	**<0.001**	**<0.001**
Grey-cheeked mangabey (*L. albigena*)	0.951	0.900	0.261	0.511	**<0.001**	**<0.001**
Thomas's rope squirrel (*F. anerythrus*)	0.947	0.978	0.278	0.113	**<0.001**	**<0.001**
Congo rope squirrel (*F. congicus*)	0.968	0.946	0.331	0.196	**<0.001**	**0.003**
Fire-footed rope squirrel (*F. pyrropus*)	0.969	0.978	0.295	0.225	0.374	**<0.001**

Shown are AUC values for models using points from the west (W) or the east (E); the percentage of the total study area predicted to be over the balance threshold; and the p-values for one-tailed binomial tests between the number of test points predicted to be suitable for the species and the fractional predicted area over the balance threshold. P-values in bold indicate models that performed significantly better than random; that is, significantly more test points were predicted to be suitable for the species than expected based on the fraction of the total area predicted to be suitable. W>E = western points were used for training, eastern for testing; E>W = eastern points were used for training, and western for testing.

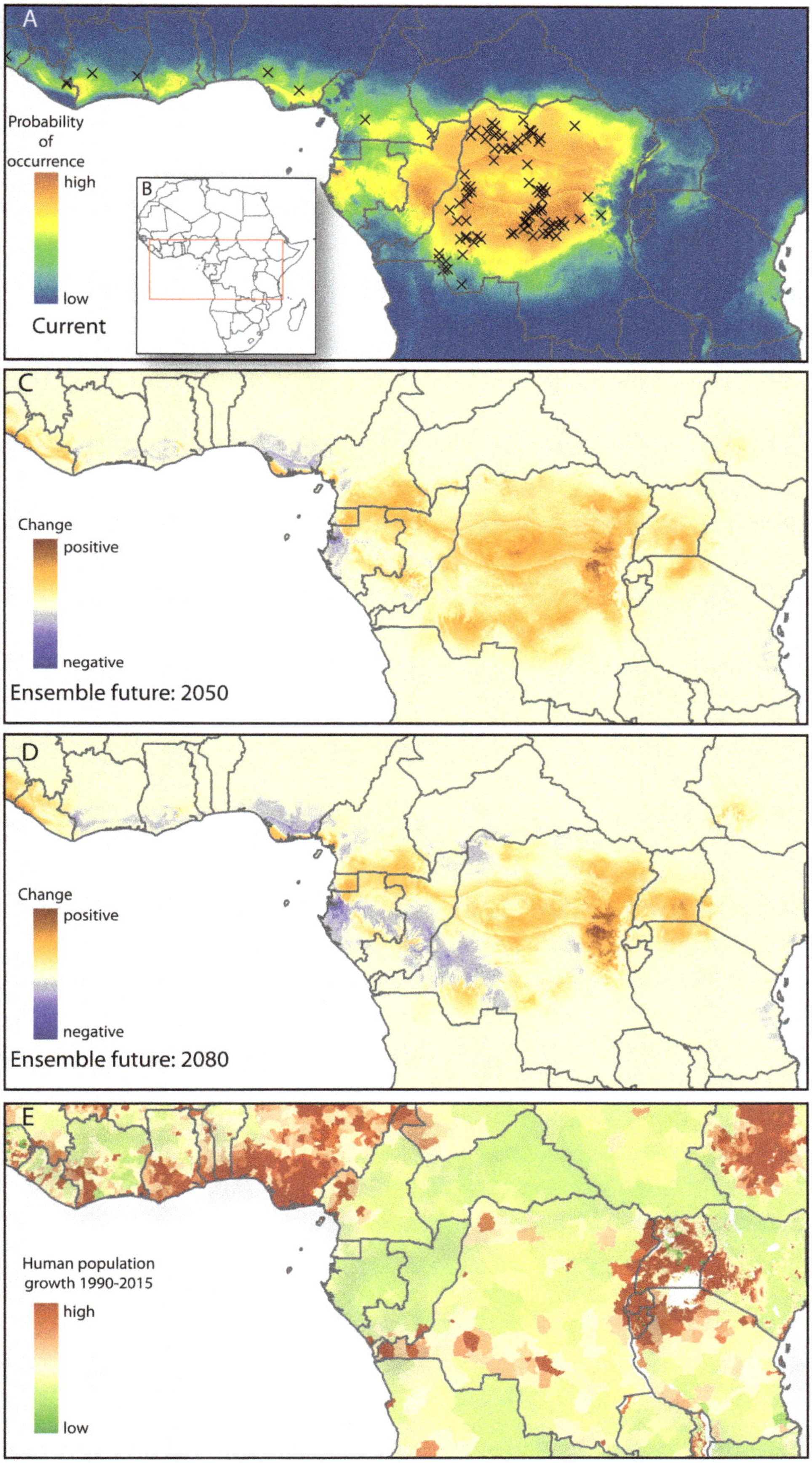
A
Probability of occurrence
high
low
Current
B
C
Change
positive
negative
Ensemble future: 2050
D
Change
positive
negative
Ensemble future: 2080
E
Human population growth 1990-2015
high
low

Figure 3. Observed and predicted human MPX occurrence. (A) Maxent prediction of human MPX occurrence under contemporary climate conditions, using reservoir species as predictor variables, with only linear 'features' (i.e. only linear coefficients are used for each predictor) allowed in the model. Colors indicate the probability of MPX occurrence, with cooler colors indicating lower probabilities and warmer colors higher probabilities (see color bar). Crosses indicate the reduced set of observed cases of MPX in humans (see Materials and Methods). (B) Study area. (C) Average projected change in probability of human MPX occurrence for eight climate change scenarios for 2050. (D) Average projected change in probability of human MPX occurrence under eight climate change scenarios for 2080. Colors in (C) and (D) indicate the change in probability of occurrence, with cooler colors indicating a decrease, and warmer colors an increase. (E) Projected human population growth from 1990–2015. The growth of the human population in the eastern DRC and Uganda, which borders North Kivu province in the eastern DRC, will be among the greatest of any region in central Africa. For areas shown in red, there has been a large increase in the human population since 1990 and further growth is forecast in the imminent future. Projections for later in the 21 century are qualitatively similar. For example, the population density of Uganda is forecast to increase 171% by 2050 [56]. A potential epicenter of future MPX outbreaks in the eastern DRC could arise from the confluence of population growth and increased MPX prevalence under climate change.

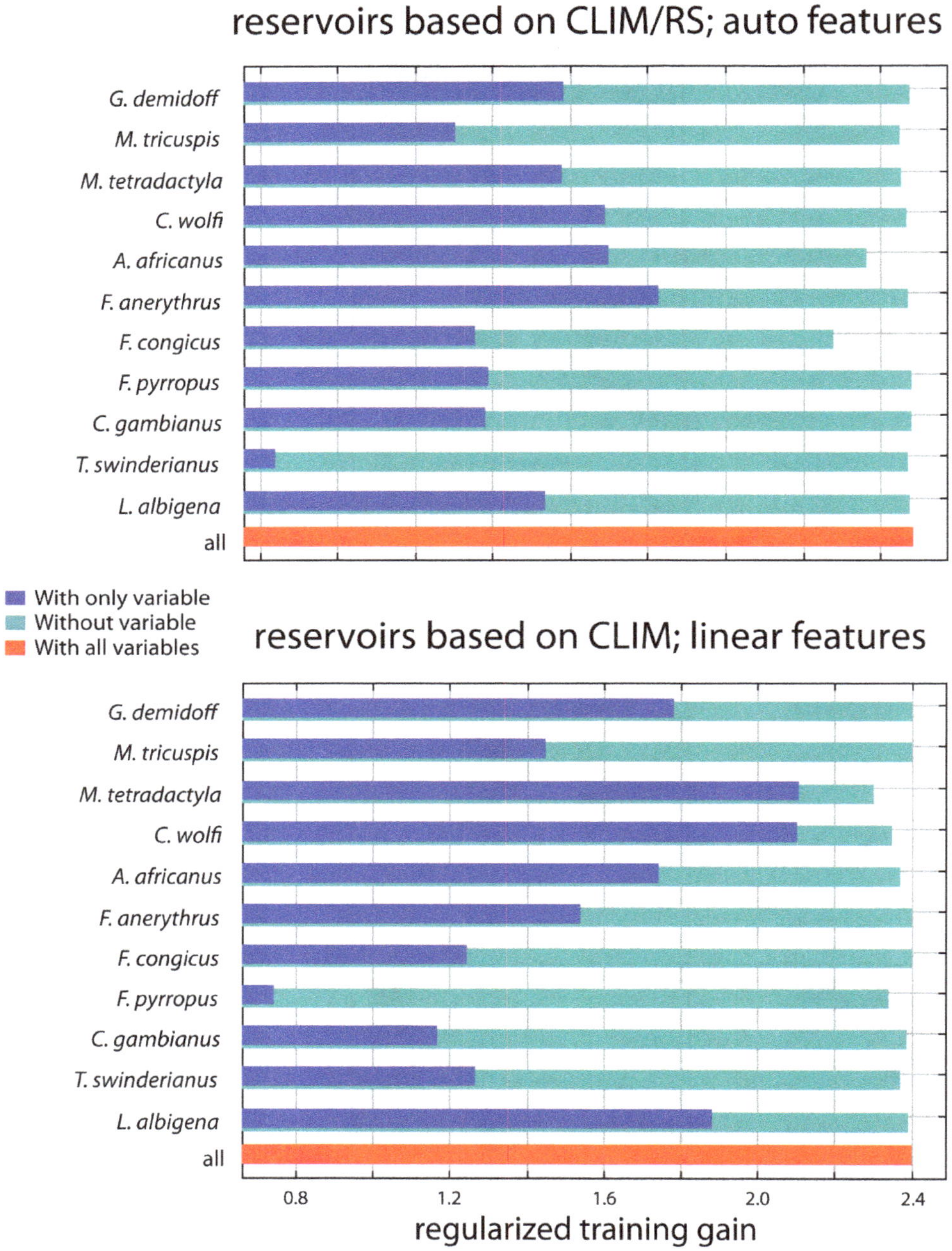

Figure 4. Variable importance of Maxent models of MPXV occurrence. Results are shown for a model that included reservoir species based on climate and remote sensing variables with all 'features' allowed (auto features, i.e. linear and quadratic coefficients can be used for each predictor, as well as step functions and interactions) (top panel), and a model that included only the reservoir species, based on climate variables, and with only linear 'features' (i.e. only linear coefficients are used for each predictor) allowed in the model (bottom panel). Dark blue bars indicate test results in which only the variable in question was entered into the model, and light blue bars in which all variables except the one in question were entered. Longer dark blue bars and shorter light blue bars indicate higher variable importance.

Table 4. Results of Maxent runs for human MPX.

Predictor variable set	N sites	AUC	Test AUC	AICc
Res(CLIM/RS)+CLIM+RS	93	0.983	0.973	2924.6
Res(CLIM/RS)	93	0.989	0.974	2793.9
Res(CLIM)+CLIM	93	0.977	0.974	2784.9
Res(CLIM)	93	0.976	0.972	2797.2
Res(CLIM) – linear features	93	0.969	0.969	**2777.3**

Results are shown for models with different sets of input predictor variables. CLIM = climate variables; RS = remote sensing variables; Res() indicate reservoir species based on the variables in brackets; linear features = only linear features allowed in the Maxent model.

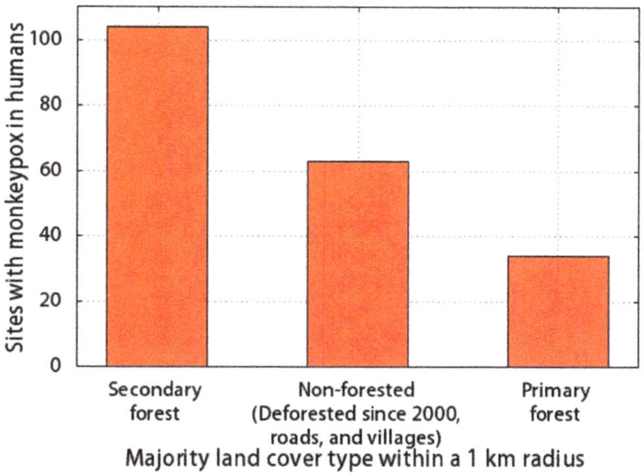

Figure 5. Land cover at sites with MPX infections in humans in Sankuru (2005–07). The land cover data are from [58].

part of our study region). The threshold-dependent testing approach suggested that models trained on western sites predicted MPXV occurrence in their corresponding eastern test sites and vice versa significantly better than random (MPXV was predicted to be present in 31 out of 45 western sites when eastern sites were used as training data, and in 48 out of 48 eastern sites when western sites were used as training; one-tailed binomial tests, $p<0.001$).

In addition, the threshold-independent test also suggested that a model based on western sites performed well in predicting suitability in eastern sites (1-tailed Wilcoxon signed rank test between models using western versus eastern sites as training sites: $Z = 120.0$, $p = 0.889$, suggesting there is no difference in predicted suitability of eastern sites between models using eastern or western sites as training data). However, eastern sites did not perform well in predicting the suitability for sites in the west (1-tailed Wilcoxon signed rank test between models using eastern versus western sites as training sites: $Z = 500.5$, $p = 0.0001$). This result may be attributed to the fact that the western sites comprise those from Central Africa, but the eastern sites do not include any from West Africa, where environmental conditions are rather different. Thus, the western sites span much of the environmental heterogeneity observed in the eastern sites, but the reverse is not true.

Future projections

To project the distribution of human MPX under future climate conditions, we first projected reservoir species distributions on climate variables from the eight climate change scenarios each for 2050 and 2080. Changes in reservoir species distributions varied between decreasing or increasing suitability and in some cases showed considerable shifts in the geographic range (Fig. S2). For instance, the size of the distribution of the tree pangolin *M. tricuspis* was projected to decrease. In contrast, changing climate was projected to result in drastic range expansions of the rope squirrel *F. pyrropus*, whereas the African brush-tailed porcupine (*A. africanus*) was projected to retain nearly the same distribution. Very little to no clamping was detected, suggesting that the projected future environmental conditions in our study region were not more extreme than any observed currently. Subsequent projections for human MPX occurrence (based on projections of current reservoir species distributions on future climate conditions) showed strong patterns of change, where some regions were projected to become less suitable, and others to become more suitable (Fig. 3c, d for multi-model ensemble change maps and Fig. S4 for projections on each climate change scenario). Interestingly, the future projections suggest an overall increase in the geographic range of human MPX for 2050–2060, whereas a subsequent decrease is projected for areas in the southwest by 2080–2090. More specifically, by the turn of the century, human MPX is projected to become less common in much of western Africa, southern Gabon, central Republic of Congo, and parts of DRC. In contrast, increased suitability may result in a range shift eastwards in DRC and into Uganda, southwestern Kenya, and northwestern Tanzania. This shift is consistently predicted for all climate change scenarios (Fig. S4). In addition, southeastern Cameroon, northern Gabon, and Equatorial Guinea are projected to become much more suitable for human MPX than under current climate conditions.

The teardrop shape in central DRC (Fig. 3c–d) is a result of high precipitation in the driest quarter (Bio17) and potentially the bull's eye effect, an artifact of the interpolation method used in the WorldClim dataset in combination with relatively sparse climate station network in the area (1–25 stations per 0.5° grid cell, which is approximately 2500 km^2); yet the sharp transitions are caused by the interaction of Bio17 and other climate variables.

Table 5. Results of model performance tests using spatial subdivisions.

Training set	Test set	N sites	AUC	Balance threshold	% area>balance threshold	P binomial test*	P Wilcoxon test*
West	East	45	0.960	0.048	0.183	**<0.001**	0.889
East	West	48	0.986	0.017	0.071	**<0.001**	**<0.001**

*The binomial test compares model performance to random performance, whereas the Wilcoxon signed rank test compares the predicted and observed logistic probabilities. Thus, good model performance is indicated by significant test results in the case of the binomial test, but non-significant results in the case of the Wilcoxon signed rank tests.

Like WorldClim, the CRU climate data set, which is constructed from ground-based weather stations, also predicts that central DRC is currently wetter than western or eastern DRC [36]. However, satellite observations suggest that central DRC is drier than the west or east, which could be due to the divergence of air near the sides of the Congo River leading to low precipitation near the river [37]. Although future field work and regional atmospheric models are needed to validate the WorldClim rainfall estimates utilized here, the use of this data appears to be justified to the extent that it represents the best ground-based precipitation estimates currently available for the DRC.

Discussion

Three-quarters of the emerging pathogens in humans are zoonotic in origin [38]. Thus, there is an urgent need to elucidate the geographic distributions of wildlife diseases. Our present analysis is a contribution toward the development of a methodology that can be used to predict the distribution of any virus with a wild mammal reservoir. This approach can potentially be adapted to a broad range of other emerging pathogens with mammalian reservoirs including hantavirus and Lassa virus [16,39]. Furthermore, the approach reported here can be utilized to design inventories to identify the most important reservoirs for a range of pathogens in tropical forests. In particular, the distribution modeling techniques developed here can focus surveillance to a subset of species from an initial large set of potential reservoir species.

Assessing the effects of present-day climate and forest clearing on MPX outbreaks in DRC

Our local-scale study in Sankuru supports the hypothesis that climate and deforestation are both important drivers of MPX insofar as the model that included covariates related to climate and forest clearing provided the best fit to the data. Here we hypothesize about two mechanisms by which forest clearing and climatic factors might affect the transmission of MPXV from wildlife to humans. First, the clearing of primary forest by humans could increase habitats that are optimal for wildlife species that carry the virus, leading to higher abundances of these species, more frequent contact between humans and wildlife carriers, and increased transmission of the virus to humans. We note that increased habitat for wildlife hosts due to the clearing of primary forests was implicated as one of the main drivers behind the range expansion of Japanese Encephalitis Virus in Asia in the mid-1990s [40]. In Sankuru, the conversion of primary forest to agricultural fields could increase the habitat for MPX hosts such as rope squirrels, which inhabit secondary forest and ecotones between agricultural fields and forests [19,20]. According to this hypothesis, human MPX infections should occur more often near secondary forests and recently deforested areas than near intact primary forest. Furthermore, we might expect that people who spend time in land that has recently been deforested or in land that represents a transition zone between forest and agricultural land would be infected with MPXV at a higher rate than other members of the population. The data on MPX cases in humans in Sankuru support this hypothesis to the extent that 83% of infections in humans were at sites surrounded either by secondary forest or formerly forested land that has been cleared in the past decade (Fig. 5, Fig. S5). In addition, boys, who typically play in agricultural areas along the perimeter of secondary forests, where they trap and eat small mammals, have the highest incidence of MPX in Sankuru [9].

Second, climatic factors could lead to increased MPX infections in humans in Sankuru because drought in forested areas could compel sylvan mammals that carry MPX to disperse into non-forested areas such as human settlements to forage. During these foraging bouts, small mammals infected with MPX could transmit the virus to villagers. If this hypothesis were correct, we would expect there to be more cases of MPX in secteurs where there has been severe drought in forested areas in the past decade. The data from Sankuru tend to support this hypothesis to the extent that Ngandu and Ukulungu, which emerged as the primary MPX hotspot, have had high water deficits in forested areas over the past decade, based on satellite observations (Fig. 2b, Text S1; Fig. S6). However, these results must be interpreted cautiously because additional ground-truthing is needed to confirm that water deficits estimated using remote sensing provide accurate estimates of canopy moisture in central DRC. At the national scale, there is a west-to-east gradient within DRC with respect to canopy water deficits measured by satellite. In western DRC, there is a significant negative correlation between canopy water deficits and the Atlantic Multidecadal Oscillation (AMO), which is a circulation pattern in the Atlantic Ocean that is correlated with climate change [41,42]. Western DRC has a distinct strain of MPXV, whereas in eastern DRC, where there is no association between climate change and canopy water deficits estimated via remote sensing, a different MPXV strain is prevalent (Fig. S7). This could indicate that different species are the primary reservoirs in different habitat types. The data presented here are suggestive of a possible relationship between forest canopy moisture, climate change, and MPX, which merits further investigation.

Finally, we note that the two mechanisms considered here, drought and deforestation, may not be exclusive. A hypothesis that awaits confirmation via future field work is that drought could trigger desiccation and tree dieback in primary forests in DRC and that this might increase secondary forest that is optimal habitat for wildlife that carry MPXV. Although drought has been shown to trigger tree die-offs in the Amazon [43], additional field campaigns are needed to test whether the conversion of primary to secondary forest is tied to drought in the Congo Basin. An alternative hypothesis is that DRC's forests may have evolved resistance to water deficits because multidecadal droughts have occurred frequently in sub-Saharan Africa during the Holocene period [13]. If our preliminary data that are suggestive of a relationship between present day climate and MPX can be confirmed, this will provide further impetus for predicting how climate change in the coming decades will trigger increases in MPX infections in humans. The development of high resolution maps of future forest cover in Africa remains an important area for future research. However, since future climate maps are already available for Central Africa [44], we developed forecasts of the distribution of MPX under a range of climate scenarios for the 2050s and 2080s.

Assessing the role of present and future climate change on MPX outbreaks at the regional scale in Tropical Africa

Our study uses the most extensive incidence data set of MPX occurrence yet developed, and is novel in using candidate reservoir species ranges to assess their potential importance for MPXV infections in humans, and in projecting shifts in the geographic distribution of human MPX under climate change. Our models for MPXV occurrence under current climate conditions, and the importance of temperature and precipitation variables are broadly consistent with those found in previous studies [23,45], providing confidence that our models under current climate conditions are useful for subsequent projection onto future climate variables, offering new insight in potential range shifts of human MPX.

Models for the occurrence of human MPX using reservoir species were nearly identical to those using only climate variables. This suggests that the MPXV and reservoir species share a unique habitat that is largely captured by environmental variables. Nevertheless, reservoir species are likely to be more biologically relevant with respect to the transmission of MPXV to humans. This is supported by the facts that MPX outbreaks have recently been recorded in Sudan [10,46], where it might be endemic, and the US, where MPXV was incidentally introduced through imported African rope squirrels (*Funisciurus* sp.), dormice (*Graphiurus* sp.), and giant pouched rats (*Cricetomys* sp.), and subsequently transmitted to prairie dogs (*Cynomys* sp.) [47,48,49]. In these areas, where MPX has not been previously recorded, environmental conditions are highly dissimilar to those seen in the African tropical forest area, yet locally abundant and immunologically naive rodent populations may be effective MPXV reservoirs once exposed to the virus. This underlines the relation between human MPX and current and potentially new reservoir species as a function of their habitat requirements and shifts due to climate change. Despite the importance of reservoir species in MPXV transmission, it is plausible that certain environmental conditions directly favor the survival or proliferation of the virus itself. For instance, specific temperature and light regimes increase the time that pox virus survives outside a host (reviewed in [50]). Experimental studies suggested that transmission of MPXV was via bodily excrements [18,19,49] and it is likely that environmental conditions affect the efficiency of these transmission mechanisms.

Important reservoir species

We used a set of candidate MPXV reservoir species to identify those that are potentially the most important. It is possible that species not included in our list of candidate reservoir species are also, or even more important reservoirs for MPXV, but our list contains most of the species that have previously been implicated in MPX occurrence (although not all of them have been shown to be infected with MPXV, and only one species (*F. anerythrus*) has been found infected in the wild [51]). In a human MPX model where reservoir species were based on both climate and remote sensing variables, presence of rope squirrels (*Funisciurus* sp.) was highly important in determining MPXV occurrence (Fig. 4). This is consistent with previous modeling [30] and serological screening results [19,20]. When reservoir species distributions were based on climate variables only, a model for human MPX suggested that the pangolin *M. tetradactyla* and the monkey *C. wolfi* were important predictor variables, as well as the rope squirrel *F. pyrropus*, which contained unique information not present in other reservoir species distributions (Fig. 4). Yet, it remains unclear whether the pangolin *M. tetradactyla* is an important reservoir species, as there is little evidence for MPXV infections [8], and contact with pangolins was not identified as a risk factor for human MPX infection [9]. We speculate that its importance could be the result of shared habitat requirements with true reservoir species. Such cross-correlations between species' distributions could result in the false identification of potential reservoirs that are in fact not a reservoir for MPXV. They are, however, unlikely to result in false negatives, because species distributions that do not correlate with the distribution of human MPX cases will all be regarded of low importance. As such, our results provide a basis for identification of suspected reservoir species that should be targeted for intensive screening efforts, including the pangolin *M. tetradactyla*.

Future projections of human MPX

Two of the key assumptions in the procedure of projecting current reservoir species distributions on future climate layers, and their subsequent use in modeling human MPX occurrence, is that current reservoir species will continue to be key reservoirs in the future, and that their environmental requirements will remain the same over the course of the next 50–100 years, which is the approximate time span used in many future climate models. These assumptions seem justified given that this period is relatively short to allow for major evolutionary responses to climate change in vertebrates. Nonetheless, the potential for MPXV to jump species boundaries into new hosts remains [52], subsequently affecting areas where our models project low MPXV suitability in the future. An additional assumption of our projections for human MPX is that the interaction between humans and reservoir species remains the same. If human populations in the future become less reliant on bush meat for protein, the number of human MPX cases may decrease – though this will be challenged by the continuing decline in population immunity as time passes since the cessation of smallpox vaccination. Conversely, if the dependence on bush meat remains at current levels or increases, it is plausible that the rise in human MPX cases [9] will continue in the future.

The consensus emerging from the eight global climate models considered here is that the geographic distribution of human MPX occurrence will change considerably across Tropical Africa. Areas in western Africa, where a less virulent strain of MPXV is present, are projected to become less suitable for the virus. Projections in West Africa are tentative, however, because only few recorded presence sites were available, and the West African MPXV strain may be associated with a different set of reservoir species, each with its own specific habitat requirements. Central regions in DRC and the Republic of Congo are also projected to experience lower levels of human MPX occurrence under future climate conditions. However, areas in the northern part of the contemporary range of human MPX will likely become more suitable as a result of changes in temperature seasonality (Bio4; mean annual standard deviation in temperature), which was an important variable in several reservoir species' niche models. In addition, projections derived from all included climate models suggest a shift eastward into areas where MPXV has not yet been recorded. Interestingly, and suggestive of the significance of our results, the incidence of human MPX in DRC has been shown to have increased over the past three decades beyond the level expected by the higher percentage of the population susceptible to smallpox, after cessation of the smallpox vaccination program in 1980 (unpublished). Although the cause of this increase remains unclear, it is concordant with the projected rise in suitability under changing climate conditions. Moreover, new reports confirm human cases of MPXV in southern Central African Republic associated with bushmeat consumption [53].

The question whether MPXV will be transmitted to humans in areas that will become more favorable to MPXV reservoir species is at least partially dependent on human behavior. For example, although our models suggest an increased risk for human MPX in Uganda in the future, Ugandan communities are generally more pastoral and less reliant on hunting animals that may act as MPXV reservoirs than communities in DRC. Hence, the spread of MPXV reservoirs into new areas may not necessarily result in higher incidence of MPX in humans, as long as contact between humans and reservoirs can be avoided. On the other hand, the projected increased suitability for MPXV in eastern DRC may have severe consequences for its human population. This is an area with high human population densities that, in contrast to communities in bordering Uganda, are largely reliant on bush meat for protein. Furthermore, the area's population is predicted to continue growing based on extrapolations from recent censuses using annual growth rate estimates (Fig. 3E; [54]). In particular, the projected population structure indicates a high proportion of individuals under 20 years old, which are at increased risk of MPX. For example, boys have the highest incidence of MPX, probably due to hunting behavior [55]. Although projections for human population growth at spatial scales smaller than the country size are highly uncertain for periods beyond 2015, it seems reasonable to suspect that growth will continue in areas where high growth rates are currently projected. According to United Nations projections, by 2050 there will be a 36% increase in male children in DRC compared to 2009 numbers [56]. In light of this, an epicenter of future MPX outbreaks and increased human-to-human transmission could potentially arise in eastern DRC. This situation is aggravated by the civil unrest in the region. First, the unstable situation may cause people to move from urban to rural areas and into the forest, where contact with animals infected with MPXV is more likely. Second, health care may be poor or non-existent. Third, surveillance in areas of civil unrest is severely limited, as was the case during the 1996 to 1997 outbreak of what was potentially human MPX [17] or chickenpox [57]. Although our future projections are for a period 40–80 years from now, there is no clear indication that the trend of increasing human MPX incidence during the past 30 years will cease in the near future. Thus, there is a need for continued and increased monitoring of human MPXV infections, as well as more detailed efforts to identify reservoir species and characterize their ecology, and to understand the risk factors underlying animal-to-human transmission. Strategies to reduce the risks of MPX transmission to humans should be developed and implemented. Unfortunately, the places where MPX occurs most frequently are in villages that are remote and difficult to access. Disease surveillance is challenging even for relatively well funded programs, and the development of a functioning health system is of utmost importance, yet complex and not easily resolved. Before now, little was known about potential range shifts of MPXV under climate change. By identifying high-risk areas, we hope that the limited resources can be utilized more effectively. At the same time, it is essential to understand the risk posed by human-to-human transmission as the population grows and becomes more susceptible.

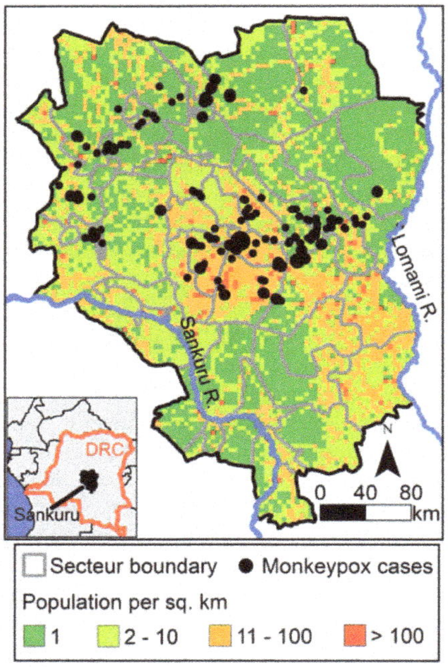

Figure 6. Sankuru, DRC. Site of the local scale study of the effects of climate and deforestation on MPX occurrence.

Our finding that deforestation has been a significant driving factor of MPX transmission in Sankuru suggests that, in addition to climate change, land use changes may also influence reservoir species distributions and abundance. The majority of the areas where human MPX is projected to increase are covered in forest, where the putative reservoir species naturally occur (Figs. S8, S9). Land conversion in these areas is rampant, where primary forest is turned into secondary forest due to timber extraction, or into agricultural fields. This type of land cover change may result in increased habitat suitability for some key reservoir species. For instance, rope squirrels (*Funisciurus* sp.), potentially one of the most important primary hosts, have been reported to be more abundant in converted areas that are heavily used by humans than in primary forest [18]. These may also be areas where other rodents are more abundant than in primary forests, due to higher availability of their major food sources, seeds and seedlings. This may have consequences for the distribution of MPXV in the future, as continued deforestation may result in range expansions or higher abundance of important reservoir species. The complex relationship between virus, reservoirs, and environmental conditions further emphasize the need for detailed studies regarding viral and host ecology.

In sum, our study contributes to a better understanding of the potential impacts of climate change and deforestation on the distribution of human MPX in Tropical Africa. Our results support previous findings that rope squirrels (*Funisciurus* sp.) may be important MPXV reservoirs. In addition, our models also suggest that the monkeys *C. wolfi* and *C. albigena*, as well as locally abundant pangolin (*M. tetradactyla*) merit investigation as potential reservoirs. Although we have not established these species as MPXV reservoirs based on active screening, our models suggest that closer investigation of these species is warranted. In contrast, Gambian rat (*C. gambianus*), greater cane rat (*T. swinderianus*), and tree pangolin (*M. tricuspis*) are not supported as likely reservoirs by our model. Our predictive maps of potential future MPX occurrence are helpful in identifying areas with particularly suitable future environmental conditions for human MPX reservoirs, which may be high-priority areas for monitoring by public health decision-makers. Areas where the probability of MPXV occurrence is projected to increase should be closely monitored. In addition, the identification of locally abundant potential reservoir species and behavioral risk factors will be helpful in assessing the human health hazards imposed by shifts in the range and prevalence of MPXV as a result of changing environmental conditions.

Materials and Methods

Assessing the effects of present-day climate and forest clearing on MPX outbreaks at the local scale in DRC

To investigate the contributions of recent climate change and deforestation to the transmission of MPX, we carried out a local scale study in Sankuru, DRC (Fig. 6). We selected Sankuru because MPX surveillance has been carried out recently in the district [9] and maps of forest clearing since 2000 are also available for the area [58]. Thus, deforestation is defined as a relatively recent event. In DRC, and over such a recent time frame, deforestation typically does not result in urbanization, but in either secondary-growth forest or agricultural landscapes. It is these kinds of landscape where close contact between humans and reservoir

Table 6. Climatic and land cover variables used to analyze MPX transmission.

Scale	Variable	Year(s)	Satellite	Reference(s)
Sankuru	Canopy moisture	2000–09	QuikSCAT	[71]
	Correlation between MWD and AMO	1998–2010	TRMM	[86]
	Deforestation	2000–10	Landsat	[58]
	Temperature of the coldest quarter	2000–10	MODIS	[87,88]
	Temperature of the warmest quarter	2000–10	MODIS	[87,88]
Tropical Africa	Vegetation greenness (NDVI)	2001–05	MODIS	[89]
	Percent tree cover	2001–05	MODIS	[70]
	Canopy moisture	2000–09	QuikSCAT	[71]
	Elevation	2000	SRTM	[90]
	Mean temperature (Bio 1)	1950–2000	Interpolated from weather stations	[67]
	Temperature range (Bio 2)	1950–2000	Interpolated from weather stations	[67]
	Temperature seasonality (Bio 4)	1950–2000	Interpolated from weather stations	[67]
	Temperature of warmest month (Bio 5)	1950–2000	Interpolated from weather stations	[67]
	Mean precipitation (Bio 12)	1950–2000	Interpolated from weather stations	[67]
	Precipitation seasonality (Bio 16)	1950–2000	Interpolated from weather stations	[67]
	Precipitation of the driest quarter (Bio 17)	1950–2000	Interpolated from weather stations	[67]

AMO = Atlantic Multidecadal Oscillation, MWD = maximum water deficit, NDVI = normalized difference vegetation index.

species is likely to be most frequent. Next, we analyzed remotely sensed data to generate maps of Sankuru's climate over the past ten years (Fig. S6), developing maps of temperature and canopy moisture (Table 6). We examined these climatic variables because temperature and wetness are important ecological drivers of rodent-borne diseases in DRC [23,59]. We also calculated the correlation between drought and the Atlantic Multidecadal Oscillation (AMO; see Text S1). This provides an indirect measure of the strength of recent climate change in Sankuru [41]. We combined the climatic data with maps of deforestation in Sankuru to estimate the relative importance of forest clearing and climate as drivers of MPX infections in humans.

Our objective was to identify regions with significantly more MPX cases in humans than would be expected by chance (hereafter "MPX hotspots"). Analyses were carried out at the level of administrative regions within Sankuru called "secteurs" (Fig. 6). The climate and deforestation data were for this purpose aggregated to the scale of secteurs in ArcGIS 9.3 (ESRI, Redlands, CA) and utilized as covariates in a statistical model implemented in the SaTScan 9.0 software package SaTScan tests the null hypothesis that the number of MPX cases in each secteur is Poisson distributed and proportional to human population size [31,32,60,61]. The alternative hypothesis is that the risk of MPX within a secteur or secteurs is elevated compared to other secteurs. SaTScan searches for secteurs that have anomalously high risk of MPX by scanning an elliptical window across Sankuru and calculating the number of observed and expected cases inside the window. A variety of ellipses with various angles and shapes were tested. The MPX primary hotspot is the window with the maximum likelihood and the secondary hotspot is the window with the second highest likelihood (see Kulldorff [60] for the likelihood function). A hotspot is an area that has a significantly higher number of disease cases than would be expected by chance, while accounting for the geographic area of the hotspot and covariates [60,62,63]. In our analyses, each hotspot typically contained several secteurs, but there was only one primary hotspot and one secondary hotspot. We compared the models that included climate, deforestation, and their interaction to a reference model with no covariates.

Assessing the role of present-day climate on MPX outbreaks at the regional scale in Tropical Africa

Next, we expanded the local scale study to map the geographic distribution of MPX throughout Tropical Africa. Like the local scale study in Sankuru, our analysis at the regional scale utilized climatic variables. However, this analysis did not consider deforestation because maps of forest clearing are not available for the entire region. In addition, the regional scale study examined the geographic distributions of wildlife hosts to predict MPX risk.

We used ecological niche modeling techniques to model reservoir species and human MPX distributions across Tropical Africa, with special emphasis on Central Africa. We created a total of 198 models with different combinations of predictor variables for candidate reservoir species and human MPX in a two-step approach (Fig. 1): 1) model reservoir species distributions using environmental data (Table 6) under current and future climate conditions, and 2) use the predicted current and projected future reservoir species distributions to model the distribution of human MPX. Because future projections are entirely based upon predicted changes in the climate, and do not take into account land cover changes, we first ran models for contemporary climate conditions excluding remotely sensed habitat characteristics to establish the current climatic determinants of reservoir species occurrence. These climate-only models were compared to models including both climate as well as remote sensing variables to assess the influence of omitting remotely sensed habitat variables, and the validity of the climate-only models and future projections. Because remote sensing variables may add valuable information to the distribution models, future projections (see below) were considered to be useful only when models using current climate broadly recovered the same distribution of reservoir species as models using both current climate as well as habitat. We expected to see that climate-only models would show a more generalized and potentially slightly broader distribution due to the relatively low native resolution and less small-scale spatial heterogeneity of the climate variables compared to remotely sensed data layers.

After modeling the distributions of candidate reservoir species, we modeled human MPX distributions under current conditions and assessed the concordance among models using the following predictor datasets (Fig. 1): 1) reservoir species distributions (based on climate and remote sensing variables) and in addition the climate and remote sensing variables (the full model), which may contain additive information on top of the reservoir species distributions that were also based on both types of data; 2) only reservoir species distributions (based on climate and remote sensing variables); 3) reservoir species distributions (based on climate variables) plus climate variables; and 4) only reservoir species distributions (based on climate variables). To model future human MPX distributions under different climate change scenarios, we projected reservoir species distributions onto future climate variables. We then used the results as input for models of human MPX under approaches 3) and 4) above, since future remote sensing variables are not available. Below, we describe the data acquisition and modeling approaches in more detail.

Human MPX occurrence data

Human MPX occurrence data for this study was partly collected during an ongoing study in DRC and partly derived from a World Health Organization (WHO) based intensified disease surveillance program during the 1970s and 1980s. Since 2001, UCLA, in collaboration with the WHO, the National Institute of Biomedical Research, and the Ministry of Health in Kinshasa have been actively conducting MPX disease surveillance in humans in DRC. As part of this screening effort, disease surveillance has been intensified since 2004. The study methods are described in detail in [9,64]. Briefly, we initiated an intensified active disease surveillance program in 12 health zones of the Sankuru district. This district was selected for intensified surveillance because the majority of reported MPX cases since 2001 were in this area, and it was a region of epidemiologic priority for the WHO from 1981–1986. The northern part of the district consists mainly of lowland tropical forest, while the southeast has a more varied terrain characterized by ecotone habitats consisting of mosaics of forest, grassland, and woody savannah. The economy of Sankuru is largely based on traditional agriculture and hunting. As a consequence, contact of people with the natural environment is intimate: the majority of villages, surrounded by traditional fields, are located in clearings in the forest, and virtually all protein is obtained from hunting locally available wild animals, of which monkeys and rodents are among the most commonly hunted species.

The surveillance effort was relatively even across the Sankuru district. Trained field teams traveled regularly in their respective health zones, encouraging local communities to report suspected cases of MPX and examining and documenting reported cases. Crusted scabs and vesicle fluids were inoculated onto MA104-cells, and presence of MPXV in samples was confirmed by sequencing the entire open reading frame of the hemagglutinin (HA) gene. The home village of each suspected MPX case was noted and its location later georeferenced by GPS. Ethical approval was obtained from the Committee on Human Research at UCLA School of Public Health and the Institutional Review Board at the Kinshasa School of Public Health (KSPH). Informed consent (and child assent in children 5–18 years of age) was obtained verbally from all participants and legal guardians of children in French and Otetela (the common local language spoken in the area). Consent was obtained verbally given the low rate of literacy in the region. Consent was documented by participants either by signature or their mark and confirmed by a witness. The verbal consent process was approved by both institutional review boards at UCLA and KSPH.

In addition to these localized data, we used locality information from human MPX cases reported by the Centers for Disease Control (CDC) and WHO in the 1970s and 1980s as part of the smallpox eradication program, and previously used in [23]. Combining data sets from the two time periods is justified in terms of our predictors, as the climate variables used are based on 50-year averages, spanning 1950–2000. Cases from the data set from the 1970s and 1980s that could not be georeferenced accurately due to incomplete or confounding location information were omitted. Location assignments for the remaining cases were accurate to at least 1 minute (~2 km) [23]. Because the combined datasets resulted in high clustering of human MPX cases in the Sankuru District, potentially biasing the distribution models to areas with high sample density, we resampled the data, and only retained sites that were at least 10 km apart, resulting in a total of 93 presence-sites. The 10 km threshold was used, because it is an order of magnitude larger than the resolution of our environmental variables, and resulted in greatly reduced clustering of human MPX positive sites. Selection of sites at distances under 10 km to be omitted from the final dataset was done at random. Because the climate variables used in the spatial predictions of reservoir species tend to vary at distances >10 km, it is unlikely that our subsampling procedure would result in a reduction of the environmental niche space included in the final dataset. Moreover, inaccuracies in georeferencing in the order of a few kilometers are unlikely to have an effect on our results.

Reservoir species data

Potential reservoir species (Table 2) were identified from previous studies showing MPXV infections in these species, or studies suggesting these species might be important in transmitting MPXV to humans [17,18,19,20]. Point locality occurrence data of potential reservoir species were collected as part of our MPX screening efforts, as well as from museum data, compiled in the online Global Biodiversity Information Facility (GBIF; http://www.gbif.org/ Accessed April 2010) and Mammal Networked Information System (MaNIS; http://manisnet.org/ Accessed April 2010) databases. Suspected MPXV-positive human subjects (see above) were asked to fill out a questionnaire, which included questions regarding recent and regular exposure to potential MPX reservoir species. For each potential reservoir species, localities from the questionnaires were added to museum data. To avoid a bias towards the Sankuru District due to high sample density in that area, we restricted the number of localities from the questionnaires to 10–15 per reservoir species. Since species identification by human subjects could not be validated, we also ran models without localities from the questionnaires. These models did not differ from those where localities from questionnaires were included, and are therefore not shown.

Museum records that were collected before 1940, or that were geo-referenced to only 0.1 decimal degree, were omitted from the final datasets used to predict the reservoir species' distributions. The use of museum records could result in inaccurate ecological niche models if they are imprecisely georeferenced or when land conversion has occurred at the sampling locality since the time of collection [65,66]. However, because our primary focus was on climate variables comprising 50-year averages (see below), and showing broad rather than small-scale spatial heterogeneity, we deemed the use of museum records appropriate. As an additional check of the accuracy of the museum records, a comparison was made to species distribution maps from the International Union for Conservation of Nature (IUCN; www.iucnredlist.org Accessed May 2010), estimated using expert knowledge. Records that were outside the ranges shown on these maps were considered to be of low confidence and were omitted from the final input data set. This procedure also at least partially mitigates issues resulting from new taxonomical insights and splitting of a single species into two or more separate species to the extent that the new taxonomy has been implemented in the IUCN database. Nevertheless, records may still include those of two closely related species. In such a case, models would likely present a wider niche than the one suitable for just one of the two sister species. However, species that are closely related may likely both be potential reservoirs of MPXV, and the possibility that models show slightly wider niches was not regarded a major concern for their subsequent use in modeling human MPX occurrence.

Few readily available museum records existed for the following potential reservoir species: the long-tailed pangolin (*Manis tricuspis*), tree pangolin (*M. tetradactyla*), Demidoff's galago (*Galagoides demidoffi*), Wolf's monkey (*Cercopithecus wolfi*), and the grey-cheeked mangabey (*Lophocebus albigena*). Although it was suggested that these species might carry MPXV, they have not been implicated as species that might act as key reservoirs in which the virus continues to exist. We resampled downloaded ecological niche models for these species from the African Mammals Databank (AMD; http://www.gisbau.uniroma1.it/amd/ Accessed April 2010) and built new distribution models using the same set of environmental variables used for the species with good coverage of museum records (see below). The AMD species distributions are the result of a combination of expert knowledge on extent of occurrence and ecological requirements, modeled species-environment relationships, and model validation at 400 sites where species presence or absence was recorded in the field. To resample the downloaded distribution models, in ArcGIS 9.3 (ESRI, Redlands, CA) we created 2500 random points across Tropical Africa, and extracted the probability values of species occurrence. For each species, we subsequently used the 100 localities with the highest probabilities to model its distribution, and visually compared the level of concordance between our results and the original AMD maps. For each species, the localities that we used fell within the top 3% of suitability.

Contemporary climate and remote sensing variables

We compiled a set of moderately high-resolution climate and satellite remote sensing variables to characterize the various habitat types in Tropical Africa (Table 6). These included eight out of a total of 19 bioclimatic variables (for the procedure to reduce the number of variables, see below) from the WorldClim database [67], which are spatially explicit estimates of annual means, seasonal extremes and degrees of seasonality in temperature and precipitation based on a 50-year climatology (1950–2000), and have been shown to represent biologically meaningful variables for characterizing habitat requirements [68].

In addition to these ground-based measurements of climate, we used satellite remote sensing data from both passive optical sensors (MODIS; https://lpdaac.usgs.gov/lpdaac/products/modis_overview/ Accessed June 2010) and active radar scatterometers (QuikScat; http://www.scp.byu.edu/data/Quikscat/SIRv2/qush/World_regions.htm Accessed June 2010) to infer a broad spectrum of ecological characteristics of the land surface. From the MODIS archive, we used the monthly Normalized Difference Vegetation Index (NDVI) to infer vegetation density [69]. In addition, we used the vegetation continuous field [70] product as a measure of the percentage of tree cover. From QuikScat (QSCAT), we obtained monthly raw backscatter measurements that capture attributes related to surface moisture and roughness [71], and from the Shuttle Radar Topography Mission (SRTM), we used elevation data. Time series of the remote sensing data sources were acquired to roughly match the period of field sampling (QSCAT and tree cover from 2001; NDVI data represent means over 2000–2004). Variables with native resolutions higher (e.g. SRTM: 30 m) or lower (e.g. QSCAT: 2.25 km) than 1 km were reaggregated to a 1 km grid cell resolution in ArcGIS 9.3 (ESRI, Redlands, CA) Spatial Analyst with the 'resample' and 'aggregate' functions respectively (Table 6). This resolution is often used in ecological niche modeling at the regional or sub-continental scales, and balances resolutions from climate data, which often have coarser native resolutions, and remote sensing data, with higher native resolutions.

To improve interpretation and avoid overfitting due to cross-correlations among environmental variables, we checked for covariance among variables, and only included those with substantial unique variance (with Pearson's correlations <0.9). Various criteria were used to decide which layers of correlated pairs were retained for further analysis. These included keeping layers that are more commonly used in distribution modeling (WorldClim) or that exhibit larger contrast/variance over the study area (QSCAT) as well as having best data quality (NDVI). To assess whether the spatial resolution (i.e. the level of aggregation) influenced the cross-correlation among environmental variables, we also checked for correlations between variables at 5 km and 25 km resolutions. The results were highly similar to those obtained from correlation analyses at 1 km resolution, and none of the variables selected based on 1 km resolution analyses were correlated with Pearson's correlations >0.9.

Projections under future climate change

To assess the potential impact of future climate change on the distribution of MPXV in Central Africa, we first projected the current relationship between reservoir species and climatic conditions onto predicted future climate layers. In a second step, we projected the current relationship between MPX occurrence and reservoir species onto the projected future reservoir species distributions. To determine the range of potential shifts in MPXV distributions under different climate change scenarios, sixteen climate models from the IPCC 4th Assessment Report with differing climate sensitivities and IPCC–SRES greenhouse gas emission scenarios were used for 2050–2060 and 2080–2090 (Table S1). These IPCC projections were statistically downscaled to 1 km resolution [67] and made available by the International Centre for Tropical Agriculture (http://ccafs-climate.org/ Accessed September 2010) [44]. As the confidence intervals around climate change scenarios tend to become broader with predictions further into the future, we included the 2050–2060 predictions. Nevertheless, as the 2080–2090 predictions represent more extreme climate change scenarios, these were also included in our study. In fact, the 2080 predictions of atmospheric CO_2 concentrations may be reached much sooner, as current emissions already exceed the trajectories of the highest scenarios [72,73,74]. Thus, projections of human MPX occurrence on the 2080–2090 climate scenarios may be relevant for purposes of our study.

A comparison among the IPCC 4th Assessment climate projections shows that the direction and magnitude of the future projections are relatively consistent in our study area. Between 14–18 out of the 21 models (IPCC 4th Assessment Report) agreed on an increase in annual precipitation in the Congo Basin, lending credibility to these projections. The multi-model ensemble (based on 21 models) from the IPCC 4th Assessment Report predicts approximately 3°C temperature increase and 0–10% increase in annual precipitation in our study region.

Distribution modeling

To model the spatial heterogeneity in reservoir species and MPXV occurrence, we used the Maximum Entropy approach, a machine learning technique implemented in Maxent 3.3.3a [75]. It relies on the assumption that the incomplete empirical probability distribution of occurrence (based on known occurrences) can be approximated with a probability distribution of maximum entropy (the Maxent distribution) subject to certain environmental constraints (based on all environmental variables in the model), and that from this distribution the potential geographic distribution of the group under study can be determined. Rather than relying on both recorded presence and absence data points, Maxent uses presence-only input data, and subsequently selects a set of random background points. This makes the algorithm well-suited for our purposes since we did not have absence data. The input data consist of a set of environmental layers for the study region and the observed disease presence localities within that region. Maxent then uses these data to estimate the environmental niche space that accurately describes the observed occurrences. Its predictions are continuous logistic probabilities with increasing values referring to more suitable habitats. In several multi-model comparisons, Maxent was shown to perform well with few point localities [76,77], and in comparisons of ecological niche modeling techniques its performance was generally rated among the highest [77,78]. We used the following settings of Maxent: 10000 background points; only linear or auto features; regularization multiplier = 3.0; maximum iterations = 500; convergence threshold = 0.00005. To assess model fit as a function of the different sets of input predictor variables, we used the Akaike Information Criterion corrected for sample size (AICc), implemented in ENMTools [79].

For modeling purposes, a location was considered to be positive for human MPX if one or more cases had been identified there. Localities of active human MPX cases used in modeling are shown in Fig. 3a. The number of sites used in each model for reservoir species and human MPX is shown in Tables 2 and 3. To examine whether predicted future environmental conditions in our study region are more extreme than those observed under current climate, we used the clamping option implemented in Maxent.

Variable importance

To assess the relative importance of predictor variables in determining the distribution of human MPX, we used the variable jackknifing procedure implemented in Maxent. Using a subset of the original variable set, new models are computed and their performance as measured by the regularized training gain (the average log probability of the presence samples, corrected for a uniform distribution with gain = 0) compared to that of the full model [75]. Two ways of selecting a subset of variables are implemented. First, each variable is used on its own, providing an assessment of the information content of the variable itself. Second, models are computed using all but one predictor variable, which is useful in assessing the amount of unique information in the omitted variable. This jackknifing approach is particularly helpful when some predictor variables (e.g., reservoir species distributions) are cross-correlated, because they will be nearly equally important in determining the range of human MPX cases.

Model performance

The area under the receiver operator curve (AUC), implemented in Maxent, was used as a first-order assessment of model performance, where an AUC score of 0.5 indicates random prediction, and a score of 1 a perfect prediction. In addition, the available data were split into training and test datasets, where the former is used to estimate parameters for ("train") the model using the settings from the full model (with the lowest AICc in the case of human MPX), and the latter to test the predictions of that model. Here, we used two different approaches to generate different training and test datasets. First, we ran additional models for each predictor data set with a random test percentage of 40% and 500 iterations for each model, as implemented in Maxent. We considered model performance high if test AUC values were >0.75 [80]. This procedure randomly selects the training data from the original dataset, and training and test sites may thus be spatially autocorrelated. To address this issue of spatial autocorrelation, in a second approach, we divided the study region in two spatial partitions, one of which was used as training data and the other as test data (Text S2). In this case, training and test datasets are unlikely to be spatially autocorrelated, in contrast to the situation when a random subset is taken to serve as test data. While there may be many ways to create a spatial subdivision of the dataset, we focused on one that generated nearly even sampling sizes in training and test datasets in a west-east split. We modeled reservoir species and human MPX occurrence across the entire region, using only the training data (east or west), and compared the projections to the corresponding observed test data.

The use of AUC values unaccompanied by other means of assessing model performance in species distribution modeling has been shown to be problematic [81,82]. Therefore, to further ensure that our model validation was robust, we used a threshold-dependent approach for the comparison of predicted versus observed occurrences for each model, in which the logistic probability output was converted to a presence-absence map based upon an empirical threshold [83]. We used the 'balance threshold' from Maxent, which balances the training omission rate and the proportional predicted area, two measures of quality of a binary prediction [75]. With this threshold, we calculated the proportion of the total area of our study region where human MPX was predicted to be present, and tested whether more test localities were predicted to be positive for human MPX than expected based upon the proportional predicted area using a one-tailed binomial test [84]. Thus, the expectation for the null (a random prediction) is that the proportion of sites predicted to be positive for MPXV is equal to the proportion of the total area where MPXV is predicted to be present. For predictions of human MPX, we also used a threshold-independent approach, where we extracted the logistic probability scores for localities that were used as training sites in one model and test sites in a second model, in which the data partitioning was reversed. That is, the logistic probability values of western sites based on a model using western sites for training were compared to those based on a model with eastern sites for training. Similarly, the probability values of eastern sites were compared between models using eastern versus western sites for training. Models were then compared using a Wilcoxon signed-rank test, expecting to see no significant differences if the training sites accurately predicted the test sites. All statistical tests were carried out in the R statistical framework [85].

As a final means to evaluate whether the modeled species distributions were meaningful, we visually compared modeled species ranges with range maps based on expert knowledge, available from the IUCN Red List (IUCN; www.iucnredlist.org Accessed May 2010), Mammal Networked Information System (MaNIS; http://manisnet.org/ Accessed April 2010), African Mammals Databank (AMD; http://www.gisbau.uniroma1.it/amd/ Accessed April 2010), and from [33] and [34]. Expert-based maps often show only the range limits of species, whereas species distribution models offer the advantage that they also show the gaps within those outer boundaries. In all cases, modeled distributions showed range limits very similar to those based on expert knowledge.

Supporting Information

Figure S1 Variable importance of Maxent predictions for reservoir species distributions. Results are shown for tests in which only the variable in question was entered into the model (dark blue bars) and in which all variables except the one in question were entered (light blue bars). Longer dark blue bars and shorter light blue bars indicate higher variable importance. Bio 1: mean annual temperature; Bio 2: daily temperature range; Bio 4: temperature seasonality; Bio 5: maximum temperature of the warmest month; Bio 12: mean annual precipitation; Bio 16: precipitation seasonality; Bio 17: precipitation of the driest quarter.

Figure S2 Predictive models for MPXV reservoir species in Tropical Africa. Maxent predictions of MPXV reservoir species occurrence under contemporary climate conditions, using climate variables as predictor, and averages for eight climate change scenarios each for the periods 2050–2060 and 2080–2090.

Figure S3 Predictive models for human MPX in Tropical Africa. Maxent predictions of human MPX occurrence under contemporary climate conditions, using different environmental variable sets as predictors: reservoir species (based on climate and remote sensing variables) plus climate and remote sensing variables; reservoir species (based on climate and remote sensing variables); reservoir species (based on climate variables) and climate variables; reservoir species (based on climate variables). For each model, all 'features' were allowed to be used in Maxent ("auto features", i.e. linear and quadratic coefficients can be used for each predictor, as well as step functions and interactions).

Figure S4 Future projections for human MPX occurrence under different climate change scenarios for 2050 (first eight panels) and 2080 (second set of eight panels). Current reservoir species distributions were first projected onto future climate variables. The resulting future reservoir species distributions were subsequently used to estimate the future human MPX distributions.

Figure S5 Spatial distribution of MPX occurrences in Sankuru showing overlap with (a) land cover types and (b) areas deforested since 2000.

Figure S6 Climatic and land cover variables used in the local scale study of MPX transmission. (a) correlation between maximum water deficit (MWD) and the Atlantic Multidecadal Oscillation (AMO), (b) temperature of the wettest quarter, (c) temperature of the coldest quarter, (d) canopy moisture defined as maximum K_u band radar backscatter minus minimum backscatter (units: dB), (e) forest cover. The polygons outlined in black in panel (a) represent secteurs identified as primary or secondary hotspots of MPX.

Figure S7 Association between climate and MPX genotypes in DRC at the national scale. MWD = maximum water deficit. AMO = Atlantic Multidecadal Oscillation.

Figure S8 Land cover map of our study area. Hatched areas indicate a predicted increase in human MPX occurrence under the multi-model ensemble climate change scenarios for 2050–2060 (horizontal) and 2080–2090 (vertical).

Figure S9 Plots of percent tree cover versus the predicted average increase in probability of human MPX occurrence for 2050 and 2080. Small increases in human MPX occurrence are predicted in both savanna and forest habitat types, but the largest increases are mainly located in forest areas.

Table S1 Intergovernmental Panel on Climate Change (IPCC) scenarios examined for Central Africa.

Acknowledgments

We thank the DRC Ministry of Health and local health workers who were responsible for specimen collection and case investigation. We also thank the World Health Organization for kindly sharing their human MPX data from the 1980s. We acknowledge support by the Deutsche Forschungsgemeinschaft and Open Access Publishing Fund of Tübingen University.

Author Contributions

Conceived and designed the experiments: HAT TF SB JOLS TBS AWR. Performed the experiments: HAT JAGS TF. Analyzed the data: HAT SA-N. Contributed reagents/materials/analysis tools: SA-N PMM SCJ NKK TLK JNF NDW RLS ML HM LLW JJM WB EO LEH AWR. Wrote the paper: HAT TF TBS AWR.

References

1. Patz JA, Campbell-Lendrum D, Holloway T, Foley JA (2005) Impact of regional climate change on human health. Nature 438: 310–317.
2. Lafferty KD (2009) The ecology of climate change and infectious diseases. Ecology 90: 888–900.
3. Pham HV, Doan HTM, Phan TTT, Minh NNT (2011) Ecological factors associated with dengue fever in a central highlands Province, Vietnam. Bmc Infectious Diseases 11.
4. Wittmann EJ, Baylis M (2000) Climate change: Effects on Culicoides-transmitted viruses and implications for the UK. Veterinary Journal 160: 107–117.
5. Gould EA, Higgs S (2009) Impact of climate change and other factors on emerging arbovirus diseases. Transactions of the Royal Society of Tropical Medicine and Hygiene 103: 109–121.
6. Brugger K, Rubel F (2009) Simulation of climate-change scenarios to explain Usutu-virus dynamics in Austria. Preventive Veterinary Medicine 88: 24–31.
7. Gilbert L (2010) Altitudinal patterns of tick and host abundance: a potential role for climate change in regulating tick-borne diseases? Oecologia 162: 217–225.
8. Breman JG (2000) Monkeypox: an emerging infection for humans? In: Scheld WM, Craig WA, Hughes JM, editors. Emerging Infections. Washington, DC: ASM Press. pp. 45–67.

9. Rimoin AW, Mulembakani PM, Johnston SC, Lloyd-Smith JO, Kisalu NK, et al. (2010) Major increase in human monkeypox incidence 30 years after smallpox vaccination campaigns cease in the Democratic Republic of Congo. Proceedings of the National Academy of Sciences 107: 16262–16267.
10. Damon IK, Roth CE, Chowdhary V (2006) Discovery of monkeypox in Sudan. New England Journal of Medicine 355: 962–963.
11. Boko M, Niang I, Nyong A, Vogel C (2007) Africa. In: Parry ML, Canziana OF, Palutikof JP, van der Linden PJ, Hanson CE, editors. Climate Change 2007: Impacts, Adaptation, and Vulnerability Contributions of Working Group II to the Fourth Assessment Report of the Intergovernmental Panel on Climate Change. Cambridge, UK: Cambridge University Press. pp. 433–467.
12. Meehl GA, Stocker TF, Collins WD, Friedlingstein P, Gaye AT (2007) Global climate projections. In: Solomon S, Qin D, Manning M, Chen Z, Marquis M et al., editors. Climate Change 2007: The Physical Science Basis Contributions of Working Group I to the Fourth Assessment Report of the Intergovernmental Panel on Climate Change. Cambridge, UK: Cambridge University Press.
13. Shanahan TM, Overpeck JT, Anchukaitis KJ, Beck JW, Cole JE, et al. (2009) Atlantic forcing of persistent drought in West Africa. Science 324: 377–380.
14. Madhi SA, Huebner RE, Doedens L, Aduc T, Wesley D, et al. (2000) HIV-1 co-infection in children hospitalised with tuberculosis in South Africa. International Journal of Tuberculosis and Lung Disease 4: 448–454.
15. Albuja S, Beau C, Farmer A, Glatz AK, Jennings E, et al. (2010) Internal Displacement. Global Overview of Trends and Developments in 2009. Geneva, Switzerland: Norwegian Refugee Council.
16. Fair J, Jentes E, Inapogui A, Kourouma K, Goba A, et al. (2007) Lassa virus-infected rodents in refugee camps in Guinea: A looming threat to public health in a politically unstable region. Vector-Borne and Zoonotic Diseases 7: 167–171.
17. Hutin YJF, Williams RJ, Malfait P, Pebody R, Loparev VN, et al. (2001) Outbreak of human monkeypox, Democratic Republic of Congo, 1996–1997. Emerging Infectious Diseases 7: 434–438.
18. Khodakevich L, Jezek Z, Messinger D (1988) Monkeypox virus: ecology and public health significance. Bulletin of the World Health Organization 66: 747–752.
19. Khodakevich L, Szczeniowski M, Manbu-ma-Disu, Jezek Z, Marennikova S, et al. (1987) The role of squirrels in sustaining monkeypox virus transmission. Tropical and Geographical Medicine 39: 115–122.
20. Jezek Z, Fenner F (1988) Human Monkeypox. Monographs in Virology Vol. 17. Basel: Karger.
21. Reynolds MG, Davidson WB, Curns AT, Conover CS, Huhn G, et al. (2007) Spectrum of infection and risk factors for human monkeypox, United States, 2003. Emerging Infectious Diseases 13: 1332–1339.
22. Lloyd-Smith JO, George D, Pepin KM, Pitzer VE, Pulliam JRC, et al. (2009) Epidemic Dynamics at the Human-Animal Interface. Science 326: 1362–1367.
23. Levine RS, Peterson AT, Yorita KL, Carroll D, Damon IK, et al. (2007) Ecological niche and geographic distribution of human monkeypox in Africa. Plos One 2: e176.
24. Reed KD, Melski JW, Graham MB, Regnery RL, Sotir MJ, et al. (2004) The detection of monkeypox in humans in the Western Hemisphere. New England Journal of Medicine 350: 342–350.
25. Beck LR, Lobitz BM, Wood BL (2000) Remote sensing and human health: New sensors and new opportunities. Emerging Infectious Diseases 6: 217–227.
26. Rogers DJ, Randolph SE (2003) Studying the global distribution of infectious diseases using GIS and RS. Nature Reviews Microbiology 1: 231–237.
27. Peterson AT (2006) Ecologic niche modeling and spatial patterns of disease transmission. Emerging Infectious Diseases 12: 1822–1826.
28. Peterson AT (2009) Shifting suitability for malaria vectors across Africa with warming climates. Bmc Infectious Diseases 9.
29. Gonzalez C, Wang O, Strutz SE, Gonzalez-Salazar C, Sanchez-Cordero V, et al. (2010) Climate Change and Risk of Leishmaniasis in North America: Predictions from Ecological Niche Models of Vector and Reservoir Species. Plos Neglected Tropical Diseases 4.
30. Fuller T, Thomassen HA, Mulembakani PM, Johnston SC, Lloyd-Smith JO, et al. (2011) Using remote sensing to map the risk of human monkeypox virus in the Congo Basin. EcoHealth 8: 14–25.
31. Root ED, Meyer RE, Emch ME (2009) Evidence of localized clustering of gastroschisis births in North Carolina, 1999–2004. Social Science & Medicine 68: 1361–1367.
32. Kulldorff M (2010) SaTScan User Guide for Version 9.0. 109 pages. Available: http://www.satscan.org. Accessed 2011 June.
33. Kingdon J (1979) East African Mammals: An Atlas of Evolution in Africa Volume 3B. Chicago: University of Chicago Press.
34. Kingdon J (1997) The Kingdon Field Guide to African Mammals. San Diego: Academic Press.
35. Elith J, Phillips SJ, Hastie T, Dudik M, Chee YE, et al. (2011) A statistical explanation of MaxEnt for ecologists. Diversity and Distributions 17: 43–57.
36. Mitchell TD, Jones PD (2005) An improved method of constructing a database of monthly climate observations and associated high-resolution grids. International Journal of Climatology 25: 693–712.
37. Asefi-Najafabady S, Saatchi S (accepted pending revisions) Impacts of climate variability on African Tropical Rainforest. Geophysical Research Letters.
38. Keesing F, Belden LK, Daszak P, Dobson A, Harvell CD, et al. (2010) Impacts of biodiversity on the emergence and transmission of infectious diseases. Nature 468: 647–652.
39. Schmaljohn C, Hjelle B (1997) Hantaviruses: A global disease problem. Emerging Infectious Diseases 3: 95–104.
40. Mackenzie JS, Williams DT (2009) The zoonotic flaviviruses of Southern, South-Eastern and Eastern Asia, and Australasia: the potential for emergent viruses. Zoonoses and Public Health 56: 338–356.
41. Baines PG, Folland CK (2007) Evidence for a rapid global climate shift across the late 1960s. Journal of Climate 20: 2721–2744.
42. Chylek P, Folland CK, Lesins G, Dubey MK (2010) Twentieth century bipolar seesaw of the Arctic and Antarctic surface air temperatures. Geophysical Research Letters 37.
43. Aragao L, Malhi Y, Roman-Cuesta RM, Saatchi S, Anderson LO, et al. (2007) Spatial patterns and fire response of recent Amazonian droughts. Geophysical Research Letters 34.
44. Ramirez J, Jarvis A (2010) Downscaling Global Circulation Model Outputs: The Delta Method. Decision and Policy Analysis Working Paper No. 1. Cali, Colombia: International Center for Tropical Agriculture (CIAT). CGIAR Challenge Program on Climate Change, Agriculture, and Food Security.
45. Ellis CK, Carroll DS, Lash RR, Peterson AT, Damon IK, et al. (2012) Ecology and geographic of human monkeypox case occurrences across Africa. Journal of Wildlife Diseases 48: 335–347.
46. Formenty P, Muntasir MO, Damon I, Chowdhary V, Opoka ML, et al. (2010) Human monkeypox outbreak caused by novel virus belonging to Congo Basin clade, Sudan, 2005. Emerging Infectious Diseases 16: 1539–1545.
47. Melski J, Reed K, Stratman E, Graham MB, Fairley J, et al. (2003) Multistate outbreak of monkeypox - Illinois, Indiana, and Wisconsin, 2003 (Reprinted from MMWR, vol 52, pg 537–540, 2003). Jama-Journal of the American Medical Association 290: 30–31.
48. Sejvar JJ, Chowdary Y, Schomogyi M, Stevens J, Patel J, et al. (2004) Human monkeypox infection: A family cluster in the Midwestern United States. Journal of Infectious Diseases 190: 1833–1840.
49. Hutson CL, Lee KN, Abel J, Carroll DS, Montgomery JM, et al. (2007) Monkeypox zoonotic associations: insights from laboratory evaluation of animals associated with the multi-state US outbreak. American Journal of Tropical Medicine and Hygiene 76: 757–767.
50. Walther BA, Ewald PW (2004) Pathogen survival in the external environment and the evolution of virulence. Biological Reviews 79: 849–869.
51. Khodakevich L, Jezek Z, Kinzanzka K (1986) Isolation of monkeypox virus from wild squirrel infected in nature. Lancet 327, 98–99.
52. Pepin KM, Lass S, Pulliam JRC, Read AF, Lloyd-Smith JO (2010) Identifying genetic markers of adaptation for surveillance of viral host jumps. Nature Reviews Microbiology 8: 802–813.
53. Nakouné E, Kazanji M (2012) Monkeypox detection in maculopapular lesions in two young Pygmies in the Central African Republic. International Journal of Infectious Diseases 16, Supplement 1: e266–e267.
54. Balk D, Brickman M, Anderson B, Pozzi F, Yetman G (2005) Mapping global urban and rural population distributions. Estimates of future global population distribution to 2015. Environment and Natural Resources Working Papers # 24. New York: United Nations Food and Agriculture Organization/Socioeconomic Data and Applications Center, Center for International Earth Science Information Network, Columbia University.
55. Rimoin AW, Mulembakani PM, Johnston SC, Lloyd-Smith JO, Kisalu NK, et al. (2010, in press) Major increase in human monkeypox incidence 30 years after smallpox vaccination campaigns cease in the Democratic Republic of Congo. Proceedings of the National Academy of Sciences.
56. United Nations Population Division (2008) World Population Prospects: The 2008 Revision Population Database. New York: Population Division of the Department of Economic and Social Affairs of the United Nations Secretariat.
57. Di Giulio DB, Eckburg PB (2004) Human monkeypox: an emerging zoonosis. Lancet Infectious Diseases 4: 15–25.
58. FACET - Initiative for remote sensing forest monitoring in Central Africa (2010) Atlas of forest cover and change 2000–2010 in the Democratic Republic of the Congo. Observatoire satellital des forets d'Afrique central, South Dakota State University, and University of Maryland.
59. Eisen RJ, Griffith KS, Borchert JN, MacMillan K, Apangu T, et al. (2010) Assessing human risk of exposure to plague bacteria in northwestern Uganda based on remotely sensed predictors. American Journal of Tropical Medicine and Hygiene 82: 904–911.
60. Kulldorff M (1997) A spatial scan statistic. Communications in Statistics: Theory and Methods 26: 1481–1496.
61. Brooker S, Clarke S, Njagi JK, Polack S, Mugo B, et al. (2004) Spatial clustering of malaria and associated risk factors during an epidemic in a highland area of western Kenya. Tropical Medicine & International Health 9: 757–766.
62. Kulldorff M, Heffernan R, Hartman J, Assuncao R, Mostashari F (2005) A space-time permutation scan statistic for disease outbreak detection. Plos Medicine 2: 216–224.
63. Kulldorff M, Nagarwalla N (1995) Spatial disease clusters - detection and inference. Statistics in Medicine 14: 799–810.
64. Rimoin AW, Kisalu N, Kebela-Ilunga B, Mukaba T, Wright LL, et al. (2007) Endemic human monkeypox, Democratic Republic of Congo, 2001–2004. Emerging Infectious Diseases 13: 934–937.
65. Graham CH, Ferrier S, Huettman F, Moritz C, Peterson AT (2004) New developments in museum-based informatics and applications in biodiversity analysis. Trends in Ecology & Evolution 19: 497–503.

66. Newbold T (2010) Applications and limitations of museum data for conservation and ecology, with particular attention to species distribution models. Progress in Physical Geography 34: 3–22.
67. Hijmans RJ, Cameron SE, Parra JL, Jones PG, Jarvis A (2005) Very high resolution interpolated climate surfaces for global land areas. International Journal of Climatology 25: 1965–1978.
68. Nix H (1986) Atlas of Elapid Snakes of Australia. Canberra: Australian Government Publishing Service.
69. Myneni RB, Hoffman S, Knyazikhin Y, Privette JL, Glassy J, et al. (2002) Global products of vegetation leaf area and fraction absorbed PAR from year one of MODIS data. Remote Sensing of Environment 83: 214–231.
70. Hansen MC, DeFries RS, Townshend JRG, Sohlberg R, Dimiceli C, et al. (2002) Towards an operational MODIS continuous field of percent tree cover algorithm: examples using AVHRR and MODIS data. Remote Sensing of Environment 83: 303–319.
71. Long DG, Drinkwater MR, Holt B, Saatchi S, Bertoia C (2001) Global ice and land climate studies using scatterometer image data. EOS Transactions AGU 82: 503.
72. Nakicenovic N, Alcamo J, Davis G, de Vries B, Fenhann J, et al. (2000) Special Report on Emissions Scenarios : a special report of Working Group III of the Intergovernmental Panel on Climate Change. Medium: ED; Size: vp. p.
73. Canadell JG, Le Quere C, Raupach MR, Field CB, Buitenhuis ET, et al. (2007) Contributions to accelerating atmospheric CO(2) growth from economic activity, carbon intensity, and efficiency of natural sinks. Proceedings of the National Academy of Sciences of the United States of America 104: 18866–18870.
74. Raupach MR, Marland G, Ciais P, Le Quere C, Canadell JG, et al. (2007) Global and regional drivers of accelerating CO2 emissions. Proceedings of the National Academy of Sciences of the United States of America 104: 10288–10293.
75. Phillips SJ, Anderson RP, Schapire RE (2006) Maximum entropy modeling of species geographic distributions. Ecological Modelling 190: 231–259.
76. Hernandez PA, Graham CH, Master LL, Albert DL (2006) The effect of sample size and species characteristics on performance of different species distribution modeling methods. Ecography 29: 773–785.
77. Wisz MS, Hijmans RJ, Li J, Peterson AT, Graham CH, et al. (2008) Effects of sample size on the performance of species distribution models. Diversity and Distributions 14: 763–773.
78. Elith J, Graham CH, Anderson RP, Dudik M, Ferrier S, et al. (2006) Novel methods improve prediction of species' distributions from occurrence data. Ecography 29: 129–151.
79. Warren DL, Glor RE, Turelli M (2008) Environmental niche equivalency versus conservatism: quantitative approaches to niche evolution. Evolution 62: 2868–2883.
80. Elith J (2002) Quantitative methods for modeling species habitat: comparative performance and an application to Australian plants. In: Ferson S, Burgman M, editors. Quantitative methods for conservation biology. Berlin: Springer. pp. 39–58.
81. Lobo JM, Jimenez-Valverde A, Real R (2008) AUC: a misleading measure of the performance of predictive distribution models. Global Ecology and Biogeography 17: 145–151.
82. Peterson AT, Papes M, Soberon J (2008) Rethinking receiver operating characteristic analysis applications in ecological niche modeling. Ecological Modelling 213: 63–72.
83. Buermann W, Saatchi S, Smith TB, Zutta BR, Chaves JA, et al. (2008) Predicting species distributions across the Amazonian and Andean regions using remote sensing data. Journal of Biogeography 35: 1160–1176.
84. Anderson RP, Gomez-Laverde M, Peterson AT (2002) Geographical distributions of spiny pocket mice in South America: insights from predictive models. Global Ecology and Biogeography 11: 131–141.
85. R Development Core Team (2004) R: A Language and Environment for Statistical Computing. Vienna: R Foundation for Statistical Computing.
86. Kummerow C, Barnes W, Kozu T, Shiue J, Simpson J (1998) The Tropical Rainfall Measuring Mission (TRMM) sensor package. Journal of Atmospheric and Oceanic Technology 15: 809–817.
87. Wan Z, Zhang Y, Zhang Q, Li ZL (2004) Quality assessment and validation of the MODIS global land surface temperature. International Journal of Remote Sensing 25: 261–274.
88. Wan ZM (2008) New refinements and validation of the MODIS Land-Surface Temperature/Emissivity products. Remote Sensing of Environment 112: 59–74.
89. Jiang ZY, Huete AR, Chen J, Chen YH, Li J, et al. (2006) Analysis of NDVI and scaled difference vegetation index retrievals of vegetation fraction. Remote Sensing of Environment 101: 366–378.
90. Farr TG, Rosen PA, Caro E, Crippen R, Duren R, et al. (2007) The shuttle radar topography mission. Reviews of Geophysics 45.

14

The Relative Impacts of Climate and Land-Use Change on Conterminous United States Bird Species from 2001 to 2075

Terry L. Sohl*

Earth Resources Observation and Science (EROS) Center, U.S. Geological Survey, Sioux Falls, South Dakota, United States of America

Abstract

Species distribution models often use climate data to assess contemporary and/or future ranges for animal or plant species. Land use and land cover (LULC) data are important predictor variables for determining species range, yet are rarely used when modeling future distributions. In this study, maximum entropy modeling was used to construct species distribution maps for 50 North American bird species to determine relative contributions of climate and LULC for contemporary (2001) and future (2075) time periods. Species presence data were used as a dependent variable, while climate, LULC, and topographic data were used as predictor variables. Results varied by species, but in general, measures of model fit for 2001 indicated significantly poorer fit when either climate or LULC data were excluded from model simulations. Climate covariates provided a higher contribution to 2001 model results than did LULC variables, although both categories of variables strongly contributed. The area deemed to be "suitable" for 2001 species presence was strongly affected by the choice of model covariates, with significantly larger ranges predicted when LULC was excluded as a covariate. Changes in species ranges for 2075 indicate much larger overall range changes due to projected climate change than due to projected LULC change. However, the choice of study area impacted results for both current and projected model applications, with truncation of actual species ranges resulting in lower model fit scores and increased difficulty in interpreting covariate impacts on species range. Results indicate species-specific response to climate and LULC variables; however, both climate and LULC variables clearly are important for modeling both contemporary and potential future species ranges.

Editor: Stephanie S. Romanach, U.S. Geological Survey, United States of America

Funding: TLS was supported by the U.S. Geological Survey's Climate and Land Use (CLU) Mission Area. The funders had no role in study design, data collection and analysis, decision to publish, or preparation of the manuscript.

Competing Interests: The author has declared that no competing interests exist.

* Email: sohl@ugs.gov

Introduction

Species distribution models (SDMs) are based on the assumption that presence at a given location is based on suitable environmental conditions to support the species' ability to find shelter, feed, and/or reproduce [1,2]. Such models have been widely used to model current species distributions, either to establish extant distributions or to understand the specific environmental variables that drive species distributions [3,4,5]. A central premise of many SDMs is that climate is a primary driving force of the distribution of species [6]. Projected climate data are frequently used with SDMs to explore potential future impacts of climate change on species distributions [7,8,9], based on the assumption that the basic physiological tolerances of species to environmental conditions are constant through time [10]. Jimenez-Valverde et al. [11] modeled typical climate conditions for 94 bird species in North America and noted the dominant signal of climate in shaping North American bird distributions. Thuiller et al. [12] modeled distributions of plants, birds, mammals, and reptiles in Europe and found that models using climate alone performed nearly as well as models that included both climate and landscape variables. Bucklin et al. [13] found that climate variables were strong predictors for contemporary species distribution modeling and that additional predictors (including land cover) were not essential.

Climate is obviously a primary driver of many SDMs. While land use and land cover (LULC) change is often used for modeling contemporary species distributions, it is not often used when examining future time frames [7]. Despite the results from the studies listed above, other studies have found that including LULC in bioclimatic models of species distribution can improve the explanatory power of SDMs [7,12]. Lee and Jetz [14] found that LULC projections were vital for future modeling, noting that loss of habitat is a high predictor of extinction for bird species. Barbet-Massin et al. [7] found that SDMs perform best if both climate and LULC are included. Sinclair et al. [15] were critical of SDMs for rarely including anthropogenic impacts on biological systems, suggesting that changing landscape patterns are likely to have at least as great an impact on species distributions as climate change. Many other studies have found that projected land use is a vital component of SDMs, with loss of predictive power when LULC is not included in the assessment [1,8,9,16].

While many state the need to use LULC data when projecting future change in species distributions, such data are often not available [17]. Riordan et al. [16] noted the disconnect between relatively high-resolution climate projections used in global assessment models and the generally coarse treatment of LULC, leading to high-resolution, projected LULC rarely being used (or available) for SDMs. Those studies that have used both projected climate and LULC data to model future species distributions have often relied on very coarse spatial-scale LULC data [1,7]. LULC is much more heterogeneous at local scales than climate. Those studies that do use LULC but only at very coarse scales miss the inherent spatial variability in LULC that is typically not found with climate data.

The goals of this paper are to examine the relative effects of climate and LULC change on bird species distributions in the conterminous United States, using maximum entropy modeling and projected climate and LULC data from 2001 to 2075. Specific research questions include:

- What are the relative influences of LULC and climate in modeling contemporary (2001) breeding bird distributions for the conterminous United States?
- What are the relative impacts of projected climate and projected LULC change on United States breeding bird distributions in the future (2075)?
- What are the specific impacts of climate and LULC on one focus species (Hooded Warbler (*Wilsonia citrina*) used to demonstrate individual species results)?
- What are the implications for the use of climate and LULC data in SDMs?

Materials and Methods

A maximum entropy model (Maxent) was used in conjunction with species presence data, current and projected climate and LULC data, and topographic data to model distributions for 50 diverse bird species in the conterminous United States. Twelve distinct modeling simulations were conducted for each species to disentangle the effects of climate and LULC on species distributions for both the "present" (2001 for this assessment) and for multiple scenarios in the future (2075). Although many supporting data were available through the year 2100, 2075 was selected as the assessed "future" year to accommodate the use of 30-year averaged climate data (as described below). The following provides a summary of data sources, the model structure, model parameterization, and the assessment framework.

Materials

The modeling approach described below required spatially explicit data on both bird species "presence", as well as environmental variables (covariates) that could be used to model species distributions. Note the goal was to assess long-term trends in changes to species distributions. Longer-term aggregate or average data values were thus used to temper the effects of seasonal or yearly variation, both for the species presence data (where distributions of single-season presence records could be impacted by unusual seasonal conditions such as drought or large disturbance events) and for the covariates (where single-season climate data, for example, may be unrepresentative of longer-term climate trends).

Species Presence Data – eBird. The source of bird species presence data was eBird, a "citizen-science" database [18,19]. eBird allows public entry of bird sightings, with recorded information on the time and date of the sighting, location, observation protocol, quantity of each species, and observer information. As a citizen science database, there are potential issues (discussed below) related to the lack of a formal sampling protocol [20], but eBird offers several potential advantages for species distribution modeling including 1) large number of sample points (millions for some species), 2) global data (although observations are currently heavily biased towards North America and Europe), and 3) observations for all seasons. eBird data have been successfully used for a number of SDM assessments [20,21,22,23]. Hochachka and Fink [20] found strong linkages between individual species and land cover using eBird data, and that the data were valuable for examining distribution patterns at multiple scales.

Fifty bird species with breeding ranges partially or completely within the conterminous United States were selected for the assessment (**Table 1**). Species were selected to ensure variability in size of breeding range, geographic region, and preferred breeding habitat. The goal was to ensure that a variety of "real-world" model applications were represented. To minimize potential effects of annual variation in species presence, data from eBird entries from 1992 to 2012 were used to establish "current" breeding records. With current and projected land cover data available for every year from 1992 to 2100 (see below), a nominal date of 2001 (middle of the 1992 to 2010 period) was used to represent contemporary species distributions and tie the 1992 to 2010 species occurrences to one specific date of land-cover conditions. Data were also filtered by season to ensure records corresponded to breeding populations; for all species a consistent June 1 to July 15 observation period was used to represent likely "breeding" presence, a reasonable assumption for the species that were assessed. Some migratory species initially included in the assessment were removed from consideration based on dispersed patterns of eBird sightings for the June 1 to July 15 period, indicating post-breeding movement had already occurred by July 15 (e.g., Long-billed Curlew (*Numenius americanus*)). One species included, the American Goldfinch (*Spinus tristis*) generally begins breeding after this period, but is considered non-migratory and still is within breeding range. For the 50 species assessed, eBird sightings for the June 1 to July 15 period corresponded well to published breeding range maps from NatureServe [23].

eBird allows users to enter one of several potential observation protocols, including "stationary count", "traveling count", or "exhaustive area count". However, regardless of observation protocol, users only enter one geographic coordinate. A single coordinate for a "traveling count" where the travel distance was substantial could result in a data point that was many kilometers from the actual observation. For exhaustive area counts with a large search area, a single coordinate may similarly be some distance from the actual observation. To eliminate potential issues with unrepresentative locations of eBird sightings, all "traveling count" sightings with a travel distance of more than 2 km were eliminated (similar to Fink et al. [23]), as were all "exhaustive area count" sightings with a search area of more than 100 hectares.

Additional potential issues with eBird data include spatial bias in presence samples [25]. eBird observations, like other citizen science data, tend to be clustered around highly populated and/or easily accessible areas [15,18,19]. Sampling bias has a much stronger effect on presence-only models (used here) than on presence-absence models, as model results end up representing both presence, as well as the density of the sampling effort [26]. Spatial filtering is an effective means to reduce bias in sample data prior to use in species distribution modeling [27,28]. For this assessment, the seasonal 1992 to 2012 observations were spatially

Table 1. Species modeled and number of eBird sample points for each.

	Species	Scientific Name	Original	Final
1	**American Goldfinch**	*Spinus tristis*	236,217	2,663
2	**Anna's Hummingbird**	*Calypte anna*	32,047	427
3	**Baird's Sparrow**	*Ammodramus bairdii*	513	48
4	**Band-tailed Pigeon**	*Patagioenas fasciata*	17,415	407
5	**Black-capped Chickadee**	*Poecile atricapillus*	131,634	1,877
6	**Blue-winged Teal**	*Anas discors*	15,288	1,243
7	**Bobolink**	*Dolichonyx oryzivorus*	28,658	1,105
8	**Brown-headed Cowbird**	*Molothrus ater*	178,324	3,996
9	**Brown Thrasher**	*Toxostoma rufum*	61,661	2,254
10	**Cactus Wren**	*Campylorhynchus brunneicapillus*	4,714	215
11	**Carolina Wren**	*Thryothorus ludovicianus*	107,244	1,893
12	**Chestnut-collared Longspur**	*Calcarius ornatus*	1,426	105
13	**Dickcissel**	*Spiza americana*	29,479	1,411
14	**Downy Woodpecker**	*Picoides pubescens*	150,261	2,925
15	**Eastern Kingbird**	*Tyrannus tyrannus*	111,057	2,956
16	**Ferruginous Hawk**	*Buteo regalis*	1,587	238
17	**Gambel's Quail**	*Callipepla gambelii*	6,307	198
18	**Grasshopper Sparrow**	*Ammodramus savannarum*	23,254	1,323
19	**Gray Partridge**	*Perdix perdix*	616	129
20	**Gray Vireo**	*Vireo vicinior*	265	43
21	**Great Blue Heron**	*Ardea herodias*	141,552	3,449
22	**Great Horned Owl**	*Bubo virginianus*	11,130	1,487
23	**Green-winged Teal**	*Anas carolinensis*	6,726	530
24	**Hooded Warbler**	*Wilsonia citrina*	15,482	773
25	**Lark Bunting**	*Calamospiza melanocorys*	3,268	355
26	**Lark Sparrow**	*Chondestes grammacus*	20,978	1,467
27	**Northern Harrier**	*Circus cyaneus*	14,795	1,231
28	**Northern Pintail**	*Anas acuta*	4,269	466
29	**Orchard Oriole**	*Icterus spurius*	41,136	1,876
30	**Painting Bunting**	*Passerina ciris*	15,294	569
31	**Pied-billed Grebe**	*Podilymbus podiceps*	23,272	1,287
32	**Pileated Woodpecker**	*Dryocopus pileatus*	48,118	1,982
33	**Pygmy Nuthatch**	*Sitta pygmaea*	11,848	322
34	**Red-eyed Vireo**	*Vireo olivaceus*	138,887	2,389
35	**Red-headed Woodpecker**	*Melanerpes erythrocephalus*	22,809	1,593
36	**Red-tailed Hawk**	*Buteo jamaicensis*	101,388	3,715
37	**Ruby-throated Hummingbird**	*Archilochus colubris*	81,241	2,090
38	**Savannah Sparrow**	*Passerculus sandwichensis*	41,214	1,435
39	**Scissor-tailed Flycatcher**	*Tyrannus forficatus*	16,571	718
40	**Sedge Wren**	*Cistothorus platensis*	7,827	478
41	**Sharp-tailed Grouse**	*Tympanuchus phasianellus*	801	121
42	**Short-eared Owl**	*Asio flammeus*	760	139
43	**Sora**	*Porzana carolina*	6,687	649
44	**Tufted Titmouse**	*Baeolophus bicolor*	129,472	2,058
45	**Vesper Sparrow**	*Pooecetes gramineus*	16,387	1,164
46	**Western Kingbird**	*Tyrannus verticalis*	45,319	2,028
47	**Western Meadowlark**	*Sturnella neglecta*	36,755	1,825
48	**Western Tanager**	*Piranga ludoviciana*	33,108	1,127

Table 1. Cont.

	Species	Scientific Name	Original	Final
49	**White-headed Woodpecker**	*Picoides albolarvatus*	4,389	137
50	**Yellow-headed Blackbird**	*Xanthocephalus xanthocephalus*	16,794	1,031

"Original" represents conterminous United States observations from 1992 to 2012, from June 1 to July 15. "Final" represents points that have had 1) spatial filtering applied to reduce points in heavily sampled areas, and 2) removal of points with long travel distances (traveling count) or large search areas (search area count).

filtered to eliminate sample points within 20 km of any other sample point. The threshold of 20 km was chosen because it more aggressively reduced sampling density in the very dense eBird database than past studies [27,28], while still maintaining adequate numbers of points for modeling. The elimination of sample points based on observation protocol or sampling density greatly reduced the number of sample points used in the assessment, often by a factor of 20 or more (**Table 1**). However, the filtering successfully eliminated the high concentration of points in heavily populated areas while maintaining a relatively large number of observations for most species (minimum of 43 points, maximum of 3,996, mean of 1,313). Only two species had fewer than 100 sample points (Gray Vireo (*Vireo vicinior*) and Baird's Sparrow (*Ammodramus bairdii*)), at 48 and 43 points, respectively. The number of sample points were considered adequate, as Wisz et al. [29] and Hernandez et al. [30] examined the effect of sample size on species distribution models and found that Maxent outperformed other modeling techniques when sample sizes were small, with "reasonable" models possible with sample sizes as small as 10.

Land-Use and Land-Cover Data. A newly available suite of LULC projections for the conterminous United States was used [31,32]. The LULC projections were produced for the conterminous United States, with annual LULC maps from 1992 to 2100 for four Intergovernmental Panel on Climate Change (IPCC) Special Report on Emissions Scenarios (SRES) [33]. The spatial resolution of the data was 250 m, with 16 LULC classes. The four modeled SRES were the A1B, A2, B1, and B2 scenarios; however, complimentary climate data were not available for the B2 scenario, so only A1B, A2, and B1 were used in this assessment (see **table 2** for characteristics of the three IPCC SRES scenarios used in this assessment). To simplify the modeling and interpretation of model results, the original sixteen LULC classes were aggregated to eight basic LULC classes (**table 3**). Aggregated 2001 LULC data served as one of the covariates when constructing the initial models. Projected 2075 LULC data provided information on LULC change for the 2075 model simulations.

Error in LULC data obtained from remote sensing sources is a concern for SDMs [4]. The LULC projections described above used the 1992 National Land Cover Database (NLCD) [34] as the mapping starting point. The projections were thus subject to not only the inherent uncertainty in projecting future LULC conditions, but also carried the legacy of any mapping error in the original 1992 NLCD. Given the lack of a rigid sampling protocol in the citizen-science eBird data, locational inaccuracies may also be a factor for the species' presence data. To reduce the effects of potential locational or mapping error in the LULC and presence data, LULC covariates used in the model were "neighborhood" measures of abundance for a given LULC class, rather than per-pixel measures. Use of a neighborhood LULC measure provided not only site-level habitat information, but also provided information on habitat in the surrounding area. Individual species have unique, scale-dependent responses to landscape structure [35,36,37]. In modeling one individual species, it would be preferable to identify the appropriate scale of analysis that captures that species' habitat preferences. However, the objective here was to identify relative influences of climate and LULC across 50 different species. Optimizing (varying) the scale of analysis for each individual species introduces another (unwanted) variable into the assessment. One set scale was thus selected to minimize the scale-dependent impacts on modeling results. In tests of multiple landscape scales for SDMs, Cunningham and Johnson [35] found that scales between 800 m and 1600 m were the most suitable a majority of 19 bird species tested. For this study, a 5×5 pixel (1,250 m×1,250 m) window around each point was chosen within which counts were tallied for each LULC class. The neighborhood counts for each LULC variable served as the LULC covariates within the modeling framework. A "LULC diversity" measure was also calculated, tallying the number of different LULC classes within each 5×5 window. The LULC diversity measure was also used as a covariate, as a measure of local landscape heterogeneity. **Table 3** summarizes the LULC covariates (as well as climate and topographic covariates described below).

Climate Data. The goal of this work was to examine long-term trends in bird species distributions in response to climate and LULC change. Global circulation models often produce climate data with monthly and yearly summaries, with year-to-year variability inherent in the output data. However, the use of average conditions was preferable to modeling with a single year of future climate projections, to minimize annual variability and focus on long-term trends [39]. A suite of global circulation models (GCMs) was used to obtain 30-year averages of climate consistent with IPCC SRES characteristics. A downscaling methodology similar to Hay et al. [39] was used to downscale coarse-scale climate data to a 4-km resolution for the conterminous United States [40], with data ultimately resampled to 250 m to match other covariates. Downscaled output was produced for six GCMs (BCCR-BCM2, CCSM3, CSIRO3.0-Mk, CSIRO-Mk3.5, INM-CM3.0, and MIROC 3.2) that provided climate data consistent with the IPCC SRES storylines (see http://www.ipcc-data.org/gcm/montly/SRES_AR4/index.html). Monthly data on average temperature, minimum temperature, maximum temperature, and precipitation were produced from each of the models. Variable averages across the six GCMs were calculated to reduce the bias present in any one individual model.

Covariates available for use in this assessment included not only yearly averages for temperature and precipitation, but also monthly averages, and monthly and annual minimum and maximum temperatures. However, not all variables were used, in order to reduce potential effects of multicollinearity. Correlation between potential climate variables was very high, particularly between the various temperature variables (Pearson correlation coefficient $r>0.90$ between nearly all paired temperature variables, such as monthly temperature and annual temperature).

Table 2. Relative socioeconomic characteristics of the three IPCC SRES scenarios used in this assessment.

	A1B	A2	B1
Primary focus	Economic growth	Economic growth	Environmental sustainability
Globalization or Regionalization	Global Convergence	Regional Development	Global Convergence
Global Population	Increase to 8.7 billion by 2050, then slow decline	Continuous increase to 15.1 billion by 2100	Increase to 8.7 billion by 2050, then slow decline
Gross Domestic Product Growth	Very High	Medium	High
Energy Use	Very High	High	Low
Energy Strategy	Balanced, fossil fuel and alternative fuels	Regionally variable, based on local resources	Push to alternative and post-fossil fuel energy
Pace of technology change	Rapid	Slow	Medium
Technology diffusion	Rapid	Slow, regional variability	Rapid
Economic equity	Homogenization, higher incomes	Fragmented, uneven, continued income gaps	Homogenization, but lower incomes than A1B
Environmental Protection	Focus on "management" of resources rather than "conservation"	Uneven environmental management, protection higher in affluent areas	Broad support for environmental conservation, efficiency gains for resource use

See Nakicenovic et al. (2000) for additional information on SRES characteristics and Sohl et al. (2014) for how these characteristics were interpreted to create the LULC projections used in this assessment.

To minimize multicollinearity effects and to simplify data analysis, only the 30-year climate averages of annual temperature and annual precipitation, averaged across the six GCMs, were used as climate covariates in this assessment.

Bradley et al. [41] noted that the use of LULC data in conjunction with climate variables often does little to improve SDM results, due to collinearity of LULC and climate data at regional scales. There was little evidence of highly correlated LULC and climate variables in this assessment. Pearson correlation coefficients were computed for all LULC and climate covariate pairs. The highest correlation was between precipitation and the shrubland count ($|r| = 0.39$), while no other LULC and climate variable pair had $|r|$ values higher than 0.29.

Topographic Data. The few studies that have used projected LULC data in conjunction with projected climate data to look at future species distributions have often restricted themselves to those two categories of data [7,8,16]. However, when developing SDMs for current conditions, modelers tend to use a wider array of input variables, with topography often playing a key role [1,42,43,44]. Because the objective of this study was to assess the relative impacts of climate and LULC in "real-world" modeling applications, topography variables were included as covariates in

Table 3. Covariates used as predictor variables within Maxent.

Variable Category	Variable Name	Description
Land Cover	Cropland Count	5×5 neighborhood count of "cropland" pixels
Land Cover	Forest Count	5×5 neighborhood count of "forest" pixels (all forest)
Land Cover	Grass Count	5×5 neighborhood count of "grassland" pixels
Land Cover	Hay Count	5×5 neighborhood count of "hay/pasture" pixels
Land Cover	Shrub Count	5×5 neighborhood count of "shrubland" pixels
Land Cover	Urban Count	5×5 neighborhood count of "urban" pixels
Land Cover	Water Count	5×5 neighborhood count of "water" pixels
Land Cover	Wetland Count	5×5 neighborhood count of "wetland" pixels (all wetland)
Land Cover	LULC Diversity	5×5 neighborhood count of the number of different LULC classes
Climate	Average Temp	Average annual temperature
Climate	Average Precip	Average annual (total) precipitation
Topography	Elevation	Elevation data from National Elevation Database
Topography	Slope	Slope data derived from National Elevation Database
Topography	Compound Topographic Index	Compound Topographic Index data derived from National Elevation Database

All data were mapped to a common geographic extent at 250-m resolution.

this assessment in recognition that SDMs often do not focus solely on LULC and climate. Three topographic variables were used, based on the USGS National Elevation Dataset for the conterminous United States [45]; 1) elevation, 2) slope, and 3) compound topographic index (a measure of "wetness" and high flow accumulation). Each variable was resampled to match the geographic extent and 250 m spatial resolution of the LULC and climate covariates.

Methods

Maximum Entropy Modeling Framework. MaxEnt model [46] (Version 3.3.1) running on a Windows desktop was used to model bird species distributions. Maxent was designed to model species distributions based on presence-only species data [26]. Maxent statistically minimizes entropy between the probability density of "presence" data, and probability density from "background" data, as defined in covariate space [26]. Maxent has been shown to be one of the most effective methodologies for modeling species distributions when presence-only data are used [2,26].

Maxent estimates suitability for a given species by fitting feature classes based on environmental covariates. The filtered eBird data for each of the 50 species served as presence points. Environmental covariates were the LULC, climate, and topographic variables described above and shown in **Table 3**. Modeled feature classes in Maxent potentially included linear, quadratic, product, hinge, threshold, and categorical [46]. Linear features model linear response to a covariate, while quadratic features model response to the variable squared. Product features model interactions between paired variables. Hinge features model piecewise constant responses, while threshold features model abrupt boundary relationships between covariates and response. Category features are binary indicators used to indicate positive or null response to each class within a categorical covariate (e.g., thematic land cover map). All variables in this assessment were presented as continuous variables, including nominally thematic LULC data that were represented as counts within a 5×5 neighborhood around each point. Categorical features were thus not used in this assessment, but the other five Maxent features were used in modeling species response to the covariates.

The most widespread method for testing model results is a random hold-out of sample data [47]; 75% of the filtered eBird samples were used for training the model while 25% were reserved for testing. Maxent uses "background" points as locations where presence was not recorded, with background points either selected at random from the geographic extent of the study area, or specifically provided by the model user. The relationship between presence and background points in Maxent can strongly influence model results. Spatial bias in the presence points can result in a selection of background points with a fundamentally different spatial distribution [27], resulting in a model that represents the sampling effort as much as species presence [48]. A number of options were available to correct for spatial bias issues [25,46,49]. Several studies have discussed the use of spatially filtering or discarding records in over-sampled efforts [27,48,50], the approach used here and described above, with Kramer-Schadt et al. [26] finding it better reduced both errors of commission and of omission compared to other methodologies. Because the eBird data were spatially filtered, no attempts were made to account for bias through other measures.

Choice of the study area extent also can influence Maxent results [51,52]. VanDerWal et al. [53] found that model performance suffered when background points were selected from either too restricted or too broad a geographic extent, in relationship to the presence points. Specifically, if background points are selected from too broad a geographic area, predictive models were dominated by coarse-scale determinants of distribution (such as climate) [53], while those that use too limited a geographic area underestimate the importance of these variables [52] To reduce the influence of a mismatch between background area and "presence" points, a consistent buffer was applied around 2001 (contemporary) input presence points to construct a unique geographic extent for each species. The buffer zone was used to definitively set the study area for each species, both for defining where background points could be selected by Maxent, and to set the complete geographic range for modeling both current and future distributions. Ideally a unique geographic region would be optimized for each species according to characteristics of the observation data [51], but to facilitate comparison across the 50 species, a consistent buffered extent was used for all species. VanDerWal et al. [53] used a 200-km buffer, but initial experimentation for this assessment found that to be too restrictive for changes in conterminous United States bird species range from 2001 to 2075, with some species' ranges shifting by more than 200 km. A 500-km buffer around input eBird points was used, restricting both the range from which background points could be selected, and restricting the prediction space for each species' range.

Remaining parameterization of Maxent largely followed model defaults. Anderson and Gonzalez [54] and Warren and Seifert [55] recommended species-specific tuning of Maxent settings, noting that the regularization value (used to restrict model "overfitting" to input data) had a large effect on results. However, Phillips and Dudik [46] tested regularization values and found that "regularization parameters which are the defaults in MaxEnt software...are well suited for a wide range of presence-only datasets." The six feature types are also selectable, yet Syfert et al. [56] found little influence on model results by varying the feature types that are used. Phillips and Dudik [46] found that using the default 10,000 background points achieved similar model results as if all possible background sites were used; the default setting was thus used. Default settings were also used that enabled "clamping" of covariate and feature values for the 2075 model simulations. With the model trained on the 2001 covariate data, the potential existed for "novel" covariate values when the model was applied in 2075, using projected climate and LULC data. For the 2075 model simulations, the enabled clamping resulted in a rescaling of both covariate and feature values if their values were higher or lower than those found in the training data. Values higher than those encountered in the training data were rescaled to the training data maximum, while values lower than those encountered in the training data were rescaled to the training data minimum. The implications of the use of clamping are provided in the discussion section.

Parameterizing Maxent as described above, initial model simulations for each species were conducted using the filtered eBird data for presence points, and the 2001 LULC, 2001 climate, and topographic variables as covariates. Twelve model simulations were made in total for each species (**Table 4**). A base model simulation was done for 2001 using all variables (simulation 1), while additional simulations were done for 2001 with climate and topography (excluding LULC) (simulation 2) or land cover and topography (excluding climate) (simulation 3). The model developed for simulation 1 was applied for 2075 to examine potential future impacts of climate change, LULC change, or both (topographic variables are static in all simulations). For each of the three IPCC scenarios, simulations were done with all 2075 variables, with projected climate but static LULC, and with

Table 4. Twelve model simulations were conducted for each species, three for 2001 and nine for 2075.

Simulation	Description	Climate (Scenario)	LULC (Scenario)	Topo Data	Scenario
1	2001 All	2001	2001	Yes	-
2	2001 Climate	2001	-	Yes	-
3	2001 LULC	-	2001	Yes	-
4*	2075 A1B All	2075 A1B	2075 A1B	Yes	A1B
5*	2075 A1B Climate Change	2075 A1B	2001	Yes	A1B
6*	2075 A1B LULC Change	2001	2075 A1B	Yes	A1B
7*	2075 A2 All	2075 A2	2075 A2	Yes	A2
8*	2075 A2 Climate Change	2075 A2	2001	Yes	A2
9*	2075 A2 LULC Change	2001	2075 A2	Yes	A2
10*	2075 B1 All	2075 B1	2075 B1	Yes	B1
11*	2075 B1 Climate Change	2075 B1	2001	Yes	B1
12*	2075 B1 LULC Change	2001	2075 B1	Yes	B1

Model simulations variously include or exclude climate and LULC covariates in order to assess the individual effects of each.
*Simulations for 2075 used the model developed for run 1 (2001 "All"), applying 2075 climate and/or LULC data from the appropriate scenario.

projected LULC but static climate. Keeping either climate or LULC static from 2001 to 2075 allowed for the examination of the relative effects of projected climate versus projected land use change on future bird species distributions. Three 2001 simulations and nine 2075 simulations were thus conducted for each of the 50 species, resulting in 600 individual model simulations.

Assessing Model Results. Several different metrics were used to assess the relative impacts of climate and LULC change on bird species distributions. The three 2001 model simulations were assessed for model fit through a comparison of Area Under the Curve (AUC) of the Receiver Operating Characteristic (ROC). AUC values represent the probability that a randomly selected "presence" site will have a higher AUC value than a randomly chosen "background" site. Comparison of AUC scores was used to examine relative impacts on model fit when LULC or climate data were excluded from the analysis. A second criterion was the relative contributions of the covariates to model results, measured by relative changes in regularized training gain between variables. This information was provided as a "percent contribution" from Maxent. A third criterion was a comparison of modeled "suitable" range for each species. Elith et al. [26] cautions against cross-species comparisons using logistic output from Maxent, as probability of presence is relative to the sampling effort for a given species. However, changes in relative range for each individual species can be identified by applying a threshold value to Maxent's logistic output, to differentiate between likely presence and absence locations. The "maximum sensitivity plus specificity" threshold was used [15,38], a thresholding technique that limits both errors of commission and errors of omission and has been found to outperform other techniques [57,58].

The 2075 simulations were evaluated by assessing changes in "suitable" breeding range as compared to 2001. Net change in range area was determined for each species by first applying the "maximum sensitivity plus specificity" threshold to modeled output and then differencing the threshold results, with comparisons of net effects of climate change alone, LULC change alone, and both climate and LULC change from 2001 to 2075 (for each scenario).

Finally, results were examined in terms of species range and relationship to the conterminous United States study area. While data sources and analyses often stop at political boundaries, species ranges obviously do not, and the use of conterminous United States borders for this work resulted in the modeling of truncated ranges for many species. Both 2001 and 2075 model results could be impacted dependent upon whether the entire range was modeled or if one or more maximum extent boundaries were artificially truncated [51,52,53]. Many SDM applications model truncated species distributions (see discussion below); assessing results on species range characteristics allowed for an examination of LULC and climate impacts across a variety of "real-world" modeling situations. For each of the assessment criteria discussed above, mean values were provided (**Table 5**) for species within the following "range classes": 1) "Single Truncated" (species with either the northern *or* southern extent artificially truncated by United States borders, 2) "Double Truncated" (species with ranges truncated at *both* the northern and southern United States border, and 3) "Whole Ranges" (species with >95% of current breeding ranges found within the conterminous United States, measured with NatureServe species distributions [24].

Results

2001 Models ("current" species' distributions)

The 2001 models were assessed for model fit using AUC scores. Values of 0.5 indicate model fit was no better than random, while increasing values above 0.5 indicated an improved model fit. **Figure 1** provides AUC scores for the three 2001 model simulations for each of the 50 species. AUC scores ranged from a low of 0.716 to a high of 0.987, with considerable variation among species, as well as among the three model simulations for a given species. Model simulations with all variables included (simuilation 1) had the highest mean AUC score, at 0.891, and the highest AUC score for each of the 50 species. AUC scores were significantly lower ($p<0.001$; paired t-test) for both simulation 2 (climate, no LULC) and simulation 3 (LULC, no climate), with mean AUC scores of 0.863 and 0.874, respectively. Results indicate significantly poorer model fit when LULC data were excluded than if climate data were excluded ($p<0.01$; paired t-test). By range class AUC scores were significantly lower when either LULC or climate data was omitted, for all range classes ($p< 0.01$, paired t-test) (**Table 5A**). AUC scores overall were similar for the Single Truncated and Whole Range classes, but were much

Table 5. Impacts on assessment variables by range class.

	Single Truncated	Double Truncated	Whole Range
5(A) -2001 MODEL FIT (AUC Score - Mean Value)			
All Variables	0.916	0.839	0.906
LULC, No Climate	0.891	0.834	0.892
Climate, No LULC	0.891	0.799	0.888
5(B) -2001 VARIABLE CONTRIBUTION (in percent)			
Climate Variables	49.5%	52.8%	52.7%
Topography Variables	12.8%	6.0%	13.9%
Land Cover Variables	37.7%	41.2%	33.4%
5(C) -2001 Range (Mean Values - Percent of conterminous United States area)			
All Variables	23.0%	42.2%	26.6%
LULC, No Climate	28.4%	43.7%	30.1%
Climate, No LULC	31.1%	57.0%	32.8%
5(D) -2075 Breeding Range (Mean Values - Percent change from 2001)			
All Variables	−9.9%	+2.6%	+12.0%
LULC, No Climate	+3.8%	+3.5%	+1.5%
Climate, No LULC	−13.0%	+1.2%	+10.2%

Values represent mean values across all species in a class. "Single Truncated" (27 species) represents species with ranges artificially truncated at either the north *or* south by the United States border. "Double Truncated" (15 species) represents species with truncated ranges that extend to or past the United States/Canada border in the north and the United States/Mexico border in the south. "Whole Range" (8 species) represents species where >95% of the current range is found within the conterminous United States.

lower on average for the Double-Truncated class. Omission of LULC resulted in the lowest overall AUC score for every species in this class. The relative impact of LULC or climate data omission was more balanced for the other two range classes and varied by species.

Figure 2 depicts Maxent-provided proportional contributions of each covariate to the regularized training gain, aggregated across all 50 species for simulation 1 (all variables modeled). The climate covariates played an important role in shaping 2001 simulations, with annual temperature and precipitation providing 51.0% of the contribution to model results. Temperature was one

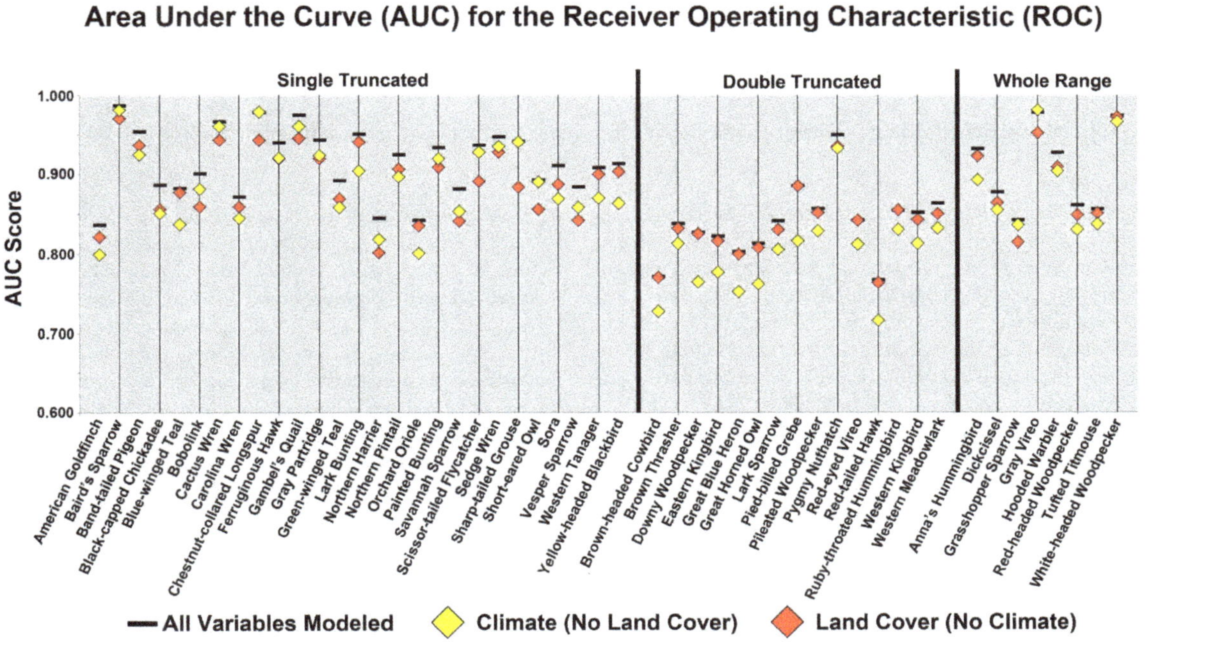

Figure 1. AUC scores for each species, for run 1 (all variables modeled), run 2 (Climate, no Land Cover), and run 3 (Land Cover no Climate). AUC scores are also parsed by range class.

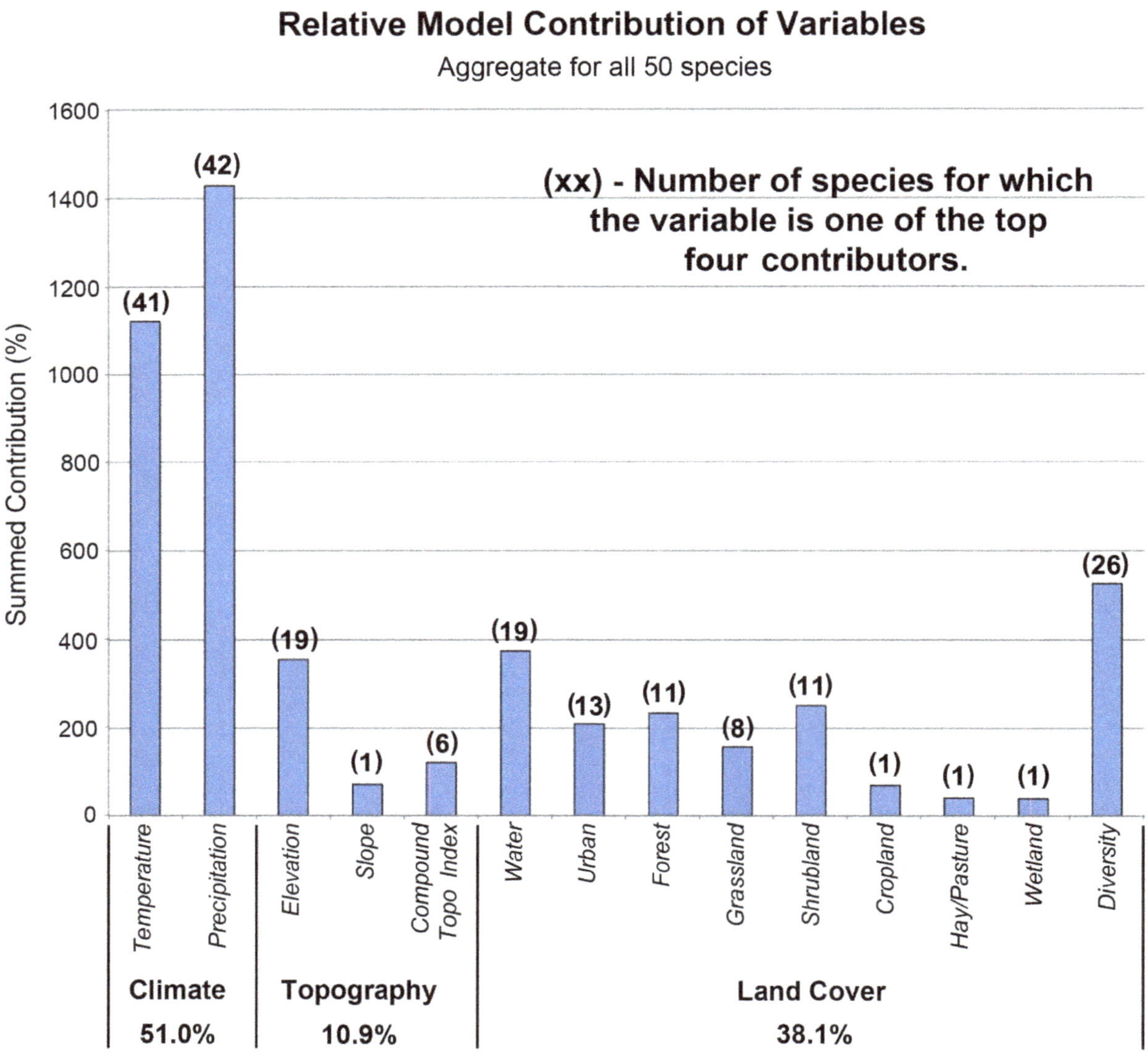

Figure 2. Proportional contributions of each covariate to the regularized training gain, aggregated across all 50 species.

of the top four contributing covariates for 41 species, while precipitation was one of the top three covariates for 42 species. LULC variables in aggregate contributed 38.1% to model results, while topographic variables contributed 10.9%. Results vary among individual species, but overall, it is clear that both climate and LULC were important contributors to model output when both were included as covariates. Results were similar when categorizing species by range class (**Table 5B**).

Figure 3 provides a comparison of modeled "suitable" area for each species, among model simulations 1, 2, and 3, using the unique maximum sensitivity plus specificity threshold criteria for each species and simulation. Values are presented as a percentage of the total land surface for the conterminous United States with Maxent logistic output values above the threshold criterion. While the predicted suitable range for a given species was sometimes similar across each of the three 2001 model simulations for a species, in many cases, the area deemed to be suitable varied dramatically depending upon what variables were used as covariates. For 36 of the 50 species, the area deemed suitable was highest in simulation 2, when only climate and topographic variables were used as covariates (LULC excluded). The area deemed suitable was nearly double in some cases (e.g., Great Horned Owl (*Bubo virginianus*), Yellow-headed Blackbird (*Xanthocephalus xanthocephalus*)) for simulation 2, as opposed to simulation1 when LULC data were also incorporated. For the other 14 species, the area deemed suitable was highest for simulation 3, when only LULC and topographic variables were used as covariates (climate excluded). Adding covariates to the model, be they LULC or climate, clearly acted to further define (and restrict) the area deemed to be suitable for species' habitation. Using climate data alone resulted in broad, overly generalized suitability ranges if LULC data were not used to help further define suitable landscapes. Results were similar when evaluating the three different range classes (**Table 5C**), with the smallest range consistently modeled when all variables were used as covariates. However, for the Double Truncated range class, the omission of LULC data from the model resulted in a much larger increase in range as compared to the other two range classes, while omitting climate data had little impact.

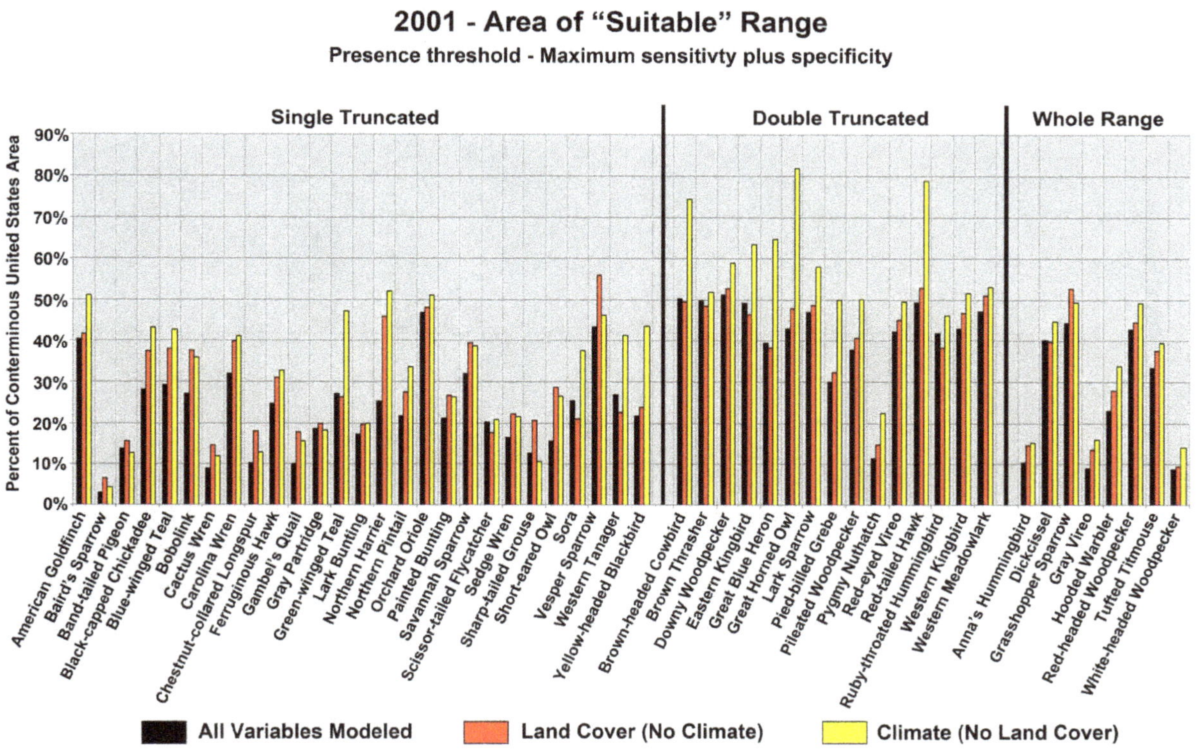

Figure 3. Area (percent of conterminous U.S. land mass) classified as suitable to support a given species, for model run 1 (all variables modeled), run 2 (Climate, no Land Cover), and run 3 (Land Cover no Climate). Suitability was determined by applying the maximum sensitivity plus specificity threshold to Maxent logistic output. Results are also parsed by range class.

2075 Models ("projected" species' distributions)

Figure 4 depicts projected changes in range for each of the 50 species, measured as change relative to the range modeled in 2001 (simulation 1), using the maximum sensitivity plus specificity threshold to differentiate between presence and absence. Range differences are provided for each of the 3 model simulations, for each of the 3 scenarios, with bar height providing the mean change in range across all three scenarios, and deviation bars providing the variation between scenarios. Depending upon species, modeled changes in range varied according to which covariates were used and between different IPCC scenarios. Changes in range varied from a near complete loss of all conterminous United States suitable range (Baird's Sparrow(*Ammodramus bairdii*)) to range expansions that nearly double the current range (Cactus Wren (*Campylorhynchus brunneicapillus*), Gambel's Quail (*Callipepla gambelii*), Gray Vireo (*Vireo vicinior*)). **Figure 4** indicates that the magnitude of projected changes in range was much more strongly impacted by projected climate change than by projected LULC change, when using a threshold to define suitability. When only LULC changed (climate static) from 2001 to 2075, changes in projected ranges from 2001 were highly significant ($p<0.001$; paired t-test) but were never more than 20% (either positive or negative). When only climate changed (LULC static) from 2001 to 2075, range changes were often quite dramatic, with 20 species showing range changes of 25% or more for a given scenario. Climate and LULC could either both influence species' distributions in the same direction, or a positive species response to one category of covariates could be offset by a negative species response to the other category.

Table 5(D) and Figure 4 show substantial differences in the relative effects of LULC and climate on 2075 model results, depending upon range class. The most dramatic overall changes in range were in the Single Truncated class, where climate change obviously had a strong effect on model results. Climate change had a much more muted impact on the Double Truncated class, with low overall changes in range. Climate had moderate to strong impacts for the Whole Range class. The impacts of LULC change were much more consistent across range classes than were the impacts of climate change.

Species Focus - Hooded Warbler (*Wilsonia citrina*)

While it is impractical to individually discuss each of the 50 modeled species, the relative impacts of climate and LULC change on one species, the Hooded Warbler (*Wilsonia citrina*), are highlighted here to demonstrate specific impacts of climate and LULC. The Hooded Warbler is a forest-dependent species that primarily breeds in the eastern United States. **Figure 5** provides 1) a map of Maxent logistic output for 2001, using simulation 1 (all covariates modeled), and 2) changes in output for each 2075 scenario, and for each 2075 model simulation. The AUC score for simulation 1 (2001) indicated a high-level of model fit (AUC = 0.927), with precipitation, temperature, and forest count (in relative order) measured as the three covariates contributing the most to model results. For simulation 1, 23.1% of the conterminous United States was deemed "suitable" (threshold) range for the Hooded Warbler. The predicted range sharply increased to 27.9% in simulation 3 (climate excluded) and 33.9% for simulation 2 (LULC excluded), a pattern seen for many species (**Figure 3**).

Changes in predicted range by 2075 indicate a strong influence of both climate and LULC change (**Figure 5**). The economically focused A1B and A2 scenarios are similar, as a changing climate

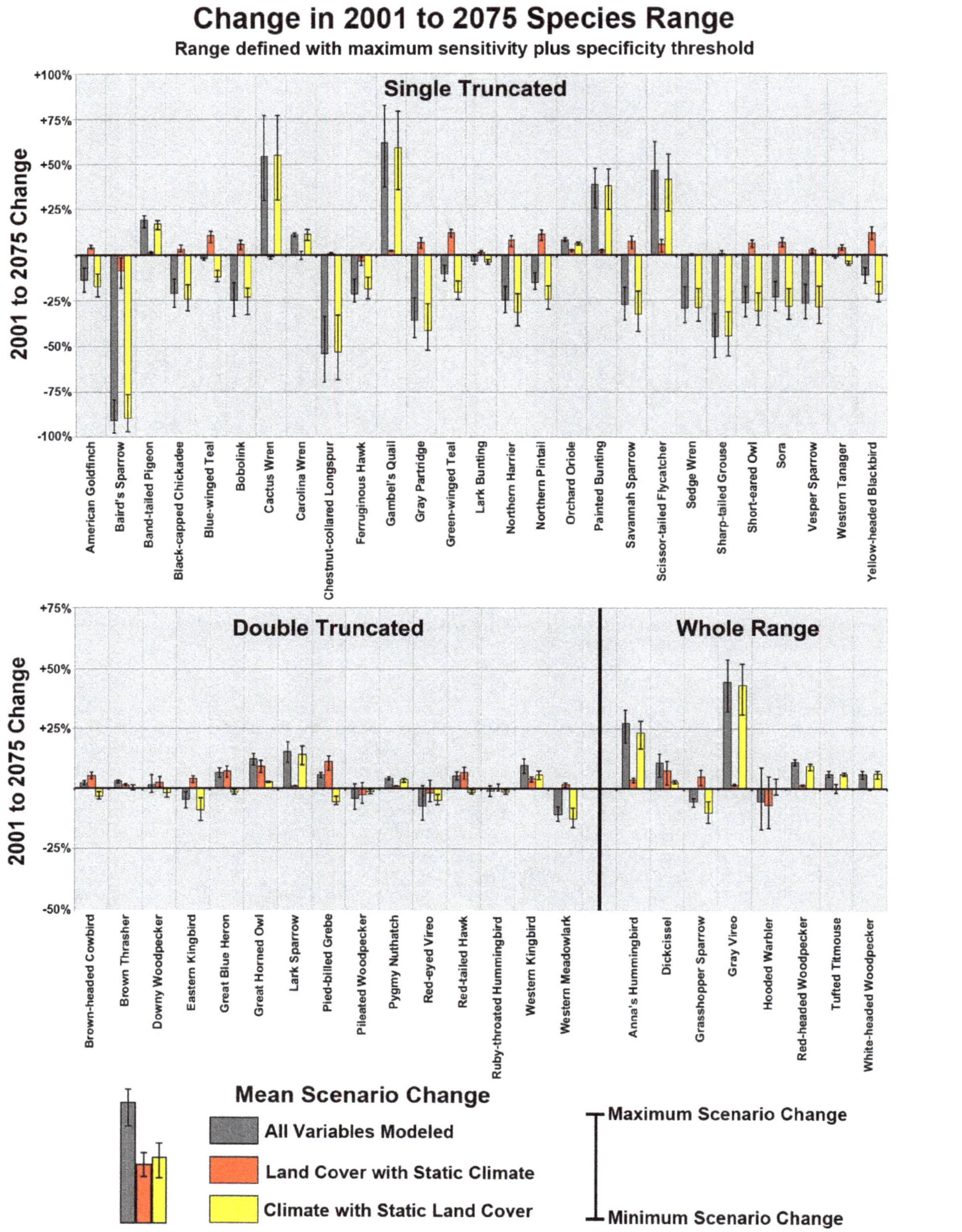

Figure 4. Changes (2001 to 2075) in area classified as suitable to support a given species. Change is presented as area change, relative to the contemporary (2001) modeled range. Bar height represents mean change across the 3 IPCC scenarios, while error bars represent scenario variability. Suitability was determined by applying the maximum sensitivity plus specificity threshold to Maxent logistic output. Results are also parsed by range class.

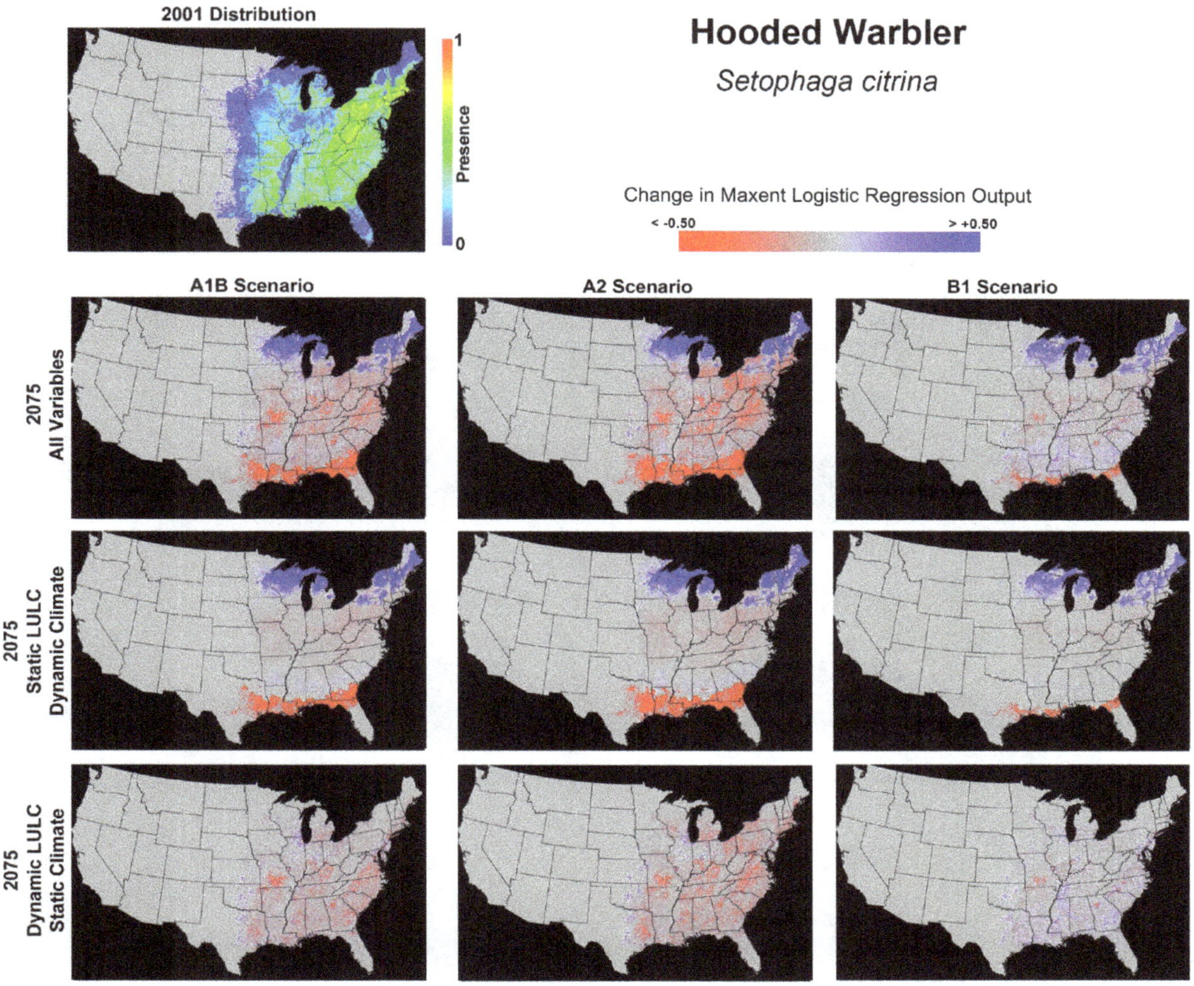

Figure 5. Maxent 2001 logistic output for the Hooded Warbler, and projected changes under each 2075 scenario and model run. Climate change results in broad northward shifts in species range across all scenarios. LULC change alters the local pattern of habitat suitability, with losses under the A1B and A2 scenarios, and general increases in the B1 scenario.

resulted in strong shifts in overall species range, with large contiguous bands of losses of range in the south and gains in the north. The effects of LULC change are more fragmented, but substantial forest loss results in local areas of decline throughout much of the eastern United States. The effects of climate change are more muted for the environmentally focused B1 scenario, with less severe shifts to the north. While local areas of forest loss do result in range declines in the B1 scenario, afforestation and forest regeneration result in higher presence scores in many locations.

Figure 6 displays modeling results for the Hooded Warbler for both 2001 and 2075 (A2 scenario) for a smaller area within their current breeding range. At this scale the relative impacts of both LULC and climate are evident for both current (2001) modeling, and for future projections. **Figure 6(a)** and **6(e)** show LULC change from 2001 to 2075, characterized by substantial expansion of urban and agricultural lands, at the expense of forest land. **Figures 6(b)**, **6(c)**, and **6(d)** show model results with LULC and topography as covariates, all variables as covariates (LULC, climate, and topography), and climate and topography as covariates, respectively. Without the use of climate data, suitability was highly heterogeneous, but higher elevation areas that currently do not support Hooded Warbler populations (e.g., parts of the upper-left quadrant) often had high suitability values even with the use of topographic information (**Figure 6(b)**). Without the use of LULC data, suitability was less heterogeneous and the cooler high-elevation areas were characterized by lower values, yet areas of dense anthropogenic land-use that are unsuitable for Hooded Warbler breeding were often characterized as highly suitable (**Figure 6(d)**). The use of both LULC and climate data, in conjunction with topographic data, resulted in a heterogeneous distribution of suitability values, capturing both the influence of cooler high-elevation areas as well as areas of dense anthropogenic land-use (**Figure 6(c)**).

For 2075 model simulations, **Figure 6(g)** shows the impacts on Hooded Warbler range when both projected climate and projected LULC are used in the model. Range expansion occurred towards higher elevations as a warming climate results in more suitable breeding conditions. LULC change, primarily urbanization and agricultural expansion, resulted in large but heterogeneous losses of breeding range, counter-balancing range gains due to climate change. **Figure 6(f)** shows a model simulation with projected 2075 LULC but a static 2001 climate. Without the use of projected climate data, the range expansion due to warming was not captured (**Figure 6(i)**). **Figure 6(h)**

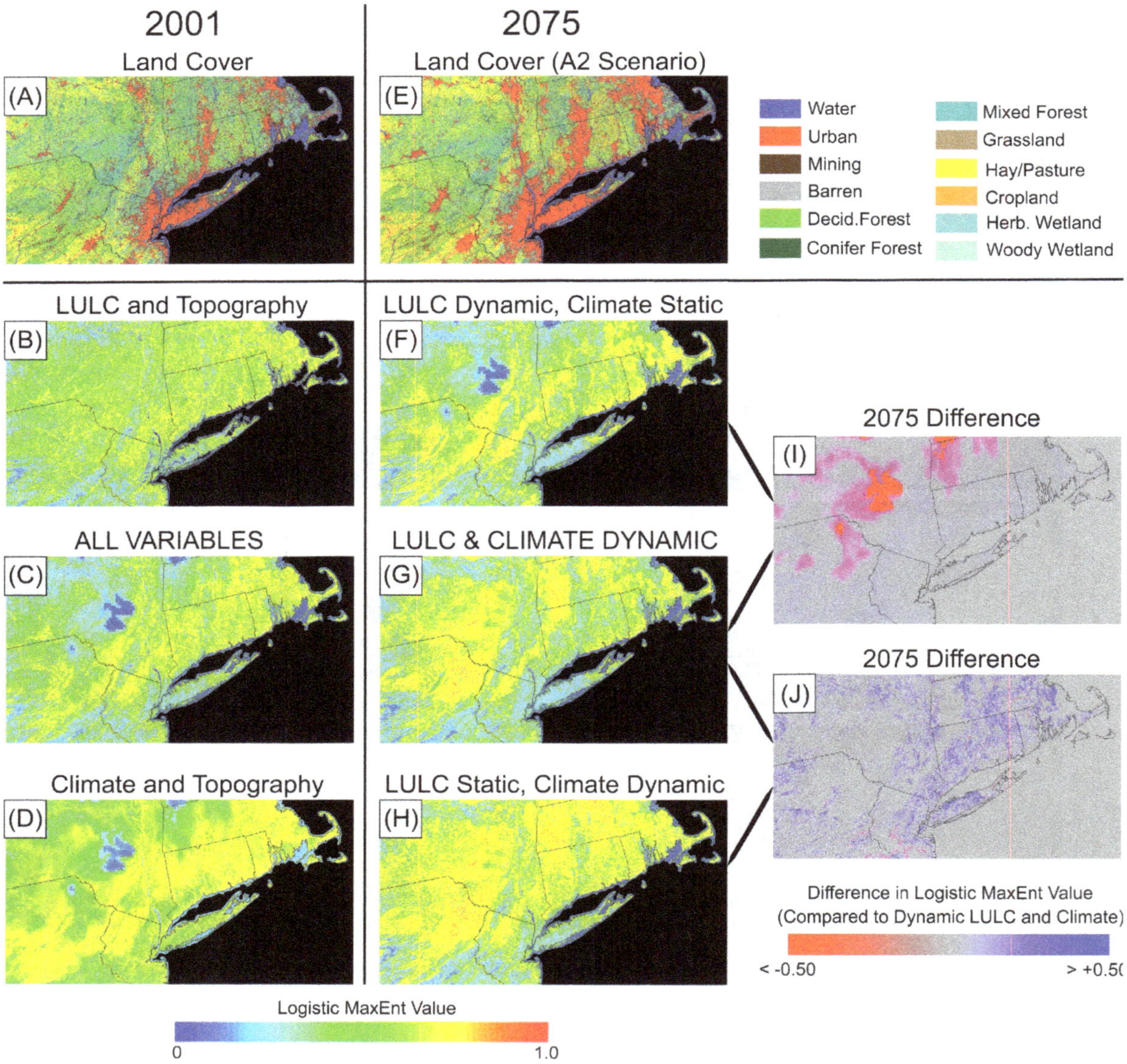

Figure 6. LULC and model results for a portion of the Hooded Warbler's range. Panels on the left (6 a–d) depict 2001 LULC, and the three model runs for 2001. Panels in the middle (6 e–h) depict 2075 LULC for the A2 scenario, and the three model runs for 2075 for that scenario. The two panels on the right depict differences in model results (compared to 6 g, when all variables were modeled) if 1) climate was held static for 2075 (6i), and 2) LULC was held static for 2075 (6j).

shows a model simulation with projected 2075 climate but a static 2001 LULC, where many climatically suitable areas were still noted as suitable for breeding despite the substantial loss of forest habitat (**Figure 6(j)**).

Specific results for all 50 species, including range maps as provided in Figure 6, are accessible through a companion website (http://landcover-modeling.cr.usgs.gov/sdm.php).

Discussion

Research Questions

What are the relative influences of LULC and climate in modeling contemporary (2001) breeding bird distributions? Clearly both climate and LULC change impact current bird species distributions, with relative impacts that are species specific. 2001 model fit was generally better with LULC simulations (climate excluded) than for climate simulations (LULC excluded), yet climate data covariates contributed more to model results than LULC data. One story that arises from these seemingly conflicting results is one of scale. Results suggest that climate data alone, without constraints afforded by the use of habitat (LULC) data, provide a "broad-brush" picture of suitability for a given species. LULC data alone excel at providing local-level insight to site-level habitat suitability. Given the inherent heterogeneity of the moderate-scale LULC data used here compared to variations in climate across geographic space, it is not surprising that climate alone offers only a general characterization of species range. For the 2001 species models, the modeled area deemed to be suitable to support a species was generally much higher when climate data were used without LULC data (**Figure 3**). The addition of LULC data to climate-based model simulations greatly restricted modeled species ranges in most cases. Prince et al. [59] similarly described climate as determining overall potential carrying capacity for a species, but noted the impact of climate change itself may be overestimated, as other factors that determine local suitability must be assessed. These results suggest that the use of climate data, without supporting LULC data, likely results in errors of commission, where climatically suitable regions are labeled as appropriate for supporting a given species, despite underlying LULC conditions that make actual presence unlikely. Araujo and Peterson [9] discussed such commission errors in bioclimatic envelope modeling, attributing overzealous predictions of range to an incomplete model; in this case, climate-only models for 2001 are "incomplete" without supporting LULC data.

Actual species range in relationship to the modeled study area influenced 2001 model results. Model fit was negatively impacted if the study area was largely contained within the actual species range. This was the case for the Double Truncated species, where ranges spanned all latitudes in the conterminous United States and were truncated at both the northern and southern borders. Climate data, in particular temperature data, thus did little to improve model fit, as species occurrences already spanned most potential climate regimes within the conterminous United States. With the resultant small impact of climate data, overall model fit suffered (**Figure 1; Table 5a**) and the addition of climate data did little to improve model results over the LULC and topography model (**Figure 1**). Modeled species range was also influenced by the relationship between actual range and study area. As noted above, the use of climate data without LULC data often resulted in errors of commission, but these errors were magnified for the Double Truncated species, with an over-prediction in suitable range as compared to the complete model with both climate and LULC data (**Figure 3**).

What are the relative impacts of projected climate and projected LULC change on breeding bird distributions? For modeled species ranges, projected changes in climate provided more dramatic shifts in future species' ranges than did projected LULC change. LULC change alone altered suitable range by no more than 20% for any species, yet climate change resulted in shifts of 50% or more for several species. Differences between the three different scenarios were often substantial, with some scenarios projecting double the range shift compared to other scenarios. However, the overall storyline was climate change impacting net changes in species range more than projected LULC change.

The relationship of the actual species range to the study area obviously affected 2075 results, with climate impacts often over- or under-estimated in relationship to LULC impacts, depending upon species. The Single Truncated class contained species where either the northern or southern extent of their actual range was artificially truncated by the borders of the study area. With a warming climate, for species with ranges truncated along their southern extent but not the north, the models thus predicted overall range expansion to the north, without capturing the (presumed) range contraction due to climate change at the species' southern range extent. Conversely, for species with ranges truncated along their northern extent but not the south, the models predicted overall range contraction, capturing contraction in the south but failing to capture (presumed) expansion in the north. By only capturing "half of the story" (i.e., either capturing range expansion in the north or range contraction in the south), these results provided an unrealistically high impacts of climate on *net* range, either positive or negative. While the impacts of climate on *net* change were thus likely overestimated for these species, *gross* change was likely underestimated, as half of the story was "missing". For the Double-Truncated class, the relative impacts of climate change to LULC change were likely underestimated. For these species, the impact of climate on species range was artificially dampened by the truncation of northern and southern range boundaries, areas where range could potentially expand or contract, respectively, due to a warming climate. For the Whole Range species, the relative impacts of climate change versus land use change vary, with net change values only providing part of the story as evidenced when assessing results for the focus species, the Hooded Warbler.

What are the specific impacts of climate and LULC on one focus species, the Hooded Warbler? The presented results for the Hooded Warbler mirrored those for many of the 50 modeled species. Net change in breeding range area showed relatively little change, while geographic patterns change dramatically. Climate change resulted in a broad overall shift in range to the north and to higher elevations, while LULC change resulted in heterogeneous, local-scale changes in habitat suitability. The breeding distributions of the Hooded Warbler have been found to be highly correlated with climate variables [60]. The species was unknown as a breeder in Canada until 1949, but with a warming climate they have started to breed in increasing numbers in extreme southern Ontario [60]. Melles et al. [60] modeled the relationship between climate and habitat covariates and the Hooded Warbler range, and found strong relationships between range expansion to the north and changes in climate over the last few decades, with habitat availability acting as a constraint on expansion. Naujokaitis-Lewis et al. [61] examined the potential impacts of climate change out to 2080 on Hooded Warblers and projected breeding range shifts to the north with characteristics dependent upon which GCM was assessed. However, they also found that land-use pressures around the Great Lakes were limiting factors to range expansion, and recommended future work that focused on "the development of more realistic (habitat) loss scenarios". The newly available LULC projections used in this work allowed for such an analysis.

As discussed, most future projections use projected climate data but ignore future LULC change. **Figure 6** clearly indicates that for a species such as the Hooded Warbler where climate change drives a broad overall shift in range to the north and to higher elevations, the modeled extent of suitable range at a local level can potentially be misrepresented without the use of projected LULC data. In this case, habitat loss due to urbanization and agricultural expansion would be missed without the use of LULC data, resulting in an over-prediction of suitable range (**Figure 6j**). Alternatively, without the use of projected LULC data, suitable range may be under-predicted if beneficial LULC change occurs (e.g., Grasshopper Sparrow (*Ammodramus savannarum*) results within the Eastern United States, where projected clearing of forest land in most scenarios resulted in more suitable habitat conditions by 2075). Exclusion of climate data can also result in a misrepresentation of modeled range. For the Hooded Warbler, range expansion to higher elevations was missed for the 2075 model excluding climate data (**Figure 6i**).

What are the implications for the use of climate and LULC data in SDMs? For contemporary species modeling or for projected changes in species range, both climate and LULC data should ideally be used. In general, model fit consistently increases with the use of both climate and LULC data, while predicted suitable range decreases. The implication is that information is missing from SDMs if both climate and LULC are not used as covariates. For modeling of current species range, areal summaries of modeled range (**Figure 3**) as well as spatially explicit maps of modeled range (**Figure 6b, c, and d**) show that

SDMs relying on climate data without LULC data provide only a broad-brush, generalized species range, while LULC data alone provide site-level information on habitat suitability while omitting climatic thresholds of unsuitability. Dependent on application, a broad-brush generalization of a species range may be adequate. However, it should be recognized that the results likely over-represent the area of suitable range and fine-scale detail is unlikely to be obtained.

Exclusion of LULC data is primarily an issue for projections of future species' range. As shown here, bioclimatic modeling where LULC information is not included or is considered static likely results in a misrepresentation of future species' range. For example, Hooded Warbler results provided here and in past studies indicate that while climate drives broad-scale shifts in range, SDMs likely misrepresent the extent of future range shifts if LULC change is not taken into account. Given how little projected LULC data is used in modeling future species distributions, quantitative estimates of range shifts are likely overestimated if habitat loss dominates projections of LULC change, or underestimated if habitat gain dominates projections of LULC change. Bioclimatic models that do not use any form of LULC information, even static LULC information for the future, likely overestimate suitable ranges (**Figure 3**).

The relationship between the study area and the actual species range also needs to be strongly considered, both in the project design and assessment phases. The methodology used here mimics that of many modeling applications. Of all the citations included in this paper where species distribution and/or probability-of-occurrence modeling was done, over two-thirds (21 of 31) of model applications assessed only partial/truncated species ranges. Many of the recommendations referenced in this paper with regard to model parameterization and handling scale issues [35,38,46,53], spatial bias and other issues with presence data [23,48,50], and relative influences of climate and LULC [1,8,12,16,59,60] were derived from studies where only partial ranges were assessed. Despite the prevalence of modeling of partial ranges, the results here indicate that caution is needed in project design, both for accurate modeling of species range, and for the interpretation of modeling results. Modeling of an entire species' range may improve model fit and enable a more direct interpretation of results, yet is often not practical due to data or processing limitations. While modeling a partial range is thus unavoidable in many cases, model results should be interpreted within the context of the overall project design and the relationship between species range and the study area. Modeling results may still be "valid" when using truncated ranges, but if the intent is to study the impacts of climate change on species distribution, for example, then the use of a "double-truncated" study area boundary would obviously be a poor choice, as the effects of climate would likely be artificially muted. If the intent is to quantify specific impacts of LULC and climate, disentangling the relative effects of LULC, climate, and other covariates would be complicated by the modeling of truncated ranges.

Comparison to Existing Research

The conceptual approach behind Barbet-Massin et al. [7] modeling of bird species in Europe and Matthews et al. [1] modeling of eastern U.S. bird species are likely the most similar work to this assessment. Each assessed a large number of species across broad geographic regions, and both incorporated projected climate and projected LULC data. Similar to Barbet-Massin et al. [7], this assessment found that LULC-based models alone predicted smaller overall shifts in future range size than did climate-based models. However, these results differ from multiple studies that discussed the relative influence of climate versus LULC, including components of Barbet-Massin et al. [7]. Barbet-Massin et al. [7] found that modeling accuracy was higher with climate-only variables than with habitat-only variables; in this study, the opposite was true in the majority of species that were assessed. Thuiller et al. [12] found that the inclusion of LULC covariates improved explanatory power of bioclimatic models, but that the "addition of land cover variables to pure bioclimatic models does not improve their predictive accuracy". In this assessment, on average, AUC scores declined more in the absence of LULC data than in the absence of climate data, while for all 50 species, 2001 model fit was improved when LULC data were included as a covariate as opposed to models with only climate and topographic data.

The differences in results may potentially be explained by 1) the difference in scale between the different assessments, 2) variations in the number of climate covariates, and 3) the use of topographic data within this assessment. Barbet-Massin et al. [7] used much coarser, 0.5-degree resolution LULC data, and noted that "such a resolution was probably too rough to precisely account for habitat factors." Thuiller et al. [12] also used a very coarse spatial resolution (50-km grid cells) and noted results may differ at finer resolutions. Bucklin et al. [13] similarly found that LULC variables provided little benefit in SDMs, but noted that both thematic and spatial resolution improvements over their LULC data source may have provided different results. Barbet-Massin et al. [7] and Thuiller et al. [12] also noted the lack of a measure of fragmentation or landscape heterogeneity in their assessments; in this study a LULC diversity measure was used to represent heterogeneity. This assessment also used only two climate covariates, while Barbet-Massin et al. [7], Thuiller et al. [12], and Bucklin et al. [13] each used eight climate covariates. Additional research is needed to assess the optimum combination of covariates and how covariate choice impacts results, particularly for heavily correlated climate covariates. The use of topographic data in this assessment also may have impacted the relative impacts of LULC and climate data. The information content provided by topographic data alone, or topographic data in combination with LULC data (e.g., "product features" within Maxent that assess 2-way interactions between covariate pairs) may partially mimic or replace the information content that is provided by climate variables [42].

Matthews et al. [1] also modeled U.S. bird species, and also used projected climate data, projected LULC data, and topographic data to assess future changes in distributions. They modeled the eastern portion of the United States at a very coarse spatial resolution (20-km grid cells), with changes in tree species representing the only modeled form of LULC change. Similar to their results, the predictive power (indicated by goodness-of-fit measures) of the models described here decreased when only climate and elevation data were used as predictor variables (i.e., LULC excluded). Even with the differences in spatial resolution, both studies found that modeling with only climate and topographic variables leads to generalized species distribution maps that lack fine-scale detail. Matthews et al. [1] noted that modeling with climate and topographic data alone makes the resultant models much more susceptible to over-prediction of future impacts of climate change on species range. It should also be noted that their use of the eastern United States as the study area resulted in artificial truncation of nearly all modeled species ranges.

Caveats and Future Research

In assessing potential future changes in species ranges, caution has been recommended when attempting to apply a contemporary model to future climate conditions [26,46]. Transferability of model results is confounded when novel conditions (i.e., specific combinations of covariates not found in the original model's training data) are found for future dates or for other geographic regions. In this assessment novel conditions were most likely to occur with higher temperatures due to climate change. However, the use of the 500-km buffer to establish the study area for each species, as well as the use of the clamping feature in Maxent, resulted in a muted influence of novel conditions on model results. Selecting background points within a 500-km buffer of a species' current range enabled the collection of background points with higher (points selected south of the breeding range) or lower (points selected north of the breeding range) average temperatures than those found in the breeding range. Thus temperature was often used as a threshold feature in species' models, with conditions modeled as unsuitable if temperature in a given location was above or below a modeled tolerance level for the species. For example, the breeding range for the Bobolink (*Dolichonyx oryzivorus*) covers much of the northern United States, but they are absent in southern areas. An examination of the 2001 model developed for the Bobolink shows average temperature used as a threshold feature. For a species such as this, with the southern end of its breeding range currently within the conterminous United States, novel conditions potentially introduced by a warming climate were unimportant, as the model already ensured exclusion of the species as a breeder in areas with temperatures above threshold values found in the training data.

Novel conditions could potentially be a problem for species with current breeding ranges extending to the United States and Mexico border. No background data south of the border were used to train the model. With a warming climate, for some species, it is likely that local temperatures in the projected climate data exceeded any temperatures found in the training data. The clamping feature in Maxent was used to control novel conditions in situations such as this, effectively rescaling novel covariate values to maximum values found in the training data. Clamping thus eliminated statistical issues with applying models into a novel prediction space, but by rescaling extreme values in the projection space, the model may effectively be dampening the impact of future change on future species distributions for 2075. Clamping of novel temperatures, for example, could result in the model incorrectly representing far southern portions of a species range as "suitable" for breeding, when in fact a temperature tolerance limit has been reached and has pushed the southern limit of the breeding range north of the United States and Mexico border.

There are additional potential caveats in interpreting results of this assessment. These results are based on one modeling methodology (Maxent), with one defined method for parameterization. Many papers have focused on the effects of different parameterization settings when using the Maxent model [27,48,56], and it was not the intention of this paper to revisit how different parameterizations affect Maxent results. The results presented here were also conducted at one specific spatial scale, with one specific suite of covariates and bird presence data. It was impractical to perform comprehensive analyses across all possible permutations of modeling frameworks, parameterization settings, spatial scales, thematic scales, temporal resolution, and data sources; results may differ for assessments where these components are altered. There was no attempt to rigorously address all potential sources of modeling uncertainty in this assessment. Conlisk et al. [38] attempted to disentangle all sources of uncertainty in SDMs, concluding that the modeling framework itself is the most important source of uncertainty. Ideally multiple models would be used to also disentangle effects of the modeling frameworks themselves, but resources were unavailable for a multi-model assessment given the large number of species, and multiple combinations of dates, covariates, and scenarios. Other potential drawbacks to the approach used here is an oversimplified representation of the driving forces behind species distributions. One final area that needs further exploration is the correction of bias for eBird data. Spatial bias was mitigated by spatially filtering the data. However, given the number and diversity of species modeled, a consistent filtering threshold of 20 km was used for all species; no attempts were made to tailor the filtering protocol to the spatial data characteristics for each species, nor were attempts made to quantify the reduction in spatial bias in this assessment. Additional potential sources of error and bias in eBird data that were not accounted for include accuracy of geographic data entry and highly variable observation and identification skills among eBird participants [62,63].

Conclusion

This work represents the first assessment of the effects of climate and LULC for bird species in the conterminous United States using both 1) newly available LULC projections of high-spatial and thematic resolution and 2) climate and LULC projections that are both consistent with IPCC SRES scenario frameworks. While modeling results clearly indicate a species-dependent determination of the relative impacts of climate and LULC change on both current and future range, it is clear that SDMs benefit by including both climate and LULC covariates. The use of climate data alone likely results in errors of commission and an over-prediction of current range. For future modeling of species range, the use of climate change information without corresponding LULC change may result in the misrepresentation of future range either positively or negatively, dependent upon whether projected LULC change was harmful or beneficial to a species. The inclusion of LULC data in SDMs 1) significantly increased measures of model fit, and 2) "tempered" predicted ranges from climate-only modeling frameworks by providing fine-scale information on local habitat suitability. When modeling future shifts in range, climate had the dominant impact on range shifts, yet LULC change was dominant for many species. Relationship of the species' range to the geographic bounds of the study area also clearly impacts whether climate or LULC has the dominant effect on modeled species range, and needs to be considered at both the design and assessment stages of a study.

All LULC projections used for this assessment are available at http://landcover-modeling.cr.usgs.gov. The computed predictor variables (covariates) used in this assessment, all range maps for 2001 and 2075 for each of the fifty modeled species, and a spreadsheet of all quantitative data reported in this paper are accessible at http://landcover-modeling.cr.usgs.gov/sdm.php. While this paper has focused on generalized results across the 50 modeled species, detailed model results for each of the 50 modeled species also are included herin. eBird data used as presence locations for this work may be obtained by through http://ebird.org.

Acknowledgments

Funding for this research was provided by the U.S. Geological Survey's Climate and Land Use Program. I thank Alisa Gallant for her work on downscaling the GCM data that were used in this assessment.

Author Contributions

Conceived and designed the experiments: TLS. Performed the experiments: TLS. Analyzed the data: TLS. Contributed reagents/materials/analysis tools: TLS. Contributed to the writing of the manuscript: TLS.

References

1. Matthews SN, Iverson LR, Prasad AM, Peters MP (2011) Changes in potential habitat of 147 North American breeding bird species in response to redistribution of trees and climate following predicted climate change. Ecography 34: 933–945.
2. Brambilla M, Ficetola GF (2012) Species distribution models as a tool to estimate reproductive parameters: a case study with a passerine bird species. Journal of Animal Ecology 81: 781–787.
3. Root T (1988) Energy constraints on avian distributions and abundances. Ecology 69(2): 330–339.
4. Thogmartin WE, Knutson MG, Sauer JR (2006) Predicting regional abundance of rare grassland birds with a hierarchical spatial count model. The Condor 108: 25–46.
5. Rahbek C, Gotelli NJ, Colwell RK, Entsminger GL, Rangel TF, et al. (2007) Predicting continental-scale patterns of bird species richenss with spatially explicit models. Proceedings of the Royal Society B (274): 165–274.
6. Pearson RG, Dawson TP (2003) Predicting the impacts of climate change on the distribution of species: are bioclimate envelope models useful? Global ecology and Biogeography 12: 361–371.
7. Barbet-Massin M, Thuiller W, Jiguet F (2012) The fate of European breeding birds under climate, land-use and dispersal scenarios. Global Change Biology 18: 881–890.
8. Jongsomjit D, Stralberg D, Gardali T, Salas L, Wiens J (2013) Between a rock and a hard place the impacts of climate change and housing development on breeding birds in California. Landscape Ecology 28: 187–200.
9. Watling JI, Bucklin DN, Speroterra C, Brandt LA, Mazzotti FJ, et al. (2013) Validating predictions from climate envelope models. PLoS ONE 8(5): e63600. doi:10.1371/journal.pone.0063600.
10. Araujo MB, Peterson AT (2012) Uses and misuses of bioclimatic envelope modeling. Ecology 93(7): 1527–1539.
11. Jimenez-Valverde A, Barve N, Lira-Noriega A, Maher SP, Nakazawa Y, et al. (2011) Dominant climate influences on North American bird distributions. Global Ecology and Biogeography 20: 114–118.
12. Thuiller W, Araujo MB, Lavorel S (2004) Do we need land-cover data to model species distributions in Europe? Journal of Biogeography 31: 353–361.
13. Bucklin DN, Basille M, Benscoter AM, Brandt LA, Mazzotti FJ, et al. (2014) Comparing species distribution models constructed with different subsets of environmental predictors. Diversity and Distributions. doi:10.1111/ddi.12247.
14. Lee TM, Jetz W (2011) Unravelling the structure of species extinction risk for predictive conservation science. Proceeding of the Royal Society B(278): 1329–1338.
15. Sinclair SJ, White MD, Newell GR (2010) How useful are species distribution models for managing biodiversity under future climates? Ecology and Society 15(1): 8. [online]. Available: http://www.ecologandsociety.org/vol15/iss1/art8/.
16. Riordan EC, Rundel PW (2014) Land use compounds habitat losses under projected climate change in a threatened California ecosystem. PLoS ONE 9(1): e86487. doi:10.1371/journal/pone.0086487.
17. Pearson RG, Dawson TP, Liu C (2004) Modelling species distributions in Britain: A hierarchical integration of climate and land-cover data. Ecography 27: 285–298.
18. Sullivan BL, Wood CL, Iliff MJ, Bonney RE, Fink D, et al. (2009) eBird: A citizen-based observation network in the biological sciences. Biological Conservation 142: 2282–2292.
19. Sullivan BL, Aycrigg JL, Barry JH, Bonney RE, Bruns N, et al. (2014) The eBird enterprise: An integrated approach to development and application of citizen science. Biological Conservation 169: 31–40.
20. Hochachka W, Fink D (2012) Broad-scale citizen science data from checklists: prospects and challenges for macroecology. Frontiers of Biogeography 4(4): 150–154.
21. Lerman SB, Nislow KH, Nowak DJ, DeStefano S, King DI, et al. (2014) Using urban forest assessment tools to model bird habitat potential. Landscape and Urban Planning 122: 29–40.
22. Hurlbert AH, Liang Z (2012) Spatiotemporal variation in avian migration phenology: Citizen science reveals effects of climate change. PLoS ONE 7(2): e31662. doi:10.1371/journal.pone.0031662.
23. Fink D, Hochachka WM, Zuckerberg B, Winkler DW, Shaby B, et al. (2010) Spatiotemporal exploratory models for broad-scale survey data. Ecological Applications 20(8): 2131–2147.
24. Ridgeby RS, Allnutt TF, Brooks T, McNicol DK, Mehlman DW, et al. (2003) Digital distribution maps of the birds of the Western Hemisphere, version 1.0 NatureServe, Arlington, Virginia USA.
25. Yackulic CB, Chandler R, Zipkin EF, Royle JA, Nichols JD, et al. (2013) Presence-only modeling using MaxEnt: when can we trust the inferences? Methods in Ecology and Evolution 4: 236–243.
26. Elith J, Phillips SJ, Hastie T, Dudik M, Chee YE, et al. (2011) A statistical explanation of MaxEnt for ecologists. Diversity and Distributions 17: 43–57.
27. Kramer-Schadt S, Niedballa J, Pilgrim JD, Schroder B, Lindenborn J (2013) The importance of correcting for sampling bias in MaxEnt species distribution models. Diversity and Distributions 19: 1366–1379.
28. Boria RA, Olson LE, Goodman SM, Anderson RP (2014) Spatial filtering to reduce sampling bias can improve the performance of ecological niche models. Ecological Modelling 275: 73–77.
29. Wisz MS, Hijmans RJ, Li J, Peterson AT, Graham CH, et al. (2008) Effects of sample size on the performance of species distribution models. Diversity and Distributions 14: 763–773.
30. Hernandez PA, Graham CH, Master LL, Albert DL (2006) The effect of sample size and species characteristics on performance of different species distribution modeling methods. Ecography 29: 773–785.
31. Sohl TL, Sleeter BM, Zhu Z, Sayler KL, Bennett S, et al. (2012) A land-use and land-cover modeling strategy to support a national assessment of carbon stocks and fluxes: Applied Geography 34: 111–124.
32. Sohl TL, Sayler KL, Bouchard MA, Reker RR, Friesz AM, et al. (2014) Spatially explicit modeling of 1992 to 2100 land cover and forest stand age for the conterminous United States. Ecological Applications 24(5): 1015–1036. Available: http://dx.doi.org/10.1890/13-1245.1.
33. Nakicenovic N, Alcamo J, Davis G, de Vries HJM, Fenhann J, et al. (2000) Special Report on Emissions Scenarios (SRES). Intergovernmental Panel on Climate Change (IPCC). Cambridge University Press, Cambridge, UK. 570 p.
34. Vogelmann JE, Howard SM, Yang L, Larson CR, Wylie BK, et al. (2001) Completion of the 1990s National Land Cover Data Set for the conterminous United States. Photogrammetric Engineering and Remote Sensing 67: 650–652.
35. Cunningham MA, Johnson DH (2006) Proximate and landscape factors influence grassland bird distributions. Ecological Applications 16(3): 1062–1075.
36. Bakker KK, Naugle DE, Higgins DF (2002) Incorporating landscape attributes into models for migratory grassland bird conservation. Conservation Biology 16(6): 1638–1646.
37. Fearer TM, Prisley SP, Stauffer DF, Keyser PD (2007) A method for integrating the Breeding Bird Survey and Forest Inventory and Analysis databases to evaluate forest bird-habitat relationships at multiple spatial scales. Forest Ecology and Management 243: 128–143.
38. Conlisk E, Syphard AD, Franklin J, Flint L, Flint A, et al. (2013) Uncertainty in assessing the impacts of global change with coupled dynamic species distribution and population models. Global Change Biology 19: 858–869.
39. Hay LE, Markstrom SL, Ward-Garrison C (2011) Watershed-scale response to climate change through the twenty-first century for selected basins across the United States. Earth Interactions 15: 1–37.
40. Wu Y, Liu S, Gallant AL (2012) Predicting impacts of increased CO^2 and climate change on the water cycle and water quality in the semiarid James River Basin of the Midwestern USA. Science of the Total Environment 430: 150–160.
41. Bradley BA, Olsson AD, Wang O, Dickson BG, Pelech L, et al. (2012) Species detection vs. habitat suitability: Are we biasing habitat suitability models with remotely sensed data? Ecological Modelling 244: 57–64.
42. Kery M, Gardner B, Monnerat C (2010) Predicting species distributions from checklist data using site-occupancy models. Journal of Biogeography 37: 1851–1862.
43. Moreno R, Zamora R, Molina JR, Vasquez A, Herrera MA (2011) Predictive modeling of microhabitats for endemic birds in South Chilean temperate forests using Maximum entropy (Maxent). Ecological Informatics 6: 364–370.
44. Johnston KM, Freund KA, Schmitz OJ (2012) Projected range shifting by montane mammals under climate change: implications for Cascadia's National Parks. Ecosphere 3(11): 97. Available: http://dx.doi.org/10.1890/ES12-00077.1.
45. U.S. Geological Survey (1999) USGS 30-Meter Resolution, One-Sixtieth Degree National Elevation Dataset for CONUS, Alaska, Hawaii, Puerto Rico, and the U.S. Virgin Islands. U.S. Geological Survey (USGS) Earth Resources Observation and Science (EROS) Center, Sioux Falls, SD.
46. Phillips SJ, Dudik M (2008) Modeling of species distributions with Maxent: new extensions and a comprehensive evaluation. Ecography 31: 161–175.
47. Bahn V, McGill BJ (2012) Testing the predictive performance of distribution models. Oikos 000: 001–011, doi:10.1111/j.1600-0706.2012.00299.x.
48. Phillips SJ, Dudik M, Elith J, Graham CH, Lehmann A, et al. (2009) Sample selection bias and presence-only distribution models: implications for background and pseudo-absence data. Ecological Applications 19(1): 181–197.
49. Dudik M, Schapire RE, Phillips SJ (2005) Correcting sample selection bias in maximum entropy density estimation. Advances in Neural Information Processing Systems 18. MIT Press, Cambridge, Massachusetts, USA. 320–330.
50. Veloz SD (2009) Spatially autocorrelated sampling falsely inflates measures of accuracy for presence-only niche models. Journal of Biogeography 36: 2290–2299.

51. Elith J, Kearney M, Phillips SJ (2010) The art of modelling range-shifting species. Methods in Ecology and Evolution 1: 330–342.
52. Barve N, Barve V, Jimenez-Valverde A, Lira-Noriega A, Maher AP, et al. (2011) The crucial role of the accessible area in ecological niche modeling and species distribution modeling. Ecological Modelling 222: 1810–1819.
53. VanDerWal J, Shoo LP, Graham C, Williams SE (2009) Selecting pseudo-absence data for presence-only distribution modeling: How far should you stray from what you know? Ecological Modelling 220: 589–594.
54. Anderson RP, Gonzalez Jr. I (2011) Species-specific tuning increases robustness to sampling bias in models of species distributions: An implementation with MaxEnt. Ecological Modelling 222: 2796–2811.
55. Warren DL, Seifert SN (2011) Ecological niche modeling in Maxent: the importance of model complexity and the performance of model selection criteria. Ecological Applications 21(2): 335–342.
56. Syfert MM, Smith MJ, Coomes DA (2013) The effects of sampling bias and model complexity on the predictive performance of MaxEnt species distribution models. PLoS One 8(2): e55158. doi:10.1371/journal.pone.0055158.
57. Liu C, Berry PM, Dawon TP, Pearson RG (2005) Selecting thresholds of occurrence in the prediction of species distributions. Ecography 28: 385–393.
58. Liu C, White M, Newell G (2013) Selecting thresholds for the prediction of species occurrence with presence-only data. Journal of Biogeography (40): 778–789.
59. Prince K, Lorrilliere R, Barbet-Massin M, Jiguet F (2013) Predicting the fate of French bird communities under agriculture and climate change scenarios. Environmental Science & Policy 33: 120–132.
60. Melles SJ, Fortin MJ, Lindsay K, Badzinski D (2011) Expanding northward: influence of climate change, forest connectivity, and population processes on a threatened species' range shift. Global Change Biology 17: 17–31.
61. Naujokaitis-Lewis IR, Curtis JMR, Tischendorf L, Badzinski D, Lindsay K, et al. (2013) Uncertainties in coupled species distribution-metapopulation dynamics models for risk assessments under climate change. Diversity and Distributions 19: 541–554.
62. Yu J, Wong WK, Hutchinson R (2010) Modeling Experts and Novices in Citizen Science Data for Species Distribution Modeling. Proceedings of the 2010 IEEE International Conference on Data Mining, (pp. 1157–1162), Washington, DC: IEEE Computer Society.
63. Dickinson JL, Zuckerberg B, Bonter DN (2010) Citizen Science as an Ecological Research Tool: Challenges and Benefits. Annual Review of Ecology, Evolution, and Systematics 41: 149–172.

15

Patterns and Variability of Projected Bioclimatic Habitat for *Pinus albicaulis* in the Greater Yellowstone Area

Tony Chang*, Andrew J. Hansen, Nathan Piekielek

Department of Ecology, Montana State University, Bozeman, Montana, United States of America

Abstract

Projected climate change at a regional level is expected to shift vegetation habitat distributions over the next century. For the sub-alpine species whitebark pine (*Pinus albicaulis*), warming temperatures may indirectly result in loss of suitable bioclimatic habitat, reducing its distribution within its historic range. This research focuses on understanding the patterns of spatiotemporal variability for future projected *P.albicaulis* suitable habitat in the Greater Yellowstone Area (GYA) through a bioclimatic envelope approach. Since intermodel variability from General Circulation Models (GCMs) lead to differing predictions regarding the magnitude and direction of modeled suitable habitat area, nine bias-corrected statistically down-scaled GCMs were utilized to understand the uncertainty associated with modeled projections. *P.albicaulis* was modeled using a Random Forests algorithm for the 1980–2010 climate period and showed strong presence/absence separations by summer maximum temperatures and springtime snowpack. Patterns of projected habitat change by the end of the century suggested a constant decrease in suitable climate area from the 2010 baseline for both Representative Concentration Pathways (RCPs) 8.5 and 4.5 climate forcing scenarios. Percent suitable climate area estimates ranged from 2–29% and 0.04–10% by 2099 for RCP 8.5 and 4.5 respectively. Habitat projections between GCMs displayed a decrease of variability over the 2010–2099 time period related to consistent warming above the 1910–2010 temperature normal after 2070 for all GCMs. A decreasing pattern of projected *P.albicaulis* suitable habitat area change was consistent across GCMs, despite strong differences in magnitude. Future ecological research in species distribution modeling should consider a full suite of GCM projections in the analysis to reduce extreme range contractions/expansions predictions. The results suggest that restoration strageties such as planting of seedlings and controlling competing vegetation may be necessary to maintain *P.albicaulis* in the GYA under the more extreme future climate scenarios.

Editor: Ben Bond-Lamberty, DOE Pacific Northwest National Laboratory, United States of America

Funding: This work was supported by the National Aeronautics and Space Administration Applied Sciences Program (Grant 10-BIOCLIM10-0034); Funder URL: http://www.nasa.gov (AJH TC). It also received support from the National Science Foundation Experimental Program to Stimulate Competitive Research (EPSCoR) Track-I EPS-1101342 (INSTEP 3); Funder URL: http://www.nsf.gov/div/index.jsp?div = EPSC (NBP TC); and the North Central Climate Science Center (G13AC00392-G-8829-1); Funder URL: http://www.doi.gov/csc/northcentral/index.cfm (AJH NBP). The funders had no role in study design, data collection and analysis, decision to publish, or preparation of the manuscript.

Competing Interests: The authors have declared that no competing interests exist.

* Email: tony.chang@msu.montana.edu

Introduction

Over the next century, it is expected that most of North America will experience climate changes related to increased concentrations of anthropogenic greenhouse gas emissions and natural variability [1]. At regional scales these changes are highly variable and can result in areas of increased mesic, xeric, or even hydric habitat conditions relative to present day. These shifting climates in turn also transform the suitable habitat for individual species that may result in changes in species composition and dominant vegetation types.

Whitebark pine (*Pinus albicaulis*) is a native conifer of the Western U.S. that is considered a keystone species in the sub-alpine environment. It provides a food source for animals such as the grizzly bear (*Ursus arctos*), red squirrel (*Tamiasciurus hudsonicus*), and Clark's nutcracker (*Nucifraga columbiana*) [2]. It also serves the ecosystem functions of stabilizing soil, moderating snow melt and runoff, and facilitating establishment for other species [2,3]. Whitebark pine has experienced a notable decline in the past two decades within the U.S. Northern Rockies due to high rates of infestation from the mountain pine beetle (*Dendroctonus ponderosae*) and infections from white pine blister rust (*Cronartium ribicola*), resulting in an 80% mortality rate within the adult population [4–7]. Given the potential loss of important ecosystem functions that whitebark pine contribute to the landscape under this mortality event, there is an emphasis to understand the climate characteristics of its habitat to identify the restoration strategies and locations that may aid the persistence of the species under future climates.

One method of understanding species response to climate change is through bioclimate niche modeling, which has become a common practice for assessing potential vegetation shifts under new environmental conditions [8–13]. Ecological niche theory proposes there exists some range of bioclimatic conditions within which a species can persist [14]. In bioclimatic niche modeling, the realized niche is modeled by empirical relationships between the presence or absence of a species and the associated abiotic, and

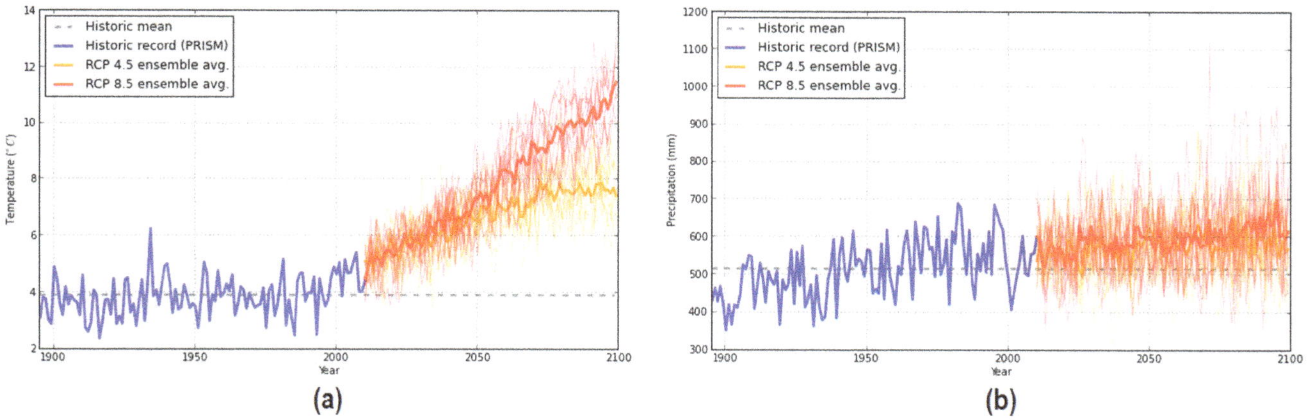

Figure 1. Historic and projected climates variables for the GYA from 1895–2099 under RCP 4.5 and 8.5 scenarios. Light shaded orange and red lines represent individual GCMs for RCP 4.5 and 8.5 respectively. Bold lines represent GCM ensemble average. (a) Mean annual temperature (b) Mean annual precipitation.

sometimes biotic, variables that describe the niche space. Bioclimatic models assume that species are in equilibrium with their environment and that the current abiotic relationships reflect a species environmental preferences which may be retained into the future [15,16]. At macro scales, bioclimatic approaches have demonstrated success at predicting current distributions of species [17,18]. Most bioclimatic models do not explicitly consider the many additional ecological factors that ultimately influence a species distribution such as dispersal, disturbance, or biotic interaction. Thus the approach does not predict where a species will actually occur in the future, but rather it predicts locations where climatic conditions will be suitable for the species.

Bioclimatic niche methodology has demonstrated utility in modeling historic ranges of species for conservation and management applications. By modeling the present day suitable habitat and then projecting those habitats into the future, bioclimatic

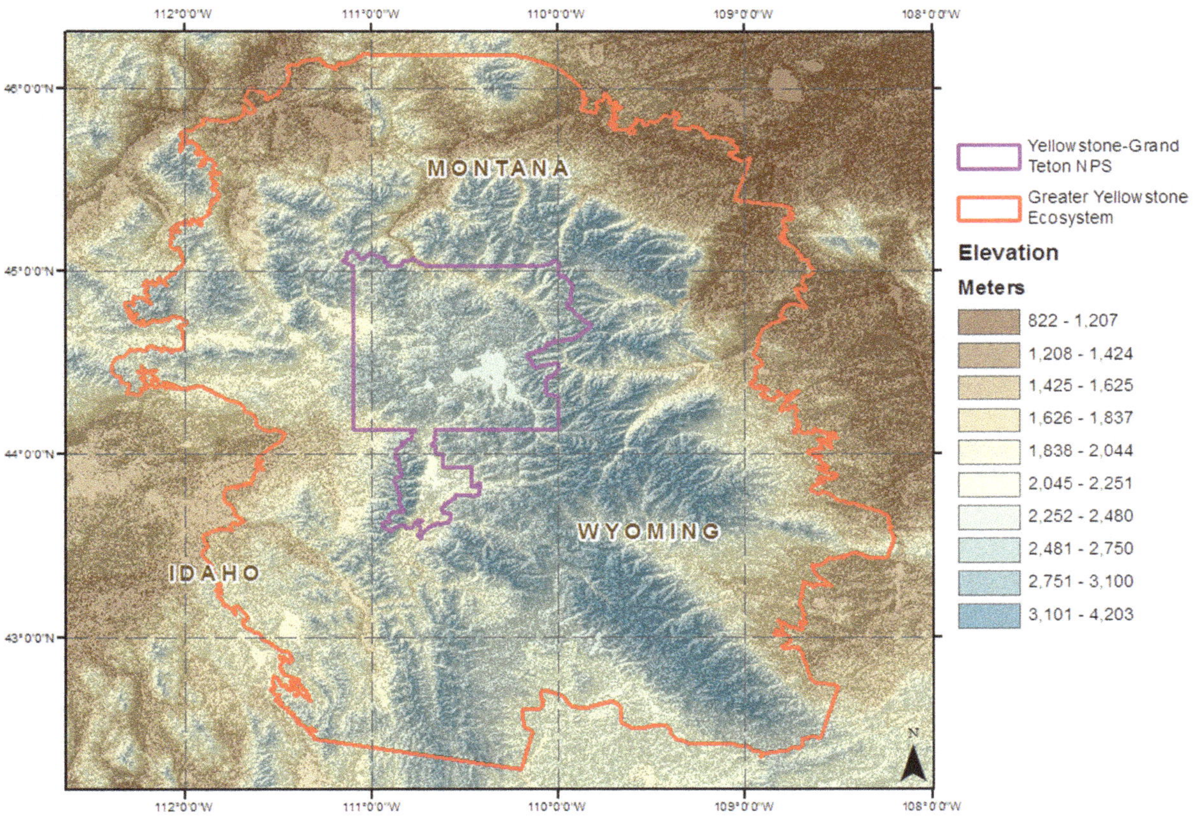

Figure 2. The Greater Yellowstone Area, representing an area of 150,700 km^2 with an elevational gradient from 522–4,206 m.

Table 1. General Circulation Models for analysis.

Name	Institute	Country
CESM1-CAM5	National Center for Atmospheric Research	US
CCSM4	National Center for Atmospheric Research	US
CESM1-BGC	National Center for Atmospheric Research	US
CNRM-CM5	Centre National de Recherche Meteorologiques	FR
HadGEM2-AO	Met Office Hadley Centre Climate Programme	UK
HadGEM2-ES	Met Office Hadley Centre Climate Programme	UK
HadGEM2-CC	Met Office Hadley Centre Climate Programme	UK
CMCC-CM	Centro Euro-Mediterraneo per Cambiamenti Climatici	ITA
CanESM2	Canadian Centre for Climate Modelling and Analysis	CAN

Selection of AR5 GCMs that represent historic climate in the U.S. Pacific Northwest region for future bioclimate habitat modeling.

niche models can serve as the first step filter for conservation action plans, such as mapping suitable species reintroduction sites or habitat reserve selection [19–21]. For *P.albicaulis*, McLane and Aitken [22] utilized bioclimate niche models to successfully implement experimental assisted migration on persisting climate habitat in British Columbia. Additionally, models of *Pinus flexis*, a closely related species of five needle pine, have been used to evaluate management options in Rocky Mountain National Park [23]. Given these examples, an effort to model and projected suitable climate habitat for *P.albicaulis* within a regional domain can provide valuable insight to land resource managers.

In this study, we present a bioclimatic habitat model for *P.albicaulis* within the Greater Yellowstone Area (GYA). Although *P. albicaulis* has a range-wide distribution that is split into two broad sections, one along Western North America: the British Columbia Coast Range, the Cascade Range, and the Sierra Nevada; and the other section in the Intermountain West that covers the Rocky Mountains from Wyoming to Alberta [2,24]; the GYA was selected as the primary geographic modeling domain for three reasons: 1) evidence that the *P. albicaulis* sub-population in the GYA is genetically distinct from other regional populations with different climate tolerances [25]; 2) the high regional investment in *P. albicaulis* conservation in the area [6]; 3) the high density of climate stations within the region. Climate within the GYA is highly heterogenous due to complex topography, and sharp elevational gradients. Current knowledge of the region expects climate to shift towards increased mean annual temperatures and earlier spring snowmelt [26,27]. This shift is expected to have an impact on the total suitable habitat area for *P. albicaulis*. Modeling at a regional scale can provide a finer resolution spatially explicit description of the bioclimatic envelope of *P. albicaulis* in the GYA.

Here we also present an opportunity to investigate the effect of future climate variability on projected species distributions. In 2013, the World Climate Research Programme Coupled Model released the new generation General Circulation Model (GCM) projections through the Coupled Model Intercomparison Project Phase 5 (CMIP5) [28]. These new GCM projections also include four possible climate futures are modeled with each GCM under the Representative Concentration Pathways (RCP) of greenhouse gas/aerosol. These RCP scenarios designate four different levels of radiative forcing (2.6, 4.5, 6.0 and 8.5 W/m^2) that may occur by the year 2099 [29]. In practice, research of future species suitable climate generally use a small suite of GCM/RCP combinations to project future climate [8,11,30]. However, internal variability in these GCMs that arise from modeled coupled interactions among the atmosphere, oceans, land, and cryosphere can result in atmospheric circulation fluctuations that are characteristic of a stochastic process [31]. Such intrinsic atmospheric circulation variations from model structure induce regional changes in air temperature and precipitation on the multi-decadal time scale [31]. For the GYA specifically, this GCM variability has be observed with mean annual temperatures projected to increase by $2-9°C$ and mean annual precipitation to change by -50 to $+$ 225 mm (Fig. 1). This suggests that magnitude and direction of projected species distributions at a regional scale can vary depending on the GCM selected and the modeled species response to more xeric or mesic future climate conditions [32].

To summarize, this study presents a bioclimate niche model for *P. albicaulis* based on historic climate observations and field sampling of *P. albicaulis* presence and absence. Using this modeled bioclimate envelope, projections of future total climate suitable habitat area under nine GCMs and two RCP scenarios will be measured. Since different GCMs may project a diverging spectrum of climates, it is expected that measures of total suitable habitat will reduce with varying degrees of area loss. It is also expect that number and size of continous patches of *P. albicaulis* habitat will reduce due to the limited available number of sub-alpine areas distributed within the landscape. This research provides an analysis of the variability of biotic response under a large suite of GCMs to provide managers/researchers with a measure of the uncertainty associated with future species distribution models. Furthermore, this analysis explicitly describes the spatial patterns of bioclimatic niches for *P. albicaulis* to gain a better understanding of topographic characteristics, such as elevation, on suitable habitat. Changes in these spatial patterns are examined through quantifying landscape patch dynamic that may result from GCM projections to understand the species trends for persistence on the landscape.

Methods

Study area

The GYA, which includes Yellowstone National Park, Grand Teton National Park, and a number of state and federally managed forests, is a mid- to high-latitude region in the Northern Rocky Mountains of western North America. Conifers are dominant in the range, with forest types composed of *Pinus contorta*, *Abies lasiocarpa*, *Pseudotsuga menziesii*, *Pinus albicaulis*, *Juniperus scopulorum*, *Pinus flexis* and *Picea engelmannii*, although the deciduous hardwood *Populus tremuloides*, is also wide spread. Plateaus and lowlands are dominated by species of *Artemisia tridentata* and open grasslands of mixed composition. The GYA study area encompasses 150,700 km^2 with an elevational gradient from 522–4,206 m that represents 14 surrounding mountain ranges (Fig. 2).

Data

Biological data. Field observations of *P. albicaulis* adult presences and absences were compiled from three data sources. First, 2,545 observations from the Forest Inventory and Analysis (FIA) program were assembled. FIA plots are located on a regular gridded sampling design with one plot at approximately every 2,500 forested hectares, with swapped and fuzzed exact plot locations within 1.6 km to protect privacy [33]. Gibson et al. [34] found that model accuracy to not be dramatically affected by data fuzzing, but to provide the most spatial accuracy, this study culled FIA field points where measured elevation were >300 m different from a 30 m USGS DEM [35]. To capitalize on additional field observations of *P. albicaulis* within the study area, and because false absences are one of the most problematic data issues in constructing bioclimatic niche models [36]; supplementary points were drawn from the Whitebark/Limber Pine Information System (WLIS) [37], and long-term monitoring plots established by the

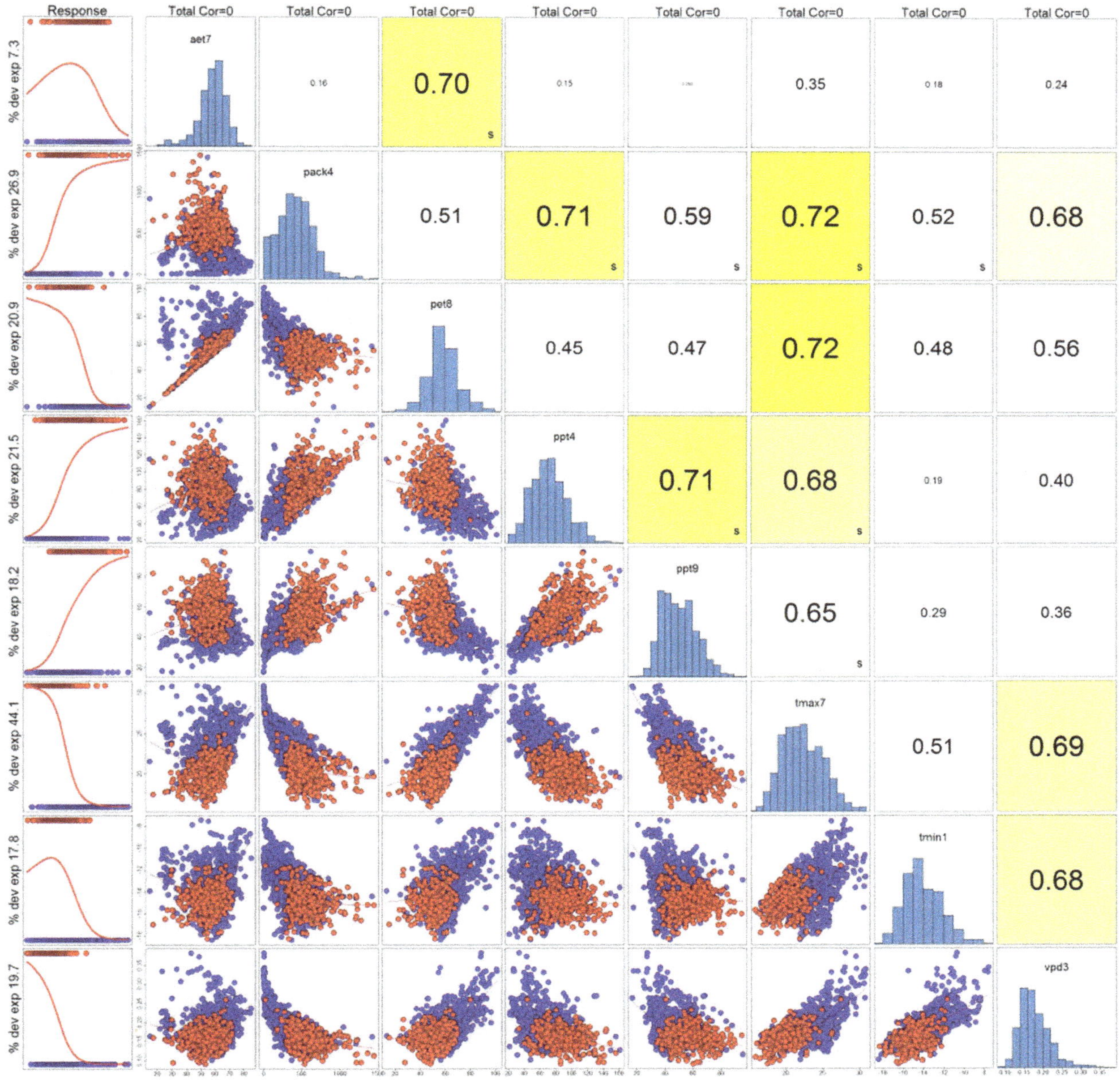

Figure 3. Selected predictor variables based on Principal Component Analysis and a maximum correlation filter of $\leq$0.75. Scatter plots represent one-to-one covariate plots where red points represent *P. albicaulis* presence, and blue points represent absence from field data. Far-left columns display logistic-regression of covariates from Generalized Additive Modeling using the Software for Assisted Habitat Modeling (SAHM [59]).

Table 2. Bioclimatic predictor variable list.

Code	Predictor Variable
tmin1	Minimum Temperature January
vpd3	Vapor Pressure Deficit March
ppt4	Precipitation April
pack4	Snow Water Equivalent April
tmax7	Maximum Temperature July
aet7	Actual Evapotranspiration July
pet8	Potential Evapotranspiration August
ppt9	Precipitation September

Final predictor variable set for Random Forest modeling. All variables were calculated as a 30-year climate mean from 1950–1980.

National Park Service Greater Yellowstone Inventory and Monitoring Network (GYRN) [38]. The presences in these two additional datasets were collocated within predictor pixels of FIA absence to correct for false absences. In doing so, only one *P. albicaulis* presence or absence record was associated per predictor pixel, thereby avoiding issues associated with sampling bias that are common when building bioclimate niche models with data from targeted surveys [39]. This compilation of data represents an effort for "completeness" as described by Kadmon et al. [40] and Franklin [36], to capture all climate conditions where a species does exist. New data sources added 119 *P. albicaulis* presences that would have been missed by using FIA data alone, for a total of 938 presences and 1,633 absences.

"Adult" class *P. albicaulis* were selected for modeling based on a recorded diameter at breast height (DBH) >20 cm. *P. albicaulis* within the Central Montana are reported to reach 100 years of age at approximately 8–12 m in height with DBHs between 15–20 cm [41]. Given previous silvicultural studies, it was assumed that 20 cm DBH *P. albicaulis* represent adult class individuals for the GYA, with potential to reproduce [24]. Furthermore, this study focused on adult size class due to difficulties distinguishing younger age class *P. albicilus* from *P. flexis*.

Historic climate data. Climate inputs for modeling were acquired from the 30-arc-second ($\sim$800 m) monthly Parameter-elevation Regressions on Independent Slopes Model (PRISM), a derived product that interpolates local station measurements across a continuous grid [42]. PRISM data includes monthly average minimum temperatures (T_{min}), maximum temperature (T_{max}), mean temperature (T_{mean}), and mean precipitation (Ppt). All monthly data were averaged for the temporal extent of 1950–1980 for bioclimatic niche model fitting. The 1950–1980 temporal extent was selected for modeling since: 1) a sufficient density of weather stations were operating by 1950 to provide a reasonable network; 2) evidence of anthropogenic warming that begins in the

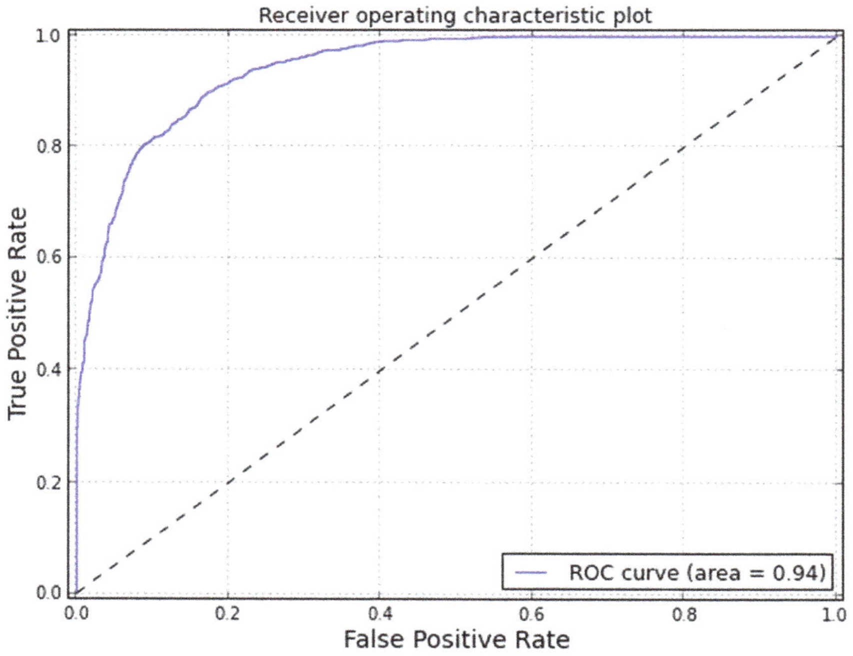

Figure 4. Area under curve for the receiver operating characteristic plot suggests adequate performance from the Random Forest modeling.

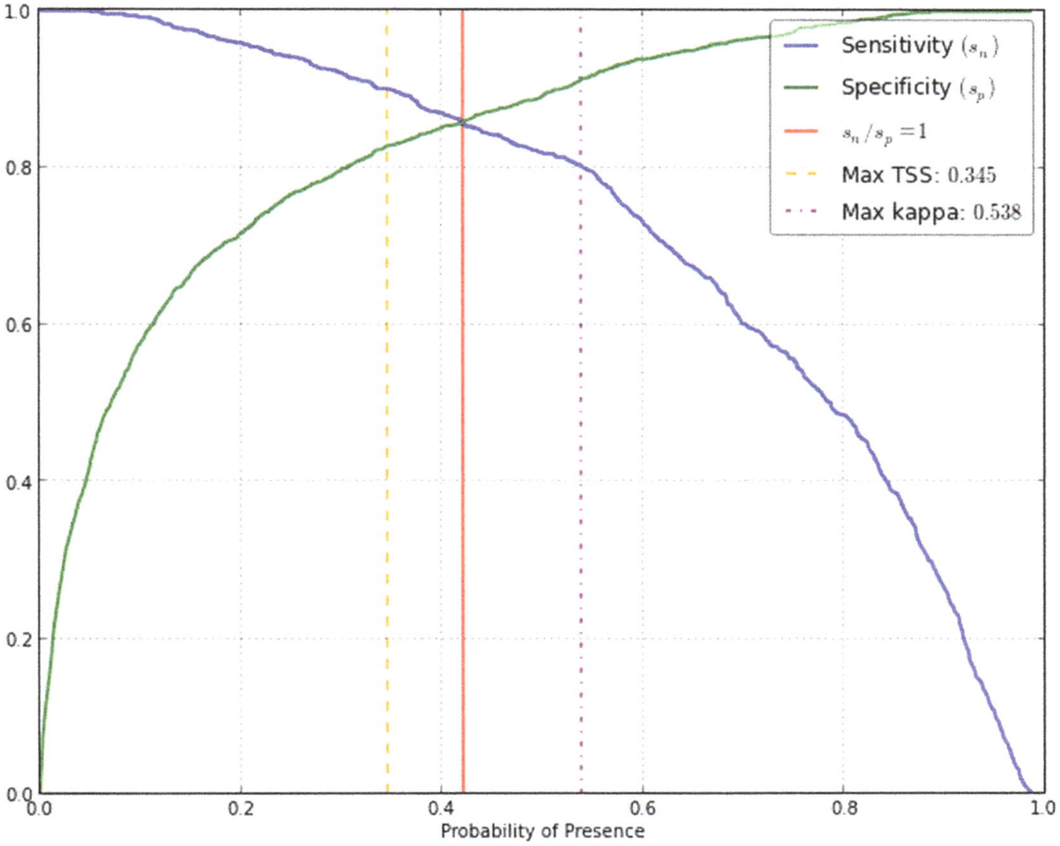

Figure 5. Threshold for probability of presence of 0.421 determined at the intersection of true positive rate (TPR) and true negative rate (TNR). Equivalent TPR and TNR, displayed a compromise between the maximum true skill statistic (TSS : 0.345) and maximum Kappa statistic (κ : 0.538).

late 1980s; 3) trees old enough to bear seeds today likely established under a similar climates to the 1950–1980 period.

Water balance. A Thornthwaite-based dynamic water balance model was used to estimate a number of variables that include actual evapotranspiration (AET) and potential evapotranspiration (PET) [43–45]. The model required only monthly mean temperatures, dew point temperatures, and precipitation (see Text S1). Water was stored as soil moisture or in surface snowpack, with the excess taking the form of evaporated vapor or loss through seepage/runoff. In addition to the climatic variables, latitude and physical characteristics of the soil were required to define water holding capacity. Soil attributes assigned by the Soil Survey Geographic (STATSGO) datasets were allocated from the Natural Resource Conservation Service at a 30-arc-second resolution to determine soil water holding capacity and estimates for soil depth [46]. All water balance variables, which include PET, AET, soil moisture, vapor pressure deficit (vpd), and snow water equivalent (pack), were averaged by month over 1950–1980 to match with historic climate data for bioclimate model fitting.

GCM data. The general circulation model (GCM) experiments conducted under CMIP5 for the Intergovernmental Panel on Climate Change Fifth Assessment Report provided future projected climate data sets for assessing the effects of global climate change. Using a Bias-Correction Spatial Disaggregation (BCSD) approach, an archive of statistically down-scaled CMIP5 climate projections for the conterminous United States at 30-arc-second spatial resolution was assembled by the NASA Center for Climate Simulation NEX-DCP30 [47]. For this analysis, a subset of the

Table 3. Confusion matrix from out-of-bag analysis.

		Validation data set	
		Presence	**Absence**
Model	Presence	763 *(81.9%)*	169 *(13.1%)*
	Absence	176 *(10.9%)*	1437 *(89.1%)*

Random Forest tree estimators displays higher OOB specificity than sensitivity. Area Under Curve (AUC) value of 0.94 suggests model has high predictive capacity for projecting future suitable bioclimate habitat.

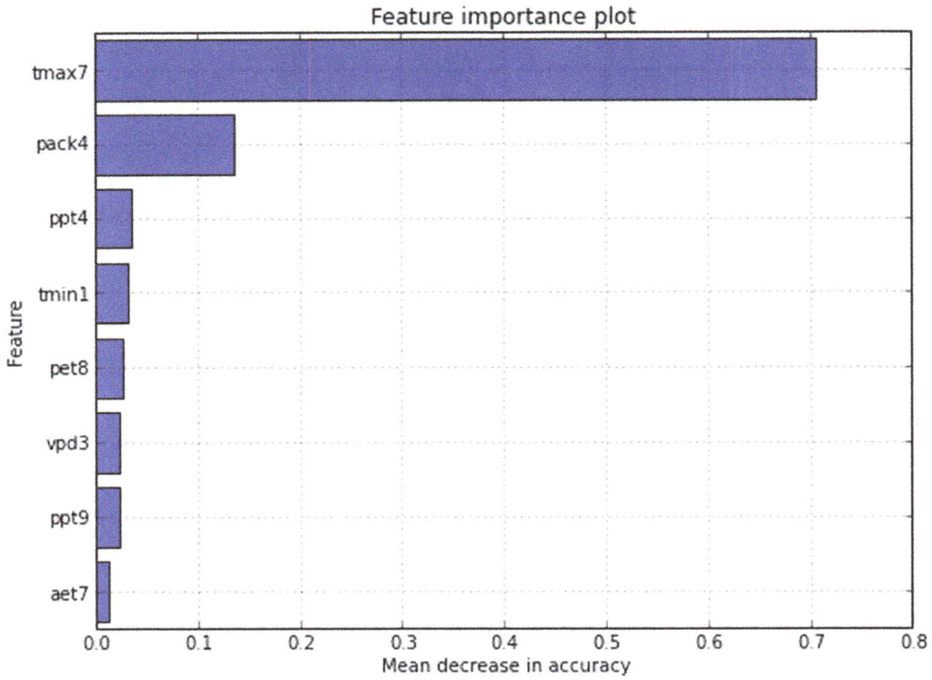

Figure 6. Random Forest out-of-bag variable importance plots find removal of maximum temperatures for July and April snow water equivalent to create the greatest reducing in model accuracy.

total GCM models available from NASA were selected that best represent the Northwestern US. Rupp et al. [48] recently presented an analysis of GCM performance versus the observed historic climate in the U.S. Pacific Northwest under 18 specified climate metrics. In their analysis, Rupp et al. ranked GCMs for accuracy using an empirical orthogonal function (EOF) analysis of the total normalized error compared to reference data. This analysis selected models with a normalized error score <0.5 as a threshold to cull the full suite of GCMs to the top nine models. These GCMs were used to project modeled *P. albicaulis* distributions into the future (Table 1). Two RCP scenarios were selected to understand effects of differing carbon futures under climate change from 2010 to 2099. RCP 4.5 was the first, representing increased radiative forcing until stabilization of greenhouse emissions between 2040–2050 and total radiative forcing of 4.5 W/m^2 by 2099. RCP 8.5 was the second, representing the "business as usual" scenario, with uncontrolled radiative forcing increasing with stabilization of 8.5 W/m^2 by 2099 [49,50].

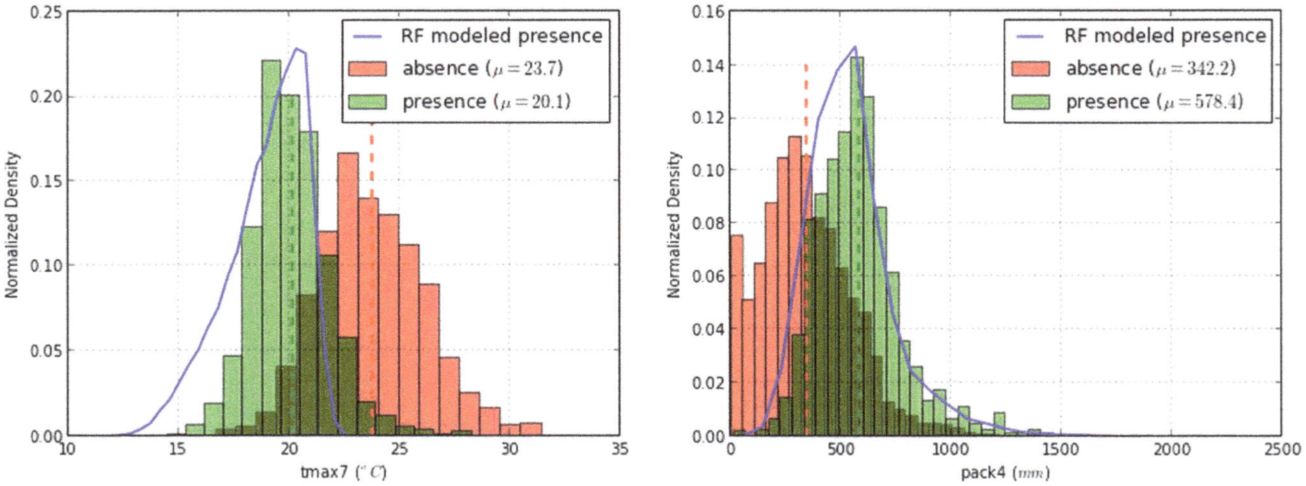

Figure 7. Modeled binary presence for *P.albicaulis* under 1980–2010 mean July maximum temperatures and mean April snow water equivalent bioclimate variables shows agreement with field presence data. Dotted lines designate climate means for corresponding *P. albicaulis* field points. Blue lines represent the distribution of Random Forest modeled presence within the GYA.

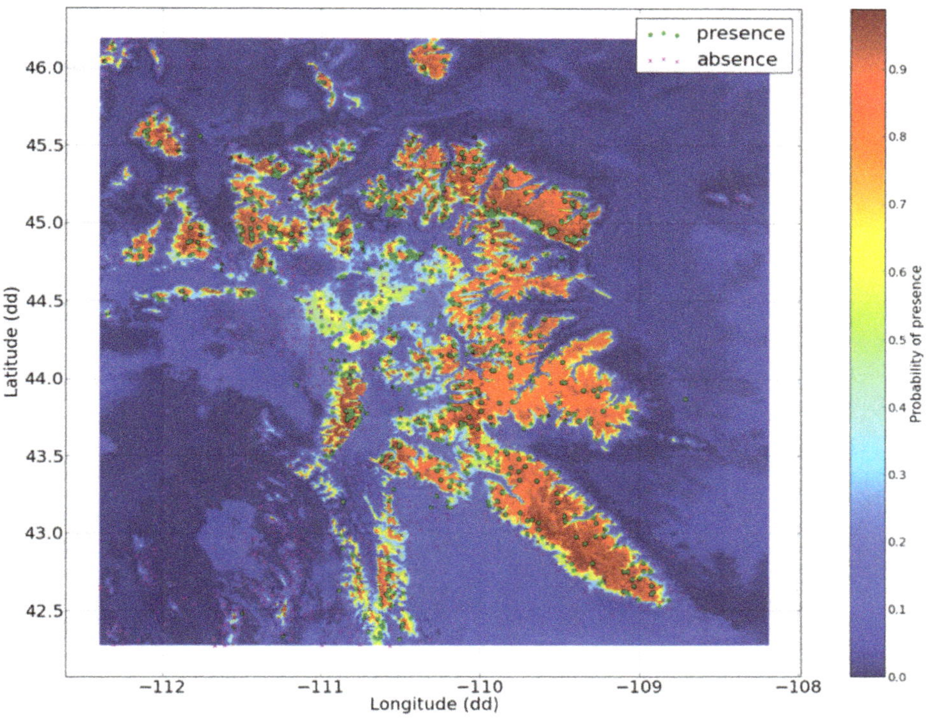

Figure 8. Probabiliy presence for *P.albicaulis* ≥20 cm DBH within the GYA for the 2010 climate period.

Modeling methods

A random forest (RF) [51] algorithm was used to create a bioclimate niche model of *P. albicaulis* in the GYA. Random forest is an ensemble learning technique that generates independent random classification trees using a subset of the total predictor variables and classifies a bootstrap random subsample of the data. These trees are aggregated and a majority vote over all trees in the random forest defines the resulting response class. This method of random trees with subsampling ensures a robust ensemble classification reducing overfitting and collinearity issues, especially with a large number of trees [9,51–53]. The python programming language (Python 3.3) and the Scikit-Learn library was used to fit the random forest model and predict current habitat niche, with parameters for number of trees ($n_{estimators} = 1000$), number of variables ($max_{features} = 4$), and node size ($min_{samplesleaf} = 20$) [54].

First pass filtering of environmental covariates was performed using Principal Component Analysis (PCA) to generate proxy sets [55–57]. After initial list was constructed, an additional filter was imposed on the variables with a 0.75 maximum correlation threshold to avoid collinearity issues (Fig. 3) [55]. Physiologically relevant variables to *P. albicaulis* presence were given precidence in final culling in cases of correlation above the specified maximum threshold. The final variable list selected were tmin1, vpd3, ppt4, pack4, tmax7, aet7, pet8, ppt9 (Table 2). The Software for Assisted Habitat Modeling (SAHM) was used to visualize correlations with the pairs function embeded in the VisTrails scientific workflow management system [58,59].

Model evaluation was performed under a variety of methods. An out-of-bag (OOB) error estimate was calculated by comparing the modeled probability of presence using approximately two-thirds of the field data, while withholding a subset of the remainder. Accuracy was evaluated by calculating: 1) the sensitivity, representing the true positive rate (TPR), 2) the specificity, representing the true negative rate (TNR), 3) the receiver operator characteristic curve (AUC). Importance of a specific predictor variable was calculated by examination of the increase in prediction error within the OOB sample when the predictor variable was permuted while others were held constant [54,60]. The rate of prediction error with permutation of a specified variable can be interpreted as the level of dependence of presence or absence response to that variable [61].

Projections for *P. albicaulis* were computed using 30 year moving climate averages for the period from 2010–2099 for both RCP 4.5 and 8.5 climate scenarios. Changes of suitable habitat area were determined using a binary classification of expected presence and absence. Binary class assignment was made under a probability of presence threshold where the ratio of sensitivity and specificity equalled 1. This method ensured an equal ability of the model to detect presence and absence. The Kappa and True Skill Statistic (TSS) were also calculated to observe how sensitivity and specificity responded under differing probability thresholds [62]. Survey plots predicted as suitable under climatic conditions in 2010 served as a reference for projections. The presence classifications were evaluated as the amount of suitable habitat changed over time, confined within specified elevational limits. To account for the need for a minimum patch size, total number of patches and median sizes using the an eight-neighbor rule (see Text S1) for patch identification were tracked over time [63].

Results

Model evaluation

The random forest model displayed an out-of-bag (OOB) error rate of 16.1% with greater errors of commission (13.1%) than omission (10.9%) (Table 3). The AUC was 0.94, displaying high

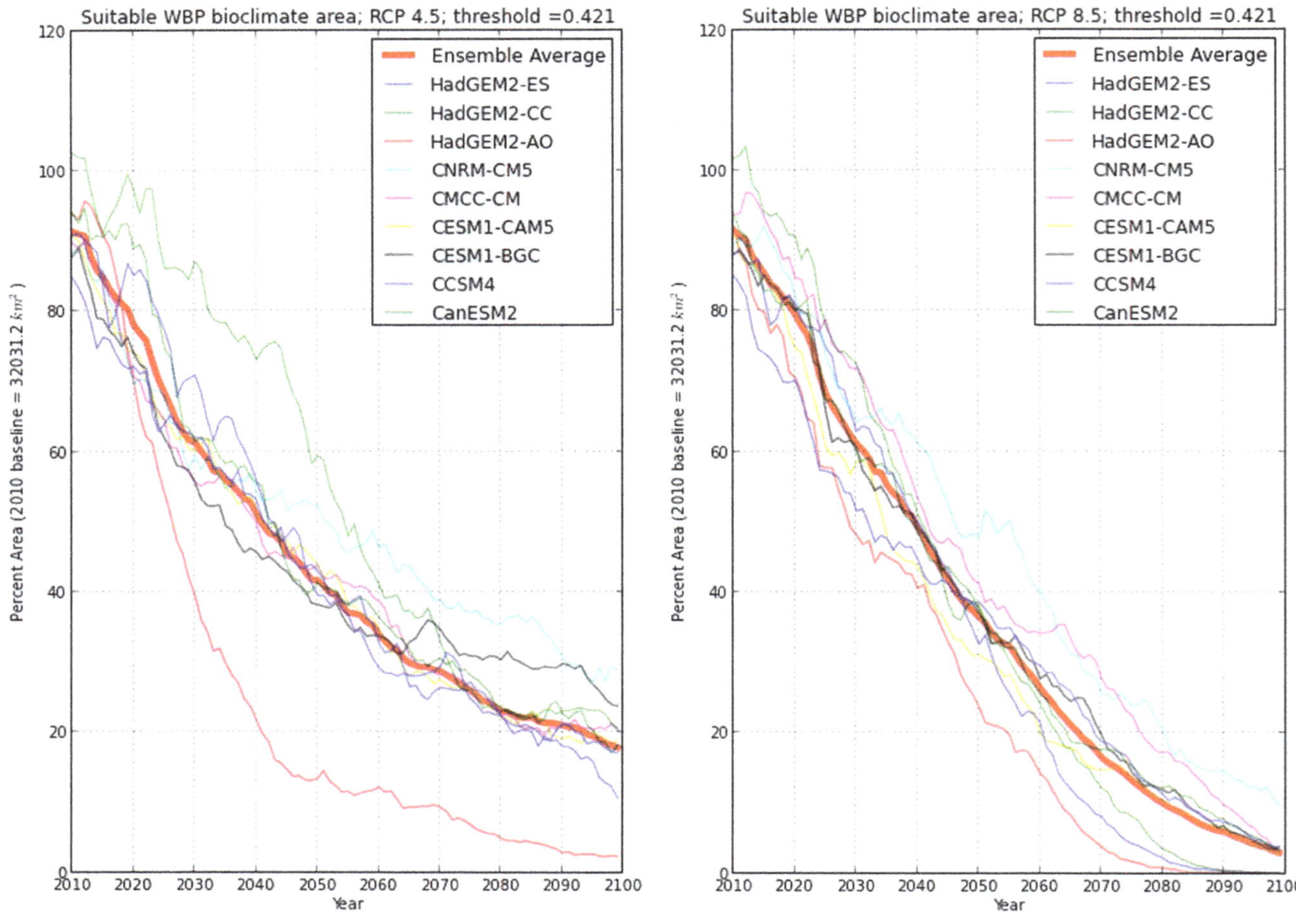

Figure 9. Bioclimate projections for *P.albicaulis* for 2010 to 2099 under 30-year moving averaged climates.

specificity and sensitivity (Fig. 4). Threshold probability of presence for a binary classification was selected at 0.421 (i.e where sensitivity = specificity). A probability threshold where TPR and TNR were equal was compared to the maximum Kappa statistic (0.538) and the maximum True Skill Statistic (TSS) (0.345) and found to be a compromise between the diagnostics (Fig. 5).

Estimates of variable importance plots revealed that permutation of maximum temperatures of summer months from all random trees resulted in a large drop in mean accuracy for distinguishing presence and absence of *P. albicaulis* (0.706 decrease in mean accuracy). This was followed by spring time snowpack (0.137 decrease in mean accuracy) (Fig. 6). Histogram plots of July maximum temperatures and April snowpack provided evidence of discrimination for presence and absence that are consistent with the modeled probability of presence for the year of calibration (Fig. 7).

Spatially explicit probability plots for the 2010 climate displayed highest probability of presence values within the ≥2500 m mountain ranges of the GYA in agreement with studies employing aerial imagery and remote sensing [4,5] (Fig. 8). Assuming that the modeled suitable bioclimate for *P. albicaulis* remains similar in the next century, the model demonstrated capacity to predict probable future *P. albicaulis* suitable habitat under projected climate conditions.

Model projections

Under both RCP 4.5 and 8.5, there was a predicted steady reduction of suitable bioclimate habitat for *P. albicaulis* over the course of this century, with RCP 8.5 displaying steeper declines than RCP 4.5 (Fig. 9). Under the RCP 4.5 and 8.5 scenarios, suitable habitat shifts from 100–85% to 2–29% by 2099, and 100–85% to 0.04–10% by 2099 respectively (Table 4).

CNRM-CM5, CMCC-CM, and CESM1-BGC projections showed the highest probabilities for suitable habitat area at the end of the century, while HadGEM2-AO, HadGEM2-ES, and HadGEM2-CC indicated the lowest probabilities. The standard deviations per year for both RCPs progressively decreased over time (Fig. 10). Among climate scenarios, standard deviations for both RCPs display low variability for the first five projection years and a rapid increase of variability peaking at 2043. For RCP 4.5, high variability existed primarily due to differing climate projections by models HadGEM2-AO and HadGEM2-CC, resulting in uncertainties in probabilities of presence fluctuating between 8 and 15% until 2068, after which variability was between 6–8%. Under RCP 8.5, standard deviations between GCMs were consistently lower than RCP 4.5. Regardless of the GCM, by 2079 the areas of suitable habitat converged to similar values.

Spatially explicit mapping of probability surfaces presented similar contractions of *P. albicaulis* habitat suitability toward the

Table 4. Projected binary *P. albicaulis* presence area within GYA to 2099.

Ensemble Average RCP 4.5	**2010**	**2040**	**2070**	**2099**
Area (km^2)	29250.9	16381.2	9151.1	5685.9
	(27134–32858)	*(6918–23359)*	*(2962–12477)*	*(763–9194)*
% Total Threshold Area*	91.3	51.1	28.6	17.8
	(85–103)	*(22–73)*	*(9–39)*	*(2–29)*
Mean Elevation (m)	2875.7	3020.2	3128.0	3217.9
	(2842–2895)	*(2938–3182)*	*(3055–3297)*	*(3114–3471)*
2.5 Percentile Elevation (m)	2356.3	2494.2	2595.0	2691.5
	(2320–2376)	*(2433–2656)*	*(2506–2758)*	*(2571–3041)*
97.5 Percentile Elevation (m)	3521.9	3603.5	3677.8	3734.6
	(3507–3530)	*(3551–3701)*	*(3636–3783)*	*(3673–3905)*
Ensemble Average RCP 8.5	**2010**	**2040**	**2070**	**2099**
Area (km^2)	29259.3	15746.0	5271.5	960.0
	(27188–32604)	*(12985–19581)*	*(1247–8850)*	*(13–3105)*
% Total Threshold Area*	91.3	49.2	16.5	3.0
	(85–102)	*(40–61)*	*(4–28)*	*(0–10)*
Mean Elevation (m)	2874.7	3022.5	3225.5	3470.5
	(2845–2893)	*(2974–3061)*	*(3116–3412)*	*(3255–3749)*
2.5 Percentile Elevation (m)	2353.1	2492.2	2691.3	3001.5
	(2322–2369)	*(2436–2547)*	*(2553–2934)*	*(2622–3401)*
97.5 Percentile Elevation (m)	3522.1	3605.7	3739.5	3908.7
	(3508–3530)	*(3576–3631)*	*(3677–3866)*	*(3775–4063)*

Summary of projection outputs under RCP 4.5 and 8.5 climate scenarios displays loss of bioclimate habitat from 2010 to 2099 (low and high probability of presence GCM summaries displayed in parentheses). Projections into 2099 under all 9 GCMs suggest rapid loss of suitable bioclimate habitat to below 70% of the current modeled distribution and shifts towards the limited high elevation zones (>3000 m). *(*Percent threshold areas calculated from the 2010 PRISM reference probabilities of presence.*)

upper elevation zones of the GYA that included the Beartooth Plateau and Wind River Ranges (Fig. 11). This implied that rapid warming may lead to conditions outside of the *P. albicaulis* niche in lower elevation areas, and limiting the species to the alpine zones. Elevational analysis of cells within threshold presence probabilities over time observed mean elevations of suitable bioclimates shifting from 2,875 to 3,218 m and 2,875 to 3,470 m for RCP 4.5 and 8.5 respectively. By 2099, ensemble averaged GCM projections displayed over 70% loss of habitat under both scenarios.

P. albicaulis patches from the 2010 baseline observed 202 patches with median patch size of ~180 km^2. Projected patch dynamics analysis denoted a quadratic relationship of patch size over time. Patch dynamics displayed a slow increase in number of *P. albicaulis* patches to a maximum at 2074 and 2057 for RCP 4.5 and 8.5 respectively, followed by a decreasing trend. RCP 4.5 patch numbers were more sporadic, displaying fluctuations across the time period compared to RCP 8.5 associated with the greater interannual climate variability amongst GCM models. Median patch size saw a steady decrease from 72–65 km^2 to 21–8 km^2 for RCP 4.5 and 8.5 respectively, for the projection period, suggesting habitat loss through fragmentation (Fig. 12).

Discussion

In this analysis, the spatiotemporal patterns for *P. albicaulis* distributions were assessed under nine climate models and two emissions scenarios. Bioclimate modeling of *P. albicaulis* illustrated that presence and absence were strongly separated by summer temperatures and spring snowpack. This was in agreement with empirical findings of *P. albicaulis* presence in cool summertime environments where July temperatures range between 4–18°C [64]. Concordantly, these cool summer regions were synonymous with late snow melt, supporting snowpack as an important feature in distinguishing presence and absence.

Future projections by all nine GCMs suggested a contraction in suitable *P. albicaulis* climate area by the end of the century to <30% of current conditions. This was consistent with the results from various other research using either niche models or hybrid process models, predicting similar amounts of *P. albicaulis* contraction [8,9,65]. Variability among projected suitable habitat areas under differing GCMs decreased as all projected maximum temperatures increased above 1°C from the 100 year historic mean. This pattern of warming convergence occurred earlier for the GCMs under the RCP 8.5 scenario than those under RCP 4.5, resulting in the observed low variability of *P. albicaulis* suitable habitat area under RCP 8.5. Despite temperature variability remaining relatively constant amongst GCMs within a RCP, once mean annual temperatures increased beyond *sim* 1°C from the historic average, all bioclimatic habitat models exhibited a pattern of contracting total area and variability. These results lead to the conclusion that explicit selection of a GCM to model under may not necessarily matter for *P. albicaulis* bioclimatic niche modeling studies, especially if the direction of change is solely of concern. However, if investigation of the magnitude of change is relevant,

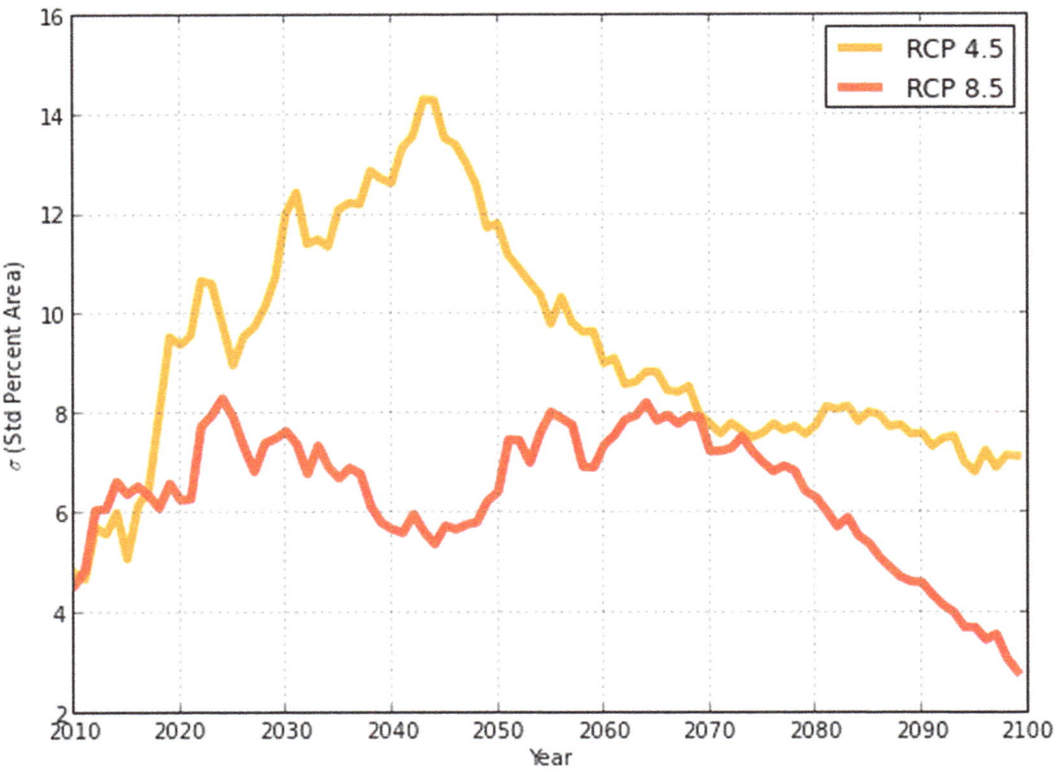

Figure 10. Evaluation of the standard deviation σ for percent suitable habitat area by RCP scenario.

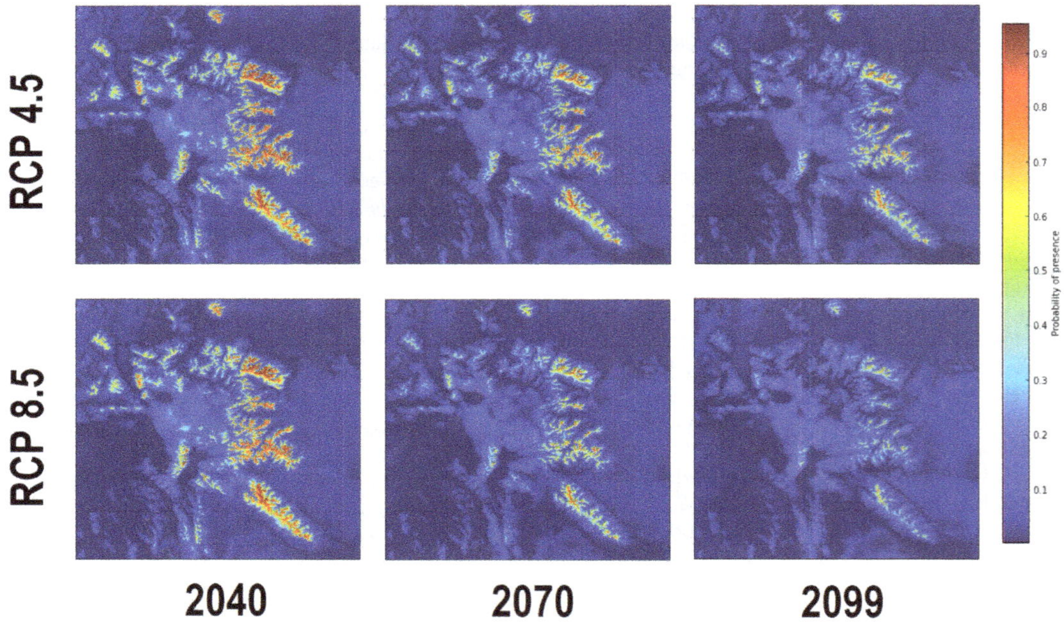

Figure 11. Spatially explicit probabilty surfaces for 2040 to 2099 suggest contraction of suitable bioclimatic habiatat for *P.albicaulis* into the $\geq$2500 m elevation zones.

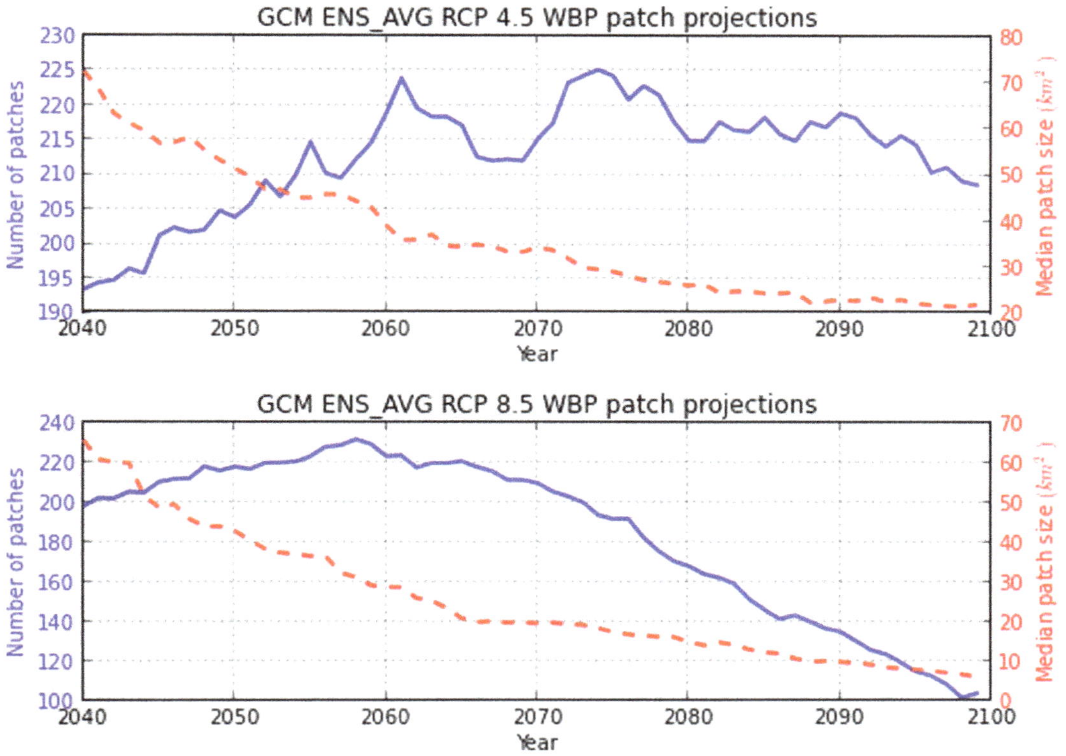

Figure 12. Patch dynamics of modeled *P.albicaulis*. Time series of *P.albicaulis* patch projections for number of patches and median patch size to 2099.

then GCM selection may directly influence the projected total suitable habitat area. This can be observed with RCP 4.5 habitat projection models differing by as much as 27% total suitable habitat area by the year 2099. Therefore arbitrary selection of a GCM for future projection modeling is likely inappropriate since it could lead to overly optimistic/pessimistic results for the species of concern.

Temporal patch dynamic analysis present an increase in fragmentation of the larger *P. albicaulis* suitable habitats over the next five decades, suggested through an increase in the total number of continuous patches but decreases in median size. This was followed by a contraction of small patches until they were almost absent from the system. Remaining habitat patches were smaller and less prevalent on the landscape by the end of the century. Reduced habitat patch size and density may reduce the likelihood for *N. columbiana* to disperse successful germinating seed caches, due to the limited size and area of suitable patch space. If changing climate habitats result in mortality within adult patches, genetic diversity may be lost resulting in a population bottleneck, thus reducing the robustness of the species to adapt to future disturbances. Experimental trials of P. albicaulis survival and fecundity under warmer and drier conditions outside the currently known range would provide greater confidence of the species ability to persist under future change. Limited analysis on seedling environmental conditions would also elucidate spatially explicit dispersal ranges and greater understanding of probable ranges for future establishment and survivorship.

Projected distributions of persistent *P. albicaulis* patches displayed a strong trend towards contraction into high elevation zones. Physiologically, there does not appear to be any upper elevation limit for *P. albicaulis* in the GYA. *P. albicaulis* in the region has been reported to survive in absolute temperatures as low as −36°C [64]. Lab experiments performed on *Pinus cembra*, a related five-needle pine residing in similar climates, were able to endure cold temperature extremes as low at −70°C without cellular tissue damage [66]. Considering the current absolute minimum temperatures the species resides in and cold tolerance of its relatives suggests that *P. albicaulis* treeline in the GYA are not limited by lower temperatures. Controlled laboratory experimentation on *P. albicaulis* tolerances to temperatures would greatly improve this physiological understanding of cold tolerance.

Elevational habitat constriction do not imply that *P. albicaulis* will be completely gone from the region, but merely the loss of suitable climate habitat. Currently pre-established adult age class individuals will likely persist, since projected conditions of increased temperatures and CO_2 concentrations physiologically indicate increased growth rates of *P. albicaulis* [67]. Furthermore, micro-refugia sites may exist in the GYA that support *P. albicaulis* survival into the future, but were failed to have been modeled due to the coarseness of 30-arc-second climate data resolution. Since this bioclimatic envelope modeling approach was parameterized by the realized niche from in-situ data, it was difficult to determine if lower elevation limits are driven by warmer climate conditions or competition for light, water, or nutrients [15,17]. For example, lower treeline limits for *P. albicaulis* maybe driven primarily by competitive exclusion from late seral species *A.lasiocarpa, P.contorta*, and *P.engelmannii*. This follows from paleoecological pollen records of competitor migration during the Early Holocene (9000–5000 yr B.P), when climate conditions were warmer and drier. Longer growing seasons allowing competitors to invade likely drove *P. albicaulis* communities +500 m in elevation [68–70]. If future climate conditions become analogous to this Early

Holocene period, invasion of competitor species will likely contract *P. albicaulis* habitat to the limited high elevation zones of the GYA, specifically the Beartooth Plateau region and Wind River range [71].

Conclusion

This analysis examined the future of *P. albicaulis* suitable climate in the GYA and explicitly addressed the question of distribution variability under 9 representative GCMs and 2 emission scenarios. Increases in temperature within the GYA will likely result in a high level of contraction of suitable climate habitat for *P. albicaulis* over the next century. This contraction was consistent for all GCM projections, with approximately 20% uncertainty in total probable area. This analysis recommends that care be taken for species distribution modeling in future studies during the selection of GCMs due to their relevance for magnitudes of change. GCM ensemble averaging may be a solution to this issue, however it should be noted that averaging should take place after an individual GCM is projected in order to maintain interannual variability.

Although other studies have examined *P. albicaulis* species distribution models [8,65,72], this study is a step forward through its focus on relevant regional scale design, expansive local datasets, inclusion of high resolution climate and dynamic water balance variables, and selective projection under the latest AR5 GCMs. It is reiterated that the bioclimate niche model approach has high utility for understanding habitat conditions through correlative relationships with environmental variables, however, it may fail to explicitly model competitive exclusion, disturbance, phenotypic plasticity, and other complex interactions that are vital in determining a species' actual presence as it experiences changes in climate [15,17,73,74]. These unmodeled factors create uncertainties suggesting that this modeling effort does not identify the full potential climatic range of *P. albicaulis* in the future. Uncertainties also exist regarding new suitable climates that may occur outside the current species range. Despite most rangewide studies confirming our results of total suitable habitat area reduction, there is potential for previously unsuitable habitat to become available under future climate change in the Northern regions [8,22,65]. Caution is therefore advised to individuals interpreting these findings. Changing climate will inevitably result in impacts on biomes and community structures. As such, mitigation and adaptation for potential futures are vital to conservation of climate sensitive species [75]. Future research that combines bioclimatic niche modeling with a mechanistic based disturbance, dispersal, and competition model will likely provide greater insight to the potential range of *P. albicaulis* in a climate changing world [76,77]. It would furthermore provide insight towards informing management options for restoration that may include controlled fire, selected thinning of competitor species, or assisted migration.

Acknowledgments

We are indebted for the insightful reviews from Richard Waring, William Monahan, and Tom Olliff. Many thanks to Marian and Colin Talbert, and Mark Greenwood for the software and statistical consultation.

Author Contributions

Conceived and designed the experiments: TC AJH NBP. Performed the experiments: TC NBP. Analyzed the data: TC. Contributed reagents/materials/analysis tools: TC AJH NBP. Wrote the paper: TC AJH NBP.

References

1. Intergovernmental Panel on Climate Change (2007) Fourth Assessment Report: Climate Change 2007: The AR4 Synthesis Report. Geneva: IPCC.
2. Tomback DF, Arno SF, Keane RE (2001) Whitebark pine communities: ecology and restoration. Island Press.
3. Callaway RM (1998) Competition and facilitation on elevation gradients in subalpine forests of the Northern Rocky Mountains, USA. Oikos 82: pp. 561–573.
4. Macfarlane WW, Logan JA, Kern W (2012) An innovative aerial assessment of greater yellowstone ecosystem mountain pine beetle-caused whitebark pine mortality. Ecological Applications.
5. Jewett JT, Lawrence RL, Marshall LA, Gessler PE, Powell SL, et al. (2011) Spatiotemporal relationships between climate and whitebark pine mortality in the greater yellowstone ecosystem. Forest Science 57: 320–335.
6. Logan JA, Macfarlane WW, Willcox L (2010) Whitebark pine vulnerability to climate-driven mountain pine beetle disturbance in the greater yellowstone ecosystem. Ecological Applications 20: 895–902.
7. Logan JA, Bentz BJ (1999) Model analysis of mountain pine beetle (coleoptera: Scolytidae) seasonality. Environmental Entomology 28: 924–934.
8. Rehfeldt GE, Crookston NL, Sáenz-Romero C, Campbell EM (2012) North American vegetation model for land-use planning in a changing climate: a solution to large classification problems. Ecological Applications 22: 119–141.
9. Rehfeldt GE, Crookston NL, Warwell MV, Evans JS (2006) Empirical analyses of plant-climate relationships for the western United States. International Journal of Plant Sciences 167: 1123–1150.
10. Thuiller W (2004) Patterns and uncertainties of species' range shifts under climate change. Global Change Biology 10: 2020–2027.
11. Iverson LR, Prasad AM, Matthews SN, Peters M (2008) Estimating potential habitat for 134 eastern US tree species under six climate scenarios. Forest Ecology and Management 254: 390–406.
12. Guisan A, Theurillat JP, Kienast F (1998) Predicting the potential distribution of plant species in an alpine environment. Journal of Vegetation Science 9: 65–74.
13. Busby J (1988) Potential impacts of climate change on Australias flora and fauna. Commonwealth Scientific and Industrial Research Organisation, Melbourne, FL, USA.
14. Hutchinson GE (1957) Concluding remarks. Cold Spring Harbor Symposia on Quantitative Biology 22: 415–427.
15. Austin M (2007) Species distribution models and ecological theory: a critical assessment and some possible new approaches. Ecological Modelling 200: 1–19.
16. Austin M (2002) Spatial prediction of species distribution: an interface between ecological theory and statistical modelling. Ecological Modelling 157: 101–118.
17. Pearson RG, Dawson TP (2003) Predicting the impacts of climate change on the distribution of species: are bioclimate envelope models useful? Global Ecology and Biogeography 12.
18. Willis KJ, Whittaker RJ (2002) Species diversity–scale matters. Science 295: 1245–1248.
19. Araújo MB, Cabeza M, Thuiller W, Hannah L, Williams PH (2004) Would climate change drive species out of reserves? an assessment of existing reserve-selection methods. Global Change Biology 10: 1618–1626.
20. Ferrier S (2002) Mapping spatial pattern in biodiversity for regional conservation planning: where to from here? Systematic Biology 51: 331–363.
21. Pearce J, Lindenmayer D (1998) Bioclimatic analysis to enhance reintroduction biology of the endangered helmeted honeyeater (lichenostomus melanops cassidix) in Southeastern Australia. Restoration Ecology 6: 238–243.
22. McLane SC, Aitken SN (2012) Whitebark pine (pinus albicaulis) assisted migration potential: testing establishment north of the species range. Ecological Applications 22: 142–153.
23. Monahan WB, Cook T, Melton F, Connor J, Bobowski B (2013) Forecasting distributional responses of limber pine to climate change at management-relevant scales in Rocky Mountain National Park. PloS ONE 8: e83163.
24. Arno SF, Hoff RJ (1989) Silvics of whitebark pine (pinus albicaulis). Intermountain Research Station GTR-INT-253.
25. Mahalovich MF, Hipkins VD (2011) Molecular genetic variation in whitebark pine (pinus albicaulis engelm.) in the inland west. In: Keane RE, Tomback DF, Murray MP, Smith CM, editors, The future of high-elevation, five-needle white pines in Western North America: Proceedings of the High Five Symposium. 28–30 June 2010; Missoula, MT. Proceedings RMRS.
26. Pederson GT, Gray ST, Ault T, Marsh W, Fagre DB, et al. (2011) Climatic controls on the snowmelt hydrology of the Northern Rocky Mountains. Journal of Climate 24: 1666–1687.

27. Westerling AL, Hidalgo HG, Cayan DR, Swetnam TW (2006) Warming and earlier spring increase western US forest wildfire activity. Science 313: 940–943.
28. Taylor KE, Stouffer RJ, Meehl GA (2012) An overview of CMIP5 and the experiment design. Bulletin of the American Meteorological Society 93.
29. Hibbard KA, van Vuuren DP, Edmonds J (2011) A primer on representative concentration pathways (RCPs) and the coordination between the climate and integrated assessment modeling communities. CLIVAR Exchanges 16: 12–13.
30. Lutz JA, van Wagtendonk JW, Franklin JF (2010) Climatic water deficit, tree species ranges, and climate change in Yosemite National Park. Journal of Biogeography 37: 936–950.
31. Deser C, Phillips AS, Alexander MA, Smoliak BV (2014) Projecting North American climate over the next 50 years: Uncertainty due to internal variability. Journal of Climate 27: 2271–2296.
32. Beaumont LJ, Hughes L, Pitman A (2008) Why is the choice of future climate scenarios for species distribution modelling important? Ecology Letters 11: 1135–1146.
33. Smith WB (2002) Forest inventory and analysis: a national inventory and monitoring program. Environmental Pollution 116: S233–S242.
34. Gibson J, Moisen G, Frescino T, Edwards Jr TC (2014) Using publicly available forest inventory data in climate-based models of tree species distribution: Examining effects of true versus altered location coordinates. Ecosystems 17: 43–53.
35. Gesch D, Oimoen M, Greenlee S, Nelson C, Steuck M, et al. (2002) The national elevation dataset. Photogrammetric engineering and remote sensing 68: 5–32.
36. Franklin J (2009) Mapping species distributions: spatial inference and prediction. Cambridge University Press.
37. Lockman IB, DeNitto GA, Courter A, Koski R (2007) WLIS: The whitebark-limber pine information system and what it can do for you. In: Proceedings of the conference whitebark pine: a Pacific Coast perspective. US Department of Agriculture, Forest Service, Pacific Northwest Region, Ashland, OR. Citeseer, pp. 146–147.
38. Jean C, Shanahan E, Daley R, DeNitto G, Reinhart D, et al. (2010) Monitoring white pine blister rust infection and mortality in whitebark pine in the Greater Yellowstone Ecosystem. Proceedings of the future of high-elevation five-needle white pines in Western North America: 28–30.
39. Edwards Jr TC, Cutler DR, Zimmermann NE, Geiser L, Moisen GG (2006) Effects of sample survey design on the accuracy of classification tree models in species distribution models. Ecological Modelling 199: 132–141.
40. Kadmon R, Farber O, Danin A (2003) A systematic analysis of factors affecting the performance of climatic envelope models. Ecological Applications 13: 853–867.
41. Weaver T, Dale D (1974) Pinus albicaulis in central Montana: environment, vegetation and production. American Midland Naturalist: 222–230.
42. Daly C, Gibson WP, Taylor GH, Johnson GL, Pasteris P (2002) A knowledge-based approach to the statistical mapping of climate. Climate Research 22: 99–113.
43. Thornthwaite C (1948) An approach toward a rational classification of climate. Geographical Review 38: 55–94.
44. Thornthwaite C, Mather J (1955) The water balance. Publication of Climatology 8.
45. Dingman S (2002) Physical hydrology. Prentice Hall.
46. National Resources Conservation Service (2014) Available: http://soildatamart.nrcs.usda.gov. Accessed 2013 Apr 3.
47. Thrasher B, Xiong J, Wang W, Melton F, Michaelis A, et al. (2013) Downscaled climate projections suitable for resource management. Eos, Transactions American Geophysical Union 94: 321–323.
48. Rupp DE, Abatzoglou JT, Hegewisch KC, Mote PW (2013) Evaluation of CMIP5 20th century climate simulations for the Pacific Northwest USA. Journal of Geophysical Research: Atmospheres 118: 10–884.
49. Gent PR, Danabasoglu G, Donner LJ, Holland MM, Hunke EC, et al. (2011) The community climate system model version 4. Journal of Climate 24: 4973–4991.
50. Moss RH, Babiker M, Brinkman S, Calvo E, Carter T, et al. (2008) Towards new scenarios for analysis of emissions, climate change, impacts, and response strategies.
51. Breiman L (2001) Random forests. Machine Learning 45: 5–32.
52. Roberts DR, Hamann A (2012) Method selection for species distribution modelling: are temporally or spatially independent evaluations necessary? Ecography 35: 792–802.
53. Lawrence RL, Wood SD, Sheley RL (2006) Mapping invasive plants using hyperspectral imagery and breiman cutler classifications (randomforest). Remote Sensing of Environment 100: 356–362.
54. Pedregosa F, Varoquaux G, Gramfort A, Michel V, Thirion B, et al. (2011) Scikit-learn: Machine learning in Python. Journal of Machine Learning Research 12: 2825–2830.
55. Dormann CF, Elith J, Bacher S, Buchmann C, Carl G, et al. (2013) Collinearity: a review of methods to deal with it and a simulation study evaluating their performance. Ecography 36: 027–046.
56. Booth GD, Niccolucci MJ, Schuster EG (1994) Identifying proxy sets in multiple linear regression: an aid to better coefficient interpretation. Research paper INT.
57. Tabachnick B, Fidell LS (1989) Using multivariate statistics, 1989. Harper Collins Tuan, PD A comment from the viewpoint of time series analysis Journal of Psychophysiology 3: 46–48.
58. Freire J (2012) Making computations and publications reproducible with vistrails. Computing in Science & Engineering 14: 18–25.
59. Morisette JT, Jarnevich CS, Holcombe TR, Talbert CB, Ignizio D, et al. (2013) Vistrails SAHM: visualization and workflow management for species habitat modeling. Ecography 36: 129–135.
60. Liaw A, Wiener M (2002) Classification and regression by randomforest. R news 2: 18–22.
61. Cutler DR, Edwards Jr TC, Beard KH, Cutler A, Hess KT, et al. (2007) Random forests for classification in ecology. Ecology 88: 2783–2792.
62. Allouche O, Tsoar A, Kadmon R (2006) Assessing the accuracy of species distribution models: prevalence, kappa and the true skill statistic (tss). Journal of Applied Ecology 43: 1223–1232.
63. Turner MG, Gardner RH, O'Neill RV (2001) Landscape ecology in theory and practice: pattern and process. Springer.
64. Weaver T (2001) Whitebark pine and its environment. In: Tomback DF, Arno SF, Keane RE, editors, Whitebark pine communities: ecology and restoration, Washington D.C, USA: Island Press.
65. Waring RH, Coops NC, Running SW (2011) Predicting satellite-derived patterns of large-scale disturbances in forests of the pacific northwest region in response to recent climatic variation. Remote Sensing of Environment 115: 3554–3566.
66. Sakai A, Larcher W (1987) Frost survival of plants. Responses and adaptation to freezing stress. Springer-Verlag.
67. Chapin III FS, Chapin MC, Matson PA, Vitousek P (2011) Principles of terrestrial ecosystem ecology. Springer.
68. Whitlock C, Shafer SL, Marlon J (2003) The role of climate and vegetation change in shaping past and future fire regimes in the Northwestern US and the implications for ecosystem management. Forest Ecology and Management 178: 5–21.
69. Whitlock C (1993) Postglacial vegetation and climate of Grand Teton and southern Yellowstone national parks. Ecological Monographs: 173–198.
70. Bartlein PJ, Whitlock C, Shafer SL (1997) Future climate in the Yellowstone national park region and its potential impact on vegetation. Conservation Biology 11: 782–792.
71. Tausch RJ, Wigand PE, Burkhardt JW (1993) Viewpoint: plant community thresholds, multiple steady states, and multiple successional pathways: legacy of the quaternary? Journal of Range Management: 439–447.
72. Bell DM, Bradford JB, Lauenroth WK (2014) Early indicators of change: divergent climate envelopes between tree life stages imply range shifts in the western united states. Global Ecology and Biogeography 23: 168–180.
73. Keane B, Tomback D, Davy L, Jenkins M, Applegate V (2013) Climate change and whitebark pine: Compelling reasons for restoration. Whitebark Pine Ecosystem Foundation Whitepaper.
74. Guisan A, Thuiller W (2005) Predicting species distribution: offering more than simple habitat models. Ecology Letters 8: 993–1009.
75. Keane RE, Tomback DF, Aubry CA, Bower EM, Campbell CL, et al. (2012) A range-wide restoration strategy for whitebark pine (pinus albicaulis): General technical report. USDA FS, Rocky Mountain Research Station RMRS-GTR-279: 108.
76. Mathys A, Coops NC, Waring RH (2014) Soil water availability effects on the distribution of 20 tree species in Western North America. Forest Ecology and Management 313: 144–152.
77. Morin X, Thuiller W (2009) Comparing niche-and process-based models to reduce prediction uncertainty in species range shifts under climate change. Ecology 90: 1301–1313.

16

Potential Climate Change Effects on the Habitat of Antarctic Krill in the Weddell Quadrant of the Southern Ocean

Simeon L. Hill[1]*, Tony Phillips[1], Angus Atkinson[2]

1 British Antarctic Survey, Natural Environment Research Council, Cambridge, United Kingdom, **2** Plymouth Marine Laboratory, Plymouth, United Kingdom

Abstract

Antarctic krill is a cold water species, an increasingly important fishery resource and a major prey item for many fish, birds and mammals in the Southern Ocean. The fishery and the summer foraging sites of many of these predators are concentrated between 0° and 90°W. Parts of this quadrant have experienced recent localised sea surface warming of up to 0.2°C per decade, and projections suggest that further widespread warming of 0.27° to 1.08°C will occur by the late 21st century. We assessed the potential influence of this projected warming on Antarctic krill habitat with a statistical model that links growth to temperature and chlorophyll concentration. The results divide the quadrant into two zones: a band around the Antarctic Circumpolar Current in which habitat quality is particularly vulnerable to warming, and a southern area which is relatively insensitive. Our analysis suggests that the direct effects of warming could reduce the area of growth habitat by up to 20%. The reduction in growth habitat within the range of predators, such as Antarctic fur seals, that forage from breeding sites on South Georgia could be up to 55%, and the habitat's ability to support Antarctic krill biomass production within this range could be reduced by up to 68%. Sensitivity analysis suggests that the effects of a 50% change in summer chlorophyll concentration could be more significant than the direct effects of warming. A reduction in primary production could lead to further habitat degradation but, even if chlorophyll increased by 50%, projected warming would still cause some degradation of the habitat accessible to predators. While there is considerable uncertainty in these projections, they suggest that future climate change could have a significant negative effect on Antarctic krill growth habitat and, consequently, on Southern Ocean biodiversity and ecosystem services.

Editor: Howard I. Browman, Institute of Marine Research, Norway

Funding: SH was funded through the Natural Environment Research Council International Opportunities Fund grant NE/I029943/1. "Coordinating International Research on Southern Ocean Ecosystems: Implementation of the ICED Programme." The funders had no role in study design, data collection and analysis, decision to publish, or preparation of the manuscript.

Competing Interests: The authors have declared that no competing interests exist.

* E-mail: sih@bas.ac.uk

Introduction

Climate warming is already producing complex spatial and seasonal changes in the Earth's habitats and ecosystems [1], [2]. Warming is expected to increase significantly over the 21st Century [3], leading to ecosystem change and potentially severe socioeconomic consequences [4]. Observed changes in the Southern Ocean include localised sea ice loss [5] and increases in summer sea surface temperatures (SSTs) of 1°C over 5 decades near the western Antarctic Peninsula [6] and 0.9°C over 8 decades at South Georgia [7]. Previous studies have identified potential relationships between climate-related variables (sea temperature, ice cover and pH) and the recruitment, survival, growth and distribution of the crustacean Antarctic krill, *Euphausia superba* [8], [9], [10]. Antarctic krill is a characteristic species of the Southern Ocean and exists within a narrow band of cold temperatures (up to ~5°C) [11], [12], [13]. It is an increasingly important fishery resource and a major prey item for a diverse suite of predators including whales, penguins, seals and fish [14], [15], [16], [17], [18], [19]. The role of Antarctic krill in supporting predators might be more significant than that of any comparable species elsewhere in the world's oceans [16].

Antarctic krill has an estimated biomass in excess of 2×10^8t [18], [20], about one-quarter of which is concentrated in about 10% of its total habitat area, specifically the Scotia Sea and southern Drake Passage (Fig. 1) [11], [12], [21]. This is where many air breathing vertebrates congregate to feed on Antarctic krill and rear offspring on islands such as South Georgia [18], [22]. For example, about 95%, 50% and 25% [23], [24], [25] of the global populations of Antarctic fur seals (*Arctocephalus gazella*), grey headed albatrosses (*Thalassarche chrysostoma*) and wandering albatrosses (*Diomedea exulans*) breed at South Georgia where Antarctic krill constitute approximately 85%, 76% and 12% of their respective diets [15], [26]. Antarctic krill is also an abundant fishery resource which, according to the Food and Agriculture Organisation of the United Nations, is underexploited [27]. The potential harvest from the Scotia Sea and southern Drake Passage is equivalent to 7% of current global marine fisheries production [28].

It is important to evaluate how further environmental change might affect Antarctic krill and consequently the biodiversity and ecosystem services of the Southern Ocean. In this study, we examine some of these effects by assessing potential changes in the habitat's ability to support Antarctic krill growth. Growth

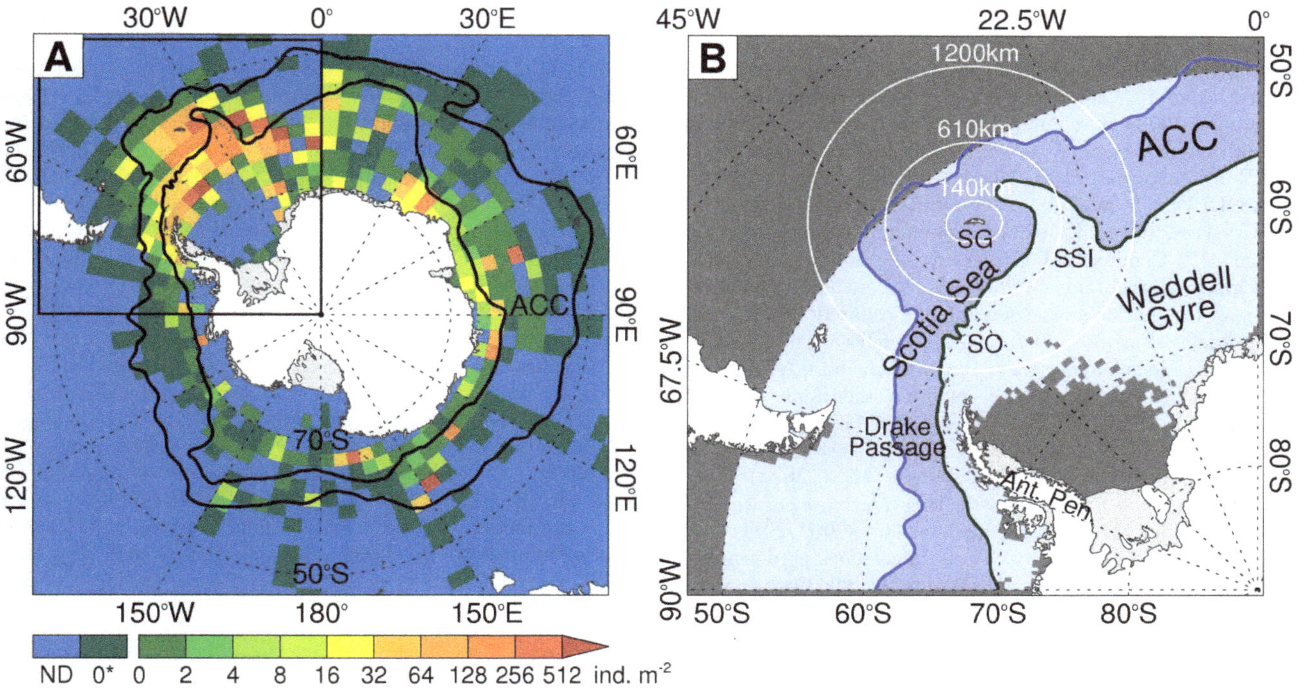

Figure 1. The distribution of Antarctic krill and the study area. (A) The observed distribution of Antarctic krill (individuals.m^{-2} within each 5° longitude by 2° latitude grid cell, ND = no data, 0* = no Antarctic krill recorded in the available data) from [12]. (Inset & B) The study area, showing the Antarctic Circumpolar Current (ACC), which is bounded to the north by the Antarctic Polar Front and to the south by the Southern boundary of the ACC (Positions from [36]). The concentric distances from South Georgia (SG) indicate the approximate foraging ranges of representative predators of Antarctic krill: Antarctic fur seals (140 km), Wandering albatrosses (610 km) and Grey-headed albatrosses (1200 km). Areas north of 50°S (shaded grey) were not included in the study. Ant. Pen = Antarctic Peninsula, SO = South Orkney Islands, SSI = South Sandwich Islands.

represents the accumulation of resources within an individual or population and is therefore a valuable indicator of overall habitat quality [29]. Individual Antarctic krill can increase their body mass by up to a factor of four within a single growing season [20] and this production of new biomass supports predators and the fishery. We assess potential habitat changes over the 21st century by combining a statistical model of Antarctic krill growth [30] with projected SST changes from the Coupled Model Intercomparison Project Phase 5 (CMIP5) multi-model ensemble [31]. The statistical model links growth to body size, SST and food availability. We focus on the quadrant between 0° and 90°W (also known as the Weddell Quadrant [32]), which encompasses the Scotia Sea and southern Drake Passage, and we assess potential changes within the ranges of predators foraging from South Georgia in more detail.

Methods

Models and Metrics

An assessment of various models for evaluating Antarctic krill growth habitat on the basis of temperature [33] favoured the use of statistical models [30], [34]. These models are based on the observed growth rates of Antarctic krill caught in a wide range of environmental conditions. We used one such model [30,12] that relates the daily increase in Antarctic krill length (daily growth rate: *DGR*, mm.d^{-1}) to sea surface temperature (*SST*, °C), food availability indicated by chlorophyll-a concentration (*CHL*, mg.m^{-3}), and starting length (*L*, mm):

$$DGR = -0.066 + 0.002L - 0.000061L^2 + \frac{0.385CHL}{0.328 + CHL} + 0.0078SST - 0.0101SST^2 \quad (1)$$

The following relationship, which was derived from the same data as equation 1 [30], converts individual Antarctic krill length (mm) to dry mass, *M* (mg):

$$\log_{10}(M) = 3.89\log_{10}(L) - 4.19 \quad (2)$$

Previous studies [12], [20] have used equations 1 and 2 to estimate Gross Growth Potential (GGP) based on spatially-resolved, monthly averages of *SST* and *CHL*. GGP is the model-predicted dry mass of an individual Antarctic krill at the end of the summer growth season divided by its dry mass at the beginning of the season. GGP is therefore a unitless quantity that indicates the habitat's ability to support Antarctic krill growth.

The CMIP5 dataset [31] provides the results of climate simulations from multiple climate models. The variations in important factors such as greenhouse gases and aerosols which were used to drive simulations to 2005 were observed values, whereas simulations from 2006 were forced with Representative Concentration Pathways (RCPs). These RCPs include representations of potential changes to factors such as greenhouse gas and air pollutant emissions and land-use, and are named according to projected radiative forcing in the year 2100 [35]. RCP2.6 has peak radiative forcing of ~3 W.m^{-2} in the first half of the 21st century, falling to ~2.6 W.m^{-2} by 2100. In this scenario, aggressive mitigation results in negative net greenhouse gas emissions by the end of the century. Under RCP4.5, greenhouse gas emissions rise until around 2040 before falling below those for the year 2000 by the end of the century, and radiative forcing stabilizes at ~4.5 W.m^{-2} around 2100. Under RCP8.5, radiative forcing reaches 8.5 W.m^{-2} in 2100 and continues to rise [31], [35].

We used results from RCP2.6, RCP4.5 and RCP8.5 to calculate projected 21st Century SST changes for the Southern Ocean in the longitudinal quadrant 0°W to 90°W. We define the Southern Ocean as the marine area south of the Antarctic Polar Front (position defined in [36], 2008 update available from: data.aad.gov.au/aadc/metadata/metadata_redirect.cfm?md = /AMD/AU/southern_ocean_fronts). We explored the implications of projected SST change for spatially-resolved GGP in the same quadrant for the area south of 50°S (the study area). 50°S is the northern limit of Antarctic krill distribution [11], [12].

We used equation 1 to calculate the length increase per week for summer growing seasons in the current (2002–2011) and projection (2070–2099) periods. These calculations used monthly averages of *SST* and *CHL*. For weeks that straddled two months, we used weighted averages of *SST* and *CHL* from those two months. We updated the Antarctic krill starting length at the beginning of each week and converted the initial and final lengths to dry mass using equation 2. We then divided final mass by starting mass to estimate GGP.

To estimate current GGP, we used monthly estimates of current SST based on observations [37], which we label $SST_{o,m,c}$. Subscript *o* indicates that the data are observations, *m* indicates the month and *c* indicates that the estimate is a climatology (i.e. a long term average). Climatologies are appropriate for variables, such as SST, that have high interannual variability [38]. Our estimates of projected GGP were based on estimates of projected SST, $SST_{p,m,y}$, from the CMIP5 results, where the subscript *p* indicates that the data are projections and *y* indicates year. To correct for bias in model estimates of SST [38], we first converted projected SST into differences from an SST climatology representing current conditions in the same model, $SST_{b,m,c}$, where subscript *b* indicates model baseline conditions. We then added the current climatology based on observations, $SST_{o,m,c}$, to calculate a bias-corrected estimate of projected SST, $SST'_{p,m,y}$:

$$SST'_{p,m,y} = SST_{p,m,y} - SST_{b,m,c} + SST_{o,m,c} \tag{3}$$

We used this bias-corrected SST estimate to calculate projected GGP.

The World Meteorological Organisation recommends calculating climatological conditions over 30 years (www.wmo.int/pages/prog/wcp/ccl/faqs.html). We therefore selected the period 1991–2020 for $SST_{b,m,c}$ and we obtained $SST_{p,m,y}$ for each year 2070–2099. The climatological period for $SST_{o,m,c}$ was restricted by data availability to 2002–2011. We obtained $SST_{b,m,c}$ and $SST_{p,m,y}$ for each available model in the CMIP5 results. Within each RCP, SST estimates for each model were calculated as the mean of the estimates for all realisations available for that model. Consequently contributions from all models were given equal weight in across-model means, regardless of the number of realisations per model.

In addition to SST data, GGP estimation requires a starting length and *CHL* estimates. We used a starting length of 40 mm, which is the observed mean length for the postlarval population of Antarctic krill [20]. Following [20] we also considered starting lengths of 30 mm and 50 mm.

We used remote-sensed, spatially-resolved *CHL* estimates ($CHL_{o,m,c}$) [39]. The climatological period for $CHL_{o,m,c}$ was restricted by data availability to 1997–2010. We varied these *CHL* estimates to assess the sensitivity of our results to assumptions about chlorophyll-a concentration. One study estimates that the Southern Ocean experienced a 10% decline in chlorophyll-a concentration over about 17 years between the 1980s and 1990s [40]. Linear extrapolation of this change suggests a 50% reduction over our eight decade projection period. We therefore decreased the *CHL* estimates by 50%. We also considered increases by the same amount.

We calculated spatially-resolved estimates of current GGP using $SST_{o,m,c}$ and $CHL_{o,m,c}$, and spatially-resolved estimates of projected GGP for each year 2070–2099 for each available CMIP5 model in each of the three RCPs. Our GGP estimates for the period 2070–2099 were resolved to grid cell (1° longitude by 0.5° latitude), model and projection year. We averaged across years and models to derive a single estimate of projected GGP for each combination of grid cell, chlorophyll-a concentration, and RCP. We then subtracted the estimated current GGP (i.e. that calculated using observed SSTs, $SST_{o,m,c}$, and observed chlorophyll-a concentrations, $CHL_{o,m,c}$) from projected GGP to estimate the GGP change between the current period and 2070–2099.

From our spatially-resolved GGP estimates, we calculated three spatially-aggregated metrics for each combination of chlorophyll-a concentration and RCP: (1) average GGP by year, (2) total GGP, and (3) growth area. Average GGP by year is the mean of those across-model, grid-cell-and-year-specific GGP estimates that were ≥1. GGP<1 indicates that the habitat does not support growth or maintenance of body size. It implies shrinkage resulting from starvation, which has been observed in Antarctic krill [41]. To estimate total GGP we first calculated, for each grid cell in each model, the across-year mean GGP for the period 2070–2099. We then calculated, for each model, the area-weighted sum of those resulting grid cell-specific estimates of GGP that were ≥1. Total GGP is the across-model mean of this sum. Weighting by grid cell area was necessary because this area changes with latitude. To estimate growth area we calculated, for each model, the total area of all grid cells in which the across-year mean GGP was ≥1. Growth area was the across-model mean of this sum.

We estimated projected change relative to current conditions in the form of relative GGP and relative growth area. We calculated these relative values by dividing total GGP and growth area by the equivalent metric calculated using observed SSTs, $SST_{o,m,c}$, and chlorophyll-a concentrations, $CHL_{o,m,c}$. Our estimate of relative GGP excludes GGP values <1 and therefore does not include further degradation of habitat that did not initially support growth. It does, however, include such habitat becoming viable for growth. We calculated relative GGP and relative growth area for our entire study area, and within the foraging ranges of representative Antarctic krill predators foraging from South Georgia.

Data

We obtained spatially-resolved monthly mean SST data at a nominal horizontal resolution of 9 km for the austral summer periods (December to March) from December 2002 to March 2011 from the archive of Aqua MODIS level 3 data [37], and chlorophyll-a concentration data for the austral summer periods from December 1997 to March 2010 from the archive of SeaWIFS data [39]. Both of these datasets are available through the NASA OceanColor website (*oceancolor.gsfc.nasa.gov*). We obtained spatially-resolved, monthly mean SST data for the period

1990–2100, for each selected RCP, from the output of multiple climate models which are available as part of the CMIP5 multi-model ensemble results [31]. The CMIP5 model data were downloaded from the distributed CMIP5 archive accessed via the Program for Climate Model Diagnosis and Intercomparison CMIP5 data portal (cmip-pcmdi.llnl.gov/cmip5/data_portal.html). We used the results that were available on 31 Jan 2012, including 14 sets of model results available for RCP2.6, 15 for RCP4.5 and 16 for RCP8.5 (Table 1). SST in these results is the average temperature of the surface layer of the modelled ocean. The depth of this layer is <20 m in all 16 models and exactly 10 m in 10 of them.

In addition to analysing habitat quality across the study area, we also extracted statistics for the areas within each of three concentric distances from South Georgia that indicate the summer foraging ranges of representative near, medium and long-range foragers. These distances were 140 km, 610 km, and 1200 km which respectively indicate the foraging ranges of Antarctic fur seals [42], wandering albatrosses [26] and grey headed albatrosses [26] during the summer offspring rearing period.

The data were initially available on different types of grid which we converted to a common grid of 1° longitude by 0.5° latitude. The remote-sensed SST and chlorophyll-a concentration data were both available on regular, fine-scale grids with a resolution of 5 arc minutes, so for each grid cell in the common grid we simply calculated the mean of the 72 constituent fine-scale grid cells. Some of the CMIP5 SST data were provided on a regular grid. We converted these data to the resolution of the common grid using bilinear interpolation [43]. For data not provided on a regular grid, we used all grid points to generate a Delaunay triangulation on an equirectangular projection [44], and we converted the data from this triangulation onto the common grid using linear interpolation [45]. We flagged as missing data any value on the common grid that was affected by a land point. The representation of coastline varies between models, so GGP estimates in coastal cells are informed by varying numbers of models. This does not affect any of our main conclusions.

The availability of remote-sensed SST and chlorophyll-a concentration data varies temporally due to the presence of cloud and ice cover, and cells with insufficient coverage appear as missing data in the Aqua MODIS and SeaWIFS monthly mean data products. Our objective was to achieve extensive spatial coverage with sufficient observations in each cell to provide representative monthly SST and chlorophyll-a concentration estimates for the summer growth season. We achieved a suitable balance of spatial and temporal coverage by including only those 1° by 0.5° cells for which data were available for a minimum of 20% of the initial fine-scale grid cells per month for at least 3 years during the climatology period. To maximise spatial data coverage, we constructed the climatologies from the full period of data availability for each data type separately, and restricted the analysis to the period January to March. The majority of krill growth occurs between December and March [11], and previous studies of current habitat quality have used this four month period but consequently had less spatial coverage [12], [20]. The application of these criteria defined the areas for inclusion and exclusion of data in our calculations, which we applied consistently to each data set that we used.

Results

The results presented in this section are for an assumed Antarctic krill starting length of 40 mm. The Supporting Information (Figs. S1, S2 and S3) compares results for different starting lengths (30 mm, 40 mm and 50 mm).

The growth model correctly identified the warmer waters north of the Antarctic Polar Front as unable to support Antarctic krill growth (Fig. 2). There was considerable spatial structure in current GGP estimates for the study area, including patches of elevated habitat quality along the coast of the Antarctic continent and around the South Orkney and South Sandwich islands. These patterns were less distinct but still apparent with changed chlorophyll-a concentrations. Fig. 2 shows extensive areas in the southern Weddell Sea and along the coast of the Antarctic Peninsula for which we did not calculate GGP because of low data availability due to frequent ice cover.

Monthly climatological SSTs from CMIP5 models for the period 1991–2020 were, on average, 2.04°C warmer than SST estimates for 2002–2011 from Aqua MODIS data, but there was reasonable spatial correlation between the two datasets ($r = 0.954$). The mean projected summer SST warming for the area south of the Antarctic Polar Front between 1991–2020 and 2070–2099 was 0.27°C, 0.56°C and 1.08°C for RCP2.6, RCP4.5 and RCP8.5 respectively. These estimates varied between years and between models (Fig. 3).

Projected GGP declines were concentrated in a band that approximates the location of the Antarctic Circumpolar Current (ACC) (Fig. 4). Most models projected significant warming of the ACC under RCP4.5 and RCP8.5. The RCP2.6 results identified an area of warming in the west Scotia Sea, but otherwise there was little agreement between the models about projected changes in GGP under RCP2.6. In the cells where ≥90% of model projections agreed on the sign of change in GGP, the projected GGP declines (with unchanged chlorophyll-a concentrations) were 16%, 25% and 37% for RCP2.6, RCP4.5 and RCP8.5 respectively. The corresponding reductions in growth area were 13%, 23% and 33%. The projections included moderate increases in habitat quality on the continental coast in the far west of the study area, but <90% of model projections agreed on the sign of change in GGP for much of this area.

The projected GGP declines (with unchanged chlorophyll-a concentrations) across all cells were 7%, 12% and 22% for RCP2.6, RCP4.5 and RCP8.5 respectively (Fig. 5). The corresponding reductions in Antarctic krill growth area were 5%, 10% and 20% for RCP2.6, RCP4.5 and RCP8.5 respectively. Inevitably a 50% reduction in chlorophyll-a concentration led to greater reductions in both GGP and growth area. When chlorophyll-a concentration was increased by 50%, the projected warming under RCPs 4.5 and 8.5 still led to significant reductions in growth area. Nonetheless, the effect of a 50% increase in chlorophyll-a concentration moderated the overall effects of warming on relative GGP.

The spatially-resolved projected GGPs for each combination of RCP and chlorophyll-a concentration had similar maxima which increased slightly (from 4.9 to 5.2) with chlorophyll-a concentration (Fig. S4). The mode of estimated current GGP was 2.1, and the modes of projected GGP with unchanged chlorophyll-a concentrations were similar (1.8 to 2.0) for all RCPs. The mode increased (from about 1.3 to about 2.5) with increasing chlorophyll-a concentration and the area with near modal values declined with increasing SST (i.e. from RCP2.6 to RCP8.5).

South Georgia is located in the band of projected GGP declines. Consequently there were pronounced negative effects within the foraging ranges of predators breeding on this island. These negative effects were greatest for those predators with the most restricted foraging ranges (Fig. 6), where the projected GGP declines (with unchanged chlorophyll-a concentrations) were 9%,

Table 1. The climate models used in this study.

Model name	Modelling group	Number of realisations			Global number of SST grid points		SST on regular grid?
		RCP2.6	RCP4.5	RCP8.5	Longitudinal	Latitudinal	
BCC-CSM1.1	Beijing Climate Center, China Meteorological Administration	1	1	1	360	232	No
CanESM2	Canadian Centre for Climate Modelling and Analysis	5	5	5	256	192	Yes
CNRM-CM5	Centre National de Recherches Météorologiques/Centre Européen de Recherche et Formation Avancée en Calcul Scientifique	1	1	5	362	292	No
CSIRO-Mk3.6.0	Commonwealth Scientific and Industrial Research Organization in collaboration with Queensland Climate Change Centre of Excellence	10	10	10	192	189	Yes
EC-EARTH	EC-EARTH consortium	2	9	9	362	292	No
GFDL-ESM2G	NOAA Geophysical Fluid Dynamics Laboratory	1		1	360	210	No
GISS-E2-R	NASA Goddard Institute for Space Studies	1	5	1	144	90	Yes
HadGEM2-CC	Met Office Hadley Centre (additional HadGEM2-ES realizations contributed by Instituto Nacional de Pesquisas Espaciais)		1	3	360	216	Yes
HadGEM2-ES		4	3	4	360	216	Yes
INM-CM4	Institute for Numerical Mathematics		1	1	360	340	No
IPSL-CM5A-LR	Institut Pierre-Simon Laplace	3	4	4	182	149	No
IPSL-CM5A-MR		1	1	1	182	149	No
MIROC5	Atmosphere and Ocean Research Institute (The University of Tokyo), National Institute for Environmental Studies, and Japan Agency for Marine-Earth Science and Technology	3	3	3	256	224	No
MPI-ESM-LR	Max-Planck-Institut für Meteorologie (Max Planck Institute for Meteorology)	3	3	3	256	220	No
MRI-CGCM3	Meteorological Research Institute	1	1	1	360	368	No
NorESM1-M	Norwegian Climate Centre	1	1	1	320	384	No

The table lists the models used in this study, identifies the number of realisations (individual model runs) available for each of the three Representative Control Pathways (RCPs 2.6, 4.5 and 8.5), and specifies the spatial resolution and grid type for each model.

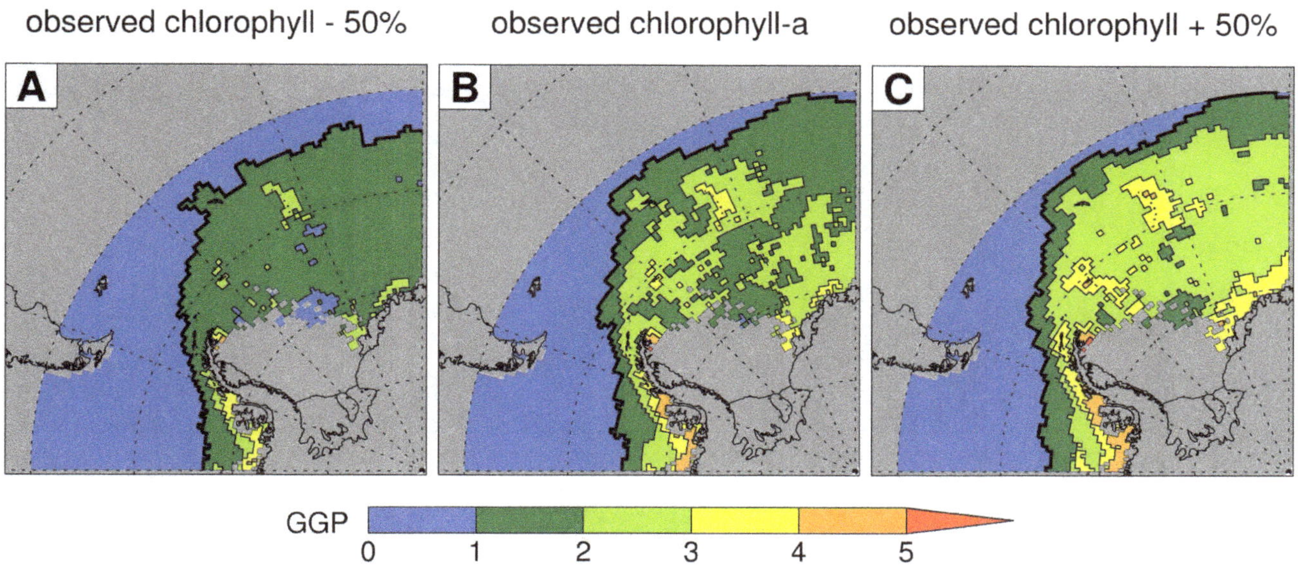

Figure 2. Current Antarctic krill summer growth habitat quality and sensitivity to chlorophyll concentration. (A) Gross Growth Potential (GGP, a unitless quantity which indicates the potential proportional increase in the mass of an individual Antarctic krill during a single summer and is therefore a measure of habitat quality) calculated for an individual with a starting length of 40 mm using observed SSTs (for the period 2002–2011), and observed chlorophyll-a concentrations (for the period 1997–2010) reduced by 50%. (B) Estimated current GGP calculated using observed SSTs, and observed chlorophyll-a concentrations. (C) Estimated current GGP calculated using observed SSTs, and observed chlorophyll-a concentrations increased by 50%. The spatial resolution is 1° longitude by 0.5° latitude and the thick black line indicates the northern extent of the growth area (the habitat that supports Antarctic krill growth with the relevant chlorophyll-a concentration). Thus, the thick black line in (B) delimits the current growth area.

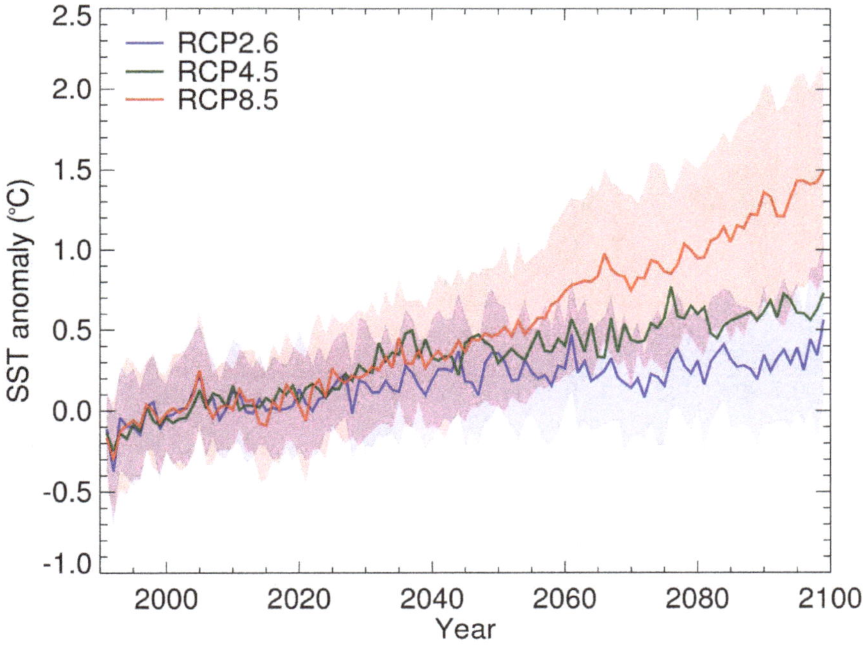

Figure 3. Projected 21st Century summer surface warming of the Southern Ocean between 0° and 90°W. Projected summer (January to March) sea surface temperature (SST) anomaly for the region between 0° and 90°W and south of the Antarctic Polar Front (Fig. 1). The SST anomaly is the within-year mean of spatially-resolved summer SSTs for a specific model realisation minus the 1991–2020 mean of spatially-resolved summer SSTs for the same model realisation. The coloured lines indicate the mean SST anomaly for 1991–2099 across all available models (Table 1) for each of three Representative Control Pathways (RCPs 2.6, 4.5 and 8.5) and the shaded envelopes indicate the between-realisation standard deviation for RCPs 2.6 and 8.5.

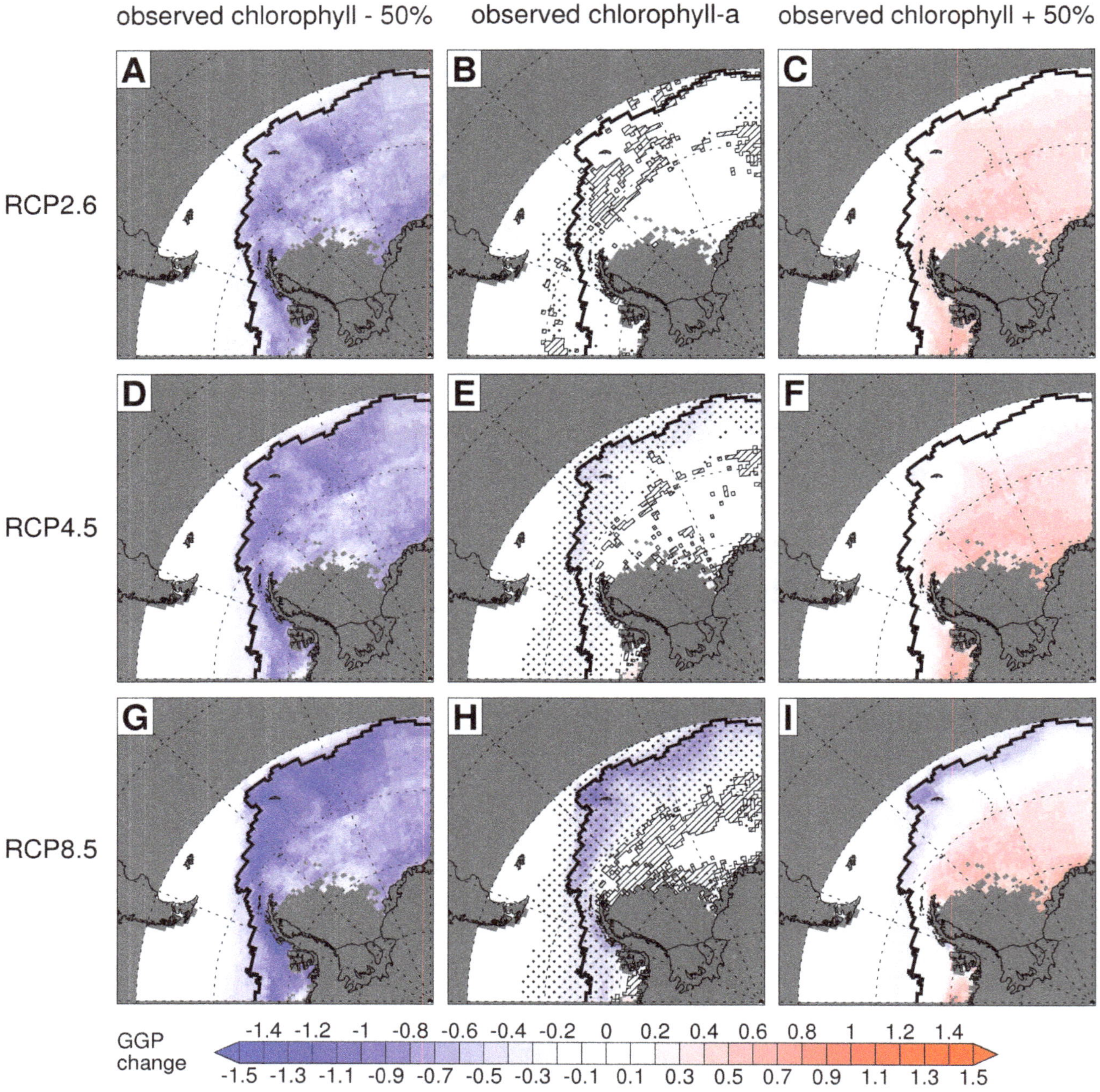

Figure 4. Spatial pattern of projected change in Antarctic krill habitat by the late 21st Century. Each panel shows the projected GGP change (GGP for the period 2070–2099 minus estimated current GGP, as shown in Fig. 2B) calculated across multiple climate models for an Antarctic krill starting length of 40 mm. The GGP values were calculated using bias-corrected SSTs from RCP2.6 (A, B & C), RCP4.5 (D, E & F) or RCP8.5 (G, H & I) and observed chlorophyll-a concentrations reduced by 50% (A, D & G), observed chlorophyll-a concentrations (B, E & H), or observed chlorophyll-a concentrations increased by 50% (C, F & I). Additional symbols (B, E & H) indicate the level of agreement between climate models. Cells where fewer than 50% of the models project significant change (t-test, $P \leq 0.05$) from the current period have no additional symbol. Cells where 50% or more of the models project significant change are highlighted with stippling if 90% or more of models agree on the sign of the change, and are highlighted with hatched lines if fewer than 90% agree. The spatial resolution is 1° longitude by 0.5° latitude and the thick black line indicates the northern extent of the current growth area (Fig. 2B).

24% and 68% and the corresponding reductions in growth area were 5%, 6% and 55% for RCP2.6, RCP4.5 and RCP8.5. The negative effects projected for RCP8.5 were apparent even with the combination of the long foraging range of grey headed albatrosses and an increase in chlorophyll-a concentration.

Analysis of variance (ANOVA) indicated that RCP, chlorophyll-a concentration and climate model each significantly influenced the results shown in Figs. 5 and 6 (e.g. for Fig. 5A three-way ANOVA suggests that each of these factors had a significant influence on the response variable, GGP: F = 187, 3182, 33,

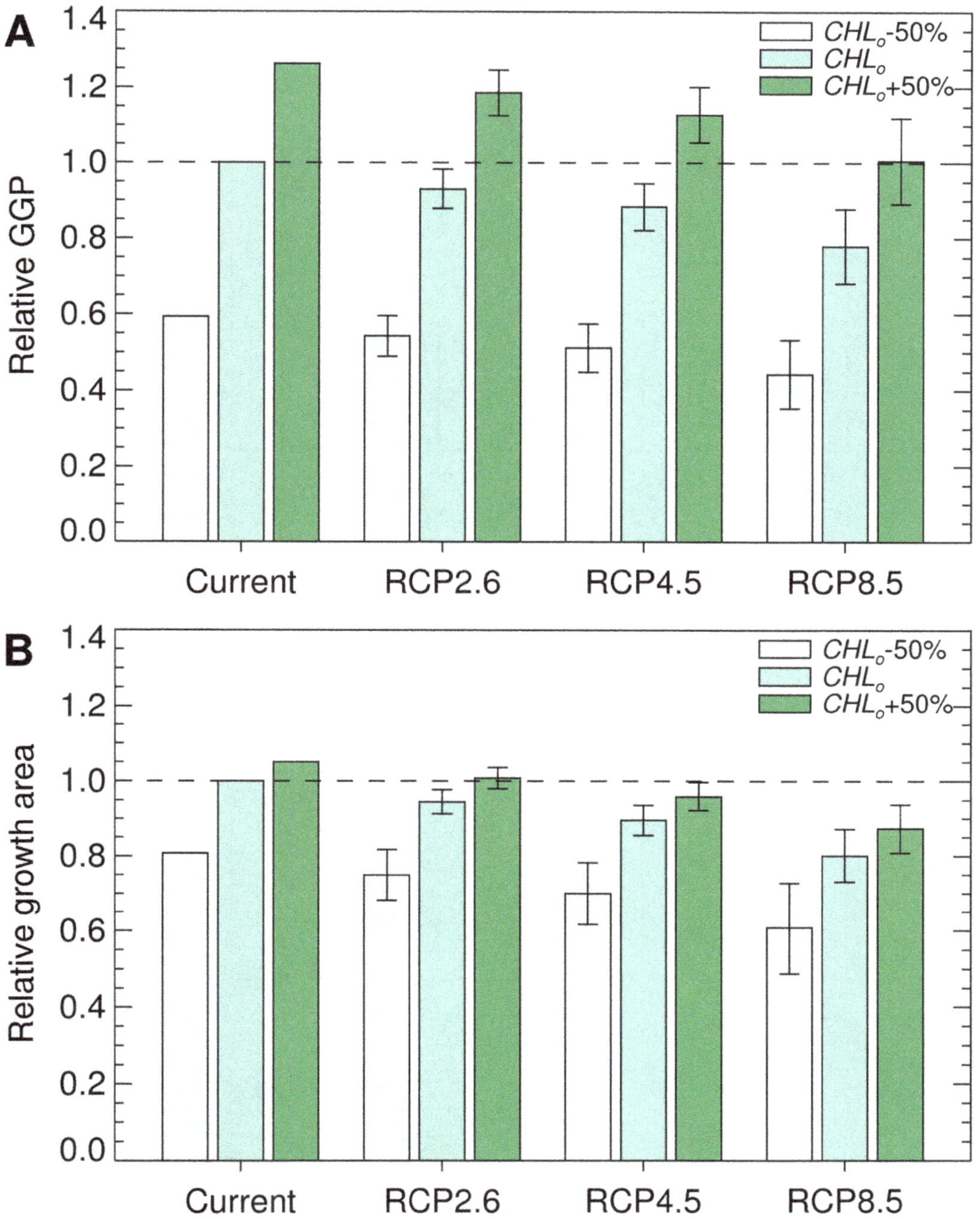

Figure 5. Projected change in Antarctic krill habitat in the study area by the late 21st Century. Relative GGP (GGP for the period 2070–2099 divided by estimated current GGP) (A) and relative growth area (growth area for the period 2070–2099 divided by estimated current growth area) (B), calculated for the study area (Fig. 1B). Results were calculated across multiple models using bias-corrected SSTs from RCP2.6, RCP4.5 or RCP8.5, and observed chlorophyll-a concentrations reduced by 50% (CHL_o-50%), observed chlorophyll-a concentrations (CHL_o), or observed chlorophyll-a concentrations increased by 50% (CHL_o+50%). The assumed Antarctic krill starting length was 40 mm and the error bars show the between-model standard deviation.

$P \ll 0.001$). There were also significant interactions between model and chlorophyll-a concentration ($F = 4$, $P < 0.001$ for Fig. 5A), and model and RCP ($F = 2$, $P = 0.026$ for Fig. 5A), which confirms the high degree of between-model variability shown in Figs. 3 and 4. Tukey multiple comparisons tests identified consistent differences ($P < 0.05$) between the results (GGP and growth area) for RCP8.5 and those for RCP2.6 or RCP4.5 when chlorophyll-a concentrations were set at observed levels or observed levels plus 50% (see Table S1, for full results).

Discussion

The projected effects of plausible SST warming on Antarctic krill growth habitat are mainly negative. Under all RCPs that we

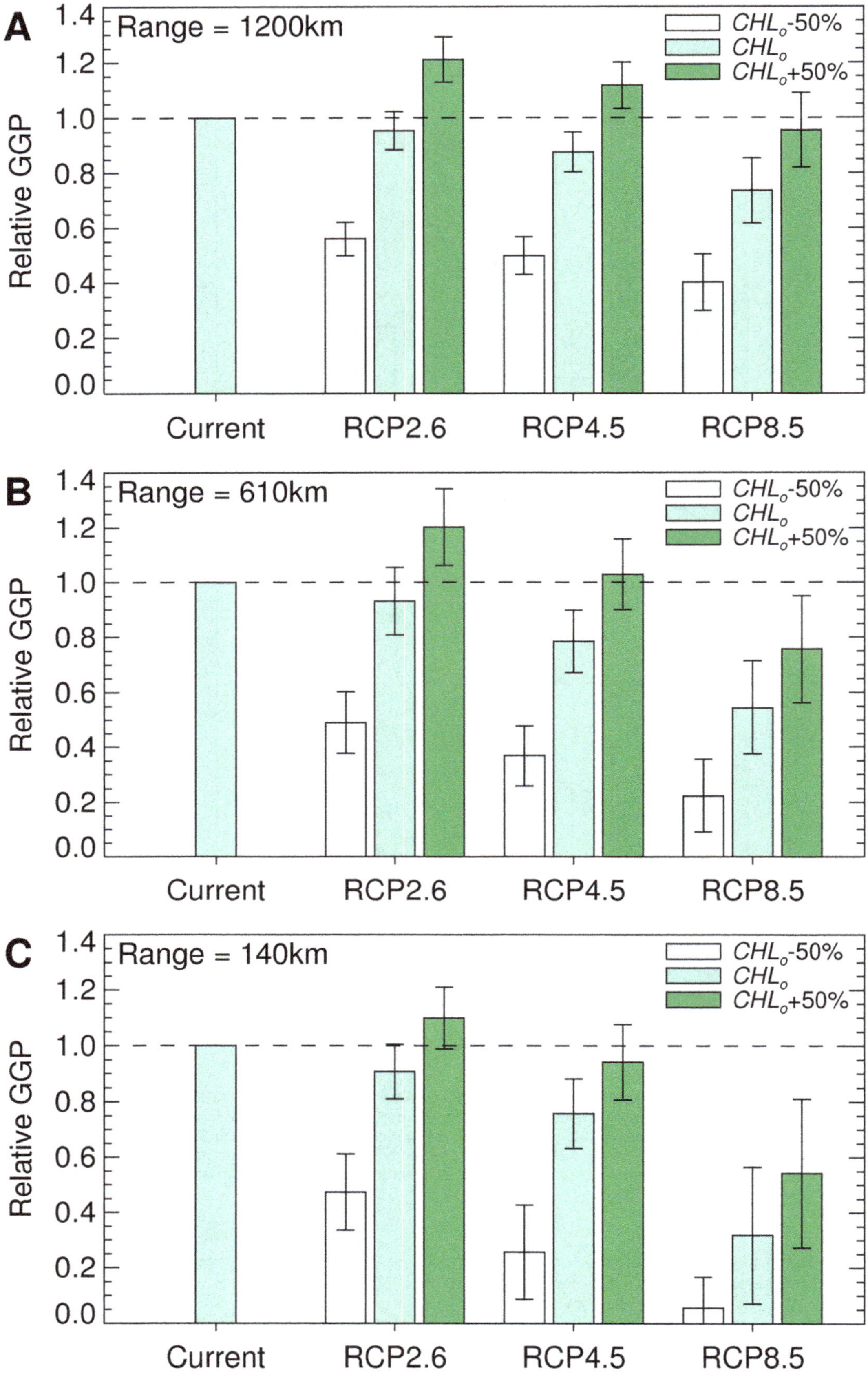
A
Range = 1200km
B
Range = 610km
C
Range = 140km
CHLo-50%
CHLo
CHLo+50%
Relative GGP
1.4
1.2
1.0
0.8
0.6
0.4
0.2
0.0
Current
RCP2.6
RCP4.5
RCP8.5

Figure 6. Projected change in Antarctic krill habitat accessible to predators foraging from South Georgia. Relative GGP (GGP for the period 2070–2099 divided by estimated current GGP), calculated within the area accessible to predators with foraging ranges of 1200 km (A), 610 km (B) and 140 km (C) from South Georgia. Results were calculated across multiple models using projected SSTs from RCP2.6, RCP4.5 or RCP8.5, and observed chlorophyll-a concentrations reduced by 50% (CHL_o-50%), observed chlorophyll-a concentrations (CHL_o), or observed chlorophyll-a concentrations increased by 50% (CHL_o+50%). The assumed Antarctic krill starting length was 40 mm and the error bars show the between-model standard deviation.

considered, the projections imply a decrease in habitat quality over the 21st century, particularly in the ACC. Our analysis suggests that these effects could be mitigated to some extent if warming leads to an overall increase in chlorophyll production. Habitat quality could improve in some marine areas close to the Antarctic continent even under the most extreme warming scenario. However this is unlikely to mitigate the negative impacts within the foraging ranges of birds and seals breeding at South Georgia.

Recent SST warming rates at South Georgia and the western Antarctic Peninsula [6], [7] are in the upper range of projected regional warming rates for the 21st century. However, some parts of the Southern Ocean have cooled over recent decades and experienced associated increases in sea ice [46]. Previous modelling studies suggest that recent warming might have already degraded Antarctic krill habitat in some areas [13], [33]. Analysis of the growth model used here with additional parameters from [34] concluded that, for the period 1970–2004, increasing temperatures probably reduced the lifetime biomass production of Antarctic krill at South Georgia but increased it at the Antarctic Peninsula [33]. Our results suggest that, in the future, increasing temperature could reduce growth in both of these areas. Reduced growth could also affect egg production as smaller females produce fewer eggs [47].

Climate affects species through their habitats. Understanding these habitat effects is a prerequisite for understanding effects on biological variables such as abundance and biomass production. There are many routes through which changing habitats can influence these variables. For example, successful completion of the Antarctic krill life cycle apparently requires spawning in water with specific depth and temperature characteristics [11], [48] and larval development under sea ice [9]. The summer months that we model encompass the main growth period of adult Antarctic krill [11], but winter processes also affect habitat suitability. Other environmental variables, such as pH, might also be critical for the sensitive larval stages [10]. Furthermore, the distribution of Antarctic krill seems to be affected by ocean currents which may transport individuals thousands of kilometres in a lifetime [49]. Thus, high quality growth habitat will only result in high biomass production if sufficient Antarctic krill arrive in the area as a result of transport or local spawning.

These multiple environmental influences on Antarctic krill abundance and biomass production have several implications. Firstly, more detailed mechanistic life-cycle and population models are needed to better assess the potential effects of climate change [9], [38]. For example, a fuller assessment of temperature effects might consider how the relationship between SST and the temperatures that Antarctic krill experience in the water column changes over time and space. Secondly, the environmental effects are likely to be more complex than a simple poleward shift in distribution. Some of the oceanographic characteristics on which Antarctic krill rely, such as the deep waters of the ACC, will not move south as the ocean warms. Coastal embayments and high latitude shelves may be reasonable refugia for growth, but they are unlikely to provide appropriate habitats for spawning [48] or connecting subpopulations [50].

Large scale analyses of ecological responses to climate change generally stress the effects of warming (e.g. [1], [2]). Polar studies also tend to emphasise warming because polar organisms are sensitive to temperature [51], [52], which is rising rapidly in some polar regions [6], [7], [53]. Nonetheless, food availability is also an important habitat characteristic which, at the physiological level, can sometimes compensate for the negative effects of temperature [29]. This is illustrated by the high Antarctic krill abundances and growth rates found at South Georgia. This is near the northern limit of the species' range and has relatively high and physiologically stressful temperatures, but it also has very high food concentrations (Fig. 1, [12]). Temperature-food interactions are therefore likely to influence the ecological effects of climate change [29]. Using models, such as ours, that explicitly include the effects of food availability is a useful step towards fuller consideration of the multiple interacting effects of climate change [38].

Previous studies have reported a 10% decline in chlorophyll-a concentration in the Southern Ocean over the 1980s and 1990s [40] and substantial localised increases and decreases in chlorophyll-a concentration at the Antarctic Peninsula in the last 30 years [54]. Such changes are associated with changes in the composition of the phytoplankton community. The main effects at the Antarctic Peninsula were an overall decline in chlorophyll-a concentration and a decrease in the abundance of diatoms relative to other phytoplankton [54]. Such changes are consistent with the expected widespread consequences of marine warming [54], [55]. A reduction in diatoms in the diet of Antarctic krill is likely to reduce both growth and reproduction [56], [57]. Therefore the most likely effects of plausible changes in chlorophyll-a concentration are in the range between our reduced and unchanged chlorophyll scenarios.

Models such as those which produced the CMIP5 results are being increasingly used to investigate climate impacts on marine species [38]. These models have many uncertainties, including regional biases and differences between models. Our results confirm a regional bias in Southern Ocean SST [58]. The representation of some Southern Ocean features, such as the ACC, has improved in the CMIP5 results compared to the previous generation of model results [59]. Nonetheless, the available models generally perform badly at reproducing sea ice conditions which, in turn, influence SST [58], [60]. The models differ markedly from each other in terms of both the magnitude and spatial distribution of projected SST changes. We have followed recommended practice for controlling and assessing the influence of these model uncertainties on our results [38]. It is clear from Fig. 4 that most models are in agreement that the ACC will experience significant warming.

Any degradation of Antarctic krill growth habitat in the ACC is likely to have consequences for predators at South Georgia. Analysis of foodweb models suggests that predators able to take advantage of copepod production might be relatively unaffected by a severe reduction in Antarctic krill availability, but that the majority of air-breathing predator populations at South Georgia would probably experience significant declines [15].

The Antarctic krill fishery took 68% of its total catch between 1980 and 2011 from the area of projected severe habitat degradation [18]. Future climate change could therefore have a significant negative effect on Southern Ocean ecosystem services as well as biodiversity. A recommendation that the Commission for the Conservation of Antarctic Marine Living Resources, which is responsible for managing the Antarctic krill fishery, should increase consideration of climate change impacts in its management decisions was made in 1992 [61] but it was not until 2009 that the Commission resolved to do so (www.ccamlr.org/en/resolution-30/xxviii-2009). We suggest that there is a need for more rapid progress in developing methods for evaluating climate change impacts in parallel with improved regional climate projections, and for adaptation to and management of the risks to the Southern Ocean ecosystem that climate change implies.

Supporting Information

Figure S1 Projected change in Antarctic krill habitat based on a starting length of 30 mm. Each panel shows the projected GGP change (GGP for the period 2070–2099 minus estimated current GGP) calculated across multiple climate models. The assumed Antarctic krill starting length was 30 mm. The GGP values were calculated using bias-corrected SSTs from RCP2.6 (A, B & C), RCP4.5 (D, E & F) or RCP8.5 (G, H & I) and observed chlorophyll-a concentrations reduced by 50% (A, D & G), observed chlorophyll-a concentrations (B, E & H), or observed chlorophyll-a concentrations increased by 50% (C, F & I). The spatial resolution is 1° longitude by 0.5° latitude and the thick black line indicates the boundaries of the growth area for that panel.

Figure S2 Projected change in Antarctic krill habitat based on a starting length of 40 mm. Each panel shows the projected GGP change calculated across multiple climate models. The assumed Antarctic krill starting length was 40 mm. Other details as Fig. S1.

Figure S3 Projected change in Antarctic krill habitat based on a starting length of 50 mm. Each panel shows the projected GGP change calculated across multiple climate models. The assumed Antarctic krill starting length was 50 mm. Other details as Fig. S1.

Figure S4 Distribution of GGP values in the results presented in Figs. 4 and 5. Each panel shows the distribution of projected GGP values for the period 2070–2099 as the percent coverage of the modelled area (coloured lines). The projected GGP values were calculated using bias-corrected SSTs from RCP2.6 (A, B & C), RCP4.5 (D, E & F) or RCP8.5 (G, H & I). The panels also show the distribution of estimated GGP values calculated using observed SSTs (for the period 2002–2011) (grey bars). Both sets of GGP values in each panel were calculated using the same chlorophyll-a concentrations: observed chlorophyll-a concentrations reduced by 50% (A, D & G), observed chlorophyll-a concentrations (B, E & H), or observed chlorophyll-a concentrations increased by 50% (C, F & I).

Table S1 Statistical comparison of scenario-specific results presented in Figs. 5 and 6. The table shows the probability (from Tukey multiple comparisons tests) that the projected GGP or growth area for each combination of chlorophyll-a concentration and RCP is significantly different from comparable results for other RCPs. We compared each result shown in Figs. 5 and 6 with the other results in the same figure panel. Comparisons which were significantly different ($P<0.05$) are highlighted in bold text.

Acknowledgments

We are grateful to Tom Bracegirdle and three anonymous referees for constructive comments and to John Turner and John King for support. We acknowledge the World Climate Research Programme's Working Group on Coupled Modelling, which is responsible for CMIP, and we thank the climate modelling groups listed in Table 1 for producing and making available their model output. The U.S. Department of Energy's Program for Climate Model Diagnosis and Intercomparison provides coordinating support for CMIP and led the development of software infrastructure in partnership with the Global Organization for Earth System Science Portals. The Antarctic krill distribution data used in Fig. 1 are available from the CCAMLR secretariat (www.ccamlr.org/en/data/ccamlr-data).

Author Contributions

Conceived and designed the experiments: SLH TP AA. Performed the experiments: TP SLH. Analyzed the data: SLH TP. Wrote the paper: SLH TP AA.

References

1. Parmesan C, Yohe G (2003) A globally coherent fingerprint of climate change impacts across natural systems. Nature 421: 37–42. doi:10.1038/nature01286.
2. Burrows MT, Schoeman DS, Buckley LB, Moore P, Poloczanska ES, et al. (2011). The pace of shifting climate in marine and terrestrial ecosystems. Science 334: 652–655. doi: 10.1126/science.1210288.
3. Solomon S, Qin D, Manning M, Chen Z, Marquis M, et al, editors (2007) Climate Change 2007: The physical sciences basis. Contribution of Working Group 1 to the Fourth Assessment Report of the Intergovernmental Panel on Climate Change. Cambridge: Cambridge University Press. 966 p.
4. Cheung WW, Lam VW, Sarmiento JL, Kearney K, Watson R, Pauly D (2009) Projecting global marine biodiversity impacts under climate change scenarios. Fish Fish 10: 235–251. doi: 10.1111/j.1467-2979.2008.00315.x.
5. Stammerjohn SE, Martinson DG, Smith RC, Yuan X, Rind D (2008) Trends in Antarctic annual sea ice retreat and advance and their relation to El Niño–Southern Oscillation and Southern Annular Mode variability. J Geophys Res-Oceans (1978–2012). doi:10.1029/2007JC004269.
6. Meredith MP, King JC (2005) Rapid climate change in the ocean west of the Antarctic Peninsula during the second half of the 20th century. Geophys Res Lett. doi: 10.1029/2005GL024042.
7. Whitehouse MJ, Priddle J, Symon C (1996) Seasonal and annual change in seawater temperature, salinity, nutrient and chlorophyll-a distributions around South Georgia, South Atlantic. Deep Sea Res Part 1 Oceanogr Res Pap 43: 425–443. doi: 10.1016/0967-0637(96)00020-9.
8. Atkinson A, Siegel V, Pakhomov E, Rothery P (2004) Long-term decline in krill stock and increase in salps within the Southern Ocean. Nature 432: 100–103. doi:10.1038/nature02996.
9. Quetin LB, Ross RM, Fritsen CH, Vernet M (2007) Ecological responses of Antarctic krill to environmental variability: can we predict the future? Antarct Sci 19: 253–266. doi:10.1017/S0954102007000363.
10. Kawaguchi S, Kurihara H, King R, Hale L, Berli T, et al. (2011) Will krill fare well under Southern Ocean acidification? Biol Lett 7: 288–291.
11. Marr JWS (1962) The natural history and geography of the Antarctic krill (*Euphasia superba* Dana). Discovery Reports 32: 33–464.
12. Atkinson A, Siegel V, Pakhomov EA, Rothery P, Loeb V, et al. (2008) Oceanic circumpolar habitats of Antarctic krill. Mar Ecol Prog Ser 362: 1–23. doi: 10.3354/meps07498.
13. Mackey AP, Atkinson A, Hill SL, Ward P, Cunningham NJ, et al. (2012) Antarctic macrozooplankton of the southwest Atlantic sector and Bellingshausen Sea: Baseline historical distributions (Discovery Investigations, 1928–1935) related to temperature and food, with projections for subsequent ocean warming. Deep Sea Res Part 2 Top Stud Oceanogr 59: 130–146. doi: 10.1016/j.dsr2.2011.08.011.
14. Atkinson A, Whitehouse MJ, Priddle J, Cripps GC, Ward P, Brandon MA (2001) South Georgia, Antarctica: a productive, cold water, pelagic ecosystem. Mar Ecol Prog Ser 216: 279–308. doi:10.3354/meps216279.
15. Hill SL, Keeble K, Atkinson A, Murphy EJ (2012) A foodweb model to explore uncertainties in the South Georgia shelf pelagic ecosystem. Deep Sea Res Part 2 Top Stud Oceanogr 59: 237–252. doi: 10.1016/j.dsr2.2011.09.001.

16. Pikitch EK, Rountos KJ, Essington TE, Santora C, Pauly D et al. (2012) The global contribution of forage fish to marine fisheries and ecosystems. Fish Fish doi: 10.1111/faf.12004.
17. Everson I (2000) Role of krill in marine foodwebs: The Southern Ocean. In Everson I, editor. Krill: biology, ecology and fisheries. Oxford: Blackwell. pp 63–79.
18. Hill SL (2013) Prospects for a sustainable increase in the availability of long chain omega 3s: Lessons from the Antarctic Krill fishery. In: De Meester F, Watson RF, Zibadi S, editors. Omega-6/3 fatty acids: Functions, sustainability strategies and perspectives (Nutrition and Health). New York: Humana Press. 267–296. doi: 10.1007/978-1-62703-215-5_14.
19. Nicol S, Foster J, Kawaguchi S (2012) The fishery for Antarctic krill–recent developments. Fish Fish 13: 30–40. doi: 10.1111/j.1467-2979.2011.00406.x.
20. Atkinson A, Siegel V, Pakhomov EA, Jessopp MJ, Loeb V (2009) A re-appraisal of the total biomass and annual production of Antarctic krill. Deep Sea Res Part 1 Oceanogr Res Pap 56: 727–740. doi:10.1016/j.dsr.2008.12.007.
21. Atkinson A, Nicol S, Kawaguchi S, Pakhomov E, Quetin L, et al. (2012) Fitting *Euphausia superba* into southern ocean food-web models: a review of data sources and their limitations. CCAMLR Sci 19: 219–245.
22. Murphy EJ, Watkins JL, Trathan PN, Reid K, Meredith MP, et al. (2007). Spatial and temporal operation of the Scotia Sea ecosystem: a review of large-scale links in a krill centred food web. Philos Trans R Soc Lond B Biol Sci 362: 113–148. doi: 10.1098/rstb.2006.1957.
23. Kovacs K, Lowry L (2008) *Arctocephalus gazella*. In: IUCN, editors. IUCN Red List of threatened species. Version 2012.2. Available: www.iucnredlist.org. Accessed 2013 March 28.
24. Butchart S, Taylor J (2012a) *Thalassarche chrysostoma*. In: IUCN, editors. IUCN Red List of threatened species. Version 2012.2. Available: www.iucnredlist.org. Accessed 2013 March 28.
25. Butchart S, Taylor J (2012b) *Diomedea exulans*. In: IUCN, editors. IUCN Red List of threatened species. Version 2012.2. Available: www.iucnredlist.org. Accessed 2013 March 28.
26. Xavier JC, Croxall JP, Trathan PN, Wood AG (2003). Feeding strategies and diets of breeding grey-headed and wandering albatrosses at South Georgia. Mar Biol 143: 221–232. doi: 10.1007/s00227-003-1049-0.
27. FAO (2005) Review of the state of world marine fishery resources. Fisheries Technical Paper 457. Rome: FAO. 235 p.
28. Grant SM, Hill SL, Trathan PN, Murphy EJ (2013) Ecosystem services of the Southern Ocean: trade-offs in decision-making. Antarct Sci. doi: 10.1017/S0954102013000308.
29. Pörtner HO (2012) Integrating climate-related stressor effects on marine organisms: unifying principles linking molecule to ecosystem-level changes. Mar Ecol Prog Ser 47: 273–290. doi:10.3354/meps10123.
30. Atkinson A, Shreeve RS, Hirst AG, Rothery P, Tarling GA, et al. (2006) Natural growth rates in Antarctic krill (*Euphausia superba*): II. Predictive models based on food, temperature, body length, sex, and maturity stage. Limnol Oceanogr 51: 973–987. doi: 10.4319/lo.2006.51.2.0973.
31. Taylor KE, Stouffer RJ, Meehl GA (2012) An overview of CMIP5 and the experiment design. Bull. Amer. Meteor. Soc. 93: 485–498. doi: 10.1175/BAMS-D-11-00094.1.
32. Hince B (2000) The Antarctic Dictionary: A complete guide to Antarctic English. Melbourne: CSIRO Publishing. 404 p.
33. Wiedenmann J, Cresswell K, Mangel M (2008) Temperature-dependent growth of Antarctic krill: predictions for a changing climate from a cohort model. Mar Ecol Prog Ser 358: 191–202. doi:10.3354/meps07350.
34. Kawaguchi S, Candy S, King R, Naganobu M, Nicol S (2006). Modelling growth of Antarctic krill. I. Growth trends with sex, length, season, and region. Mar Ecol Prog Ser 306: 1–15.
35. Van Vuuren DP, Edmonds J, Kainuma M, Riahi K, Thomson A, et al. (2011) The representative concentration pathways: an overview. Clim Change 109: 5–31. doi: 10.1007/s10584-011-0148-z.
36. Orsi AH, Whitworth T, Nowlin WD (1995) On the meridional extent and fronts of the Antarctic Circumpolar Current. Deep Sea Res Part 1 Oceanogr Res Pap 42: 641–673. doi: 10.1016/0967-0637(95)00021-W.
37. Feldman GC, McClain CR (2007) Ocean Color Web, MODIS-Aqua Reprocessing 2010.0. NASA Goddard Space Flight Center. Eds. Kuring N, Bailey SW. January 2012. oceancolor.gsfc.nasa.gov.
38. Stock CA, Alexander MA, Bond NA, Brander KM, Cheung WW (2011) On the use of IPCC-class models to assess the impact of climate on living marine resources. Prog Oceanogr 88: 1–27. doi: 10.3410/f.8898957.9452055.
39. Feldman GC, McClain CR (2007) Ocean Color Web, SeaWiFS Reprocessing 2010.0. NASA Goddard Space Flight Center. Eds. Kuring N, Bailey SW. October 2011. oceancolor.gsfc.nasa.gov.
40. Gregg WW, Conkright ME, Ginoux P, O'Reilly JE, Casey NW (2003). Ocean primary production and climate: Global decadal changes. Geophys Res Lett, 30. doi: 10.1029/2003GL016889.
41. Nicol S, Stolp M, Cochran T, Geijsel P, Marshall J (1992) Growth and shrinkage of Antarctic krill *Euphausia superba* from the Indian Ocean sector of the Southern Ocean during summer. Mar Ecol Prog Ser 89: 175–181.
42. Staniland IJ, Boyd IL (2003) Variation in the foraging location of Antarctic fur seals (*Arctocephalus gazella*) and the effects on diving behavior. Mar Mammal Sci 19: 331–343. doi: 10.1111/j.1748-7692.2003.tb01112.
43. Press WH, Flannery BP, Teukolsky SA, Vetterling WT (1992) Numerical recipes in C: The art of scientific computing, Second Edition. Cambridge: Cambridge University Press. 1020 p.
44. Lee DT, Schachter BJ (1980) Two algorithms for constructing a Delaunay triangulation. Int J Comput Inf Sci 9: 219–242. doi:10.1007/bf00977785.
45. Coxeter HSM (1969) Introduction to geometry, Second Edition. Oxford: Wiley. 496 p.
46. Parkinson CL (2004) Southern Ocean sea ice and its wider linkages: insights revealed from models and observations. Antarct Sci 16: 387–400. doi: 10.1017/S0954102004002214.
47. Tarling GA, Cuzin-Roudy J, Thorpe SE, Shreeve RS, Ward P, Murphy EJ (2007). Recruitment of Antarctic krill *Euphausia superba* in the South Georgia region: adult fecundity and the fate of larvae. Mar Ecol Prog Ser 331: 161–179. doi: 10.3354/meps331161.
48. Hofmann EE, Hüsrevoğlu YS (2003) A circumpolar modeling study of habitat control of Antarctic krill (*Euphausia superba*) reproductive success. Deep Sea Res Part 2 Top Stud Oceanogr 50: 3121–3142. doi: 10.1016/j.dsr2.2003.07.012.
49. Thorpe SE, Murphy EJ, Watkins JL (2007). Circumpolar connections between Antarctic krill (*Euphausia superba* Dana) populations: Investigating the roles of ocean and sea ice transport. Deep Sea Res Part 1 Oceanogr Res Pap 54: 792–810. doi: 10.1016/j.dsr.2007.01.008.
50. Siegel V (2005) Distribution and population dynamics of *Euphausia superba*: summary of recent findings. Polar Biol 29: 1–22. doi: 10.1007/s00300-005-0058-5.
51. Pörtner HO, Farrell AP (2008) Physiology and climate change. Science 322: 690–692. doi: 0.1126/science.1163156.
52. Peck LS, Webb KE, Bailey DM (2004) Extreme sensitivity of biological function to temperature in Antarctic marine species. Funct Ecol 18: 625–630. doi: 10.1111/j.0269-8463.2004.00903.x.
53. Steele M, Zhang J, Ermold W (2010) Mechanisms of summertime upper Arctic Ocean warming and the effect on sea ice melt. J Geophys Res-Oceans (1978–2012), 115(C11). doi: 10.1029/2009JC005849.
54. Montes-Hugo M, Doney SC, Ducklow HW, Fraser W, Martinson D, et al. (2009) Recent changes in phytoplankton communities associated with rapid regional climate change along the western Antarctic Peninsula. Science 323: 1470–1473. doi: 10.1126/science.1164533.
55. Moline MA, Blackwell SM, Chant R, Oliver MJ, Bergmann T, et al. (2004) Episodic physical forcing and the structure of phytoplankton communities in the coastal waters of New Jersey. J Geophys Res-Oceans (1978–2012) 109(C12): C12S05. doi: 10.1029/2003JC001985.
56. Ross RM, Quetin LB, Baker KS, Vernet M, Smith RC (2000) Growth limitation in young *Euphausia superba* under field conditions. Limnol Oceanogr 45: 31–43.
57. Schmidt K, Atkinson A, Venables HJ, Pond DW (2012) Early spawning of Antarctic krill in the Scotia Sea is fuelled by "superfluous" feeding on non-ice associated phytoplankton blooms. Deep Sea Res Part 2 Top Stud Oceanogr 59: 159–172. doi: 10.1016/j.dsr2.2011.05.002.
58. Sallée JB, Shuckburgh E, Bruneau N, Meijers A, Wang Z, et al. (2013) Assessment of Southern Ocean water mass circulation and characteristics in CMIP5 models: historical bias and forcing response. J Geophys Res-Oceans 118: 1830–1844. doi: 10.1002/jgrc.20135.
59. Meijers AJS, Shuckburgh E, Bruneau N, Sallee JB, Bracegirdle TJ, et al. (2012) Representation of the Antarctic Circumpolar Current in the CMIP5 climate models and future changes under warming scenarios. J Geophys Res-Oceans (1978–2012) 117(C12). doi: 10.1029/2012JC008412.
60. Turner J, Bracegirdle TJ, Phillips T, Marshall GJ, Hosking S (2013). An initial assessment of Antarctic sea ice extent in the CMIP5 models. J Climate 26: 1473–1484 doi: 10.1175/JCLI-D-12-00068.1.
61. Everson I, Stonehouse B, Drewry DJ, Barker PF (1992) Managing Southern Ocean krill and fish stocks in a changing environment [and Discussion]. Philos Trans R Soc Lond B Biol Sci 338: 311–317. doi: 10.1098/rstb.1992.0151.

17

Investigation of Climate Change Impact on Water Resources for an Alpine Basin in Northern Italy: Implications for Evapotranspiration Modeling Complexity

Giovanni Ravazzani[1]*, Matteo Ghilardi[1], Thomas Mendlik[2], Andreas Gobiet[2], Chiara Corbari[1], Marco Mancini[1]

1 Politecnico di Milano, Piazza Leonardo da Vinci, Milan, Italy, **2** Wegener Center for Climate and Global Change and Institute for Geophysics, Astrophysics, and Meteorology, University of Graz, Graz, Austria

Abstract

Assessing the future effects of climate change on water availability requires an understanding of how precipitation and evapotranspiration rates will respond to changes in atmospheric forcing. Use of simplified hydrological models is required beacause of lack of meteorological forcings with the high space and time resolutions required to model hydrological processes in mountains river basins, and the necessity of reducing the computational costs. The main objective of this study was to quantify the differences between a simplified hydrological model, which uses only precipitation and temperature to compute the hydrological balance when simulating the impact of climate change, and an enhanced version of the model, which solves the energy balance to compute the actual evapotranspiration. For the meteorological forcing of future scenario, at-site bias-corrected time series based on two regional climate models were used. A quantile-based error-correction approach was used to downscale the regional climate model simulations to a point scale and to reduce its error characteristics. The study shows that a simple temperature-based approach for computing the evapotranspiration is sufficiently accurate for performing hydrological impact investigations of climate change for the Alpine river basin which was studied.

Editor: João Miguel Dias, University of Aveiro, Portugal

Funding: This work was supported by ACQWA EU/FP7 project (grant number 212250) "Assessing Climate impacts on the Quantity and quality of WAter" [link:http://www.acqwa.ch]. The funders had no role in study design, data collection and analysis, decision to publish, or preparation of the manuscript.

Competing Interests: The authors have declared that no competing interests exist.

* Email: giovanni.ravazzani@polimi.it

Introduction

According to the Fifth Assessment Report (AR5) of the United Nations Intergovernmental Panel on Climate Change (IPCC) [1], for average annual Northern Hemisphere temperatures, the period 1983–2012 was very likely the warmest 30-year period of the last 800 years. Climate change has significant implications for the environment [2], [3], water resources [4], and human life in general [5], which have motivated a multitude of scientific investigations over the past two decades [6], [7], [8], [9]. One of the expected impacts of climate change is a modification of water availability, due to the strict interaction between the climate system and the hydrological cycle. Therefore, an accurate assessment of the future effects of climate change requires an understanding of how precipitation and evapotranspiration rates will respond to changes in atmospheric forcing. The most common approach used to assess the hydrologic impact of global climate change involves climate models as input of hydrological models. In particular, the climate models simulate the climatic effects of increasing atmospheric concentrations of greenhouse gases, while the hydrological models are used to simulate the hydrological impacts of climate change [10]. River discharges, and their temporal distributions, are strongly affected by high mountainous areas [11], [12], which are particularly sensitive to global warming [13], [14]. The quality of hydrological impact investigations, even of larger catchments, thus depends on the capability to model those specific processes in mountainous regions.

The extreme complexity of the processes involved in the hydrology of mountainous areas, and the great spatial variability of meteorological forcings and river basin characteristics, require the use of physically based and spatially distributed hydrological models to simulate the transformation of rainfall into runoff [15], [16], [17]. Recent advances have made physically based hydrologic models more complex through the inclusion of more sophisticated land surface models, which compute the water and energy balances between the land surface and the atmosphere [18], [19]. This should improve the predictive skill, and facilitate the estimation of parameter values based on physiological

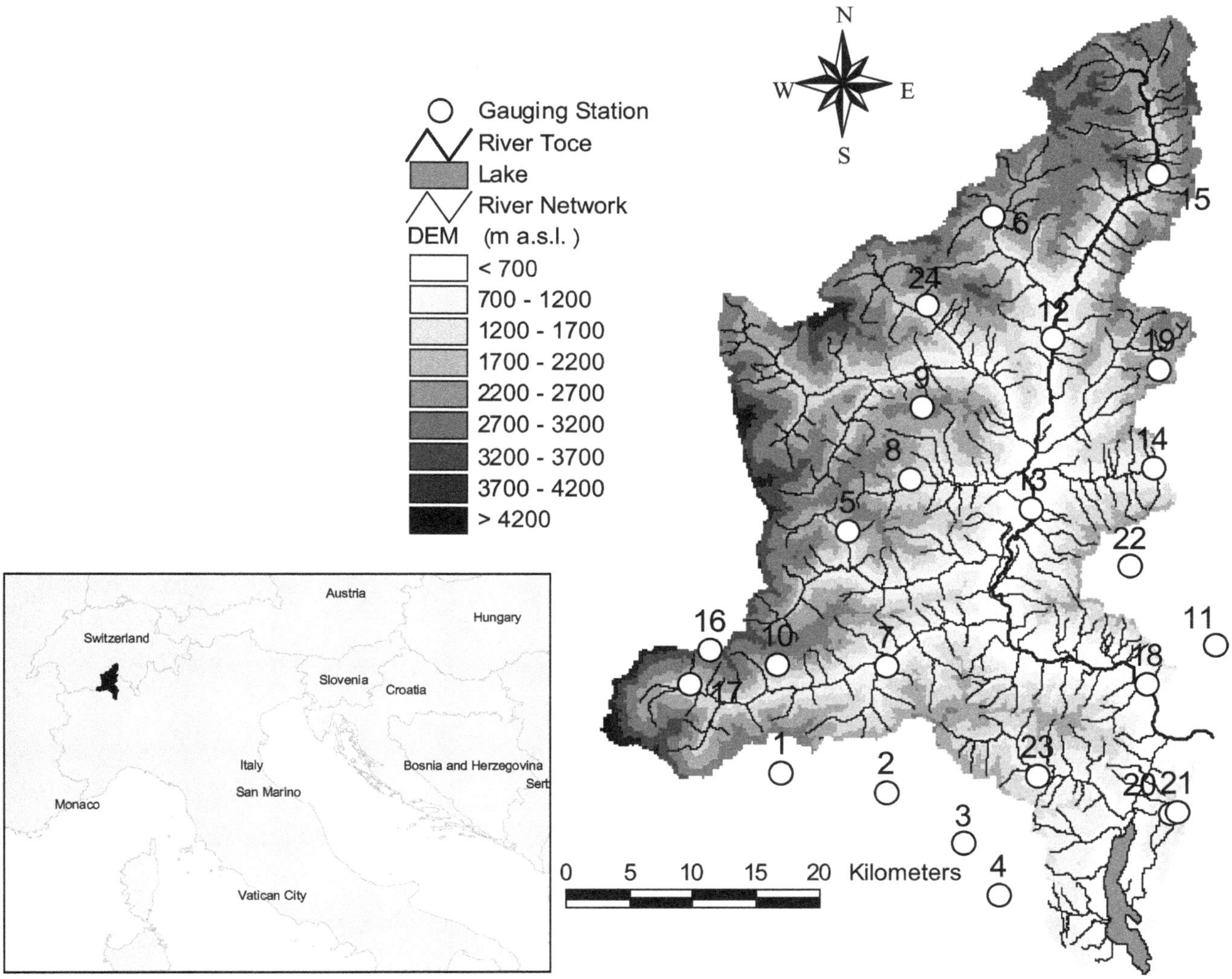

Figure 1. Localization of the stations on a DEM of the Toce watershed.

characteristics or measurements. Conversely, more complex models suffer from computational requirements, which can limit their applicability when simulating long time series such as those required for climate change impact analyses. Moreover, in addition to precipitation and temperature data, more sophisticated models require, as an input, a complete dataset of meteorological forcings, including solar radiation, wind speed, and relative humidity. These variables may not be available, at proper spatial and temporal resolutions, to accurately capture the dynamics of the hydrological processes in mountainous areas [20]. As a consequence, the hydrological model used for the analysis of climate change impacts should be a compromise between its accuracy and its simulation time. This requires an assessment of the reliability of simplified hydrological models in contrast to the more sophisticated land surface models.

The main objective of this study was to quantify the differences between a simplified hydrological model, which computes the hydrological balance based on precipitation and temperature only, and an enhanced version of the model, which solves the energy balance to compute the actual evapotranspiration. The study was performed in three steps: first, the hydrological models were calibrated and validated against the river discharge measured in the control period; second, the hydrological models driven by climatic forcings were evaluated for their performance in reproducing the water balance components during the control period; and third, climate change impacts was assessed computing the differences between the hydrological variables simulated for the decade spanning 2041–2050 and those of the control period.

The structure of the paper is as follows. In section 2, description of the study area, and data and mathematical models used are presented. In section 3.1, the hydrological models driven by meterological forcings, are evaluated in reproducing the daily streamflow; in section 3.2 the hydrological models, driven by modelled climatic forcings, are evaluated in reproducing the hydrological aspects of the control period; in section 3.3 the climate change impacts on hydrological processes are presented. In the last section, conclusions are drawn.

Table 1. Availability of data at the stations used in the hydrological analysis.

ID	*Name*	*Precipitation (mm)*	*Temperature (°C)*	*Radiation (w/m²)*	*Wind Speed (m/s)*	*Relative Humidity (%)*	*Discharge (m³/s)*
1	Carcoforo		X				
2	Fobello	X	X				
3	Sabbia	X	X				
4	Varallo	X	X			X	
5	Alpe Cheggio		X				
6	Alpe Devero	X	X			X	
7	Anzino	X	X				
8	Pizzanco	X	X			X	
9	Lago Paione	X	X	X		X	
10	Ceppo Morelli		X				
11	Cicogna	X	X				
12	Crodo	X	X			X	
13	Domodossola	X	X	X	X	X	
14	Druogno	X	X			X	
15	Formazza	X	X			X	
16	Passo Moro	X	X			X	
17	Pecetto	X	X				
18	Candoglia	X	X				X
19	Larecchio	X	X			X	
20	Baita CAI	X	X				
21	Mottarone	X	X			X	
22	Mottac	X	X				
23	Sambughetto	X	X				
24	Varzo	X	X			X	

Data and Methods

Study area

The Toce watershed is a typical glacial basin, with steep hillslopes bounding a narrow valley located primarily in the north Piedmont region of Italy, and partially in Switzerland (10% of the total area), and with a total drainage area of approximately 1,800 km² (Fig. 1). Its elevation ranges from 193 m above sea level (a.s.l.) at the outlet to approximately 4,600 m a.s.l. at the Monte Rosa crest. The average elevation is 1,641 m a.s.l. Geographic coordinates of basin outlet are: 8.49027° longitude, 45.94028° latitude.

The land cover is composed of forests (70%), bare rocks (9%), agricultural land (7%), natural grassland (6%), urban centers (4%), bodies of water (3%), and glaciers and perpetual snow (1%). The Toce lithology has five main classes: augean gneiss (49%), micaceous schists (27%), calcareous schists (11%), grindstones (7%), and granites (6%). The steep hillslopes, forming the most significant area of the basin, are mostly covered by trees on thin soil layers resting on bedrock. The soil depth increases in the downstream narrow alluvial region where an unconfined aquifer interacts with the river course. Fourteen major dams are located within the Toce watershed, with a total effective storage capacity of approximately 151×10^6 m³ [21].

A digital elevation model was available at a 200×200 meter resolution as retrieved from 1:10,000-scale topographic maps [22]. The digital land use map was derived by coupling the CORINE (Coordination of Information on the Environment) land cover map [23] for the Italian portion, with the Swiss land use map (Arealstatistik) for the small portion of the basin located in Switzerland. Both the maps were derived from remote sensing observations [21]. Most of the parameter maps were produced during the European Union research project RAPHAEL (Runoff and Atmospheric Processes for flood HAzard forEcasting and controL), whose objective was to improve flood forecasting in the complex mountain watershed [24], [25].

The meteorological and hydrologic data were collected hourly by a telemetric monitoring system of the Regione Piemonte flood warning system. The data were available from January 1, 2000 to December 31, 2010 at the stations shown in Table 1 and Fig. 1 for the rainfall, air temperature, short wave solar radiation, air humidity, wind speed, and river discharges at Candoglia (1,534 km² basin area). The mean value of the maximum annual flood peak is 944 m³/s, and the average discharge is 64 m³/s.

Two hydrological models with increasing complexity

Two distributed hydrological models were used for simulating the water balance components of the Toce river basin: the FEST-WB (Flash–flood Event–based Spatially distributed rainfall–runoff Transformation, including Water Balance [15], [26]) and the FEST-EWB (Flash–flood Event–based Spatially distributed rainfall–runoff Transformation, including Energy and Water Balance [27], [18]). The main difference between them is in the computation of evapotranspiration. The FEST-WB model derives the actual evapotranspiration by rescaling the potential evapotranspiration using a simple empirical approximation, where the

Table 2. Meteorological forcings and parameters used as input to the FEST-WB and FEST-EWB models.

Input	*Unit*	*FEST-WB*	*FEST-EWB*
Precipitation	mm	X	X
Temperature	°C	X	X
Solar Radiation	W/m^2		X
Wind Speed	m/s		X
Relative Humidity	%		X
Saturated Hydraulic Conductivity	m/s	X	X
Residual Moisture Content	-	X	X
Saturated Moisture Content	-	X	X
Wilting Point	-	X	X
Field Capacity	-	X	X
Pore Size Index	-	X	X
Curve Number	-	X	X
Soil Depth	m	X	X
Vegetation Fraction	%	X	X
Crop Coefficient	-	X	
Leaf Area Index	m^2/m^2		X
Albedo	-		X
Minimum Stomatal Resistance	s/m		X
Vegetation Height	m		X

potential evapotranspiration is computed based only on air temperature measurements. By contrast, the FEST-EWB model computes the actual evapotranspiration by solving the system of water mass and energy balance equations. The differences in the input parameters and meteorological forcings are listed in Table 2.

Six principal components can be identified in all models (Fig. 2): 1) the flow paths and channel network definition; 2) the spatial interpolation of meteorological forcings; 3) the simulation of snow pack and glacier dynamics; 4) the estimation of losses and soil moisture updating; 5) the runoff and base flow routings, including the effect of artificial reservoirs; and 6) the groundwater and hyporheic exchanges with streamflow.

For further details on distributed hydrological models and their applications, the reader can refer to [28], [29], [30], [31], [32], [33], [34].

Actual evapotranspiration in the FEST-WB hydrological model. The global actual evapotranspiration rate is given by:

$$ET = f_{bs}E_{bs} + f_v T \tag{1}$$

where E_{bs} is the actual rate of bare soil evaporation, T is the actual rate of transpiration, and f_{bs} and f_v are the fraction of the bare soil and the vegetation area, respectively ($f_{bs} + f_v = 1$). The actual rates of the bare soil evaporation and transpiration are computed as a fraction of the potential evapotranspiration, PET:

$$E_{bs} = \alpha(\theta)PET \tag{2a}$$

$$T = \beta(\theta)PET \tag{2b}$$

where

$$\alpha(\theta) = 0.082\theta + 9.173\theta^2 - 9.815\theta^3 \tag{3a}$$

$$\beta(\theta) = \begin{cases} 0 & if \quad \theta \le \theta_{wp} \\ 1 & if \quad \theta \ge \theta_{fc} \\ \dfrac{\theta - \theta_{wp}}{\theta_{fc} - \theta_{wp}} & if \quad \theta_{wp} < \theta < \theta_{fc} \end{cases} \tag{3b}$$

and where θ (-),θ_{fc} (-), and θ_{wp} (-) are current water content, field capacity, and wilting point, respectively.

The potential evapotranspiration is given by

$$PET = K_c PET_0 \tag{4}$$

where K_c is the crop coefficient [35] retrieved from satellite images [36], [37], and PET_0 is the reference potential evapotranspiration that is computed with a temperature-based equation specifically developed for the Alpine environment [38]:

$$PET = K_c PET_0$$

$$PET_0 = (0.817 + 0.00022 \cdot z) \cdot HC \cdot R_a \cdot (T_{\max} - T_{\min})^{HE} \left(\frac{T_{\max} - T_{\min}}{2} + HT\right) \tag{5}$$

where z is the elevation (m a.s.l.), R_a is the extraterrestrial radiation (mm·day^{-1}), T_{max} is the daily maximum air temperature

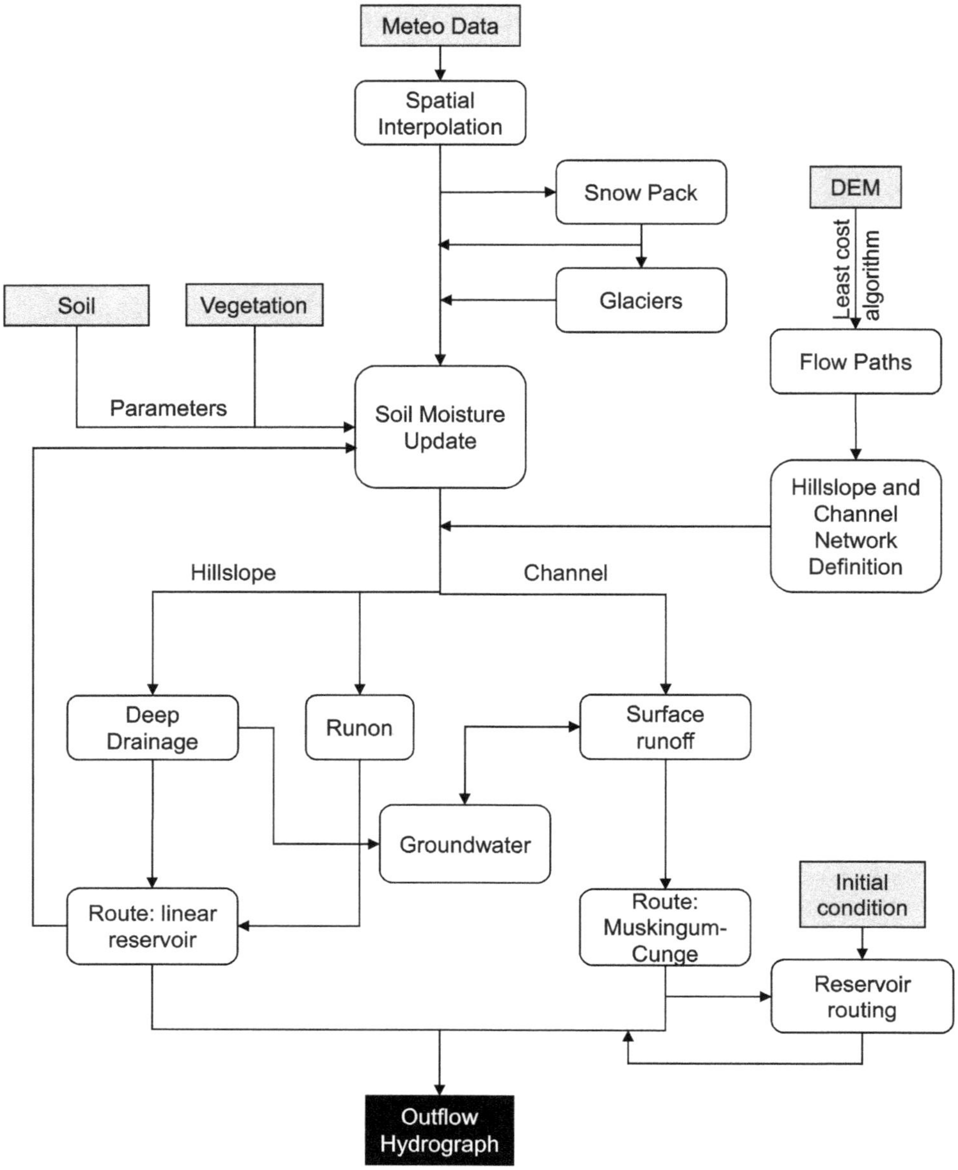

Figure 2. Scheme of the primary features common to the FEST-WB and FEST-EWB distributed-hydrological models.

(°C), T_{min} is the daily minimum air temperature (°C), HC is the empirical coefficient ($HC = 0.0023$), HE is the empirical exponent ($HE = 0.5$), and HT is to convert units of Fahrenheit to Celsius ($HT = 32/1.8 = 17.8$) [39].

Actual evapotranspiration in FEST-EWB hydrological model. In the FEST-EWB model, the actual evapotranspiration is computed by solving the energy balance equation at the ground surface expressed as

Table 3. Root mean square error (*RMSE*), m^3/s, and Nash and Sutcliffe efficiency (η) for the FEST-WB and FEST-EWB driven by observed meteorological forcings.

Index	*FEST-WB*	*FEST-EWB*
RMSE	26.7	29.5
η	0.81	0.76

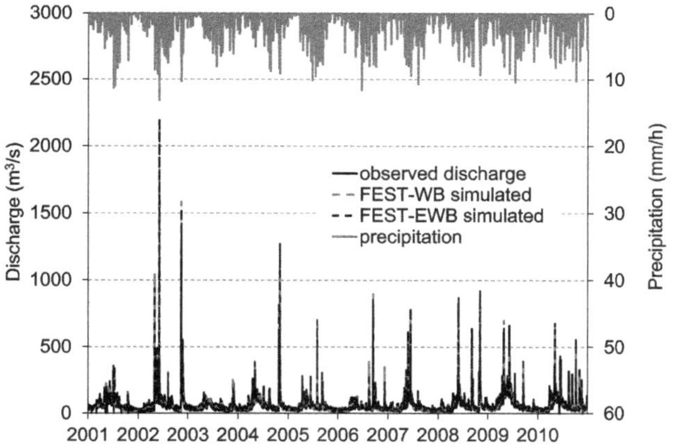

Figure 3. Comparison between the simulated and observed hourly discharge from the FEST-WB and FEST-EWB hydrological models.

$$R_n - G - (H_S + H_C) - (LE_S + LE_C) = \frac{\Delta W}{\Delta t} \qquad (6)$$

where R_n ($W{\cdot}m^{-2}$) is the net radiation, G ($W{\cdot}m^{-2}$) is the soil heat flux, H_s and H_c ($W{\cdot}m^{-2}$) and LE_s and LE_c ($W{\cdot}m^{-2}$) are the sensible heat and latent heat fluxes for the bare soil (s) and canopy (c), respectively, and $\Delta W/\Delta t$ ($W{\cdot}m^{-2}$) assembles the energy storage terms. These terms are often negligible, especially with a low spatial resolution at the basin scale; however, the contribution of these terms can be significant at the local scale [40], [41]. LE_c is a function of the canopy resistance, which is expressed as a function of the leaf area index, while LE_s is a function of the soil resistance [18]. In this study leaf area index was retrieved from satellite images.

All of the terms of the energy balance depend on the land surface temperature (LST), which allows the energy balance equation to be solved by finding the thermodynamic equilibrium temperature which closes the equation using the Newton-Raphson method:

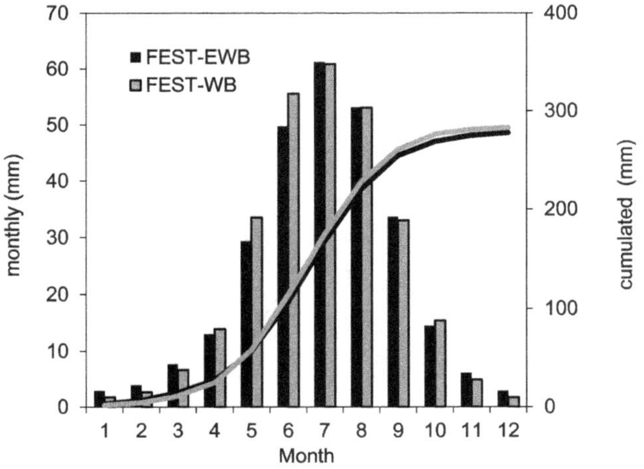

Figure 4. Mean monthly and cumulated actual evapotranspiration as computed by the FEST-WB and FEST-EWB hydrological models driven by meteorological observations.

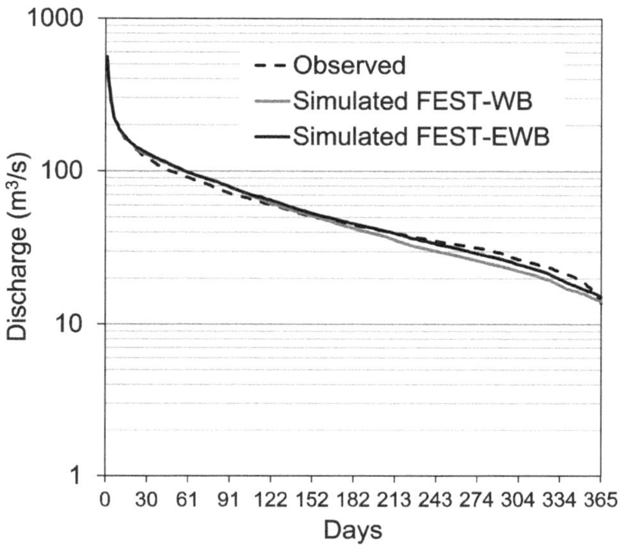

Figure 5. Mean flow duration curves for 2001–2010 from the observed discharges and those simulated by the FEST-WB and FEST-EWB hydrological models driven by meteorological observations.

$$LST_n = LST_{n-1} + \frac{f_t(LST_{n-1})}{f_t'(LST_{n-1})} \qquad (7)$$

where LST_n is the actual value, LST_{n-1} is the value at the previous iteration, $f_t(LST_{n-1})$ is the energy balance function, and $f_t'(LST_{n-1})$ is its derivative. The solution is acceptable when $\left|\frac{f_t(LST)}{f_t'(LST)}\right| < tolerance$ and $f_t(LST) < tolerance$, with *tolerance* equal to 0.001.

The FEST-EWB model has been proven to make accurate projections of the actual evapotranspiration against the energy and mass exchange measurements acquired by an eddy covariance station [18] and at the agricultural district scale against ground and remote sensing information [27].

Calibration and validation of the hydrological models

The calibration of the snow module parameters was performed in a previous study described by [42] and [43]. Given that the first assigned values (based upon measured values or reference literature or an educated guess) provided satisfactory results in terms of time series discharge simulation, no other parameters were calibrated. The performance of the model was assessed by comparing the daily simulated and observed discharge at Candoglia in the period from 2001 to 2010. The year 2000 was treated as the period for the model initialization. The performance of the models was assessed through two goodness of fit indices, the Root Mean Square Error (RMSE) and the Nash and Sutcliffe [44] efficiency (η), defined as follows:

$$RMSE = \left[\frac{\sum_{i=1}^{n}(Q_{sim}^{i} - Q_{obs}^{i})^2}{n}\right]^{0.5} \qquad (8)$$

Table 4. Daily mean (T), maximum (T_{max}) and minimum (T_{max}) temperature and mean annual precipitation (P) observed and simulated by error corrected REMO and RegCM3 climate models for control period (2001–2010).

	T (°C)	T_{max} (°C)	T_{min} (°C)	P (mm)
observed	4.17	8.14	0.52	1412.75
REMO	4.06	8.33	0.46	1399.25
RegCM3	4.09	7.04	1.33	1339.04

$$\eta = 1 - \frac{\sum_{i=1}^{n} \left(Q_{sim}^{i} - Q_{obs}^{i}\right)^2}{\sum_{i=1}^{n} \left(Q_{obs}^{i} - \overline{Q_{obs}}\right)^2} \qquad (9)$$

where n is the total number of time steps, Q_{sim}^{i} is the ith simulated discharge, Q_{obs}^{i} is the ith observed discharge, and $\overline{Q_{obs}}$ is the mean of the observed discharges.

At-site bias-corrected climate-scenario forcings

For the meteorological forcing of future scenarios, two different regional climate models (RCMs) were used, the REMO [45] and the RegCM3 [46]. Both models cover Europe on a 25×25 km grid, in the same simulation period (1951–2100). Moreover, they are driven by the same global ocean-atmosphere-coupled model, ECHAM5 [47], using the observed greenhouse gas concentrations between 1951 and 2000 and IPCC's (Intergovernmental Panel on Climate Change) greenhouse gas emission scenario A1B [48] between 2001 and 2100. Both were produced within the EU FP6 Integrated Project ENSEMBLES (http://www.ensembles-eu.org/) and can be downloaded from http://ensemblesrt3.dmi.dk on a daily basis. Hourly and 3-hourly data were provided directly by the Max Planck Institute for Meteorology and the Abdus Salam International Center for Theoretical Physics. In comparison to the larger ensemble of regional simulations for Europe, the REMO and RegCM3 models represent moderate warming (below average) and near-average precipitation changes [49].

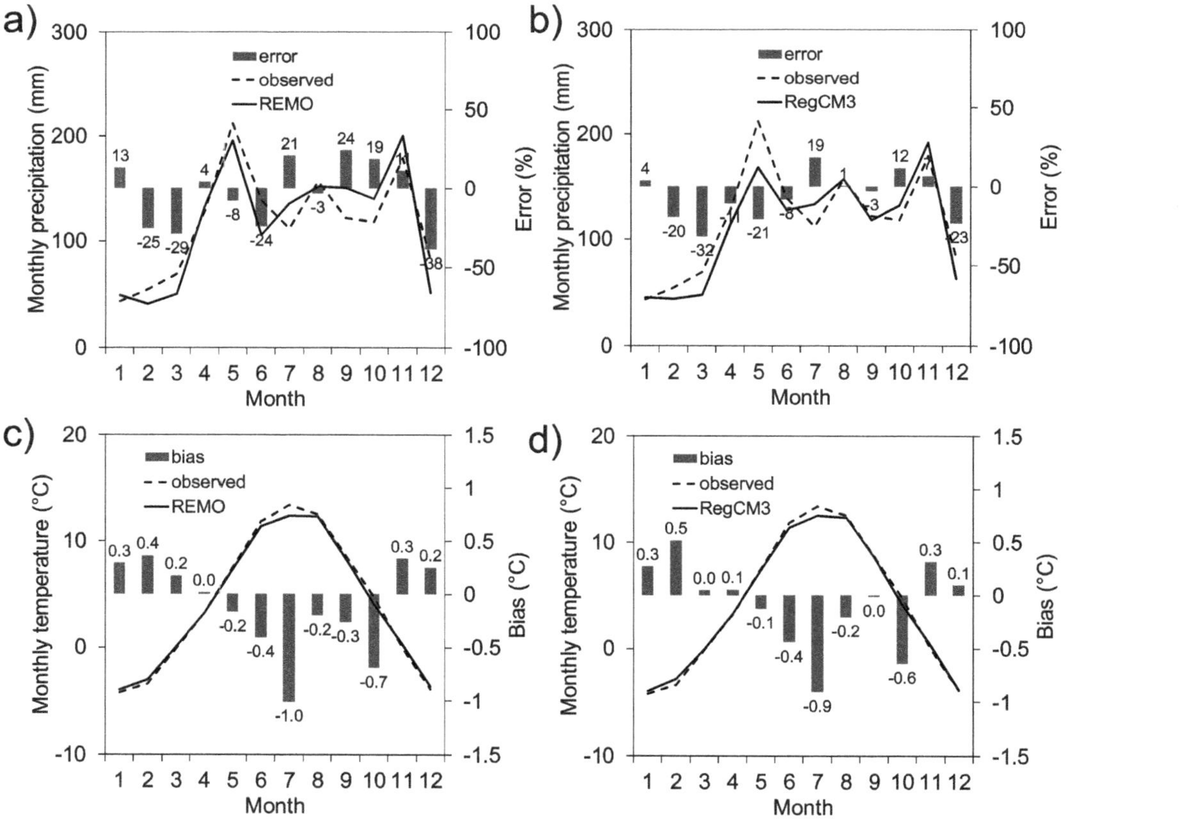

Figure 6. Mean monthly precipitation and temperature for 2001–2010 as simulated by the REMO and RegCM3 regional climate models and their deviations versus the observations.

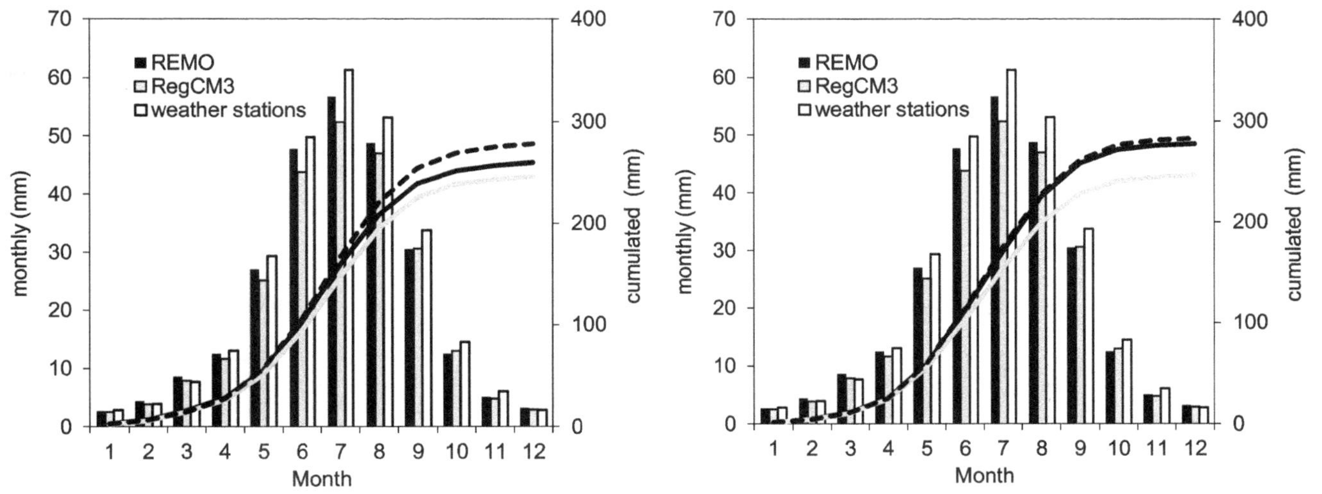

Figure 7. Mean monthly and cumulated actual evapotranspiration as computed for 2001–2010 by the FEST-EWB (left) and FEST-WB (right) hydrological models driven by the REMO and RegCM3 regional climate models and the weather observations during the control period (2001–2010).

A quantile-based error-correction approach (quantile mapping) has been used to downscale the RCM simulations to a point scale and to reduce its error characteristics. The potential of quantile mapping for correcting GCM data has already been demonstrated in previous hydrological studies [50], [51], but its application to regional climate simulations is somewhat recent [52], [53], [54], [55], [56], [57]. In this study, the quantile mapping applied observational stations data to climate data, from the regional climate models, on a daily basis. It adapted the modelled time series to the observed empirical cumulative frequency distribution [58]. The method and its application were discussed by [59] and [54] as to what concerns daily temperature and precipitation, and by [60] as regards other meteorological variables such as relative humidity, global radiation, and wind speed. All the variables used in this study were error corrected and downscaled to a station basis. A 31-day moving window in the calibration period, centered on the day to be corrected, was used for constructing the empirical cumulative frequency distribution for that particular day of the year. This enabled an annual cycle-sensitive correction as well as a sufficiently large sample size. A point-wise implementation, which fits a separate statistical model for each observational station, was chosen to account for the regionally varying errors. Grid cell averages (3×3) of the raw RCM data were used as predictors with respect to the effective resolution of the RCM, which is below the grid-resolution.

The calibration period for the error correction ranged from 01-01-2000 to 12-31-2009. No error correction was performed for stations with less than 9 years of observational data (>10% missing data), because the climate variability could not be expected to be properly covered by only a few years of data. Quantile mapping

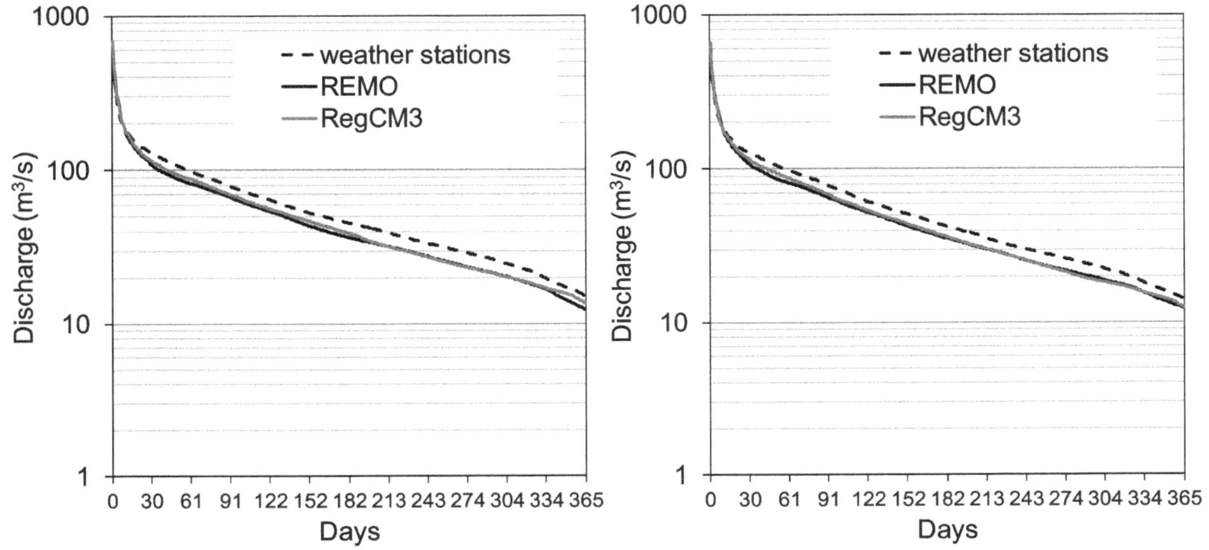

Figure 8. Mean flow duration curves simulated by the FEST-EWB (left) and FEST-WB (right) hydrological models driven by meteorological observations and the simulated climatic forcings by the REMO and RegCM3 regional climate models.

Table 5. Daily mean (*T*), maximum (T_{max}) and minimum (T_{max}) temperature and mean annual precipitation (*P*) simulated by REMO and RegCM3 climate models for decade 2041–2050.

	T (°C)	T_{max} (°C)	T_{min} (°C)	*P* (mm)
REMO	5.35	9.01	1.34	1678.54
RegCM3	5.21	7.88	2.16	1578.85

assumes that the same statistical relations of the observed and modelled climate hold within the calibration period, as well as in future scenario periods. It must be kept in mind that even a calibration period of 9–10 years can be affected by decadal climate variability, which can degrade the results of the error correction applied to the future scenario period.

The resulting daily scenarios were further refined to a 3-hourly time series using the sub-daily data from the RCMs. For the air temperature, the differences between the 3-hourly RCM data and their daily-mean values were added to the corresponding corrected-daily values. The ratios of the 3-hourly RCM data and daily precipitation values were multiplied by the corrected daily values. Similarly, for the global radiation, the wind speed, and the relative humidity, the ratios of the 3-hourly RCM data and their daily mean values were multiplied by the corrected daily value. In the case of relative humidity values exceeding 100%, all of the values from the day were multiplied by a factor to shrink the daily maximum value to 100%.

Climate-scenario dataset used in this analysis is included as supplemental file (Dataset S1).

Results and Discussion

Comparison of the models driven by meteorological observations

Table 3 shows the results in reproducing the daily streamflow by the FEST-WB and FEST-EWB models forced by meteorological observations. The goodness of fit indices of the two models are comparable, with the FEST-WB model displaying a slightly greater Nash and Sutcliffe efficiency and lower RMSE. In Fig. 3, a comparison between the FEST-WB and FEST-EWB models for the simulated and observed hourly discharge is shown for the

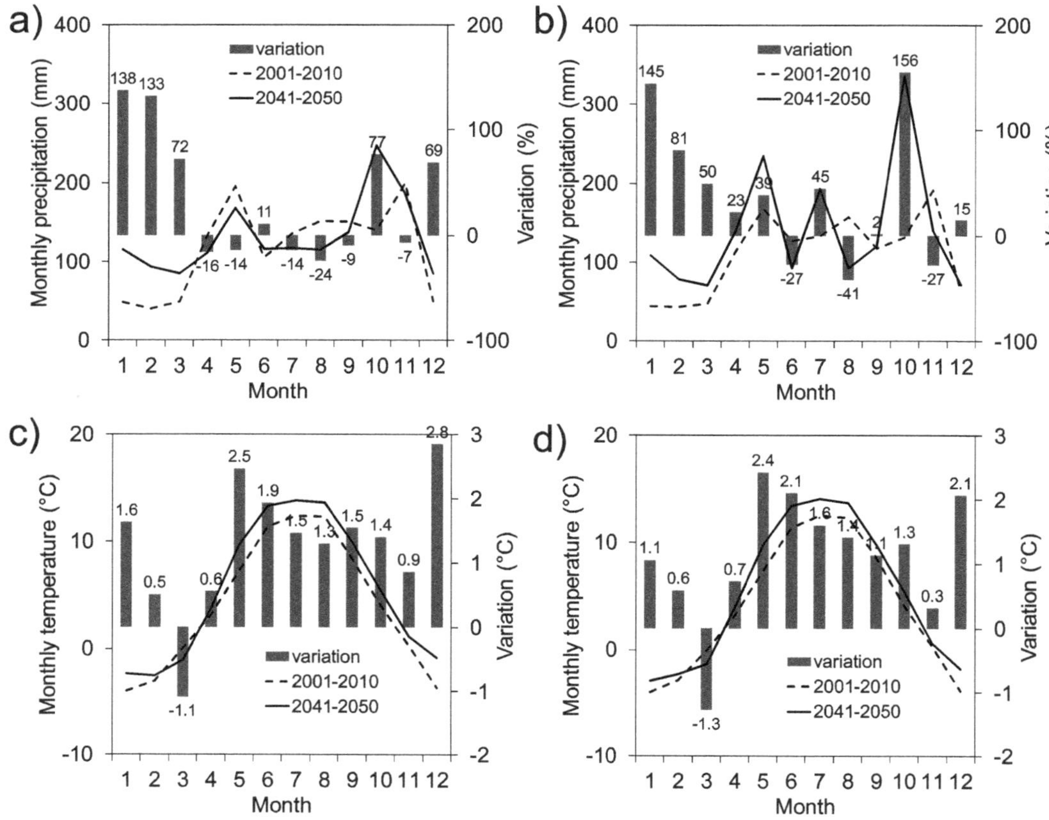

Figure 9. Mean monthly precipitation and temperature for the period 2041–2050 as projected by REMO and RegCM3 regional climate models versus the control period (2001–2010): (a) precipitation by REMO; (b) precipitation by RegCM3; (c) temperature by REMO; and (d) temperature by RegCM3.

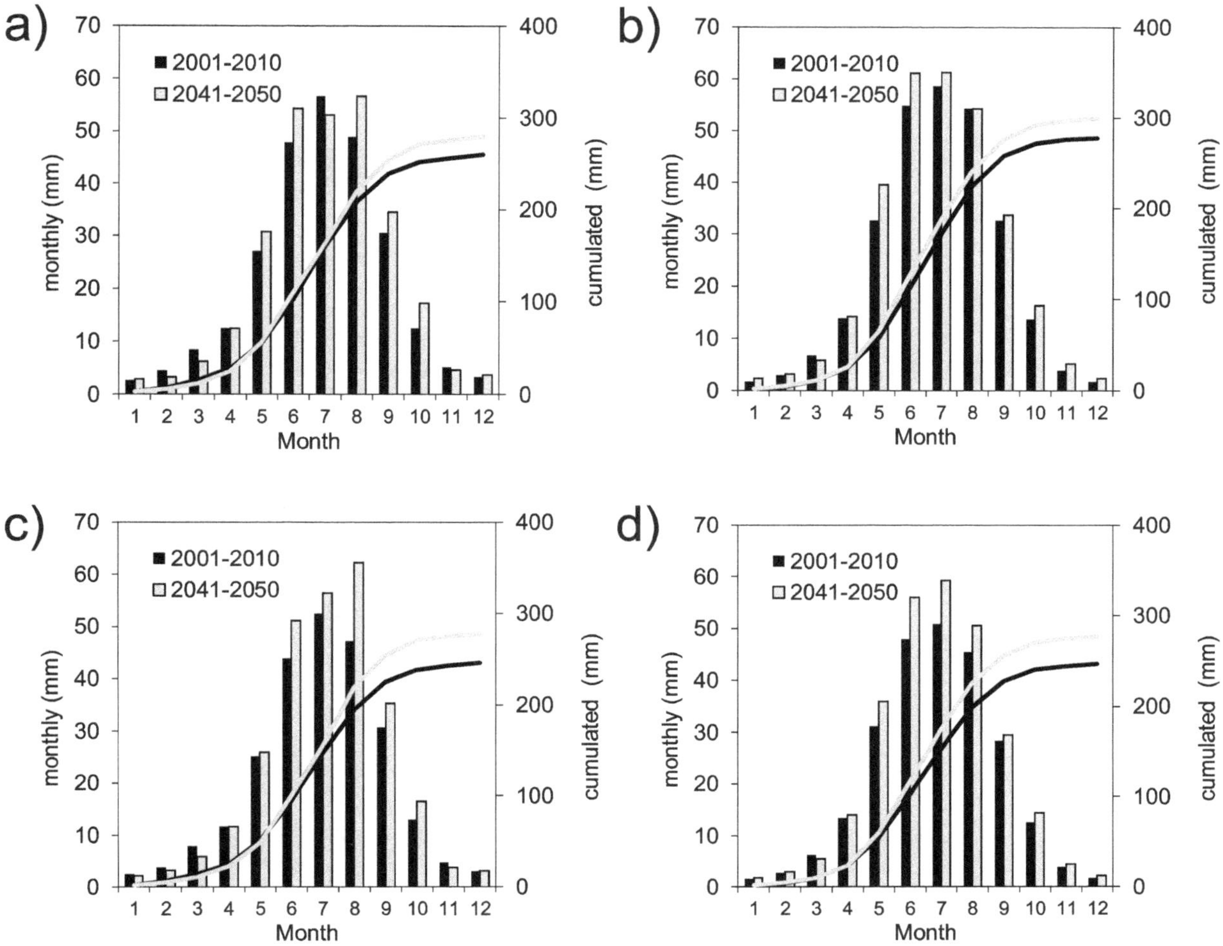

Figure 10. Mean monthly and cumulated actual evapotranspiration for the period 2041–2050 as simulated by the FEST-WB and FEST-EWB hydrological models driven by the REMO or RegCM3 regional climate models versus the control period (2001–2010): (a) FEST-EWB driven by REMO; (b) FEST-WB driven by REMO; (c) FEST-EWB driven by RegCM3; and (d) FEST WB driven by RegCM3.

period from 2001 to 2010. The differences were almost negligible and the time series overlapped.

The mean monthly and cumulated actual evapotranspiration values, as computed by the FEST-WB and FEST-EWB models driven by meteorological observations, are shown in Fig. 4. There is a general agreement between the two different approaches in computing the evapotranspiration with the exceptions of May and June, when the FEST-WB model had values 14% and 12% higher, respectively, than the FEST-EWB model. On an annual basis, the FEST-WB model had a 1.8% higher value than the FEST-EWB model.

In Fig. 5, the mean flow duration curves simulated by the two models are compared to those observed. Good agreement was seen for both the higher and lower discharges. The difference between the FEST-WB and FEST-EWB simulations was due to the different evapotranspiration losses reported by the two models.

Comparison of models in reproducing the hydrological aspects of the control period

Before assessing the climate change impacts, we performed an analysis of the hydrological models, driven by modelled climatic forcings, in reproducing the hydrological aspects of the control period (2001–2010). Table 4 shows the mean annual precipitation and the average daily mean, maximum, and minimum temperatures observed and simulated by the calibrated REMO and RegCM3 climate models during the control period. The REMO model resulted in a 0.11°C and 0.95% underestimation in reproducing the temperature and precipitation, respectively, while the RegCM3 model resulted in a 0.08°C and 5.2% underestimation.

The two climate models displayed larger differences in reproducing the daily maximum and minimum temperature. The REMO model produced errors of 0.2°C and −0.1°C in reproducing the maximum and minimum daily temperatures, respectively, while the RegCM3 model produced errors of −1.1°C and 0.8°C, respectively. This can be explained by the fact that only the daily mean values were error corrected, while the diurnal cycle was superimposed as the models simulated it without further correction (Section 2.4).

In Fig. 6, the mean monthly precipitation and temperature, as simulated by the REMO and RegCM3 climate models, are compared to those observed. Regarding precipitation, the two models underestimated it in February, March, May, June, and December and overestimated it in July, October, and November. Nevertheless, the overall behavior, such as the peaks in the spring and autumn, was well captured. The two climatic models

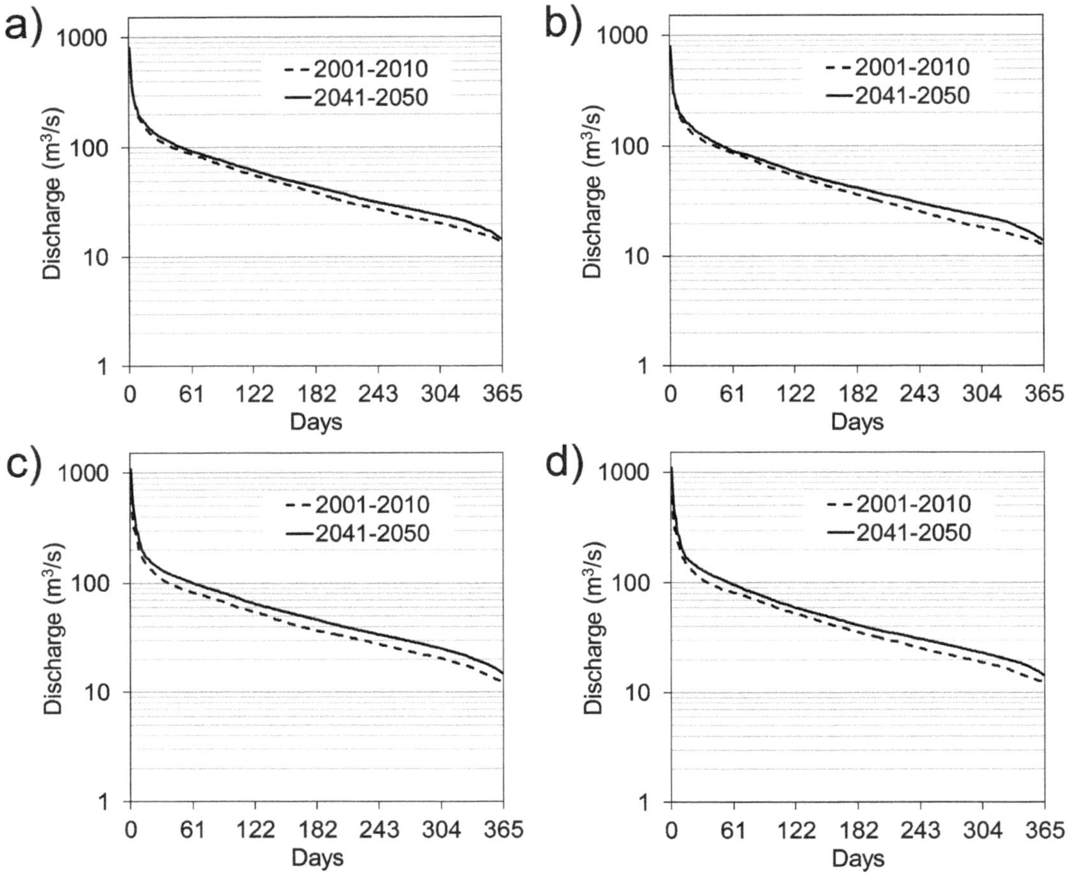

Figure 11. Mean flow duration curve for the period 2041–2050 as simulated by the FEST-WB and FEST-EWB hydrological models driven by the REMO or RegCM3 regional climate models versus the control period (2001–2010): a) FEST-EWB driven by REMO, b) FEST-WB driven by REMO, c) FEST-EWB driven by RegCM3, and d) FEST WB driven by RegCM3.

displayed the same results in reproducing the monthly temperatures, that is, they underestimated it from May to October and overestimated it in the other months of the year, but the discrepancies were under the acceptable limits.

Since the evaluation period differs from the calibration period of the bias correction, such minor discrepancies should be expected. An increase of these errors in future applications should also be expected. However, this increase can be regarded as limited, given that the bias correction is relatively stable, as has been demonstrated by [61] and [60].

Fig. 7 shows the mean monthly and cumulated actual evapotranspiration as computed by the FEST-EWB and FEST-WB hydrological models driven by REMO, RegCM3, and meteorological observations during the control period. Both the FEST-EWB and FEST-WB models simulated greater evapotranspiration when driven by REMO than by RegCM3. The FEST-WB equation to compute the evapotranspiration is very sensible to the daily temperature range. This implies that the evapotranspiration computed by the FEST-WB model, driven by REMO, matches the evapotranspiration computed by the FEST-WB model, driven by the meteorological observations, as the REMO model is more accurate than the RegCM3 model in reproducing the daily minimum and maximum temperatures.

Fig. 8 shows the mean flow duration curves simulated by the FEST-EWB and FEST-WB hydrological models driven by meteorological observations and the REMO and RegCM3 simulated climatic forcings. There was good agreement between the REMO and RegCM3 driven simulations. The discharges driven by the RegCM3 model were generally greater than those driven by REMO, particularly for the high durations, due to the underestimation of the evapotranspiration component.

Projected changes in the hydrological processes

The impacts of climate change on hydrological processes were assessed by comparing the results from the FEST-EWB and FEST-WB models, driven by the REMO and RegCM3 climate models, for the decade spanning 2041–2050 to those of the control period. The simulations were performed assuming no variation in the spatial distribution of the vegetation, or the beginning and duration of the growing season. This means that the crop coefficient functions for the future period were unmodified, and that the monthly leaf area index maps were derived as an average from the 2001–2010 remotely sensed maps.

Table 5 shows the average daily mean, maximum, and minimum temperatures and the mean annual precipitation simulated by the REMO and RegCM3 models for the 2041–2050 decade. The REMO and RegCM3 models simulated an increase in mean temperature of 1.28°C and 1.12°C, respectively, and an increase in the mean annual precipitation of 12.83% and 25.35%, respectively.

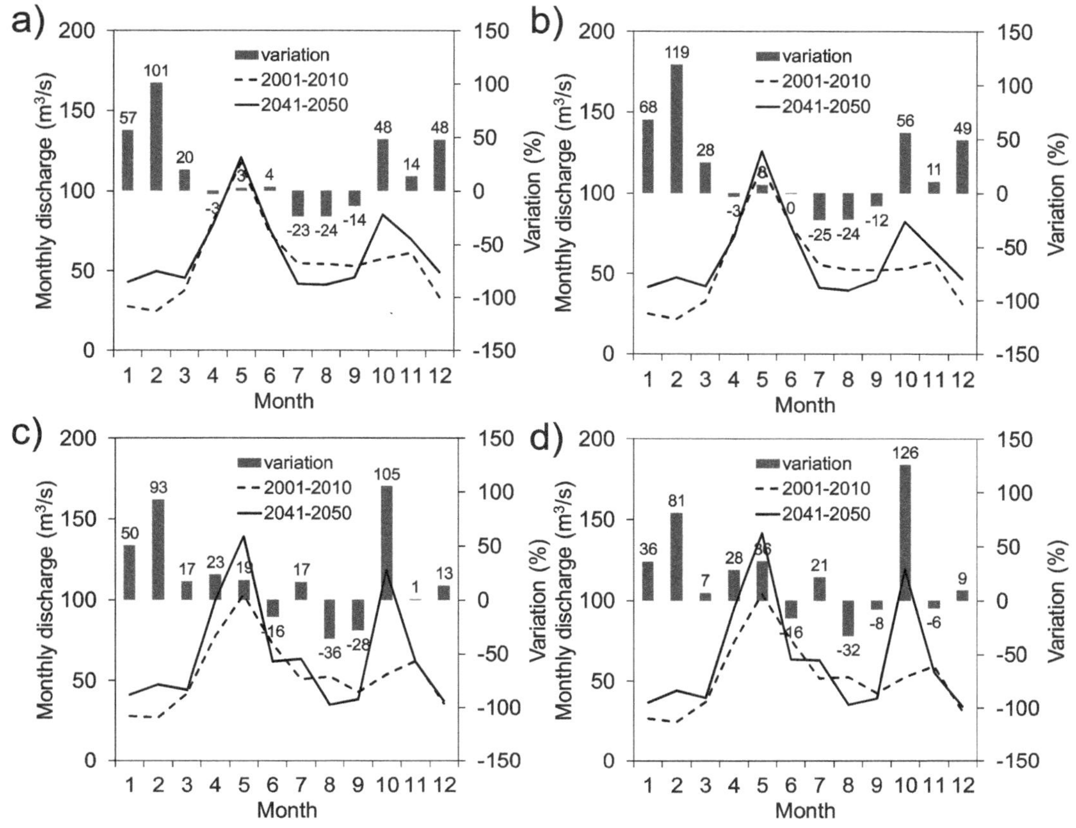

Figure 12. Mean monthly discharge for the period 2041–2050 as simulated by the FEST-WB and FEST-EWB hydrological models driven by the REMO or RegCM3 regional climate models versus the control period (2001–2010): a) FEST-EWB driven by REMO, b) FEST-WB driven by REMO, c) FEST-EWB driven by RegCM3, and d) FEST WB driven by RegCM3.

The mean monthly precipitation and temperature as simulated by REMO and RegCM3 for the control period (2001–2010) and for the 2041–2050 decade are shown in Fig. 9. The precipitation increase was mostly concentrated in the winter period and October, in which the RegCM3 model projected a precipitation increase of 156%. During the summer, the two climate models projected a significant decrease in precipitation, as much as −41% in August by the RegCM3 model. The temperature was generally predicted to increase more significantly during the summer, late spring, and winter, while a decrease was expected in March. These results are consistent with findings depicted in the IPCC AR4 for Central Europe [62] and with a recent review on expected climate change in the Alpine region [14].

Fig. 10 shows the mean monthly and cumulated actual evapotranspiration as computed by the FEST-EWB and FEST-WB models driven by REMO and RegCM3 for the control period (2001–2010) and the 2041–2050 decade. All of the simulations showed an increase in the evapotranspiration, in agreement with the increase in air temperature. For a given climate model, the FEST-EWB and FEST-WB hydrological models projected similar modifications to the evapotranspiration.

In Fig. 11, the mean flow duration curves simulated by the FEST-EWB and FEST-WB hydrological models, driven by REMO and RegCM3 for the 2041–2050 decade, were compared to those of the control period (2001–2010). A general increase of the discharge was projected for the flow duration, in agreement with the significantly increased annual precipitation not compensated by increased evapotranspiration. For a given climate model the FEST-EWB and FEST-WB hydrological models predicted similar modifications to the flow duration curve.

In Fig. 12, the mean monthly discharge for the 2041–2050 decade, simulated by the FEST-EWB and FEST-WB hydrological models driven by the REMO and RegCM3 climate models, is compared to that of the control period (2001–2010). The seasonal shift observed in the precipitation was reflected in the projected monthly discharge, with a significant increase expected in October and in the winter period and a significant decrease expected in summer.

Conclusions

This study investigated the role of climatic forcing availability, and thus hydrological model complexity, on the assessment of climate change impacts on the water resources for the Toce river basin. Two distributed hydrological models were used to simulate the water balance components of the Toce river basin: the FEST-WB model, which implements a simple temperature-based method for computing the evapotranspiration; and the FEST-EWB model, which computes evapotranspiration by solving energy and water balance equations that require temperature, net radiation, wind speed, and relative air humidity as meteorological forcings. Both the FEST-WB and FEST-EWB models performed well in reproducing the daily discharge of the 2001–2010 period and the hourly discharge for major flood events. The difference in computing the evapotranspiration was approximately

2% on an annual basis. Moreover, there was general agreement between the two hydrological models in reproducing the mean annual flow-duration curve.

An analysis of the hydrological models, driven by the climatic forcings modelled by the REMO and RegCM3 climate models, in reproducing the hydrological aspects of the control period showed that the FEST-WB model was more sensitive to the daily temperature range in simulating the evapotranspiration. The evapotranspiration differences impacted the flow duration curve, but the two hydrological models achieved good agreement.

The impact of climate change on the hydrological processes was assessed by comparing the results from the FEST-EWB and FEST-WB models, driven by REMO and RegCM3 for the 2041–2050 decade, to those of the control period (2001–2010). The REMO and RegCM3 climate models simulated increased mean temperatures, and increased mean annual precipitations. The precipitation increase was primarily concentrated during October and the winter period. The two climate models predicted a significant decrease in precipitation during the summer. This reflects an increase in the evapotranspiration and the discharges for all of the durations in the flow duration curves. The seasonal shift observed in the precipitation was reflected in the monthly discharge. Indeed, a significant increase in the discharge was expected in October and the winter period, while a significant decrease was expected in the summer. Obtained results are generally consistent with findings depicted in the IPCC assessment reports.

In general, this study showed that despite the simple temperature-based approach for computing evapotranspiration, the FEST-WB model is robust and sufficiently accurate to perform hydrological impact studies of climate change for the Alpine river basin that was investigated. The bias introduced by the approximations from the method used to compute the evapotranspiration was less than the uncertainty associated with climate models.

Acknowledgments

We thank ARPA Piemonte for providing observed meteorological data. The authors thank the two anonymous reviewers for their helpful comments that contributed to improve the paper.

Author Contributions

Conceived and designed the experiments: GR AG MM. Performed the experiments: GR MG TM AG. Analyzed the data: GR CC MG. Wrote the paper: GR CC AG TM.

References

1. IPCC (2013) Climate Change 2013: The Physical Science Basis. Contribution of Working Group I to the Fifth Assessment Report of the Intergovernmental Panel on Climate Change [Stocker, T.F., D. Qin, G.-K. Plattner, M. Tignor, S.K. Allen, J. Boschung, A. Nauels, Y. Xia, V. Bex and P.M. Midgley (eds.)]. Cambridge University Press, Cambridge, United Kingdom and New York, NY, USA, 1535 pp.
2. Loarie SR, Carter BE, Hayhoe K, McMahon S, Moe R, et al. (2008) Climate Change and the Future of California's Endemic Flora. PLoS ONE 3(6): e2502. doi:10.1371/journal.pone.0002502.
3. Shrestha UB, Gautam S, Bawa KS (2012) Widespread Climate Change in the Himalayas and Associated Changes in Local Ecosystems. PLoS ONE 7(5): e36741. doi:10.1371/journal.pone.0036741.
4. Dile YT, Berndtsson R, Setegn SG (2013) Hydrological Response to Climate Change for Gilgel Abay River, in the Lake Tana Basin - Upper Blue Nile Basin of Ethiopia. PLoS ONE 8(10): e79296. doi:10.1371/journal.pone.0079296.
5. Beniston M (2004) Climatic Change and its Impacts. An Overview Focusing on Switzerland. Kluwer Academic Publishers: Dordrecht/The Netherlands and Boston/USA (now Springer Publishers).
6. Xu C-Y, Wide'n E, Halldin S (2005) Modelling hydrological consequences of climate change - progress and challenges. Adv Atmos Sci 22 (6): 789–797.
7. Soncini A, Bocchiola D (2011) Assessment of future snowfall regimes within the Italian Alps using general circulation models. Cold Reg Sci Technol 68(3): 113–123.
8. Diolaiuti GA, Maragno D, D'Agata C, Smiraglia C, Bocchiola D (2011) Glacier retreat and climate change: Documenting the last 50 years of Alpine glacier history from area and geometry changes of Dosdè Piazzi glaciers (Lombardy Alps, Italy). Prog Phys Geog 35(2): 161–182.
9. Ficklin DL, Stewart IT, Maurer EP (2013) Climate Change Impacts on Streamflow and Subbasin-Scale Hydrology in the Upper Colorado River Basin. PLoS ONE 8(8): e71297. doi:10.1371/journal.pone.0071297.
10. Jiang T, Chen YD, Xu C, Chen X, Chen X, et al. (2007) Comparison of hydrological impacts of climate change simulated by six hydrological models in the Dongjiang Basin, South China. J Hydrol 336: 316–333.
11. Verbunt M, Gurtz J, Jasper K, Lang H, Warmerdam P, et al. (2003) The hydrological role of snow and glaciers in alpine river basins and their distributed modeling. J Hydrol 282; 36–55.
12. Null SE, Viers JH, Mount JF (2010) Hydrologic Response and Watershed Sensitivity to Climate Warming in California's Sierra Nevada. PLoS ONE 5(4): e9932. doi:10.1371/journal.pone.0009932.
13. Beniston M, Uhlmann U, Goyette S, Lopez-Moreno JI (2011) Will snow-abundant winters still exist in the Swiss Alps in an enhanced greenhouse climate? Int J Climatol 31: 1257–1263.
14. Gobiet A, Kotlarski S, Beniston M, Heinrich G, Rajczak J, et al. (2014) 21st century climate change in the European Alps–A review. Sci Total Environ 493(0): 1138–1151, doi:http://dx.doi.org/10.1016/j.scitotenv.2013.07.050.
15. Rabuffetti D, Ravazzani G, Corbari C, Mancini M (2008) Verification of operational Quantitative Discharge Forecast (QDF) for a regional warning system – the AMPHORE case studies in the upper Po River. Nat Hazard Earth Sys 8: 161–173.
16. Corbari C, Ravazzani G, Martinelli J, Mancini M (2009) Elevation based correction of snow coverage retrieved from satellite images to improve model calibration. Hydrol Earth Syst Sc 13(5): 639–649.
17. Viviroli D, Zappa M, Schwanbeck J, Gurtz J, Weingartner R (2009) Continuous simulation for flood estimation in ungauged mesoscale catchments of Switzerland – Part I: Modelling framework and calibration results. J Hydrol 377: 191–207.
18. Corbari C, Ravazzani G, Mancini M (2011) A distributed thermodynamic model for energy and mass balance computation: FEST-EWB. Hydrol Process 25(9): 1443–1452.
19. Montaldo N, Albertson JD (2001) On The Use Of The Force-Restore SVAT Model Formulation For Stratified Soils. J Hydrometeorol 2(6): 571–578.
20. Senatore A, Mendicino G, Smiatek G, Kunstmann H (2011) Regional climate change projections and hydrological impact analysis for a Mediterranean basin in Southern Italy. J Hydrol 399: 70–92.
21. Montaldo N, Mancini M, Rosso R (2004) Flood hydrograph attenuation induced by a reservoir system: analysis with a distributed rainfall-runoff model. Hydrol Process 18 (3): 545–563.
22. Regione Piemonte (1997) DTM 1/10000. *Servizio Cartografico (Cartographic Service)*: Torino (Italy).
23. CEC (2000) CORINE land cover. Technical guide, Commission of the European Communities, Luxembourg.
24. Bacchi B, Ranzi R (2003) Hydrological and meteorological aspects of floods in the Alps: an overview. Hydrol Earth Syst Sc 7(6): 784–798.
25. Montaldo N, Ravazzani G, Mancini M (2007) On the prediction of the Toce alpine basin floods with distributed hydrologic models. Hydrol Process 21: 608–621.
26. Pianosi F, Ravazzani G (2010) Assessing rainfall-runoff models for the management of Lake Verbano. Hydrol Process 24(22): 3195–3205.
27. Corbari C, Sobrino JA, Mancini M, Hidalgo V (2010) Land surface temperature representativeness in a heterogeneous area through a distributed energy-water balance model and remote sensing data. Hydrol Earth Syst Sc 14: 2141–2151.
28. Ravazzani G, Mancini M, Giudici I, Amadio P (2007) Effects of soil moisture parameterization on a real- time flood forecasting system based on rainfall thresholds. IAHS Publ. 313: 407–416.

29. Ravazzani G, Mancini M, Meroni C (2009) Design hydrograph and routing scheme for flood mapping in a dense urban area. Urban Water J 6(3): 221–231.
30. Ravazzani G, Rametta D, Mancini M (2011) Macroscopic Cellular Automata for groundwater modelling: a first approach. Environ Modell Softw 26(5): 634–643.
31. Ravazzani G, Gianoli P, Meucci S, Mancini M (2012b) Indirect estimation of design flood in urbanized river basins using a distributed hydrological model. J Hydrol Eng 19(1): 235–242.
32. Ceppi A, Ravazzani G, Salandin A, Rabuffetti D, Montani A, et al. (2013) Effects of temperature on flood forecasting: analysis of an operative case study in Alpine basins. Nat Hazard Earth Sys 13(4): 1051–1062.
33. Gaudard L, Romerio F, Dalla Valle F, Gorret R, Maran S, et al. (2013) Climate change impacts on hydropower in the Swiss and Italian Alps. Sci Total Environ, in press, doi: 10.1016/j.scitotenv.2013.10.012.
34. Ravazzani G (2013) MOSAICO, a library for raster based hydrological applications. Comput Geosci 51: 1–6.
35. Allen RG, Pereira LS, Raes D, Smith M (1998) Crop evapotranspiration – guidelines for computing crop water requirements – FAO Irrigation and Drainage Paper 56. FAO.
36. D'Urso G, Menenti M (1995) Mapping crop coefficients in irrigated areas from Landsat TM images, in Paris Europto, (Ed.) European Symposium on Satellite Remote Sensing II, SPIE, Intern Soc Optical Engineering vol. 2585, Bellingham, USA, 41–47.
37. Corbari C, Ravazzani G, Galvagno M, Cremonese E, Mancini M (2014) Assessing crop coefficient for natural vegetated area using satellite data and eddy covariance stations. J Hydrol Eng, submitted.
38. Ravazzani G, Corbari C, Morella S, Gianoli P, Mancini M (2012a) Modified Hargreaves-Samani equation for the assessment of reference evapotranspiration in Alpine river basins. J Irrig Drain Eng 138(7): 592–599.
39. Hargreaves G H (1994) Defining and using reference evapotranspiration. J Irrig Drain Eng 120(6): 1132–1139.
40. Jacobs AFG, Heusinlveld BG, Holtslag AAM (2008) Towards closing the energy surface budget of a mid-latitude grassland. Boundary Layer Meteorol 126: 125–136.
41. Meyers TP, Hollinger SE (2004) An assessment of storage terms in the surface energy balance of maize and soybean. Agr Forest Meteorol 125: 105–115.
42. Boscarello L, Ravazzani G, Rabuffetti D (2014) Integrating glaciers dynamics raster based modelling in large catchments hydrological balance: the Rhone case study. Hydrol Process 28(3): 401–1560. doi: 10.1002/hyp.9588.
43. Pellegrini M (2011) Accuracy of MODIS snow cover images and calibration of hydrological model in alpine river basins, Politecnico di Milano, Milan, Master thesis.
44. Nash JE, Sutcliffe JV (1970) River flow forecasting through the conceptual models, Part 1: A discussion of principles. J Hydrol 10 (3): 282–290.
45. Jacob D (2001) A note to the simulation of the annual and inter-annual variability of the water budget over the Baltic Sea drainage basin. Meteorol Atmos Phys 77: 61–73.
46. Pal JS, Giorgi F, Bi X, Elguindi N, Solmon F, et al. (2007) Regional climate modeling for the developing world: The ICTP RegCM3 and RegCNET. Bull Am Met Soc 88: 1395–1409.
47. Roeckner E, Baeuml G, Bonaventura L, Brokopf R, Esch M, et al. (2003) The Atmospheric General Circulation Model ECHAM5. Part 1: Model Description, Report 349, Max Planck Institute for Meteorology (MPI), Hamburg.
48. Nakicenovic N, Alcamo J, Davis G, de Vries B, Fenhann J, et al. (2000) IPCC Special Report on Emissions Scenarios, Cambridge University Press, Cambridge, United Kingdom and New York, NY, USA.
49. Heinrich G, Gobiet A, Prein AF (2011) Uncertainty of Regional Climate Simulations in the Alpine Region. EGC Report to the EU Nr. 01/2011, 59 pp, Wegener Center, University of Graz, Graz, Austria.
50. Dettinger MD, Cayan DR, Meyer MK, Jeton AE (2004) Simulated hydrologic responses to climate variations and change in the Merced, Carson, and American river basins, Sierra Nevada, California, 1900–2099. Climatic Change 62: 283–317.
51. Wood AW, Leung LR, Sridhar V, Lettenmaier DP (2004) Hydrologic Implications of Dynamical and Statistical Approaches to Downscale Climate Model Outputs. Climatic Change 62: 189–216.
52. Dobler A, Ahrens B (2008) Precipitation by a regional climate model and bias correction in Europe and South Asia. Meteorologische Zeitschrift 17: 499–509.
53. Piani C, Haerter JO, Coppola E (2010) Statistical bias correction for daily precipitation in regional climate models over Europe. Theor Appl Climatol 99: 187–192, doi: 10.1007/s00704-009-0134-9.
54. Themeßl MJ, Gobiet A, Heinrich G (2012) Empirical-statistical downscaling and error correction of regional climate models and its impact on the climate change signal. Climatic Change 112(2): 449–468. doi: 10.1007/s10584-011-0224-4.
55. Amengual A, Homar V, Romero R, Alonso S, Ramis C (2012) A statistical adjustment of regional climate model outputs to local scales: application to Platja de Palma, Spain. J Clim 25(3): 939–957.
56. Déqué M (2007) Frequency of precipitation and temperature extremes over France in an anthropogenic scenario: model results and statistical correction according to observed values. Glob Planet Chang 57: 16–26.
57. Boé J, Terray L, Habets F, Martin E (2007) Statistical and dynamical downscaling of the Seine basin climate for hydro-meteorological studies. Int J Climatol 27: 1643–1655.
58. Wilks DS (1995) Statistical Methods in Atmospheric Science, Volume 59 of International Geophysics Series. Academic Press: San Diego, London.
59. Themeßl M, Gobiet A, Leuprecht A (2011) Empirical-statistical downscaling and error correction of daily precipitation from regional climate models. Int J Climatol 31(10): 1530–1544. doi:10.1002/joc.2168.
60. Wilcke RA, Mendlik T, Gobiet A (2013) Performance and Physical Consistency of Multi-Variable Downscaling and Error-Correction of Regional Climate Models. Climatic Change 120:871–887.
61. Maraun D (2012) Nonstationarities of regional climate model biases in European seasonal mean temperature and precipitation sums. Geophys Res Lett 39: L06706. doi:10.1029/2012GL051210.
62. Christensen JH, Hewitson B, Busuioc A, Chen A, Gao A, et al. (2007) Regional climate projections. In: climate change 2007: the physical science basis. Contribution of Working Group I to the Fourth Assessment Report of the Intergovernmental Panel on Climate Change. Cambridge University Press, Cambridge, United Kingdom and New York, NY, USA.

18

Sensitivity Analysis of CLIMEX Parameters in Modeling Potential Distribution of *Phoenix dactylifera* L.

Farzin Shabani*, Lalit Kumar

Ecosystem Management, School of Environmental and Rural Science, University of New England, Armidale, New South Wales, Australia

Abstract

Using CLIMEX and the Taguchi Method, a process-based niche model was developed to estimate potential distributions of *Phoenix dactylifera* L. (date palm), an economically important crop in many counties. Development of the model was based on both its native and invasive distribution and validation was carried out in terms of its extensive distribution in Iran. To identify model parameters having greatest influence on distribution of date palm, a sensitivity analysis was carried out. Changes in suitability were established by mapping of regions where the estimated distribution changed with parameter alterations. This facilitated the assessment of certain areas in Iran where parameter modifications impacted the most, particularly in relation to suitable and highly suitable locations. Parameter sensitivities were also evaluated by the calculation of area changes within the suitable and highly suitable categories. The low temperature limit (DV2), high temperature limit (DV3), upper optimal temperature (SM2) and high soil moisture limit (SM3) had the greatest impact on sensitivity, while other parameters showed relatively less sensitivity or were insensitive to change. For an accurate fit in species distribution models, highly sensitive parameters require more extensive research and data collection methods. Results of this study demonstrate a more cost effective method for developing date palm distribution models, an integral element in species management, and may prove useful for streamlining requirements for data collection in potential distribution modeling for other species as well.

Editor: Dafeng Hui, Tennessee State University, United States of America

Funding: These authors have no support or funding to report.

Competing Interests: The authors have declared that no competing interests exist.

* E-mail: fshabani@myune.edu.au

Introduction

Species distribution models, ecological niche models (SDMs and ENMs) [1–3] and general bioclimatic models such BIOCLIM [4], are now acknowledged as essential tools in predicting a variety of future scenarios. Potential species distribution changes are one such application of these modeling tools [5]. Drawing on the distribution of the species and environmental data, a profile is compiled, relating distribution to variable environmental factors, a method termed the 'environmental envelope approach' [6]. The founding principle on which this approach is based is that the primary determinant of range and potential range of plants and other poikilotherms is climate [7]. The environmental envelope falls within the parameters of the upper and lower tolerances of a species, which are used in the modeling process to create a habitat map that describes the environmental suitability of each location or potential location [6]. The environmental envelope approach has been the basis for the development of CLIMEX [8] and a number of other models [9], designed to model current or future distribution of a species [10], using data based on the environmental factors inherent within the natural distribution area of the species, to project levels of suitability for other previously uncolonized regions [11]. This modeling is valuable for the identification of potential localities where the species could be successfully introduced and survive, as well as for estimating the impact of a potential threat.

Despite widespread usage of these models, there are many challenges relating to inaccuracies of prediction [12] which may limit the usability of the output. One cause of inaccuracy in the modeling of species distribution relates to the inherent assumption that there is equilibrium between species and environment [6]. Such inaccuracies are at their most extreme in the modeling of distributions when a species has only recently been established in a particular locality. This becomes most pertinent where invasive species are not yet in equilibrium with the new environment and full distribution coverage is not yet fully established due to the particular dispersal rate of the species [13].

Further inaccuracies in output can be linked to data and calibration methods used for the establishment of parameters [12]. However sensitivity analysis provides a technique which may be used to better understand and thus eliminate the impacts of inaccuracy and error in output [14]. This type of analysis is invaluable for establishing which parameters have the greatest impact on the modeled results [15]. Species distribution software, such as CLIMEX, may show greater sensitivity in particular parameters than in others, which may impact on the projections themselves. Analysis of parameter sensitivity levels is vital for testing hypotheses regarding the impact of climate variables on distribution, and in addition assists in the understanding of which climatic factors cause the greatest impact on the populations of a particular species [16]. A global biome model's sensitivity to parameter value inaccuracies derived from literary sources was tested by Hallgren and Pitman [17], who showed that in most parameters the model demonstrated insensitivity but was sensitive in the case of photosynthesis-related parameters.

To help explain its relationships of climatic response, CLIMEX uses the documented geographical distribution of a particular species, from which it then projects potential climatic responses in other localities and under different scenarios of climate change. CLIMEX focuses on the distribution data of a species, in order to show more clearly the climatic conditions that support or restrict its growth [18]. It draws on various types of information to model a species' distribution potential, including current distribution, phenology and empirical observations of the effects of temperature and soil moisture on growth response. Reviewing climate-based software for estimating the potential distributions of species, Kriticos and Randall [9] rated CLIMEX as most suitable for performing weed risk assessments in that it supports fitting the model to global distribution of the plant, includes a mechanism to create climate change scenarios, and provides a view of the ecological response of the plant to climate. Another aspect favoring CLIMEX is that it indicates when factors other than climate are responsible for the limitation of geographical distribution, such as biotic interactions. It is essential that all parameters are biologically logical and that every possible record of positive locality has been included in the parameter-fitting exercise. Non-inclusion of localities of estimated suitability from the point of view of climatic conditions in the data indicates that other factors may be altering the distribution patterns used for positive data extrapolations for new regions, as is generally essential in the case of biotic invasions [18].

The current study has used CLIMEX in the development of a baseline model for the species *Phoenix dactylifera* L. (date palm), a crop of major economic value in Iran, Iraq, Egypt and Saudi Arabia, as well as in Spain and Turkey [19–24], the result of this modelling have been published in [10,25–27]. Dates have been essential in supporting humans living in desert regions and thus the cultivation of the plant has had a major impact on Middle Eastern history. That dates are important is evidenced by great nutritional value of the plant (minerals, protein, vitamins, carbohydrates, fats, salts, and dietary fiber), its productivity and extensive yield life of up to 100 years [21,28–35].

This study analyses parameter uncertainty and the effect thereof in quantifying date palm response to temperature, soil moisture and cold stress changes. Thus it identifies the parameters of functional importance to provide a greater understanding of the climatic factors that most impact on the plant's distribution. The Taguchi method [36] was utilized to provide the general framework for the recognition of parameter uncertainty and evaluation of its resultant effect on estimations and decisions. The study output provides an indication of those parameters requiring an accurate fit of the detailed data collection, as well as those that are relatively change insensitive and therefore requiring less investment into research and the collection of data.

Materials and Methods

Current Date Palm *(P. dactylifera)* Distribution in Iran

Satellite data were employed for the determination of date palm plantation sites in Iran. Data were collected from Landsat images with image resolution of 30 m [37], as well the Global Biodiversity Information Facility (GBIF) [38], Missouri Botanical Gardens' database (MBG) [39] and additional date palm literature resources in CAB Abstracts database [40]. The GBIF and MBG databases contain occurrence records of many plant species for the whole world. Further date palm literature supplementation was also utilized [10,21,29,30,41,42]. A total of 145 records for *P. dactylifera* were obtained from the above sources, 19 of which had no geographical coordinates and were thus removed, leaving a working total of 126 records. To verify the point data accuracy, background satellite images were used to zoom in on each point for verification of the date palm presence in that locality. It should be mentioned that this section draws on a recent study that projected suitable regions for date palm cultivation in Iran under different climate change scenarios by 2030, 2050, 2070 and 2100 [25,26].

CLIMEX Software

The CLIMEX Version 3 software package [8,42,43] employs a model of eco-physiological growth with the inherent assumption that in a favorable season a population achieves a positive growth rate while an unfavorable season implies negative population growth [43]. Using the geographic range as reference, the parameters that describe response to climate of the species are inferred [44]. These parameters are in turn applied to the new climates to project the potential range of the species in alternative regions and different climate scenarios [45]. An index of annual growth (GI_A) rates the population growth potential when climatic conditions are favorable. Conversely, the stress indices (cold, wet, hot or dry) indicate the survival probabilities in unfavorable conditions [43]. The annual growth index combines the temperature index (TI) and the moisture index (MI) representing the species' required levels of temperature and soil moisture for growth. Four parameters, minimum limit, optimum lower, optimum upper and maximum limit are applicable to the temperature and moisture indices, which are multiplied to produce the weekly growth index and the yearly average of this represents the annual growth index (GI_A). The stress indices have two parameters, the threshold value and stress accumulation rate. The annual stress accumulation is exponential and once this value equals 1, the species will no longer be able to persist in that geographic region [43]. Growth and stress indices are calculated weekly and then integrated into the annual climatic suitability index, the ecoclimatic index (EI) on a scale between 0 and 100. A zero EI value denotes an unsuitable habitat where survival of the species is impossible; EI values between 1 and 10 denote marginal habitats; while values between 10 and 20 denote support for substantial populations and values greater than 20 denote high suitability [27,44].

The methodology described in Sutherst and Maywald [8] was used to fit the growth and stress parameters. An in-depth account of these parameters is available in Sutherst and Maywald [8]. The Climate Research Unit (CRU), Norwich, UK, global meteorological dataset at 0.5° resolution [46] was supplied with the CLIMEX version used. This includes data from many locations worldwide, based on long-term monthly average maximum and minimum temperatures, rainfall, and relative humidity between the hours of 09:00 and 15:00, between 1961 and 1990. This meteorological dataset formed the basis for initial parameter-fitting.

Model Framing

CLIMEX model depicting climatic suitability for *P. dactylifera* was created using native and exotic distributions of date palm from a variety of data sources (Figure 1), [21,23,27,29,30,47–56]. Table 1 summarizes all CLIMEX parameters. A detailed justification of these and their derivations is available in Shabani et al. [10].

Taguchi Method

The Taguchi method applies orthogonal arrays to study many decision variables in relation to few experiments [57]. In other

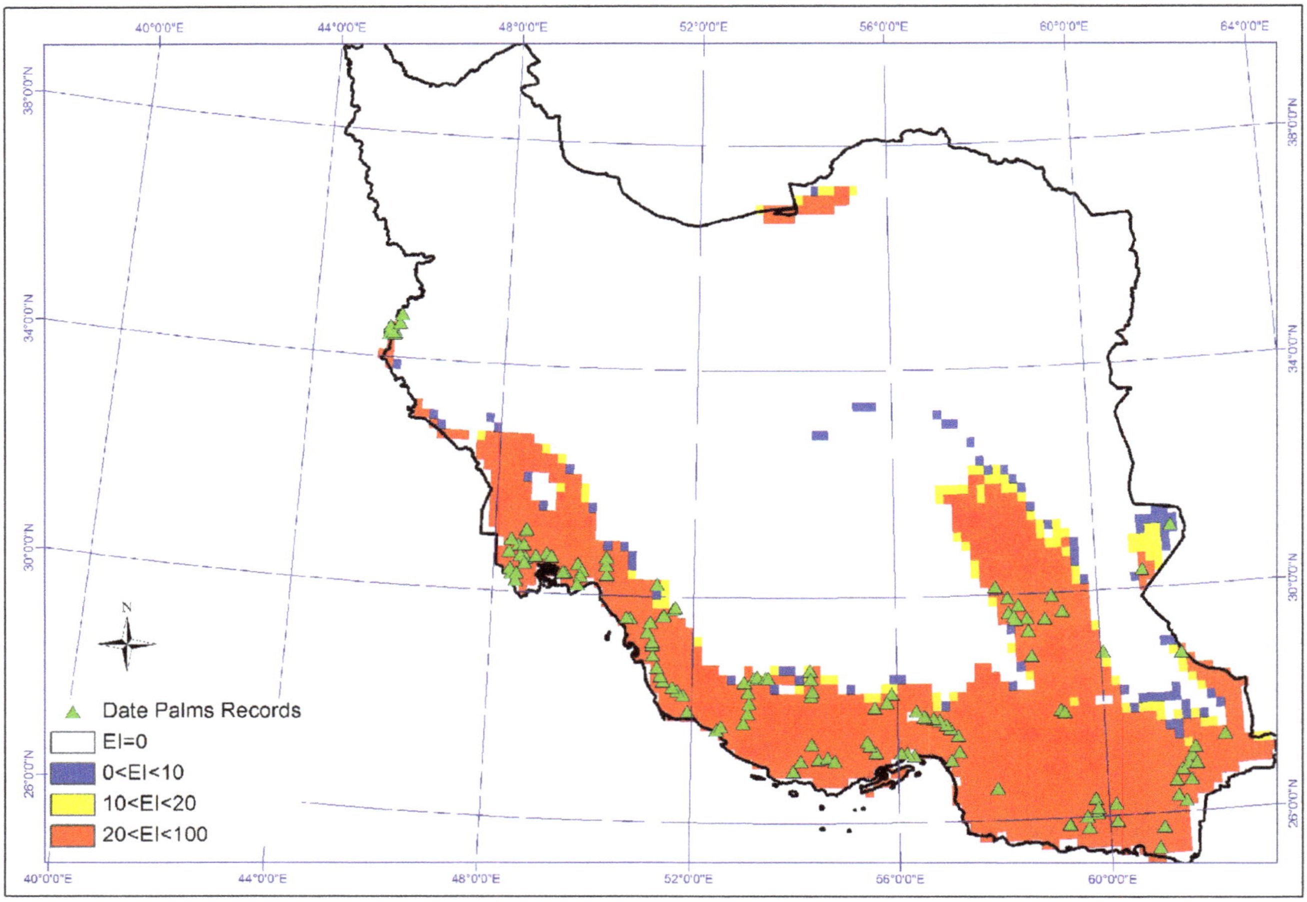

Figure 1. The current distribution of *P. dactylifera* L. and its potential distribution based on CLIMEX outputs at a country scale. (EI = 0, 0<EI<10, 10<EI<20 and 20<EI<100 means unsuitability, marginality, suitability and high suitability for date palm growth respectively).

Table 1. CLIMX parameter values used for *P. dactylifera* modeling.

Index	Parameter	Code	Values
	Limiting low temperature	DV0	14°C
Temperature	Lower optimal temperature	DV1	20°C
	Upper optimal temperature	DV2	39°C
	Limiting high temperature	DV3	46°C
	Limiting low soil moisture	SM0	0.007
Moisture	Lower optimal soil moisture	SM1	0.013
	Upper optimal soil moisture	SM2	0.81
	Limiting high soil moisture	SM3	0.9
Cold Stress	Cold stress temperature threshold	TTCS	4°C
	Cold stress temperature rate	THCS	−0.01/week
Heat Stress	Heat stress temperature threshold	TTHS	46°C
	Heat stress accumulation rate	THHS	0.9/week
Wet Stress	Wet stress threshold	SMWS	0.9
	Wet stress rate	HWS	0.022/week

Table 2. Factors and levels of potential distribution of *P. dactylifera* L. based on CLIMEX.

	Level 1	Level 2	Level 3
SM0	0.005	0.007	0.01
SM1	0.011	0.013	0.017
SM2	0.5	0.81	1
SM3	0.6	0.9	1
DV0	10	14	18
DV1	15	20	26
DV2	30	39	45
DV3	40	46	50
TTCS	2	4	5
THCS	−0.05	−0.01	0
TTHS	40	46	50
THHS	0.7	0.9	1
SMWS	0.6	0.9	1
HWS	0.018	0.022	0.029

words, this statistical method efficiently decreases the number of iterations required in an optimization process [58,59]. Compared to other optimization methods, including the Genetic Algorithm and the Particle Swarm Optimization, this method is easy to implement and can converge to the global optimum solution quickly [60]. Although the Taguchi method has been applied in many fields, including chemical and mechanical engineering [58], this method has rarely been used in climatic modeling studies. In the Taguchi method factors are divided into two main categories: controllable factors and noise factors [61]. Noise factors include those over which the experimenter cannot exert direct or exact control. Since noise factor elimination is impractical and sometimes even impossible, the Taguchi method aims to minimize noise effects and to calculate optimal levels of vital controllable factors, incorporating the concept of robustness [62]. Taguchi also proportions the significance of specific factors in terms of effects on the objective function [36]. Taguchi also proposes transformation of reiteration data to another value as a variation measure [63]. This transformation is based on the signal-to-noise (SN) ratio and explains why a particular parameter design is described as robust [62]. The term 'signal' represents the desired values (mean response variables), while 'noise' represents the undesired values (standard deviations), with the aim of maximizing the signal-to-noise ratio. Furthermore, objective functions are classed by Taguchi into three groups: 1 smaller the- better; 2 larger-the-better; and 3 nominal-best. A major feature of Taguchi method is the generation of sensitivity results within a single scale for different parameters, enabling users to compare sensitivity of specified parameters.

We established 14 Taguchi method factors, the same as the number of CLIMEX parameters and chose 54 runs, which is the maximum number of iterations and gives maximum accuracy, with 2×1 and 3×25 Array in a mixed level design to optimize sensitivity analysis accuracy.

A Taguchi-based sensitivity analysis was performed to quantify *P. dactylifera* L. response to temperature, soil moisture and cold, wet and heat stress parameter changes. Here the parameter values were based on Table 2, and in a few cases where the values fell outside the CLIMEX range, the values were adjusted. Adjusted models were then re-run after each parameter value change and thereafter the area of suitable and highly suitable categories were calculated for the baseline and for the final adjusted models, to assess the different parameters' sensitivity levels. In summary, 12 out of 54 different sets of control factors derived by the Taguchi method for potential distribution of *P. dactylifera* L. based on CLIMEX parameters is summarized in Table 3.

Incremental model EI values from the sensitivity analysis were plotted and compared with those of the baseline model. Where an altered incremental model parameter was found to be highly sensitive, a large change in EI value was expected. However where a parameter was not highly sensitive, we expected the baseline and incremental model EI values to be similar. Suitability changes were also assessed by area mapping, where suitability levels changed in terms of the suitable or highly suitable categories, for high level sensitivity parameters. All validation data changes related to parameter changes were further assessed by noting all changes in the number of occurrence records falling within each of the suitability categories.

Results

Historical Climate

A comparison of the model for climate suitability with the date palm distribution for Iran shows consistency in the correlation of the modeled EI with current distribution of *P. dactylifera*. Suitable climatic conditions for *P. dactylifera* lie between 48° E and 52° E, 57° E and 60° E, with large areas between 25° N and 29° N in central Iran (Figure 1).

A sample set of temperature parameter changes from the baseline model and the resultant impact on distribution are shown in Figure 2, while the final sensitivity analysis results, using the adjusted values of the Taguchi method, are illustrated in Figure 3. In the Taguchi method, the flat sensitivity lines indicate that that particular variable is not sensitive.

The potential distribution of *P. dactylifera* L. baseline model was highly sensitive to change in DV2 (low temperature limit) and DV3 (high temperature limit). In other words, changing DV2 and DV3 values produced markedly increased or decreased suitable and highly suitable areas. These changes produced a northward shift in the distribution of the species, creating larger areas in

Table 3. A sample set of control factors derived by Taguchi method for potential distribution of *P. dactylifera* L. based on CLIMEX parameters.

Sample Runs	Parameters														Areas of suitability (km^2)
	DV0	DV1	DV2	DV3	SM0	SM1	SM2	SM3	TTCS	THCS	TTHS	SMWS	HWS	THHS	
1	1	1	1	1	1	1	1	1	1	1	1	1	1	1	373977
2	1	1	2	2	2	2	2	2	1	1	1	1	1	1	1435284
3	1	2	2	2	3	3	1	1	1	1	2	2	3	3	1190953
4	1	2	2	2	3	3	1	1	2	2	3	3	1	1	427252
5	1	3	1	2	1	3	2	3	3	1	3	2	1	2	1345530
6	1	3	2	3	2	1	3	1	2	3	2	1	3	1	323326
7	1	3	2	3	2	1	3	1	3	1	3	2	1	2	1364426
8	1	3	3	1	3	2	1	2	3	1	3	2	1	2	1235830
9	2	2	2	3	1	2	1	3	3	1	2	3	2	1	1176781
10	2	2	3	1	2	3	2	1	2	3	1	2	1	3	539052
11	2	3	1	3	2	3	1	2	2	1	3	1	2	3	1045823
12	2	3	3	2	1	2	3	1	1	3	2	3	1	2	315978

central and northern Iran. Western Iran showed a similar trend, in areas that were originally unsuitable or marginally suitable for growth of date palm. Comparing sensitivity also showed that in CLIMEX, DV1 was more sensitive to change than DV0, but not as much as DV2 and DV3 (Figure 3). In other words, suitable and highly suitable areas changed less rapidly with DV1 adjustment from the baseline. Figure 3 also demonstrates that the values chosen for the baseline model DV2 and DV3 parameter were accurate, since the SN ratio mean was optimum.

In regard to the four soil moisture parameters, the upper optimal soil moisture level (SM2) proved most sensitive to changes (Figure 3). A small increase or decrease in this parameter projected large areas that could become suitable or highly suitable in the modeled suitability. SM0, SM1 and SM3 also proved sensitive to changes (Figure 3). Figure 3 also shows that level one in the parameters of SM0, SM1 and SM2 produced the highest SN ratio mean, while conversely the optimum mean of SN ratios for SM3 was observed at level three (Figure 3). In other words, SM0, SM1, SM2, and SM3 changes produced markedly changes in date palm distribution.

The cold stress temperature threshold (TTCS), cold stress temperature rate (THCS), heat stress accumulation rate (THHS) heat stress temperature threshold (TTHS) and the wet stress rate (HWS) demonstrated insensitivity to change. The alteration in the wet stress threshold (SMWS) thus conversely reflects the changes in EI when this parameter is modified (Figure 3).

Discussion

This study delineates the relationship in Iran between climate and *P. dactylifera* L. distribution. CLIMEX and the sensitivity analysis based on the Taguchi method illustrated informatively specific parameters that had the most effect on the modeled distribution of *P. dactylifera* L. Results demonstrate that distribution of *P. dactylifera* L. is highly sensitive to DV3 and DV2 (the limiting high and upper optimal temperature) changes as well as in SM3 and SM2 (the limiting high soil moisture and the upper optimal moisture) parameters (Figure 3). Additionally, date palm distribution was slightly sensitive to SM1 and SM0 (lower optimal soil moisture and lower limit soil moisture) and SMWS (wet stress threshold) parameters. These results contrasted with the results of the sensitivity analysis for *Lantana camara* L. [64], rated as one of the ten most destructive weeds in the world [65]. Taylor [64] showed that the distribution of lantana was highly sensitive to DV0, DV3 and SM0 parameter changes. Additionally, Taylor [64] showed that DV0 and DV3 changes had the substantially modified suitable and highly suitable lantana location in Australia. The total difference in climatic parameter requirements for these two species is possibly responsible for such markedly differences. For example, if the soil is not waterlogged for prolonged periods, lantana is tolerant of up to 3000 mm of rainfall annually [66,67] while documentation shows that 78.74 mm of rainfall over an 8-day period caused over 50% loss in yields of date palm, while 86.36 mm over 10 days resulted in a 15% loss on date palm farms in some countries [10]. Another difference in essential climatic parameters between lantana and date palm is upper optimal temperature, being 39°C and 30°C respectively [11,64].

Generally, where an area indicated decreased suitability, this was reflected in a decrease in the number of highly suitable category records, while records increased in the unsuitable and marginal categories. In certain cases, there was an increase of records in the suitable category, suggesting that some highly suitable areas were rendered just suitable by adjustment of parameters. Conversely, an increased suitable area generally

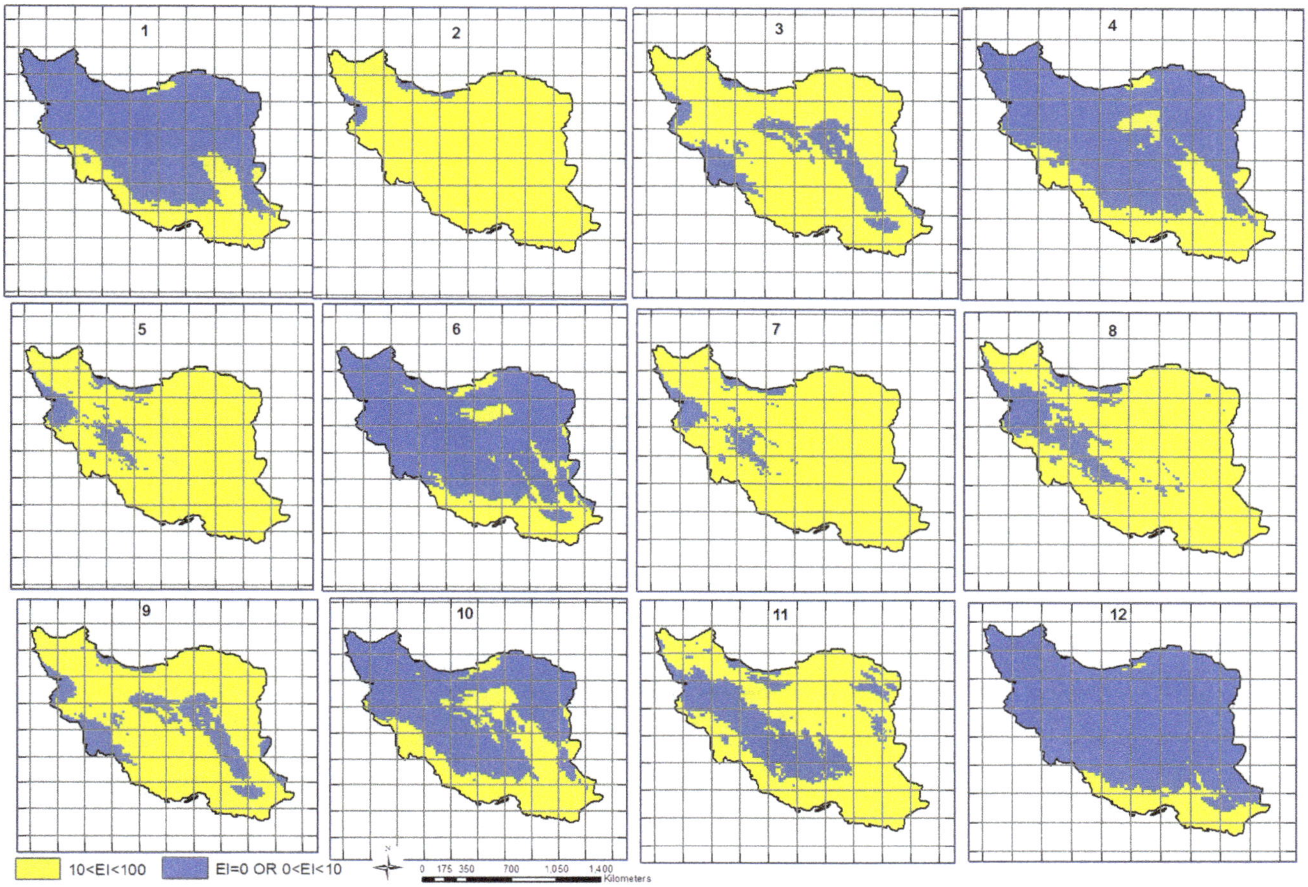

Figure 2. 12 sample results out of 54 different control factors based on Taguchi method for *P. dactylifera* L. Area changes in suitable and highly suitable categories when sensitivity analysis was taken. The values for the various parameters used are those given in table 3.

reflected in increased records in the highly suitable and suitable categories and decreased records in the unsuitable and marginal categories. Here, a decrease in records in the suitable category suggests that certain suitable areas became highly suitable with changes in parameter value.

The SM3, SM2 and SM0 moisture parameters were the most sensitive and showed greater shifts in distribution when altered, compared to SM1. A high level of sensitivity of this species within these parameters was demonstrated by these results. The cold stress temperature threshold (TTCS) and cold stress temperature rate (THCS) also demonstrated minimal levels of sensitivity to change. Acute specific changes in the sensitivity analysis to parameter values were facilitated by an in-depth understanding of the relationship of the species with climatic factors, based on the extensive date palm distribution in Iran and the relevant available research data. Table 3 gives the changes in area of suitability as the parameters are changed. Sample runs 2, 5, 7 and 8 (See Table 3 for detailed parameter settings) produced the highest suitable areas, while sample runs 1, 6 and 12 had least area.

Distribution predictive modeling is a useful tool for control and management of a species. Such models provide potential distribution mapping of a species, allowing policy makers at the national and international level to make well-informed decisions regarding management of their market. Predictive modeling techniques, such as CLIMEX, derive data on climatic requirements of a target species from species' geographic distribution data, which supports parameter-fitting in the development of the model. However, a particular species of interest may have greater sensitivity to certain climatic factors than to others, and these varying sensitivity levels may complicate distribution predictive modeling. Those parameters highly sensitive to change will have a greater impact on the output of the model than the relatively insensitive parameters. Sensitivity analysis highlights parameters of greater or lesser relative importance, useful for improving data collection [68]. It is for good reason that formal sensitivity analyses are advocated as the most effective method for the evaluation and refining of improvements to model input data [69].

For the continuation of model output development and the collecting of data for fitting of highly sensitive parameters with greater accuracy, further research is essential. Conversely, the cost effectiveness of the collection of additional data for relatively insensitive parameters may not be valid, if the model output improvements are minimal. However research towards the quantification of parameter sensitivity and the resultant refinement of model outputs are useful as a means to improving confidence in parameter estimates [70], leading to the most cost effective management strategies. Toward this end, perhaps this study's most encouraging finding is that eight of the fourteen parameters tested impacted strongly on potential date palm distribution, namely the DV1, DV2, DV3 and SM0, SM1, SM2, SM3 and SMWS parameters. Thus, where resources and support are limited, it would be most expedient to research and incorporate a wider

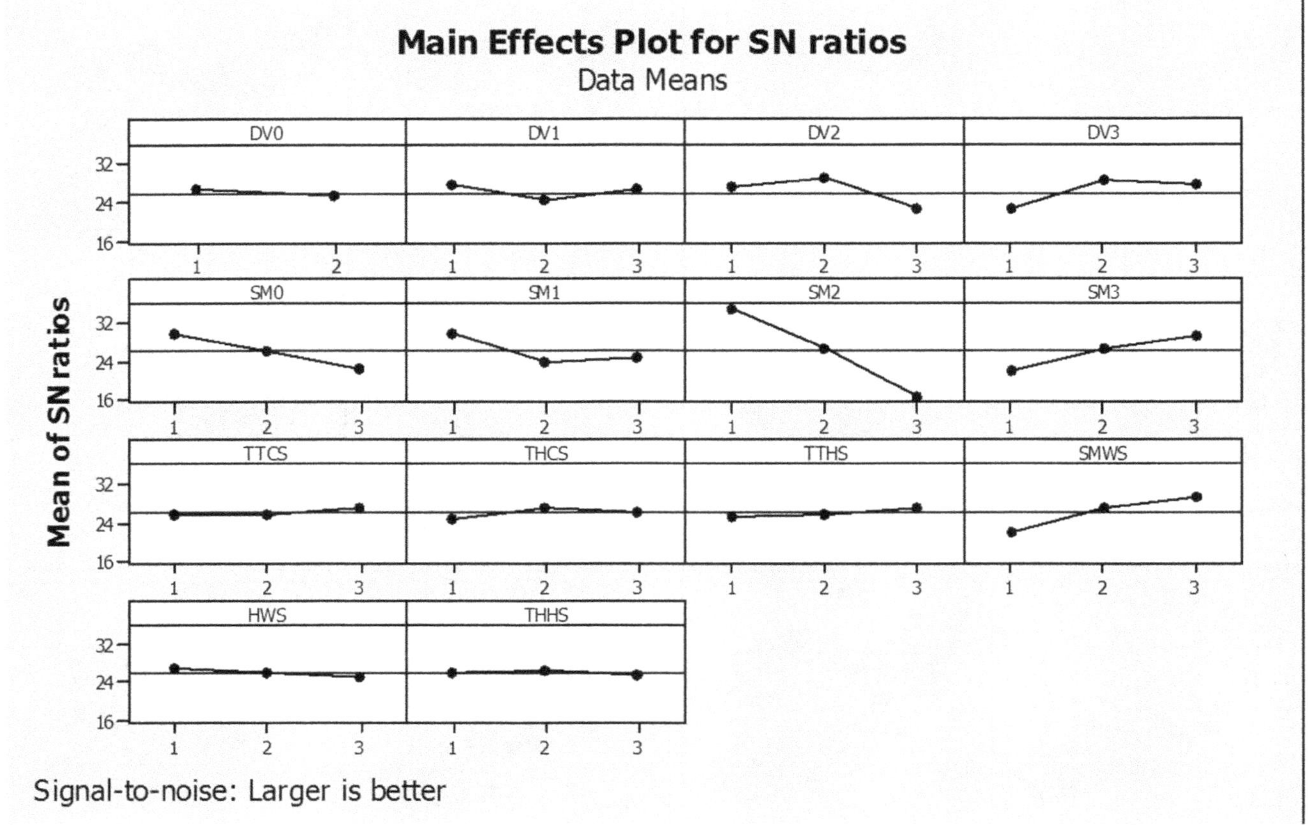

Figure 3. Sensitivity analysis of the 14 parameters in CLIMEX for *P. dactylifera* L.

range of alternative data sources towards fitting these eight parameters with increased accuracy.

The sensitivity analysis results also signal the need for caution regarding regional and global models of climate change in projecting date palm distributions, that all models and climate change scenarios in the future include ranges for both temperature and rainfall variation. The Special Report on Emissions Scenarios (SRES) A1F1 fossil intensive scenario projects an increase in global average temperatures of 2.4 to 6.4°C, making this of vital interest [71]. The lower end of this spectrum projects an increase of 1.1 to 2.9°C based on the SRES B1 scenario. This projected increase in the average global temperature would have drastic implications for date palm, in terms of the level of sensitivity its distribution has demonstrated regarding changes to the upper and lower temperature tolerance limits (DV2 and DV1).

CLIMEX's central assumption is that the primary determinant of the geographical distribution of a species is climate. Thus dispersal potential, biotic interactions, soil type, land-use and disturbance activities and other non-climatic factors are not explicitly incorporated in the modeling process. These factors can however be incorporated after the climate modeling process has been executed [26]. Additionally the inclusion of native as well as exotic distribution data should include any effects from the release of natural enemies [72] apparent in the exotic range of species such as date palm. This more clearly defines its fundamental niche [73]. Thus, the methodologies that refine the data necessary for date palm potential distribution modeling tools are invaluable. We used a sensitivity analysis for the identification of the CLIMEX parameters and consequent climate aspects that most influenced date palm potential distribution in Iran. This approach is extremely useful in the streamlining of the requirements for data collection for the modeling of potential distribution.

Acknowledgments

We thank Mohammad Esmaeili and Dr. Shahin Shadlou during conduction of this study.

Author Contributions

Conceived and designed the experiments: FS. Performed the experiments: FS. Analyzed the data: FS LK. Contributed reagents/materials/analysis tools: FS LK. Wrote the paper: FS LK.

References

1. Soberon J, Peterson A (2005) Interpretation of models of fundamental ecological niches and species distributional areas. Biodiversity Informatics 2: 1–10.
2. Guisan A, Zimmerman NE (2000) Predictive habitat distribution models in ecology. Ecological Modelling 135: 147–186.
3. Pearson RG, Dawson TP (2003) Predicting the impacts of climate change on the distribution of species: Are bioclimate envelope models useful? Global Ecology and Biogeography 12: 361–371.
4. Beaumont LJ, Hughes L, Poulsen M (2005) Predicting species distributions: use of climatic parameters in BIOCLIM and its impact on predictions of species' current and future distributions. Ecological Modelling 186: 251–270.

5. Peterson AT, Papes M, Kluza DA (2003) Predicting the potential invasive distributions of four alien plant species in North America. Weed Science 51: 863–868.
6. Barry S, Elith J (2006) Error and uncertainty in habitat models. Journal of Applied Ecology 43: 413–423.
7. Andrewartha HG, Birch LC (1954) The distribution and abundance of animals. Chicago: University of Chicago Press. 782 p.
8. Sutherst RW, Maywald G (1985) A computerized system for matching climates in ecology. Agriculture Ecosystems & Environment 13: 281–299.
9. Kriticos DJ, Randall RP (2001) A comparison of systems to analyze potential weed distributions. In: Groves RH, Panetta FD, Virtue JG, editors. Weed Risk Assessment. Collingwood: CSIRO Publishing. pp. 61–79.
10. Shabani F, Kumar L, Taylor S (2012) Climate change impacts on the future distribution of date palms: a modeling exercise using CLIMEX. PLoS ONE 7: e48021.
11. Shabani F, Kumar L, Esmaeili A (2013) Use of CLIMEX, Land use and Topography to Refine Areas Suitable for Date Palm Cultivation in Spain under Climate Change Scenarios. Journal of Earth Science & Climatic Change 4.
12. Hanspach J, Kühn I, Schweiger O, Pompe S, Klotz S (2011) Geographical patterns in prediction errors of species distribution models. Global Ecology and Biogeography 20: 779–788.
13. Robertson MP, Villet MH, Palmer AR (2004) A fuzzy classification technique for predicting speciesdistributions: application using invasive alien plants and indigenous insects. Diversity and Distributions 10: 461–474.
14. Burgman MA, Lindenmayer DB, Elith J (2005) Managing landscapes for conservation under uncertainty. Ecology 86: 2007–2017.
15. Hamby DM (1994) A review of techniques for parameter sensitivity analysis of environmental models. Environmental Monitoring & Assessment 32: 135–154.
16. Olfert O, Hallett R, Weiss RM, Soroka J, Goodfellow S (2006) Potential distribution and relative abundance of swede midge, Contarinia nasturtii, an invasive pest in Canada. Entomologia Experimentalis et Applicata 120: 221–228.
17. Hallgren WS, Pitman AJ (2000) The uncertainty in simulations by a Global Biome Model (BIOME3) to alternative parameter values. Global Change Biology 6: 483–495.
18. Sutherst RW, Bourne AS (2009) Modelling non-equilibrium distributions of invasive species: a tale of two modelling paradigms. Biological Invasions 11: 1231–1237.
19. Ferry M, Gomez S (2002) The Red Palm Weevil in the Mediterranean Area, Available at "http://www.palms.org/palmsjournal/2002/redweevil.htm". Journal of the International Palm Society 46.
20. Gómez-Vidal S, Lopez-Llorca LV, Jansson HB, Salinas J (2006) Endophytic colonization of date palm (*Phoenix dactylifera* L.) leaves by entomopathogenic fungi. Micron 37: 624–632.
21. Tengberg M (2011) Beginnings and early history of date palm garden cultivation in the Middle East. Journal of Arid Environments 5: 1–9.
22. Ahmed M, Bouna Z, Lemine F, Djeh T, Mokhtar T, et al. (2011) Use of multivariate analysis to assess phenotypic diversity of date palm (*Phoenix dactylifera* L.) cultivars. Scientia Horticulturae 127: 367–371.
23. Bokhary H (2010) Seed-borne fungi of date-palm, *Phoenix dactylifera* L. from Saudi Arabia. Saudi Journal of Biological Sciences 17: 327–329.
24. Shabani F, Kumar L, Esmaeili A, Saremi H (2013) Climate change will lead to larger areas of Spain being conducive to date palm cultivation. Journal of Food, Agriculture & Environment 11.
25. Shabani F, Kumar L, Taylor S (2014) Projecting date palm distribution in Iran under climate change using topography, physicochemical soil properties, soil taxonomy, land use and climate data. Theoretical and Applied Climatology 115.
26. Shabani F, Kumar L, Taylor S (2013) Suitable regions for date palm cultivation in Iran are predicted to increase substantially under future climate change scenarios. Journal of Agricultural Science: 1–15.
27. Shabani F, Kumar L (2013) Risk levels of Invasive *Fusarium oxysporum* f. sp. in Areas Suitable for Date Palm (*Phoenix dactylifera*) Cultivation Under Various Climate Change Projections. PLoS ONE 8: e83404.
28. Chao C, Krueger R (2007) The Date Palm (Phoenix dactylifera L.): Overview of Biology, Uses, and Cultivation. Journal of Hortscience 42: 1077–1083.
29. Eshraghi P, Zarghami R, Mirabdulbaghi M (2005) Somatic embryogenesis in two Iranian date palm. African Journal of Biotechnology 4: 1309–1312.
30. Mahmoudi H, Hosseininia G (2008) Enhancing date palm processing, marketing and pest control through organic culture. Journal of Organic Systems 3: 30–39.
31. Al-Shahib W, Marshall RJ (2003) The fruit of the date palm: its possible use as the best food for the future? International journal of food sciences and nutrition 54: 247–259.
32. Kassem H (2012) The response of date palm to calcareous soil fertilisation. Journal of soil science and plant nutrition 12: 45–58.
33. Reilly D, Reilly R (2012) Gurra downs, date palms, Available at: http://www.gurradowns.com.au/Ourplantation.php, Accessed 19 March 2013.
34. Botes A, Zaid A (2002) Date palm cultivation, Available at: http://www.fao.org/DOCREP/006/Y4360E/y4360e07.htm#bm07.2, Accessed 20 January 2013.
35. Khayyat M, Tafazoli E, Eshghi S, Rajaee S (2007) Effect of nitrogen, boron, potassium and zinc sprays on yield and fruit quality of date palm. Am Eurasian J Agric Environ Sci 2: 289–296.
36. Nalbant M, Gökkaya H, Sur G (2007) Application of Taguchi method in the optimization of cutting parameters for surface roughness in turning. Materials & design 28: 1379–1385.
37. EarthExplorer (2012).
38. Global Biodiversity Information Facility (2014) Available at: http://www.gbif.org/, Accessed 11 April 2013.
39. Missouri Botanical Garden (2012) Available at: http://www.missouribotanicalgarden.org/, Accessed 25 June 2013.
40. CAB Direct (2013) http://www.cabdirect.org/web/about.html, Accessed 7 January 2012.
41. Rahnema A (2013) Iranian date palm institution, available at: http://khorma.areo.ir/HomePage.aspx?TabID = 15022&Site = khorma.areo&Lang = en-US, Accessed 23 April 2013.
42. Shayesteh N, Marouf A, Amir-Maafi M (2010) Some biological characteristics of the Batrachedra amydraula Meyrick (Lepidoptera: Batrachedridae) on main varieties of dry and semi-dry date palm of Iran. Julius-Kühn-Archiv: 151.
43. Sutherst RW, Maywald G, Kriticos DJ (2007) CLIMEX version 3: user's guide. In: Ltd HssP, editor. Melbourne. pp. 10–126.
44. Sutherst RW, Maywald G (2005) A climate model of the red imported fire ant, Solenopsis invicta Buren (Hymenoptera: Formicidae): Implications for invasion of new regions, particularly Oceania. Environmental Entomology 34: 317–335.
45. Kriticos DJ, Sutherst RW, Brown JR, Adkins SW, Maywald GF (2003a) Climate change and the potential distribution of an invasive alien plant: Acacia nilotica ssp indica in Australia. Journal of Applied Ecology 40: 111–124.
46. New MG, Hulme M, Jones MO (1999) Representing 20th century space–time climate variability. Part I. Development of a 1961–1990 mean monthly terrestrial climatology. Journal of Climate 12: 829–856.
47. Jain S, Al-Khayri J, Dennis V, Jameel M (2011) Date Palm Biotechnology: Springer Netherlands. 743 p.
48. Shayesteh N, Marouf A, Amir-Maafi M (2010) Some biological characteristics of the Batrachedra amydraula Meyrick (Lepidoptera: Batrachedridae) on main varieties of dry and semi-dry date palm of Iran. 10th International Working Conference on Stored Product Protection.
49. Abbas I, Mouhi M, Al-Roubaie J, Hama N, El-Bahadli A (1991) *Phomopsis phoenicola* and *Fusarium equiseti*, new pathogens on date palm in Iraq. Mycological Research 95: 509.
50. Auda H, Khalaf Z (1979) Studies on sprout inhibition of potatoes and onions and shelf-life extension of dates in Iraq. Journal of Radiation Physics and Chemistry 14: 775–781.
51. Heakal MS, Al-Awajy MH (1989) Long-term effects of irrigation and date-palm production on Torripsamments, Saudi Arabia. Geoderma 44: 261–273.
52. Al-Senaidy M, Abdurrahman M, Mohammad A (2011) Purification and characterization of membrane-bound peroxidase from date palm leaves (*Phoenix dactylifera* L.). Saudi Journal of Biological Sciences 18: 293–298.
53. Markhand G (2010) Fruit characterization of Pakistani dates. Available at: http://www.pakbs.org/pjbot/PDFs/42%286%29/PJB42%286%293715.pdf, Accessed 9 January 2012. Pakistan Journal of Botany 42: 3715–3721.
54. Hasan S, Baksh K, Ahmad Z, Maqbool A, Ahmed W (2006) Economics of growing date palm in punjab, pakistan. International Journal of Agriculture and Biology 8: 1–5.
55. Elhoumaizi M, Saaidi M, Oihabi A, Cilas C (2001) Phenotypic diversity of date-palm cultivars (*Phoenix dactylifera* L.) from Morocco. Genetic Resources and Crop Evolution 49: 483–490.
56. Marqués J, Duran-Vila N, Daròs J (2011) The Mn-binding proteins of the photosystem II oxygen-evolving complex are decreased in date palms affected by brittle leaf disease. Plant Physiology and Biochemistry 49: 388–394.
57. Yang W, Tarng Y (1998) Design optimization of cutting parameters for turning operations based on the Taguchi method. Journal of Materials Processing Technology 84: 122–129.
58. Waldner F (2013) Implementation of Taguchi's Method for optimal design using Orthogonal Arrays theory: University of Trento.
59. Unal R, Dean EB (1991) Taguchi approach to design optimization for quality and cost: an overview. Annual conference of the international society of parametric analysts: 13–23.
60. Weng W-C, Yang F, Elsherbeni AZ (2007) Linear antenna array synthesis using Taguchi's method: A novel optimization technique in electromagnetics. Antennas and Propagation, IEEE Transactions on 55: 723–730.
61. Ghani J, Choudhury I, Hassan H (2004) Application of Taguchi method in the optimization of end milling parameters. Journal of Materials Processing Technology 145: 84–92.
62. Phadke MS (1995) Quality engineering using robust design: Prentice Hall PTR.
63. Lee K-H, Eom I-S, Park G-J, Lee W-I (1996) Robust design for unconstrained optimization problems using the Taguchi method. AIAA journal 34: 1059–1063.
64. Taylor S, Kumar L (2012) Sensitivity Analysis of CLIMEX Parameters in Modelling Potential Distribution of *Lantana camara* L. PLoS ONE 7: e40969.
65. Sharma GP, Raghubanshi AS, Singh JS (2005) Lantana invasion: An overview. Weed Biology and Management 5: 157–165.
66. Day MD, Broughton S, Hannan-Jones MA (2003) Current distribution and status of *Lantana camara* and its biological control agents in Australia, with recommendations for further biocontrol introductions into other countries. BioControl 24: 63N–76N.
67. Thaman R (1974) *Lantana camara*: Its introduction, dispersal and impact on islands of the tropical Pacific ocean. Micronesica 10: 17–39.
68. Merow C, LaFleur N, Silander Jr JA, Wilson AM, Rubega M (2011) Developing dynamic mechanistic species distribution models: predicting bird-mediated spread of invasive plants acrosss Northeastern North America. The American Naturalist 178: 30–43.

69. Johnson CJ, Gillingham MP (2008) Sensitivity of species-distribution models to error, bias, and model design: An application to resource selection functions for woodland caribou. Ecological Modelling 213: 143–155.
70. Van Klinken RD, Lawson BE, Zalucki MP (2009) Predicting invasions in Australia by a Neotropical shrub under climate change: the challenge of novel climates and parameter estimation. Global Ecology and Biogeography 18: 688–700.
71. Intergovernmental Panel on Climate Change IPCC (2007) Climate change 2007: synthesis report. summary for policymakers.
72. Keane RM, Crawley MJ (2002) Exotic plant invasions and the enemy release hypothesis. Trends in Ecology & Evolution 17: 164–170.
73. Wharton TN, Kriticos DJ (2004) The fundamental and realized niche of the Monterey Pine aphid, Essigella californica (Essig)(Hemiptera: Aphididae): implications for managing softwood plantations in Australia. Diversity and Distributions 10: 253–262.

PERMISSIONS

All chapters in this book were first published in PLOS ONE, by The Public Library of Science; hereby published with permission under the Creative Commons Attribution License or equivalent. Every chapter published in this book has been scrutinized by our experts. Their significance has been extensively debated. The topics covered herein carry significant findings which will fuel the growth of the discipline. They may even be implemented as practical applications or may be referred to as a beginning point for another development.

The contributors of this book come from diverse backgrounds, making this book a truly international effort. This book will bring forth new frontiers with its revolutionizing research information and detailed analysis of the nascent developments around the world.

We would like to thank all the contributing authors for lending their expertise to make the book truly unique. They have played a crucial role in the development of this book. Without their invaluable contributions this book wouldn't have been possible. They have made vital efforts to compile up to date information on the varied aspects of this subject to make this book a valuable addition to the collection of many professionals and students.

This book was conceptualized with the vision of imparting up-to-date information and advanced data in this field. To ensure the same, a matchless editorial board was set up. Every individual on the board went through rigorous rounds of assessment to prove their worth. After which they invested a large part of their time researching and compiling the most relevant data for our readers.

The editorial board has been involved in producing this book since its inception. They have spent rigorous hours researching and exploring the diverse topics which have resulted in the successful publishing of this book. They have passed on their knowledge of decades through this book. To expedite this challenging task, the publisher supported the team at every step. A small team of assistant editors was also appointed to further simplify the editing procedure and attain best results for the readers.

Apart from the editorial board, the designing team has also invested a significant amount of their time in understanding the subject and creating the most relevant covers. They scrutinized every image to scout for the most suitable representation of the subject and create an appropriate cover for the book.

The publishing team has been an ardent support to the editorial, designing and production team. Their endless efforts to recruit the best for this project, has resulted in the accomplishment of this book. They are a veteran in the field of academics and their pool of knowledge is as vast as their experience in printing. Their expertise and guidance has proved useful at every step. Their uncompromising quality standards have made this book an exceptional effort. Their encouragement from time to time has been an inspiration for everyone.

The publisher and the editorial board hope that this book will prove to be a valuable piece of knowledge for researchers, students, practitioners and scholars across the globe.

LIST OF CONTRIBUTORS

Hannah Slater and Edwin Michael
Department of Infectious Disease Epidemiology, Imperial College London, London, United Kingdom

Anna Drake, Christine A. Rock, Sam P. Quinlan, Michaela Martin and David J. Green
Center for Wildlife Ecology, Department of Biological Sciences, Simon Fraser University, Burnaby, British Columbia, Canada

Ana Luz Márquez and Raimundo Real
Biogeography, Diversity and Conservation Research Team, Dept. of Animal Biology, Faculty of Sciences, University of Malaga, Malaga, Spain

Antonio-Román Muñ
Biogeography, Diversity and Conservation Research Team, Dept. of Animal Biology, Faculty of Sciences, University of Malaga, Malaga, Spain
Fundación Migres, N-340, Km. 96 Huerta Grande, Pelayo, Algeciras, Spain
Depto. de Didáctica de las Matemáticas, de las Ciencias Sociales y de las Ciencias Experimentales, Faculty of Education Sciences, University of Malaga, Malaga, Spain

Jeanne L. Nel, David C. Le Maitre, Belinda Reyers, Greg G. Forsyth, Andre K. Theron, Patrick J. O'Farrell, Lara van Niekerk and Laurie Barwell
Natural Resources and the Environment, Council for Scientific and Industrial Research (CSIR), Stellenbosch, Western Cape, South Africa

Brian W. van Wilgen
Natural Resources and the Environment, Council for Scientific and Industrial Research (CSIR), Stellenbosch, Western Cape, South Africa
Centre for Invasion Biology, Department of Botany and Zoology, Stellenbosch University, Stellenbosch, Western Cape, South Africa

Deon C. Nel
World Wide Fund for Nature (WWF), Cape Town, Western Cape, South Africa

Sally Archibald
School of Animal, Plant and Environmental Sciences, University of the Witwatersrand, Johannesburg, Gauteng, South Africa
Natural Resources and the Environment, Council for Scientific and Industrial Research (CSIR), Pretoria, Gauteng, South Africa

Jean-Marc Mwenge Kahinda, Francois A. Engelbrecht and Evison Kapangaziwiri
Natural Resources and the Environment, Council for Scientific and Industrial Research (CSIR), Pretoria, Gauteng, South Africa

Adrian J. Das and Nathan L. Stephenson
Western Ecological Research Center, United States Geological Survey, Three Rivers, California, United States of America

Alan Flint
California Water Science Center, United States Geological Survey, Sacramento, California, United States of America

Tapash Das
Climate Atmospheric Science and Physical Oceanography, Scripps Institution of Oceanography, La Jolla, California, United States of America

Phillip J. van Mantgem
Western Ecological Research Center, United States Geological Survey, Arcata, California, United States of America

Amanda J. Chunco
Department of Geography, University of North Carolina at Chapel Hill, Chapel Hill, North Carolina, United States of America

Todd Jobe
Signal Innovations Group, Inc., Durham, North Carolina, United States of America,

Karin S. Pfennig
Department of Biology, University of North Carolina at Chapel Hill, Chapel Hill, North Carolina, United States of America

Hannah Slater
Department of Infectious Disease Epidemiology, Imperial College London, St. Mary's Campus, Norfolk Place, London, United Kingdom

Edwin Michael
Department of Biological Sciences, University of Notre Dame, Notre Dame, Indiana, United States of America

Colin S. Shanley and David M. Albert
The Nature Conservancy, Juneau, Alaska, United States of America

Holger Hoffmann and Thomas Rath
Biosystems Engineering, Institute for Biological Production Systems, Leibniz Universität Hannover, Hannover, Germany

Cleo Bertelsmeier and Franck Courchamp
Ecologie, Systématique & Evolution, Univ. Paris Sud, Orsay, France

Benoît Guénard
Biodiversity and Biocomplexity Unit, Okinawa Institute of Science and Technology, Okinawa, Japan

Jennifer A. Curtis
U. S. Geological Survey, California Water Science Center, Eureka, California, United States of America

Lorraine E. Flint
U. S. Geological Survey, California Water Science Center, Sacramento, California, United States of America

Alan L. Flint
U. S. Geological Survey, California Water Science Center, Placer Hall, California, United States of America

Jessica D. Lundquist
University of Washington, Department of Civil and Environmental Engineering, Seattle, Washington, United States of America

Brian Hudgens
Institute for Wildlife Studies, Arcata, California, United States of America

Erin E. Boydston
U. S. Geological Survey, Western Ecological Research Center, Thousand Oaks, California, United States of America

Julie K. Young
U. S. Department of Agriculture, Wildlife Services, National Wildlife Research Center and Utah State University, Wildland Resources Department, Logan, Utah, United States of America

Michael R. Guttery
Department of Forest and Wildlife Ecology, University of Wisconsin, Madison, Wisconsin, United States of America

David K. Dahlgren
Kansas Department of Wildlife and Parks, Hays, Kansas, United States of America

Terry A. Messmer and Pat A. Terletzky
Department of Wildland Resources, Utah State University, Logan, Utah United States of America

John W. Connelly
Idaho Department of Fish and Game, Blackfoot, Idaho, United States of America

Kerry P. Reese
Department of Fish and Wildlife Sciences, University of Idaho, Moscow, Idaho, United States of America

Nathan Burkepile
Northland Fish and Game, Whangarei, New Zealand

David N. Koons
Department of Wildland Resources and the Ecology Center, Utah State University, Logan, Utah, United States of America

Trevon Fuller and Julia A. G. Shiplacoff
Center for Tropical Research, University of California Los Angeles, Los Angeles, California, United States of America

Henri A. Thomassen
Center for Tropical Research, University of California Los Angeles, Los Angeles, California, United States of America
Department of Comparative Zoology, University of Tübingen, Tübingen, Germany

Thomas B. Smith
Center for Tropical Research, University of California Los Angeles, Los Angeles, California, United States of America

Salvi Asefi-Najafabady
School of Life Sciences, Arizona State University, Tempe, Arizona, United States of America
Institute of the Environment and Sustainability, University of California Los Angeles, Los Angeles, California, United States of America

Prime M. Mulembakani5, Timothée L. Kinkela and Emile Okitolonda
Kinshasa School of Public Health, Kinshasa, Democratic Republic of Congo

Seth Blumberg and James O. Lloyd-Smith
Fogarty International Center, National Institutes of Health, Bethesda, Maryland, United States of Americ
Department of Ecology and Evolutionary Biology, University of California Los Angeles, Los Angeles, California, United States of America

Sara C. Johnston
United States Army Medical Research Institute of Infectious Diseases, Fredrick, Maryland, United States of America

Neville K. Kisalu
Department of Microbiology, Immunology, and Molecular Genetics, University of California Los Angeles, Los Angeles, California, United States of America

Joseph N. Fair and Matthew LeBreton
Global Viral Forecasting, San Francisco, California, United States of America, Stanford University, Program in Human Biology, Stanford, California, United States of America

Nathan D. Wolfe
Global Viral Forecasting, San Francisco, California, United States of America, Stanford University, Program in Human Biology, Stanford, California, United States of America

Robert L. Shongo
Ministry of Health, Kinshasa, Democratic Republic of Congo

Hermann Meyer
Bundeswehr Institute of Microbiology, Munich, Germany

Linda L. Wright
The Eunice Kennedy Shriver National Institute of Child Health and Human Development, Bethesda, Maryland, United States of America

Jean-Jacques Muyembe
National Institute of Biomedical Research, Kinshasa, Democratic Republic of Congo

Wolfgang Buermann
Center for Tropical Research, University of California Los Angeles, Los Angeles, California, United States of America
Department of Atmospheric and Oceanic Sciences, University of California Los Angeles, Los Angeles, California, United States of America

Lisa E. Hensley
Medical Countermeasures Initiative, Silver Spring, Maryland, United States of America

Anne W. Rimoin
Department of Microbiology, Immunology, and Molecular Genetics, University of California Los Angeles, Los Angeles, California, United States of America
Department of Epidemiology, School of Public Health, University of California Los Angeles, Los Angeles, California, United States of America

Terry L. Sohl
Earth Resources Observation and Science (EROS) Center, U.S. Geological Survey, Sioux Falls, South Dakota, United States of America

Tony Chang, Andrew J. Hansen and Nathan Piekielek
Department of Ecology, Montana State University, Bozeman, Montana, United States of America

Simeon L. Hill, Tony Phillips and Angus Atkinson
British Antarctic Survey, Natural Environment Research Council, Cambridge, United Kingdom
Plymouth Marine Laboratory, Plymouth, United Kingdom

Giovanni Ravazzani, Matteo Ghilardi, , Chiara Corbari and Marco Mancini
Politecnico di Milano, Piazza Leonardo da Vinci, Milan, Italy

Thomas Mendlik and Andreas Gobiet
Wegener Center for Climate and Global Change and Institute for Geophysics, Astrophysics, and Meteorology, University of Graz, Graz, Austria

Farzin Shabani and Lalit Kumar
Ecosystem Management, School of Environmental and Rural Science, University of New England, Armidale, New South Wales, Australia

Index